Geographic Information Research
Bridging the Atlantic

Geographic Information Research

Bridging the Atlantic

EDITED BY

MASSIMO CRAGLIA and **HELEN COUCLELIS**

ORGANISING COMMITTEE

HANS-PETER BÄHR, KEITH CLARKE,
HELEN COUCLELIS, MASSIMO CRAGLIA,
JEAN-PAUL DONNAY, HARLAN ONSRUD
and FRANÇOIS SALGÉ

Taylor & Francis
Publishers since 1798

UK   Taylor & Francis Ltd, 1 Gunpowder Square, London EC4A 3DE
USA Taylor & Francis Inc., 1900 Frost Road, Suite 101, Bristol, PA 19007

British Library Cataloguing in Publication Data

A catalogue record for this book is available from the British Library.
ISBN 0-7484-0594-1 (paperback)

Library of Congress Cataloging in Publication Data are available

Cover design by Hans Jacobs and Robin Uleman
Typeset in Times 10/12pt by Keyset Composition, Colchester, Essex
Printed in Great Britain by T. J. Press (Padstow) Ltd

Contents

Foreword

In his brilliant essay *La Filosofia della Statistica*, written in 1827, Melchiorre Gioja defines statistics as 'the sum of notions relative to a country, that can be useful to each and the majority of its citizens, in the course of their daily business and also to their government, which is their representative, agent and solicitor'. In the same work, however, he warns against the unprofessional collection of information by inept officials, lashing against 'the many queries that bumblers called secretaries send from the capital to the provinces. Queries that never produced other than three effects: 1. Fear that the government is looking for additional taxes, and therefore opportunistically false answers. 2. Ridicule consequent to the silliness, inconsistency and lack of precision of the questions. And therefore answers untrue out of despise. 3. Heaps of paper which encumber uselessly the archives, if the government mistrusts the answers. Serious mistakes if it uses them. Not to mention the time wasted by provincial and municipal officers who must prepare the answers.'

Modern social scientists are still trying to fulfil the mission of providing their fellow citizens with notions that can help them in 'the conduct of their daily business', and the availability of contemporary sophisticated technologies such as Geographical Information Systems provides powerful aids towards that. However, the need to use carefully assessed data has not faded away. GIS, and other similar tools belong to a more general family of projections, namely devices destined to represent a real and sometimes complex object on a surface. The relation between the object to be represented or projected varies on a continuum. At one end of the continuum there is the familiar 'Chinese shadows', in which the purpose of the projection is to transform the real object – hands, fingers, fists – into an imaginary one – rabbits, dogs or ducks – thus achieving the maximum of distortion. At the opposite end there are the blueprints used by architects and engineers in which, through the rules of geometrical perspective and projection, one can fully derive an object or a construction from a series of flat charts. This results in the maximum of precision and correspondence between the object and its projection. Somewhere in between there is a host of more complex representation systems, such as X-ray slides. Here the projection is precise but not immediately evident. A double decoding process is needed, represented normally by the joint work of the doctor and the radiologist. The physician has functional knowledge about the body, but the radiologist has technical knowledge about the transformation rules of the X-ray. He/she can interpret the correspondence between the actual masses in the body and the shadows projected by X-rays on the film, given specific conditions of radiation intensity and direction and film sensitivity. Neither skill can fully operate without the other, lest the risk of serious mistakes.

GIS applied to social phenomena belong to this range of projections. The technicality of the tool, the functional meaning of the data used, and the particular coding system in which they are collected must be known. Take the familiar production of choropleth maps with population density as an example. If the density of Milan and Rome is mapped, one might conclude that Rome is a much less urbanised commune than Milan. This is true if one is satisfied with an average figure, but very misleading if one is a road planner. Rome in fact is a peculiarly large commune (1504 sq. km, without the Vatican State) with a surface of little less than ten times that of Milan. But almost two thirds of this surface is countryside, and in fact farmed. Thus the actual net demographic density is what one can see by travelling in Rome, and not very different from that of Milan or other comparable cities. In order to interpret that particular patch of colour produced by the printer of the GIS, one needs to combine the technical skill of how data are produced and projected with the functional knowledge of the municipal organisation in Italy.

This is why the Standing Committee on Social Sciences (SCSS) of the European Science Foundation has supported the establishment of the GISDATA scientific programme. We consider this as crucial in helping social scientists understand and utilise the technology which is becoming increasingly useful for furthering our knowledge of social systems.

A further reason why the SCSS is particularly keen about GISDATA is the extent of the collaboration with the NSF and in particular with its Social, Behavioral and Economic Sciences (SBE) Division. SCSS has a long-established policy of cooperation with social scientists in the US and their institutions. Our Committee membership has included participation from the US Social Science Research Council for many years, and now I am happy to say that Dr Cora Marrett sits in the SCSS in her capacity as Assistant Director of the SBE. In addition to these more general and institutional reasons, I would like to add a personal note of satisfaction for the cooperation with NCGIA, and with Helen Couclelis and Michael Goodchild, as an old-timer of the University of California at Santa Barbara.

For these reasons, on behalf of the ESF/SCSS, I am delighted that the First Summer Institute for young scientists on geographic information held in Maine in 1995, and this book which has resulted from it, has provided such tangible evidence of the close and fruitful collaboration between institutions and scientists across the Atlantic.

PROFESSOR GUIDO MARTINOTTI
Chairman of the Standing Committee of the Social Sciences of the
European Science Foundation

The word 'statistics' derives from the Latin term for state. The Greeks sub-sequently used the expression to mean the study of political facts and figures. It is instructive that the original statisticians made the boundaries of the nation-state their analytical borders. So, too, have many social scientists. These scientists, significant users of and contributors to statistical approaches, have toiled valiantly to reach the social and political bedrock of national systems.

But these are times for broadening the scope of such analyses. First, many of the concerns that captivate social scientists have no national bounds. The movements of people and of commerce, sweeping changes in transformation and information technology – and similar developments – modify social systems on a global and not just a domestic scale. To develop explanations based only on events within a nation would leave much of the landscape untouched. Second, the mark of any science is its search for general principles. The theories that social scientists propound and appraise – about human behaviour or institutional arrangements – should apply across political environments, not just to unique localities. The interest then in advancing social science makes it imperative to search for recurring tendencies and patterns. Finally, so rapid has been progress in social science research that knowledge of social, political, and economic systems is now widely dispersed around the globe. The elaboration of general principles requires ever greater communication about research designs, theoretical structures, and study findings.

It is the recognition of the changing nature of international research that led the National Science Foundation (NSF) to join with the European Science Foundation in supporting the First Summer Institute on Geographic Information upon which this volume is based. So swift have been the development and deployment of geographic information systems on both sides of the Atlantic that the time and subject seemed ripe for greater international collaboration. The need for improved conceptual frameworks is so important, that failure to consolidate the knowledge that has been developed across the Atlantic would reduce the possible contributions, to policy and theory, of research that uses geographic information systems. Hence, the Directorate for the Social, Behavioural, and Economic Sciences of NSF regards its investment in the Summer Institute as a contribution to the storehouse of wisdom about the significance of geographic setting to social life. This volume represents only one outcome from the collaboration that the NSF hopes to promote between social and behavioural scientists in the United States and our peers elsewhere.

We in the Directorate are especially proud to have supported an activity targeted towards young scientists. We contend that international experience at the beginning of careers builds a habit of outreach. In addressing the problems with which social scientists grapple and to discern the bases of human organisation, such a habit is indispensable to contemporary social science.

DR CORA MARRETT
Assistant Director of the Social and Behavioural Science Division of the
National Science Foundation

Preface

In the summer of 1995, 52 researchers met at the Wolfe's Neck Conference Centre of the University of Southern Maine, USA, for the First International Summer Institute in Geographic Information. The meeting was sponsored jointly by the European Science Foundation (ESF) GISDATA programme and the US National Science Foundation (NSF) through the National Center for Geographic Information and Analysis (NCGIA). The event was extraordinary in a number of ways. First, the participants came in equal numbers from both sides of the Atlantic, a very unusual feature compared with most international geographic information meetings which tend to be dominated by either European or American participants. Second, the duration of the Institute, which included six full days of meetings and one day for technical excursions, allowed for considerable extent and depth of interaction among the participants. Third, the majority of participants were at the early stages of their scientific career, as they were completing or had recently completed their doctoral research, and were all selected on the basis of open competitions, one in Europe and one in the USA, for the high quality and originality of their work. Being, in research terms, at the coal face in their particular field, they could capitalise on the extensive feedback and close interaction with colleagues from other countries doing research in the same area, and with senior instructors who are among the recognised leaders in that field. At the same time, while specialist input is obtained by interacting with colleagues from a similar background, it is often the case that innovative ideas and new ways of looking at issues emerge from interacting with those from different professional and cultural traditions. This volume includes most of the papers presented at the Institute, revised following peer review, as well as direct feedback from colleagues both during and after the meeting. In many instances the revised papers reflect not only the reviewers' specialist comments, but also the increment of wisdom gained from many hours of face-to-face discussions with other participants often coming from a different disciplinary perspective and working in a different area.

This Summer Institute, and the next one taking place in Berlin in 1996, are the flagships of the collaboration between the GISDATA scientific programme of the European Science Foundation (Arnaud *et al.*, 1993) and the NCGIA (Abler, 1987). The topical emphases of the Institute reflected those of the first six GISDATA specialist meetings held in 1993–4 and current or recent NCGIA

research initiatives in closely related areas. A similar approach is being pursued for the Second Summer Institute which covers the second half of the research agenda of GISDATA and the related research initiatives undertaken by the NCGIA. In many respects, the similarities between the areas identified by GISDATA and the NCGIA as being of highest research priority is indicative of the current concerns of the field as a whole, while the intense collaboration between the two programmes results in a critical mass of researchers addressing the issues with fewer duplications of effort.

The groundwork for the First International Summer Institute in Geographic Information was laid out in October 1994, when the members of the organising committee from both sides of the Atlantic met at the headquarters of the NSF in Arlington, Virginia. That meeting was also attended by John Smith, Scientific Secretary at ESF, as well as by several NSF officials. In keeping with its high-level mandate, the panel determined that the Summer Institute should meet three concurrent goals:

- promotion of basic research in geographic information;
- human resource development; and
- development of international cooperation between US and European scientists.

This volume is the tangible evidence that the first goal was met. The very high quality of most of the papers presented at the Institute, and the willingness of an international publisher of the reputation of Taylor & Francis to publish them, attest to the research contribution that was made. The significance of these papers, individually or as pieces of larger units, is further discussed in the Introduction. The achievement of the other two goals is better evaluated in hindsight. Here we can only outline the means by which the organising committee strived to maximise the value of the Summer Institute for the select international group of young (and not-so-young) scientists who participated.

A critical aspect of the success of the Institute was in bringing together early-career scientists with a substantial subset of the best known senior researchers in geographic information. Of the 52 participants, 31 were young scholars selected on the basis of the two parallel open competitions, and 21 were

Table 1 Themes of the First ESF/GISDATA-NSF/NCGIA Summer Institute on Geographic Information, Wolfe's Neck, Maine, 1995

NSF/NCGIA		ESF/GISDATA
Representing spatial data quality	1	GIS and multimedia
Formalising cartographic knowledge	2	Generalisation
Human and organisational factors in GIS implementation	3	GIS diffusion in Europe
Collaborative spatial decision support	4	GIS and spatial analysis
Cognitive perspectives on spatio-temporal reasoning	5	Spatial conceptual models for geographic objects with undefined boundaries
National spatial data infrastructure	6	European data availability

internationally known researchers, most of whom had been directly involved in either the GISDATA specialist meetings or the NCGIA research initiatives upon which the Institute's topical emphases were drawn. In their evaluations of the Institute, the young researchers gave their most glowing ratings to the active presence and day-long availability of several of the 'living legends' in the field. These senior scientists gave keynote presentations, taught mini-workshops, led or participated in the *ad hoc* research project teams that were formed, gave constructive comments to the junior paper presenters, argued vigorously among themselves, co-judged the team projects submitted, guided the two field trips to the Orono (Maine) NCGIA facility and the MIT labs in Cambridge (Massachusetts), and were enthusiastic participants in the jolly 'happy hours' that closed each hard day's work.

The Institute's programme was set up to balance plenary and small-group sessions, lectures and hands-on challenges, academic papers and practical workshops, structured and unstructured events. Above all, it sought to encourage the thorough mix of nationalities, research perspectives, and levels of academic experience, for the purpose of helping prepare tomorrow's experts for the increasingly international and multidisciplinary working environments they are likely to function in. The key device used at the Institute to foster that kind of collaboration was the formation of five research teams, each working on a preliminary research proposal on a preassigned 'Big Question'. These questions were deliberately chosen by the organisers to be orthogonal to the thematic emphases of the Institute, so that no one present was an expert in any of them. Rather, the criteria for picking these topics were (a) that they have an international dimension, (b) that they transcend the social/physical division and (c) that they be important enough to go beyond the headlines of the day. Not everyone felt comfortable with having to work outside his or her own, often narrow area of expertise, and at the time some even resented being pushed into what they saw as a dilettantish frame of mind – so contrary to the highly specialised training he or she had received. Still, some very creative and imaginative ideas resulted from that exercise, and most participants subsequently praised the value of the unconventional teamwork. Indeed, there were very few things (mosquitoes aside) that the 1995 Summer Institute participants did not like about it. No doubt, the secluded, bucolic setting of the Wolfe's Neck conference facility on the Maine coast, with the sounds and smells of grazing cows on the one side and genteel lawn, garden and woods on the other, contributed to making this a 'total' experience for all involved.

Of all that multifaceted happening that was the Summer Institute, this book can only reflect the formal academic dimension. The young scholars' papers were solicited and submitted in the following 12 paired research areas, each of which was also addressed in a keynote paper by one of the instructors (Table 1). These six double themes represent some of the key research topics identified by GISDATA and the NCGIA in their respective research agendas. Specialist meetings and other research activities have already been held on them, or will be in the near future.

The Institute was organised by a small international panel including Keith Clarke, Helen Couclelis, and Harlan Onsrud from the US, and François Salgé, Hans-Peter Bähr, Jean-Paul Donnay, and Massimo Craglia from Europe. Local arrangements were handled superbly by Kathleen Hornsby of NCGIA at Orono,

Maine, while Sandi Glendinning and LaNell Lucius at the Santa Barbara, California NCGIA office dealt with the more general logistics of the Institute. We are extremely grateful to all those involved in the organisation of this meeting, whose commitment and enthusiasm have been critical to its success. Michael Goodchild and Ian Masser spearheaded the NCGIA-GISDATA collaboration of which the two Summer Institutes are the most visible products. The organisers are equally grateful to the sponsors of these meetings, the European Science Foundation and the National Science Foundation which were represented at Wolfe's Neck by Dr Guido Martinotti, Chairman of the ESF Standing Committee for the Social Sciences, and Dr Cora Marrett, Head of the NSF Social and Behavioural Science Division, respectively. The importance of this initiative for the future researchers and teachers in the GIS field cannot be overemphasised.

Finally, a particular debt of gratitude is owed to all the manuscript referees who have at times been put under a lot of pressure to provide feedback within a very short time; and to Christine Goacher at Sheffield who has undertaken the gruelling task of standardising the format of all the chapters while never losing her good sense of humour.

MASSIMO CRAGLIA
HELEN COUCLELIS

References

ABLER, R. (1987). The National Science Foundation National Center for Geographic Information and Spatial Analysis, *International Journal of GIS*, **1**(4), 303–26.

ARNAUD, A., CRAGLIA, M., MASSER, I., SALGÉ, F. and SCHOLTEN, H. (1993). The research agenda of the European Science Foundation's GISDATA scientific programme, *International Journal of GIS*, **7**(5), 463–70.

Contributors

William Albert
Department of Geography, Boston University, 675 Commonwealth Avenue, Boston MA 02215, USA, walbert@crsa.bu.edu

Theo Arentze, Aloys Borgers and Harry Timmermans
Urban Planning Group, Technische Universiteit Eindhoven, Den Dolech 2, PO Box 513, 5600 MB Eindhoven, THE NETHERLANDS, t.a.arentze@bwk.tue.nl

Dimitris Assimakopoulos
Department of Town and Regional Planning, University of Sheffield, Sheffield S10 2TN, UK, d.g.assimakopoulos@sheffield.ac.uk

Bijan Azad
COMAP, Ministry of Finance, Riad El-Solh Square, Beirut, LEBANON, pit@inco.com.lb

Alexander Baklanov and Tatjana Makarova
Institute of North Ecology Problems, Kola Science Centre 14 Fersman Str., Apatity, Murmansk Region, 184 200, RUSSIA

Kate Beard
NCGIA, Department of Spatial Information Science and Engineering, University of Maine, 5711 Boardman Hall, Orono ME04469-5711, USA, beard@spatial.maine.edu

Lars Bodum
Department of Development and Planning, Aalborg University, Fibigerstraede 11, DK 9220 Aalborg, DENMARK, ibo@i4.auc.dk

Nina Bullen
MIDAS – Manchester Computing, University of Manchester, Oxford Road, Manchester, M13 9PL, UK, zzaanib@cs6400.mcc.ac.uk

Barbara Buttenfield
Department of Geography, University of Colorado, Campus Box 260, Boulder, Colorado 80309-0260, USA, babs@whitney.Colorado.edu

Antonio Câmara
Environmental System Analysis Group, New University of Lisbon, 2825 Monte da Caparica, PORTUGAL, asc@mail.fct.unl.pt

Sue Collins
Sheffield Centre for Geographic Information and Spatial Analysis, Geography and Planning Building, University of Sheffield, Sheffield S10 2TN, UK, s.collins@sheffield.ac.uk

Helen Couclelis
NCGIA, 3510 Phelps Hall, University of California, Santa Barbara CA 93106, USA, cook@geog.ucsb.edu

Massimo Craglia
Department of Town and Regional Planning, University of Sheffield, Sheffield S10 2TN, UK, GISDATA@sheffield.ac.uk

Michael Domaratz
Federal Geographic Data Committee, US Geological Survey, 590 National Center, Reston, VA 22092, USA, mdomarat@usgs.gov

Daniel Dorling
Department of Geography, University of Bristol, Bristol BS8 1SS, UK, danny.dorling@bristol.ac.uk

John Evans
Massachusetts Institute of Technology, 77 Massachusetts Avenue, Room 9-514, Cambridge MA 02139 USA, jdevans@mit.edu

Mark Finco
Department of Geography, University of Utah, Salt Lake City, UT 84112 USA, mark.finco@geog.utah.edu

Peter Fisher
Department of Geography, University of Leicester, Leicester LE1 7RH, UK, pff1@le.ac.uk

Alexandra Fonseca
CNIG, Rua Braamcamp, n° 82, 5E, 1200 Lisboa, PORTUGAL, xana@helios.cnig.pt

Christopher Giertsen, Oddmar Sandvik and Rune Torkildsen
Christian Michelsen Research, Fantoftvegan 38, 5036 Bergen-Fantoft, NORWAY, chrisgie@cmr.no

Michael Goodchild
*NCGIA, 3510 Phelps Hall, University of California, Santa Barbara CA 93106,
USA*, good@goodrs.geog.ucsb.edu

Jean-Philippe Lagrange
CEGN, 16 bis Avenue Prieur de la Côte d'Or, 94114 Arcueíl Cedex, FRANCE,
lagrange@ctme.etca.fr

Xavier R. Lopez
*School of Information Systems and Management, University of California,
Berkeley,* CA 94720, USA, xavier@sims.berkeley.edu

Marianne MacDonald
*Urban and Regional Planning Program, University of Colorado, Denver
Campus Box 126, PO Box 173364, Denver CO 80217 USA,*
mmacdona@carbon.cudenver.edu

David Mark
*NCGIA and Department of Geography, State University of New York at
Buffalo, Buffalo NY 14261, USA*, dmark@geog.buffalo.edu

Ian Masser
*Department of Town and Regional Planning, University of Sheffield, Sheffield
S10 2TN, UK*, i.masser@sheffield.ac.uk

Robert McMaster
*Department of Geography, University of Minnesota, Minneapolis, MN 55455,
USA*, mcmaster@cassini.geog.umn.edu

Terje Midtbø
*Department of Surveying and Mapping, Norwegian Institute of Technology,
N-7034 Trondheim, NORWAY*, terjem@iko.unit.no

Larissa Nazarenko
*Institute of Ocean Sciences, 9860 West Saanich Road, PO Box 6000, Sidney,
British Columbia, CANADA*, lara@ios.bc.ca

Zorica Nedović-Budić
*Department of Urban and Regional Planning, University of Illinois, 611 East
Lorado Taft Drive, Champaign, IL 61820, USA*, budic@uiuc.edu

David Olson
*City University of New York Graduate Center, 33 West 42 Street, Room 1444,
New York, NY 10036, USA*, dolson@email.gc.cuny.edu

Harlan J. Onsrud
*NCGIA, Department of Spatial Information Science and Engineering, University
of Maine, 5711 Boardman Hall, Orono ME04469-5711, USA,*
onsrud@spatial.maine.edu

Dimos Pantazis
University of Liège, Department of Geography, 7 Place du 20 Août, B-4000 Liège, BELGIUM, pantazis@geo.ulg.ac.be

Jeffrey K. Pinto
School of Business, Penn State University, Erie, PA 16563, USA, jkp4@psuvm.psu.edu

Corinne Plazanet
Institut Geographique National, 2 Avenue Pasteur, 94160 Sainte-Mandé, FRANCE, plazanet@cogit.ign.fr

Jonathan Raper
Department of Geography, Birkbeck College, 7–15 Gresse Street, London W1P 1PA, UK, j.raper@geog.bbk.ac.uk

Sigrid Roessner
GeoForschungs ZenTrum, Potsdam, Section 1.5 – Remote Sensing, Telegrafenberg A17, D14473 Potsdam, GERMANY, roessner@gfz-potsdam.de

François Salgé
Director CERCO/Groupe MEGRIN, abs IGN, 2 Ave. Pasteur, BP 68, 94160 Saint Mandé, FRANCE, salge@megrin.ign.fr

Michela Spagnuolo
IMA-CNR, via de Marini 6 Torre di Francia, 16149 Genova, ITALY, spagnuolo@ima.ge.cnr.it

Sabine Timpf
TU Vienna, Department of Geoinformation, E127/1 Gusshausstrasse 27-29, 1040 Vienna, AUSTRIA, timpf@geoinfo.tuwien.ac.at

Nancy Tosta
Puget Sound Regional Council, 1011 Western Avenue, Suite 500, Seattle, Washington, WA 98104–1035, USA
tosta@psrc.wa.com

David Unwin
Department of Geography, Birkbeck College, 7-15 Gresse Street, London W1P 1PA, UK, d.unwin@geog.bbk.ac.uk

Andrej Včkovski
Department of Geography, University of Zürich, Winterthurerstrasse 190, CH-8057 Zürich, SWITZERLAND, vckovski@gis.geogr.unizh.ch

Kongjian Yu
Swa Group, 580 Broadway, Suite 200, Laguna Beach, CA 92651, USA, Yukongjian@aol.com

May Yuan
*Department of Geography, University of Oklahoma, 100 E. Boyd Street, Room
684, Norman OK 73019, USA*, myuan@uoknor.edu

The **European Science Foundation** (ESF) is an association of its 59 member research councils, academies, and institutions devoted to basic scientific research in 21 countries. The ESF assists its Member Organisations in two main ways: by bringing scientists together in its Scientific Programmes, Networks and European Research Conferences, to work on topics of common concern; and through the joint study of issues of strategic importance in European science policy.

The scientific work sponsored by ESF includes basic research in the natural and technical sciences, the medical and biosciences, the humanities and social sciences. The ESF maintains close relations with other scientific institutions within and outside Europe. By its activities, ESF adds value by cooperation and coordination across national frontiers and endeavours, offers expert scientific advice on strategic issues, and provides the European forum for fundamental science.

GISDATA is one of the ESF Social Science scientific programmes and focuses on issues relating to Data Integration, Database Design, and Socio-Economic and Environmental applications of GIS technology. This four-year programme was launched in January 1993 and is sponsored by ESF member councils in 14 countries. Through its activities the programme has stimulated a number of successful collaborations among GIS researchers across Europe.

The US **National Science Foundation** (NSF) is an independent federal agency of the US Government. Its aim is to promote and advance scientific progress in the United States. In contrast to other federal agencies which support research focused on specific missions (such as health or defence), the NSF supports basic research in science and engineering across all disciplines. The Foundation accounts for about 25 per cent of Federal support to academic institutions for basic research.

The agency operates no laboratories itself but does support National Research Centres and certain other activities. The Foundation also supports cooperative research between universities and industry and US participation in international scientific efforts.

The National Center for Geographic Information and Analysis (NCGIA) is one of the National Research Centres funded by the NSF. It is a consortium comprised of the University of California at Santa Barbara, the State University of New York at Buffalo, and the University of Maine. The Centre serves as a focus for basic research activities relating to geographic information science and technology. It is a shared resource fostering collaborative and multidisciplinary research with scientists across the United States and the world.

CHAPTER ONE

Introduction

MASSIMO CRAGLIA

The title of this book, *Geographic Information Research: Bridging the Atlantic*, captures its two most salient features: the extent to which it focuses on a broad range of issues which are central to current research in geographic information, and in which technology is only one aspect, and the degree to which the book builds in equal measure on the different cultural and research traditions existing on both sides of the Atlantic. This aspect is critical as emphasised by both Guido Martinotti and Cora Marrett in their Foreword. Whilst past collaborative research efforts and academic outputs have tended to involve only a few individuals or institutions, this is the first time that the European Science Foundation and the (US) National Science Foundation, who are key players in their respective continents for the promotion of basic research, have joined hands to address common issues within the social sciences. It is also very significant that geographic information research was chosen as the area for this first joint effort as it confirms the perceived strategic importance of this field on both sides of the Atlantic (see Chapters 2 and 3).

As the book demonstrates the research issues addressed are global in nature and it is therefore extremely important that the richness of the field in terms of its multidisciplinarity and the variety of research traditions it can draw upon are fully exploited whilst minimising the duplication of efforts. At the same time, it is clear that in some research areas, particularly those less technical in nature and drawing from a larger pool of disciplines, there are differences in paradigms and approaches across Europe and the USA which will only benefit from increased dialogue and comparison of findings to advance the field further.

With this in mind, the objectives of this book are to review some of the key research areas in geographic information, identify the state of the art and issues where further progress is needed, present some of the latest research being undertaken in these areas across both sides of the Atlantic, and provide insights as to where the field as a whole might be going.

The book is divided into six sections: two are broadly concerned with issues of data integration (geographic data infrastructures, GIS diffusion and implementation), two are more technical and conceptual in nature (generalisation, concepts and paradigms), and two reflect to a larger extent the application-driven nature of GIS technology (spatial analysis and multimedia). These six themes represent some of the key research topics identified by GISDATA and the NCGIA in their research agendas and on which collaborative specialist meetings and initiatives have already been held or are in progress. Each section is introduced by one or more chapters highlighting the key research issues and includes further chapters exploring some of these issues in greater depth with complementing perspectives from European and American researchers.

The first two sections have in the past been regarded as 'soft' areas of research by comparison with the more technical issues dominating the GIS research agenda. In this volume they are placed up front in recognition of the important issues they raise not only for the research community but also for policymakers and society as a whole. As data infrastructures and diffusion of information technology are increasingly perceived as strategic for economic and social development on both sides of the Atlantic, it is a particular duty of GI researchers to raise awareness of the opportunities but also warn about the possible dangers of unchecked developments in these fields that may be detrimental to personal privacy and reinforce the power of a few at the expense of the many. It is also worth pointing out that to date the pool of researchers investigating these issues is relatively small compared with the technical or application-driven areas. Therefore, placing them at the fore is intended as a signal of their critical importance.

1.1   Part One. Geographic data infrastructure

In Chapter 2, Nancy Tosta and Michael Domaratz discuss some of the key issues related to the development of the US National Spatial Data Infrastructure or NSDI. The authors describe the main components of the NSDI, the National Digital Geospatial Data Framework, the Clearinghouse, and Metadata Standards, and the (largely organisational) issues being faced in implementing it. What is worth noting, particularly from a European perspective, is the highest political backing that this initiative has through the commitment of both President and Vice-President of the USA, and the extent to which the NSDI has become a model for other initiatives in Europe and the Far East.

The very rapid developments that have also taken place in Europe to make a case for a European Geographic Information Infrastructure are analysed by Ian Masser and François Salgé in Chapter 3. These developments, which are largely driven by government and industry and coordinated by the European Commission DG XIII (Information Markets and Exploitation of Research), are compared by the authors with the equivalent developments in the USA and to the priority areas identified by the European GI research community represented by the GISDATA programme. From this comparison the authors argue that much more research is needed on four key issues which are underestimated by current policy documents: coordination in the acquisition and maintenance of geographic data, copyright, the implications of the increasing commodification of geographic information, and the protection of privacy and confidentiality.

The different approaches pursued by governments across the Atlantic in respect to government-held information are explored and evaluated by Xavier Lopez in Chapter 4. Using the dissemination of geographic information as a test bed for evaluating the impacts of national information policies, the author argues that putting government information into the public domain and relinquishing copyright is the most appropriate approach for fostering democratic control and the development of a vibrant private market based on value-added products. Implementing this approach is however not easy, particularly in a complex institutional environment such as Europe, nor unproblematic in respect to the trade-offs between revenue generating and quality of the data provided.

In Chapter 5, John Evans examines the factors that contribute to the development of effective infrastructures for information sharing. Using two examples of multiagency initiatives in the USA, the author highlights the importance of multiple levels of connectivity among the partners, the feasibility of non-intrusive approaches to data sharing, and assesses the strength and weaknesses of current networked tools as the basis for the sharing infrastructure. The author concludes that key issues for effective sharing are organisational and human rather than technical including agreement on superordinate goals and well-developed teamwork.

In the final chapter of Part One, Daniel Dorling uses the case study of the New Social Atlas of the UK he has produced to illustrate the important issues raised by the availability of very large sets of micro-spatial data (that is census data for some 10 000 spatial units). Although data at such detailed level is not yet available at supra-national scale, the techniques developed by the author to visualise the data in a meaningful way are undoubtedly an important contribution to the fields of cartographic visualisation, spatial analysis, and the social sciences as a whole.

1.2 Part Two. GIS diffusion and implementation

Ian Masser and Massimo Craglia in Chapter 7 provide an overview and a comparative evaluation of GIS diffusion in local government in nine European countries based on a series of studies and surveys undertaken in each country by a local team of experts. The authors argue that local government is a particularly interesting area to study GIS diffusion because it covers a wide range of applications in a variety of settings. It also has a strong national dimension which is reflected in the expectations that people have of their local authorities, the tasks they carry out, and the professional cultures that have developed to support them. From the analysis of the data Masser and Craglia suggest that two key factors account for most of the differences among countries. The first of these measures links digital data availability and GIS diffusion, while the second reflects the professional cultures surrounding GIS in local government and the GIS applications developed.

The role of professional cultures and networks in the diffusion process is further explored by Dimitris Assimakopoulos in Chapter 8. Taking advantage of the small size of the country and the relatively early stages of GIS diffusion in Greece, the author examines this process through the evolution of the scientific and technological communities in that country. Adopting this novel approach, the author develops a theoretical model which is like a cognitive map of the diffusion journey plotted on the real network of key individuals, teams and the scientific and professional groupings who socially shape the adoption and implementation process of GIS in Greece.

Whilst these first two chapters of Part Two deal essentially with the adoption of innovations, the following three chapters focus on the complex issues of implementation. In their overview of the field (Chapter 9) Jeff Pinto and Harlan Onsrud highlight the lack of consensus among researchers on what constitutes 'successful' implementation. Building on the management information systems literature, the authors put forward a parsimonious model of system implementation success that takes into account a wide range of factors and synthesises them into a framework which the authors argue can be used by researchers and managers in assessing accurately their system's efficacy for organisational operations.

In Chapter 10, Bijan Azad explores the role and activities of the GIS champion(s) in a case study of GIS implementation in a large USA organisation. This is a valuable contribution as the role of the champion is often seen as critical in the literature on system and policy implementation, and more generally in guiding an organisation through change, and yet what it is that champions actually do is often unclear. The author identifies some of the key activities undertaken by the champion(s) in his case study and demonstrates the value of conducting further work in this important area which will extend the knowledge base and contextualise the role of the champion in different settings.

In the final chapter of this section, Zorica Nedović-Budić takes a broader look at the implementation process by focusing on the mutual adaptation of technologi-cal innovations and organisations. The author summarises previous research on organisational change and the diffusion of new technology and reviews the issue of organisational-technology fit. Through four case studies of GIS implementation in USA local government agencies, the author demonstrates the impact of organisational contextual factors on the process of GIS initiation and implementa-tion, and advocates the use of the socio-technical system design: that is, an approach focusing not only on the technology but also on the social setting.

1.3 Part Three. Generalisation

Whilst Parts One and Two of the book emphasise in different ways the importance of social and political processes at the level of the individual, organisation, and society for the diffusion, implementation, and use of information technology and infrastructures, it is also clear that continued research into improving both techniques and technology is vital. One area of research which has received much attention given the central role of cartography and surveying in the historical development of GIS is that of cartographic and model generalisation. The considerable progress made in this field has also been possible because of the intense collaboration among European and American researchers in this area fostered *inter alia* by both the GISDATA programme and the NCGIA through their specialist meetings and research initiatives. These meetings and initiatives are the starting point for the two overview chapters by Jean-Philippe Lagrange, Robert McMaster and Barbara Buttenfield. They are followed by four chapters addressing some of the issues identified in the overview as deserving further work, while Kate Beard concludes this section by focusing on a critical and related topic, that of data quality, which is also one of the themes of the second GISDATA-NCGIA Summer Institute taking place in Berlin in 1996.

In Chapter 12, Jean-Philippe Lagrange provides a broad and candid overview of the field of generalisation highlighting areas of significant progress as well as those where there is much work yet to be done. In particular, key issues where further research is needed include ideas and techniques to analyse objects in respect to their nature and shape to arrive at fully automated algorithms; conflict resolution, which remains one of the most serious stumbling blocks to larger automation, and understanding the role of human experts who are still indispensable to control and sequence the generalisation process. This requires further research into cognition, perception, mental planning etcetera. Lastly, further work is needed into quality monitoring and measurements to validate tools and solutions, monitor tool application and evaluate intermediate results. Robert McMaster and Barbara Buttenfield in Chapter 13 focus specifically on formalising cartographic knowledge – on the acquisition and representation of the knowledge necessary to develop an expert system for map generalisation.

Although significant progress has been made in this area as reviewed by the authors with the development of numerous prototype systems, full production systems are not yet widespread. Moreover, there are still areas such as cartographic symbolisation where little has been done beyond developing a conceptual framework. This overview chapter nicely complements, from an American perspective, the previous one by Lagrange which is more Europe-centred.

In Chapter 14, Sabine Timpf proposes a multiscale hierarchical spatial model for cartographic data in which objects are stored with increasing detail and can be used to compose a map at a particular scale. This approach leads to a multiscale cartographic forest when applied to cartographic mapping. The author explains the structure of this forest which is based on a trade-off between storage and computation requirements whilst methods for selecting cartographic objects to be rendered are based on the principle of equal information density.

The efficient handling of large data sets is explored by Terje Midtbø in Chapter 15 through the use of Delaunay triangulation. The author uses a qualified selection of points to extract representative information from the data set, and by inserting or removing the points thus selected he is able to use models of changeable resolution depending on the analysis to be performed. When surface models become extensive a block-based management of the data is proposed in which a central feature is the ability to merge blocks together. Finally, the author draws parallels between Delaunay networks of two and three dimensions and argues that incremental Delaunay tetrahedrisation and qualified point selection pave the way for generalised models of tetrahedral networks.

Michela Spagnuolo in Chapter 16 proposes a methodology to reconstruct surfaces from irregularly distributed points. This methodology is based on the premise that all information explicitly or implicitly contained in the available data set should be exploited to delineate a morphological structure on which the final surface model has to be adapted. Using the example of a project in Antarctica in which measures of the sea floor are available in parallel vertical sections (profiles), the author illustrates a prototype system in which a shape-driven generalisation is performed on the profiles using wavelets techniques to preserve the main features of the profiles. Then the structural features of the sea floor are searched through the set of simplified profiles as patterns of similar shapes across adjacent profiles. Finally, the recognised features are used as a skeleton for building the sea floor model defined as a triangulation constrained to the shape features.

In Chapter 17, Corinne Plazanet highlights the importance of modelling the geometry of linear features for their generalisation. The author proposes a hierarchical process for segmenting lines and qualifying the sections detected according to different criteria and levels. The aim of this research is to guide the choice of sequences of generalisation operations and algorithms and provide basic tools for assessing the quality of the results of generalisation alternatives in terms of shape maintenance. Together with the previous chapter by Spagnuolo, this chapter demonstrates the progress that has been made in the analysis of features and features' shapes, one of the critical areas of research in generalisation.

Data quality is the focus of the final chapter of this section (Chapter 18) by Kate Beard who argues that in spite of the increasing emphasis on data quality from producers, users and vendors, there is as yet no agreement on how best to distribute information on data quality, how to use it, and how software developers can support both producers and consumers. Research on analysing the accuracy and understanding the uncertainty of spatial data has progressed, but little of this work has been incorporated in operational systems. With this in mind, the author proposes a conceptual model for representing data quality and addresses three cases: a static deterministic model, a dynamic deterministic model, and a stochastic model.

1.4 Part Four. Concepts and paradigms

Whilst Part Three has illustrated the extent of progress that can be achieved when a relatively well-socialised group of international researchers from similar backgrounds work together, the following Part Four on conceptual issues demonstrates the richness and multidisciplinarity of GI research which makes it such a challenging environment to be working in. The two introductory chapters by Peter Fisher and David Mark provide a challenging analysis of the limitations of current paradigms in spatial data processing and an overview of cognitive perspectives in spatial and spatio-temporal reasoning respectively. These are followed by five chapters developing the central themes of how we acquire knowledge about geographic space and how to develop new models for representing and manipulating space and time in a GIS environment.

In Chapter 19, Peter Fisher argues that learning the paradigms related to an area of scientific endeavour is the essential rite of passage from student to practitioner and that the role of the researcher is to probe the usefulness of those paradigms and contribute to the progression of the subject area from one paradigm to another. In his chapter, he reviews and identifies the many limitations of some of the historical paradigms permeating the area of GI research especially the Boolean map, the layer-based view of the world, the raster-vector debate, objects and fields, and the metric of space. The author concludes that recent developments in the field may lead to shifts towards new paradigms although only time will tell whether these changes will actually take place.

David Mark paints a broad picture in Chapter 20 of the field of spatial cognition arguing that an understanding of this field is critical in explaining and predicting human behaviour in geographic space. This understanding may not only contribute to developing fundamental theory in geographic information science, but is also at the base of improving the design of information systems to support human decision making in geographic space. The author suggests that a philosophical stance termed experiential realism is a useful background for spatial cognition research and that the categorisation of kinds of things, attributes, processes, and relationships is central to this research area. To make progress in this field, Mark argues that a variety of disciplines including psychology, linguistics, artificial intelligence and others from the cognitive science domain can provide useful inputs, but that it is critical that formal definitions of all the assumptions underlying the models used are provided to allow differences in the empirical results to be accounted for, leading ultimately to unified models of human spatial and spatio-temporal cognition.

The need to develop comprehensive models of spatial cognition to account for the differences among individuals and human societies is illustrated by William Albert in Chapter 21. The author presents the results of a series of tests aimed at verifying two hypotheses: that visual memory is a good predictor of both route and survey knowledge, and that spatial orientation is a better predictor of survey knowledge than route knowledge as it requires the development of a frame of reference which is essential in many survey level spatial tasks. The tests confirm only in part these hypotheses, but above all show that differences in individual cognitive abilities account to a large extent for the total variance in performance in the spatial judgement tasks tested, thus confirming the need for further research in this field.

May Yuan proposes in Chapter 22 a conceptual data model for representing semantic, temporal, and spatial information in GIS. The key features of the model are that GI is modelled by objects in three domains (temporal, semantic, and spatial) with topological constraints used as the primary control of data integrity to construct and manage these objects. Different ways of linking the objects in the three domains are presented so that the GIS is able to represent reality from locational-centred, entity-centred, and time-centred perspectives. Whilst demonstrating the substantial progress made, the author also argues that further research is needed in three areas: geographic semantics; defining geographic entities from different perspectives; and associating language as well as logic system within and among the domains of semantics, space, and time.

In Chapter 23, Dimos Pantazis reviews some of the deficiencies of the classic formalisms for database design, such as entity/relationship, which are also commonly applied to design spatial databases. These deficiencies are particularly evident when modelling different types of geographic data, their spatial reference, and their topological relationships. With this in mind, the author illustrates a new formalism named CON.G.O.O. (Conception Géographique Orientée Objet), the main advantages of which are the ability to classify geographic objects and classes into different types, distinguish different types of relationships between objects, identify and represent all the topological relationships between all the geographic objects of the database with only two concepts, adjacency and superimposition, establish relationships not only between objects but also between objects and classes and objects and layers, and classify attributes and treatments in object attributes and treatments, and classes attributes and treatments encapsulating then these characteristics within the objects and classes. Pantazis argues that this new formalism provides models that facilitate communication between users, programmers and analysts of geographic databases because of their capacity to describe easily very complex environments.

Mark Finco focuses in Chapter 24 on the role of pattern recognition in inductive geographic investigation. As research moves in increasingly rich-data environments, such as in remotely sensed imagery, the author develops Openshaw's (1994) views that new styles of analysis are needed which are data driven and exploratory, looking at data for explanation while minimising the influence of pre-set theories and biases. Against this background, Finco illustrates a method for pattern recognition in N-dimensional remotely sensed imagery based on cellula automata and artificial life. The Artificial Explorer he is developing uses the same conceptual model of human beings when mapping the space around them. Therefore, it measures geographic space (what is 'close') and compares attributes' characteristics (what is 'similar') for every cell in the data set. Progress on the Explorer is well advanced and Finco expects to produce both tangible products (for example, computer code) and a better understanding of what pattern is and how it might be recognised at various scales using decentralised techniques.

The final chapter of this section (Chapter 25) is by Andrej Včkovski who describes the concepts and implementation of an object-oriented data repository, virtual data sets, that extends a data set with access methods that simulate continuity. The starting point for the discussion are the incompatibilities emerging when a continuous field is sampled and therefore discretised. The author argues that when a data set is used for a variety of different applications, it is likely that the initial structure of the data set will not fit the needs of every application. Therefore, there is a need for a generalised representation of a field, one that not only contains a set of sampled values but also derived information, that is a representation that models the field. On this basis, Včkovski discusses in some depth the key issues related to the sampling of continuous fields and then presents his object-oriented virtual data set. This is a representation containing information which allows field values at unsampled locations to be calculated on request, thus simulating continuity. In the implementation proposed (in Java language), the calculations are hidden to the user who gets a 'virtual' data in the sense that is data without persistence or secondary storage but only created at the query stage.

1.5 Part Five. GIS and spatial analysis

The last two sections of the book reflect to a larger extent the application-driven nature of GIS technology and focus on GIS and Spatial Analysis and GIS and Multimedia. Part Five includes nine chapters and is therefore somewhat larger than the other sections. This reflects the enormous interest and amount of research in this area as analysing spatial data is perceived (together with making maps) as one of the main features of GIS and a central activity of geography as a discipline. Given that it draws from such a large pool of research activities, it is inevitable that the chapters in this section are less homogeneous than in other sections of the book, with some more technical in nature and others being more descriptive. This is by no means a negative connotation as it testifies to the variety of domains to which GIS methodology can be applied.

The section is introduced in Chapter 26 by David Unwin who identifies three major assumptions in the current literature, namely that something called spatial analysis can be defined, that it is useful in work using GIS technology, and that it *will* be applied. Unwin distinguishes between four views of spatial analysis:

spatial data manipulation, a basic feature of all GIS and a relatively unproblematic one, spatial statistical analysis, which is something few GIS if any are good at and often requires the coupling of GIS with other software, a geography-based and data-driven view of spatial analysis which finds its origins in the quantitative revolution of the 1960s and 1970s, and finally a spatial modelling view to be found in decision-support systems and dynamic modelling environment. The author focuses in particular on the second and third view, that of 'GISable' statistics and assesses the potentials and limitations of current approaches. Above all he concludes, one of the key problems is that GIS is being taught as a technology without paying any attention to its theoretical underpinnings '. . . thus we remain ossified with a rag bag of functions that could have been (and were) implemented in software over 20 years ago. If we cannot get our act together on this it seems unlikely that we will persuade our colleagues in other disciplines who use GIS, still less the GIS vendors, to do anything about providing the necessary tools [to overcome current deficiencies]'.

The chapter by Nina Bullen provides a good example of what Unwin defined as spatial statistics which requires the loose coupling of specialist software with GIS in view of the deficiencies of current systems. Bullen presents her doctoral research into modelling the geographical variations of house prices at several spatial scales simultaneously. Whilst previous research by the author and her colleagues had developed multilevel models of individual properties nested into clearly defined, but not necessarily the most appropriate, existing spatial units (for example administrative units or travel-to-work areas), her present work focuses on identifying more germane spatial units. For this purpose, a specialist software for multilevel modelling is coupled with a GIS to map the results of the model and the zoning algorithm developed by Openshaw and Rao (ZDES) is utilised to design a set of spatial units which optimise region homogeneity. In addition to being of considerable technical interest, this chapter shows the opportunities offered by GIS (in combination with other software), for the analysis of property markets, an area where GIS methods are as yet rarely used.

In Chapter 28 Sue Collins provides a very interesting example of a GIS application relevant to epidemiologists and urban planners through her work in modelling urban air pollution to assess areawide levels of risk. Using a practical case study, the author illustrates her approach which has been developed within a GIS environment utilising a combination of dispersion modelling, adapted to work within the GIS, and spatial interpolation techniques. The validation of the results confirm the effectiveness of this approach which generates estimates very close to the observed values. Given the increasing awareness of the relationships between pollution, and more generally the urban environment, and health, the work presented in this chapter is of considerable academic and practical interest.

A very different but equally interesting application is described by Sigrid Roessner in Chapter 29 who integrates data from different sources to recreate the landscape in the central Kenya rift to estimate the extent of soil erosion over the last 3 million years. Her procedure involved creating a digital terrain model of the present surface using topographic maps, remotely sensed data, and field surveys. The idealised uneroded initial surface was created by interpolating the points representing relicts of the old plateau surface. Once both surfaces were created, it was possible to calculate the denudation rates with a higher degree of reliability than using standard methods such as long-term extrapolation of present short-term measure, thus proving the advantages of a GIS-supported approach.

The two following chapters by David Olson (Ch. 30), and Alexander Baklanov, Tatjana Makarova and Larissa Nazarenko (Ch. 31) are less technical in nature but of great interest as they show how relatively simple analyses performed with a GIS can be extremely useful for social science and environmental policy-making. Given the relatively little in-roads made by GIS in the social sciences, with the exceptions of geography and possibly town-planning, the work presented by Olson in analysing voter turnout at the 1993 elections for Mayor of New York City provides an excellent example of the value of GIS for political analysts. It also demonstrates the synergy that can be achieved between researchers in different fields when comparing their similar analytical needs and methods. The chapter by Baklanov, Makarova and Nazarenko describing the development of an ecological atlas for the Kola peninsula makes fascinating reading for the subject area (assessing the radioactive pollution created by decades of nuclear testing and dumping at sea and inland), the nightmare that data gathering and collating must have been in such a sensitive area, and not least the pleasure of having a group of Russian researchers presenting their work to the international scientific community in such a candid way.

The Red Stone National Park in China is the case study for the following chapter by Kongjian Yu (Chapter 32) who integrates GIS with the approach of security patterns in landscape planning. Yu considers landscape planning as a procedure of defence enacted by different interest groups and argues that defending by Security Patterns (strategic portions and positions) may increase the efficiency of safeguarding the processes of our concern. The author defines Security Patterns, proposes alternative change models based on SPs, and discusses decision-making based on this approach. In the case study, the defending procedure among ecologists (defenders of ecological processes), tourists (defenders of visual perceptions), and farmers (defenders of agricultural conversion processes) are simulated, thus illustrating the issues raised in the first part of the chapter. Finally, Yu concludes that integrating the SP approach with GIS has considerable potential for supporting decision-making in landscape changes.

Spatial analysis for decision support is the topic of the remaining two chapters of this section. In Chapter 33 Theo Arentze and his colleagues describe the spatial decision-support system they are developing for retail planning. The system includes two components: a GIS for the management, analysis, and visualisation of the spatial data provided, and a windows-based programme called Location Planner for decision support. The latter allows users to construct a model of the retail system out of available model components to suit specific needs. Different levels are available in the system including viewing of data (domain level), performing what-if analyses (inferential level), solving well-structured problems (task level), and structuring complex problems (strategic level). The authors argue that the proposed system design improves the flexibility and interactive properties of spatial DSS and discuss the implications for improving the decision-support capabilities of existing GIS.

While the system developed by Arentze and colleagues is essentially a stand-alone, Marianne MacDonald moves one step further by looking at the use of decision-support systems in the context of collaborative (but often conflictual) decision making. Using the example of waste disposal which requires the interaction of different interests and levels of expertise, the author identifies in Chapter 34 the challenges, strength and weaknesses of DSS in collaborative planning, providing a very useful contribution to the field. Of particular interest is the concept of a facilitator who needs to understand both the technology and group dynamics to make the most effective use of the system.

1.6 Part Six. GIS and multimedia

The concluding section of the book addresses the issues of integrating multimedia and GIS. It is placed at the end not because it is perceived as being of lesser importance but, on the contrary, because many of the issues raised in the previous sections are present with renewed challenges in this research area. This is clearly indicated in the research agenda put forward by Jonathan Raper following his comprehensive overview of the spatial multimedia field (Chapter 35) which addresses recent developments in GIS and multimedia integration, hypermedia spatial databases, hypermaps, cartographic visualisation and animation, spatial user environments and usability, simulated spatial environments, virtual reality and three-dimensional GIS, spatial multimedia explorers and travel support, and computer-based learning for spatial data. The research agenda includes data sourcing, access and pricing, data integration, information structuring, user interaction, and spatial analysis of multimedia data types and simulation. Therefore, the previous sections of the book dealing with data integration, conceptual issues, and spatial analysis all have major contributions to make in this field, and vice versa.

The three chapters that follow address a range of issues identified in the research agenda. In Chapter 36, Lars Bodum is particularly concerned with the development of systems that will be used by urban planners. His starting point therefore is a survey of what planners actually do, what kind of technology they use and how they use it. This highlights that even in a country as technologically advanced as Denmark the use of GIS and spatial analytical tools in the town halls is at best limited. From this finding, Bodum suggests that the future of GIS will be much more chaotic and individualistic than imagined. Each planner will have his or her own idea of what GIS is and how to use it, and this necessitates a much more flexible and modular system able to integrate different data types and functions in an imaginary desktop. The electronic desktop metaphor is at the base of the prototype illustrated by Bodum who also advocates the development of an electronic local plan with which the planners can interact.

According to Alexandra Fonseca and Antonio Câmara, Environmental Impact Assessment (EIA) is an activity particularly suited to the utilisation of spatial multimedia. In Chapter 37 they argue that EIA is a multidimensional process, comprising several variables and phenomena, with complex inter-relationships varying in space and time. Moreover, the results of EIA have to be communicated to the public in a simple and clear way. With this in mind, they suggest that the incorporation of images, video and sound within a spatial information system might increase its ability to deal with the multiple aspects of environmental problems and that features such as navigation through the data, analysis capabilities and the communicability of the results might be considerably improved. The chapter reviews the field, indicates the opportunities created by distributed geo-referenced information systems based on the use of the World-Wide Web, and illustrates the applications of spatial multimedia to EIA through the case study of their work for the Expo'98 in Lisbon.

In the final chapter (Chapter 38), Christopher Giertsen and his colleagues propose an architecture and example of a general software for 3D visualisation and animation of Geographic Information. This architecture includes an open interface where new software modules can be easily attached and applied to create graphical objects representing the GIS-held data. The output from this module can either be a set of 2D textures to be mapped on to a terrain surface or a set of 3D objects to be positioned relative to the terrain surface. A user interface is also developed where different applications of generalised model generators are made available through templates to avoid duplications of tasks. The advantages of this approach are discussed by the authors, who then present methods for animation design where users can create an image sequence by specifying a virtual camera motion, selecting static scenes as background, and selecting and synchronising dynamic scene components.

1.7 Summary

As the many chapters of this book show, the field of GI research is extremely varied, multidisciplinary and dynamic with established areas and research traditions such as cartography, surveying, and geography finding new opportunities and challenges while completely new research issues develop, moving the field towards uncharted waters. This is most evident in the first section of the book where the transition from analogue to digital geographic databases is opening new opportunities for data integration and analysis but also for job creation through the development of value-added service and increased international competitiveness in the information society (Commission of the European Communities, 1994 and 1995; Federal Geographic Data Committee, 1994). Although these opportunities have raised the issue of data infrastructures and increased technological diffusion to the top of the political agenda both in the USA and Europe, recent research starts questioning the possible implications of unrestrained technological development and the bias in favour of innovations, whatever their nature, prevailing in society and the research establishments (for example Wegener and Masser, 1996; Pickles, 1995). As Masser and Salgé argue in Chapter 3, key research issues which have a new dimension and urgency in the digital age include privacy, copyright, access, and ultimately power in the information society. These are big questions of no easy resolution, but also ones to which the research community must give its full attention with empirical studies of opportunities and successes but also of the potential dangers and failures to counter the prevailing pro-innovation bias (Rogers, 1993).

This same pro-innovation bias has contributed to underestimating the difficulties of implementing IT (and GIS) in real-life organisations postulating instead that once an innovation has been adopted it *will* be used. However, as the chapters in Part Two show, adopting new technologies and practices is only the first step in a complex process of adaptation and reinvention which is also influenced by national, professional and organisational cultures so that it may not necessarily result in successful implementation and effective use. Steering this process to arrive at the desired results is by no means easy. One of the more worrying aspects as clearly illustrated by Pinto and Onsrud in Chapter 9 is the dangerous circularity in this research area as we seem to learn little from past experiences in fields such as Management Information Systems implementation and organisational change, which have long traditions of research and much to teach GIS managers and researchers alike. Instead, we are still searching for consensus about definitions and methods and often remain entrenched in camps holding opposing views without making much progress. Therefore, this is an area where increased international collaboration and the exchange of results is critical in advancing the field.

A good example of a research area where greater international collaboration is in place and progress is being made is that of generalisation discussed in Part Three of the book. This is partly due to the greater opportunities for socialisation of the researchers in this field who tend to share common backgrounds in cartography and surveying, and the critical role played by a few large organisations, the national cadastral and mapping agencies, in funding research on problems common to all. The field is nevertheless full of major challenges created by the transition from analogue maps to digital geographic databases, as illustrated by Lagrange and McMaster and Buttenfield in Chapters 12 and 13. Much research is still too focused on cartographic generalisation while the major issues of object-oriented, purpose-dependent generalisation and database requirements (that is model generalisation) need progressing much further. Assessing and validating existing tools and results and developing robust measures of data quality particularly in model-oriented generalisation are other key areas of research where the interactions between cartographers and experts from other disciplines need to be strengthened. Therefore, it may be argued that whilst initial progress in a research area is helped by the international collaboration of researchers sharing similar reference points and background, to overcome major thresholds a more multidisciplinary approach is needed (see also Muller *et al.*, 1995).

This is best illustrated by Part Four of the book where geographers, computer scientists, linguists, psychologists, and philosophers all have a major part to play in developing our understanding in and ability to formalise space and time (see also Frank and Kuhn, 1995). The paradigms that have dominated the GIS field for the last 20 years including the Boolean map, the layered view of the world, raster and vector data models, and crisp measurable entities may all be shifting towards new realms as their underlying positivist thinking is increasingly challenged and found wanting. The view that different cultures (but what cultures: national, ethnic, professional?) view the world differently including such 'invariants' as natural objects and their spatial relations requires a rethink of current ideas and the tools that represent them. This necessitates *inter alia* further research into cognition and semantics and novel approaches to modelling the world we live in as highlighted by the chapters in this section.

Part Five on GIS and spatial analysis makes three important points. First, that whatever the deficiencies and problems identified in the previous sections, available tools and methods can already provide useful results that further our understanding of existing processes and may lead to more informed decision-making. These applications should not be undervalued and, if anything, should lead to greater dialogue with other social sciences disciplines on the opportunities opened up by the increased availability of digital spatial data and handling technologies. Second, that the field is rich in contributions and research endeavours suggesting that considerable progress on some of the outstanding issues in the field is likely. Third, and by no means last, is a critical point made by Unwin at the end of Chapter 26, that while as researchers we strive all the time to further the field, we must not forget that as educationalists we should focus the minds of our students less on the technology and the latest fashion and much more on what type of analysis is needed, what tools are appropriate, and above all what the results mean. This will then enrich both research and practice to a much greater extent than is now the case.

The concluding section of the book deals with a topic, spatial multimedia, that has the potential to blow away a lot of the existing 'cobwebs'. A famous cartoon some time ago showed a fish being swallowed by a larger fish which is being swallowed by a larger fish, and so on. Maybe in the case of GIS and multimedia we are witnessing a similar development that may profoundly change the technology, the database design, the spatial data-handling techniques, in other words our current view of GIS. Certainly, as the chapters in this section highlight, spatial multimedia creates enormous opportunities for addressing some of the issues identified in all the previous sections including our ability to represent space and time, generalise spatial databases, convey quality, perform analyses, explore data, and model complex processes. It is very significant though for Raper to suggest that amidst all the breathtaking technological developments, data issues are even more important than technical ones: '. . . how will confidentiality and copyright be preserved in this new environment; and, if preserved will (can?) spatial multimedia develop?'(p. 539). By highlighting the challenges that new technologies pose to intellectual property and confidentiality, the book comes to a full circle as these were two of the key issues identified by Masser and Salgé in Chapter 3 requiring urgent attention by the research community and the policy-making bodies. It also justifies once more the focus of the title and the book on geographic information rather than the technology as this is where the major research and ethical issues rest.

What the book as a whole demonstrates clearly is that research on geographic information is truly global in character by cutting across a wide range of disciplines and addressing conceptual, methodological, technical, ethical, and political issues alike. As such there are equally large benefits in addressing these challenges on a global scale through an ever increasing international collaboration which reduces the duplication of efforts, creates a critical mass of researchers addressing similar issues, and through the comparison of their results such as those included in this book gives the better prospects for advancing the field.

References

COMMISSION OF THE EUROPEAN COMMUNITIES (1994). *Europe's Way to the Information Society: An Action Plan*, COM (94) 347 final, Brussels: Commission of the European Communities.

COMMISSION OF THE EUROPEAN COMMUNITIES (1995). *GI 2000: Towards a European Geographic Information Infrastructure*, Luxembourg: Commission of the European Communities DG XIII.

FEDERAL GEOGRAPHIC DATA COMMITTEE (1994). *The 1994 Plan for the National Spatial Data Infrastructure: Building the Foundation of an Information-Based Society*, Reston, VA: Federal Geographic Data Committee.

FRANK, A. and KUHN, W. (Eds) (1995). *Spatial Information Theory: A Theoretical Basis for GIS*, Berlin: Springer-Verlag.

MULLER, J. C., LAGRANGE, J. P., and WEIBEL, R. (1995). *GIS and Generalisation: Methodology and Practice*, London: Taylor & Francis.

OPENSHAW, S. (1994). Two exploratory space-time attribute pattern analysers relevant to GIS, in Fotheringham, S. & Rogerson, P. (Eds), *Spatial Analysis and GIS*, pp. 83–104, London: Taylor & Francis.

PICKLES, J. (1995). *Ground Truth: The Social Implications of Geographical Information Systems*, New York: Guilford Press.

ROGERS, E. (1993). The diffusion of innovations model, in Masser, I. and Onsrud, H. (Eds), *Diffusion and Use of Geographic Information Technologies*, pp. 9–24, Dordrecht: Kluwer.

WEGENER, M. and MASSER, I. (1996). Brave new GIS worlds, in Masser, I., Campbell, H. & Craglia, M. *GIS Diffusion: The Adoption and Use of GIS in Local Government in Europe*, Chapter 3, London: Taylor & Francis.

PART ONE

Geographic Data Infrastructure

CHAPTER TWO

The US National Spatial Data Infrastructure

NANCY TOSTA and MICHAEL DOMARATZ

2.1 Introduction

Over the last two decades, the use of geospatial data collection and management tools has become common in many organisations. These tools include geographic information systems, global positioning systems, map digitising and scanning equipment, and image processing systems. They provide government and private sector organisations the ability to produce and manage their own geospatial information.

Concurrent with the spread of these tools is the recognition of the value of spatial analyses in the understanding of societal issues. Most environmental, and many social and economic issues are related to specific locations. Developing responses to these issues often requires consideration of many themes of data. Many organisations are recognising the powerful analysis potential provided by geospatial-data processing tools using locational information to integrate numerous variables and attributes.

The expanded base of users of these technologies, coupled with the large number and different themes of data needed to address ever more complex environmental, economic, and social questions, has led to increasingly distributed production of digital geospatial data, as well as rapidly growing demands for accurate digital data. As recently as two decades ago, few organisations, public or private, had the ability to produce analogue maps or map-based information, or to integrate and analyse map-based information. Today, thousands of organisations at all levels of government and in the private, non-profit, and academic sectors produce a variety of themes of digital geospatial data at different resolutions and levels of accuracy over many different geographic areas. Despite these capabilities, however, few organisations can produce and maintain all the data they need to take full advantage of the technology.

The proliferation of digital data sets, the number of organisations with limited ability to produce needed data, the considerable power of geospatial data processing tools to integrate data, and the complexity of modern-day life have created an opportunity to develop an infrastructure to share the data needed to support complex decision-making. In the USA, the National Spatial Data

Infrastructure (NSDI) provides a base or structure of relationships among data producers and users that facilitates data sharing. More formally, it is 'the technology, policies, standards, and human resources necessary to acquire, process, store, distribute, and improve utilisation of geospatial data' (Executive Office of the President, 1994, p. 17671).

2.2 The concept of a National Spatial Data Infrastructure (NSDI)

In the early decades of using geographic information systems, individual groups in separate divisions of an organisation implemented the technology to solve different problems. This approach resulted in unique combinations of hardware, software, and data, each of which supported a unique application. Upon examination, most organisations realised that this 'stovepipe' approach was very inefficient because of the nature of spatial data and the integrating capability of geographic information systems. Increasingly, agencies have come to recognise the value of developing a more open approach in which groups can share information that are commonly needed, as well as the ability to exchange information with groups outside the organisation. For example, instead of a planning department, highway department, and tax office each maintaining separate street centre-line databases, one street file can be developed and shared across the departments. Responsibility for updates and maintenance are defined clearly among the three offices. This approach places more emphasis on the development of standards that enable data sharing, and developing working relationships among the participants.

This approach takes advantage of the fact that aspects of geographic space are common to many government and non-government applications and that efficiencies in data development can be achieved by sharing responsibility for the data. Additionally, the approach responds to other changes taking place in organisations, including downsizing, increased distribution of authority and responsibility in flatter organisations, and enhanced interest in partnerships. Finally, the approach encourages the development of multifaceted collections of knowledge needed to cooperatively tackle complex issues such as transportation planning or community sustainability. The NSDI extends this approach to a national scale. The concept recognises that many government agencies, at all levels, as well as private sector and non-profit organisations, share interests in and are developing data related to common pieces of geography. The major goals of the NSDI are to establish simpler mechanisms to find data and to build partnerships to share data over geographic areas that are of mutual interest. The approach will result in reduced duplication in data-collection efforts (to minimise expenditures while enhancing data available), and development of a common view of shared places. The NSDI efforts anticipate that the ability to integrate and examine data nationally can be accomplished at less cost through data sharing, and that the availability of data valuable for local decision making will be improved.

In the USA, the Federal Government recognises that the concept of the National Spatial Data Infrastructure provides an approach to reinvent means of providing government services and sharing public responsibilities. In a September 1993 report, the NSDI was listed as one of the initiatives to reinvent Federal government, stating that '(I)n partnership with state and local governments and private companies we will create a National Spatial Data Infrastructure' (Gore, 1993). This recommendation was one of several that focused on establishing accountability and control at lower levels in Federal agencies, as well as at state and local government agencies.

The initial steps to implement the NSDI recommendations were carried out through the issuance of Executive Order 12906, 'Coordinating Geographic Data Acquisition and Access: The National Spatial Data Infrastructure,' by President Clinton in April 1994. Executive orders are the means by which the President directs Federal agencies to implement various actions or policies. The order directed Federal agencies and the interagency Federal Geographic Data Committee (FGDC) to carry out tasks to improve access to federally-held data, plan better means of creating and maintaining data that are commonly needed and to develop and improve standards to implement the NSDI (Executive Office of the President, 1994).

The Executive Order also had a significant effect in increasing the level of awareness about the value, use, and management of geospatial data among Federal agencies. It raised the political visibility of geospatial-data collection, management, and use, both nationally and internationally. The Executive Order created an environment within which new partnerships among levels of government and organisations are not only encouraged, but required.

In the longer term, the success of the NSDI is based on a complex web of partnerships among public and private organisations. These partnerships vary across the nation, based on organisations' interests in data over any specific location. Participation by different levels of government and different sectors is something of a self-selecting process. The Federal Government has promoted the concept of the NSDI on a national basis, but as the NSDI evolves, the roles of local and state governments are becoming more pronounced. Many state agencies have taken on the task of assimilating and integrating Federal data sets with local data to create useful statewide databases. Local agencies increasingly are collecting accurate and current spatial data. The NSDI is an excellent example of the adage, 'think globally, act locally'.

2.3 Activities to develop the NSDI

The NSDI may be thought of as an umbrella of policies, standards and procedures under which organisations and technologies interact to foster more efficient use, management and production of geospatial data. Initial activities to implement the NSDI concentrate on standards' development, data-sharing mechanisms (referred to as clearinghouse) and approaches to sharing data production responsibilities (referred to as framework). Priorities for these actions were established through a series of meetings with interested parties from the public and private sectors, and are outlined in the '1994 Plan for the National Spatial Data Infrastructure' (Federal Geographic Data Committee, 1994a) and the Executive Order.

2.3.1 Standards

Standards are required to facilitate data sharing. The success of a voluntary standard ultimately is measured by the breadth of its acceptance and use within a community. While the development of standards must account for many technical issues, it is often not recognised that there are important economic, institutional and behavioural aspects to the development and adoption of standards. The development and use of standards can be contentious and the process for developing standards must acknowledge these factors.

In the FGDC, a dozen subcommittees that deal with different themes of data, such as transportation, vegetation, cadastral, and soils, are developing standards for data collection and content, data presentation and data management to facilitate data sharing. A Standards Working Group links the subcommittees' standards activities and adjudicates differences. All standards developed by the FGDC undergo an extensive public review process that includes nationally advertised comment and testing phases and solicitation of comments from state and local government agencies, private sector firms and professional societies. This process is time consuming, but provides value in the improvements to the standard identified by broader comment and testing and enhanced likelihood that the standards will be used by geospatial data producers, users and managers throughout the community. To date, a metadata standard was developed and adopted in 1994; a cadastral standard completed national review in mid-1995 and is in the final adoption phase; a wetlands classification standard has just completed national review; and a cultural and demographic profile for metadata, a transportation profile for data transfer, a vegetation classification and an accuracy standard for spatial data collection and representation are all in the queue for national review.

2.3.2 Clearinghouse

The second activity facilitates access to data, with the goals of minimising duplication and aiding the development of collaborative data development and maintenance. This effort is to develop a National Geospatial Data Clearinghouse by using the rapidly expanding information infrastructure to facilitate access to geospatial data. The clearinghouse is a referral service to discover who has what data. Designed with the decentralised distribution of data producers and users in mind, the clearinghouse is comprised of a set of information stores that use computer hardware, software and telecommunications to link producers and users. The clearinghouse is geographically distributed in nature; there is not a centralised 'warehouse' of geospatial data. Thousands and potentially millions of data holdings will be catalogued and accessed across electronic networks.

There are at least three elements conceived to be necessary in the operation of the clearinghouse: metadata, Internet connectivity and distributed search and query software tools. Metadata can be used internally by an agency to manage data, it can be useful as a description of the data when it is transferred, and it can be provided as a 'catalogue' entry when made accessible to the Internet. The Content Standards for Digital Geospatial Metadata (Federal Geographic Data Committee, 1994b) establishes terms and definitions to provide a consistent means to describe the quality and characteristics of geospatial data. Data producers nationwide, at all levels of government and within the private sector, are being encouraged to describe the data they produce with this standard. The Executive Order mandated that Federal agencies use the standard to document all new data created after January 1995.

The Internet is the second required element of the clearinghouse. It has become the most pervasive electronic network in the world with tens of millions of users and double digit rates of growth monthly. For the clearinghouse, the Internet provides the means for organisations seeking data to retrieve metadata that describe the holdings of data producers. The Internet also may provide access to data holdings if permitted by the data producer. Federal agencies, as well as other producers, are being encouraged to provide access to the Internet so that they can search the clearinghouse and to establish Internet sites to make metadata, if not the data themselves, accessible.

The third element of the clearinghouse is the use of software tools for searching and querying data on the network. New tools are continually being developed and offered to the public over the Internet. Examples include tools for 'surfing' the World-Wide Web that provide access to 'home pages' with hypertext links between sources of data. Many organisations have established sites that permit potential users to browse through metadata. The FGDC is supporting the development and testing of enhancements to tools, based on the Z39.50 standard, used to conduct spatial searches across many sites of metadata. These advances should improve the efficiency and effectiveness of users searching for spatial data on the Internet.

Most Federal agencies and a number of state and local governments have established sites for disseminating data on the Internet. Organisations serving data on the Internet discover that thousands of files are being accessed monthly, a figure that is often a large jump above the volume of data distributed through other means. For example, the US Fish and Wildlife Service reported that, in the first month of operation, approximately 29 000 digital maps from the National Wetlands Inventory were retrieved. The US Geological Survey reported a similar volume of interest, with 40 000 files of spatial data retrieved in the first 3 months of operation (Federal Geographic Data Committee, 1994c).

2.3.3 Framework

The third activity area is the development of a framework of commonly used data based on shared responsibilities for data creation and maintenance. These data are those that seem to be required most commonly by most geographic information system users, including digital orthoimagery, geodetic control, elevation, transportation, hydrology, governmental boundaries and cadastral or ownership information. They form the backdrop for display, organisation or analysis of other data. Although these data often fill only a supporting role in an application, many organisations spend a great deal of time and money digitising or seeking them. A common approach to build and maintain these data could free an organisation's resources to concentrate on more pressing applications.

During 1994, the FGDC formed a Framework Working Group that consisted of representatives of federal, state and local government agencies. The approach outlined by the group proposed the linking of various existing data-collection activities over specific geographic areas, recognising that, in most instances, local governments are collecting higher resolution data and have more concerns about keeping their data current (Federal Geographic Data Committee, 1995). The framework concept recognises the potential critical role of local organisations in the development and maintenance of data. This approach provides an opportunity for state or Federal agencies to obtain more current and accurate data, potentially at lower costs than collecting the data themselves, by collaborating with local government organisations. The group proposed mechanisms to establish new responsibilities for data collection and maintenance which emphasised the collaboration of organisations that are interested in a specific geographic place.

The concepts outlined in the group's report are currently being examined in a series of pilot studies. The pilots will help develop standards, identify institutional issues and provide the foundation for programmatic changes and funding proposals that will contribute to the development of framework data. These pilot discussions will result in operational guidelines for building framework data that can be more easily integrated in a multitude of applications.

2.4 Encouraging participation in the NSDI

The success of the NSDI depends on a higher level of cooperation and collaboration among a more diverse and widely dispersed set of organisations than has been seen in the spatial data community in the USA. A lack of knowledge about the effort, a lack of understanding about the benefits of participation and a concern about risk in undertaking new approaches to doing business slow down the spread of the concept. To remove barriers and encourage broader participation in the effort, the FGDC has used a variety of means to promote the development of the NSDI.

Political support for the initiative has helped to provide visibility. The issuance of the Executive Order, and the fact that the President's Office of Management and Budget originally established the FGDC to coordinate spatial data collection activities among Federal agencies and develop the NSDI (Office of Management and Budget, 1990), provide a continuing mandate and high-level support from an oversight agency. The willingness of the Secretary of the Interior to chair the FGDC has contributed political visibility and a sense of importance to the effort that would not otherwise exist. His involvement has encouraged other Federal agencies to commit resources at higher political levels. This attention by the Federal Government has helped some state and local governments to make local needs for spatial data production and coordination more visible. This attention has also hastened the use of standards such as those for metadata and improved the accessibility of data.

Opportunities to participate in the development of the NSDI and to learn more about the effort provide the means to capitalise on the interest generated by this political support. Several mechanisms are being used to develop these opportunities and to make them widely available in the community.

The ability to involve the large number of interested individuals and organisations in the effort is a challenge but encouraging the establishment of groups with smaller geographic areas of interest is one means to increase the opportunities to participate and benefit. Geographic information councils that exist or are being formed in many states are a good example of these groups. Most councils represent a cross-section of geographic data interests in the state, including state and local governments, regional Federal agencies, the private sector and academia. These groups can be important vehicles for the development and promulgation of standards and for outreach on data-sharing activities. The FGDC has established a process to formally recognise these councils as partners in the development of the NSDI. As of February 1996, seven councils have been formally recognised and approaches are being developed for representatives of these councils to meet with the FGDC.

The process established by the FGDC for developing and approving standards is another mechanism for participating in the NSDI. Standards may be proposed by anyone to the FGDC Standards Working Group. A suggested standard enters a formal process of review by various levels of the FGDC and national review through advertisements in the Federal Register, major trade journals and conference presentations. Organisations can also offer comments and recommendations of standards as part of the public review and testing period.

The FGDC established a Competitive Cooperative Agreements Programme to encourage participation and experimentation in the development of the NSDI. An annual announcement is made of the availability of funds to support collaborative efforts on metadata implementation, data clearinghouse development, framework data, standards development and implementation and other issues. State and local governments and other non-Federal entities are encouraged to form partnerships with other organisations to enhance access to their data. The proposals received are competitively ranked and awarded. Nine projects were funded in 1994, 22 in 1995 and as many as 30 may be funded in 1996.

Education is also vital to the success of the NSDI. The FGDC conducts numerous training courses, workshops and presentations about the various activities including use of the metadata standard, clearinghouse development, framework concepts and overall design of the NSDI. These courses are offered at national, state, regional, and user group conferences. The FGDC and its various subcommittees publish a newsletter and several reports a year on technical issues. These can also be accessed electronically from the FGDC home page on the World-Wide Web (http://www.fgdc.gov).

2.5 Questions in the ongoing evolution of the NSDI

An endeavour such as the NSDI does not come about overnight, nor does it happen without question and contention. The technology of GIS is relatively immature and constantly changing, which adds an element of guesswork to plans and initiatives. In many ways, rather than a grand plan, the NSDI is a vision of data sharing and access that is being defined as the future unfolds. Numerous questions have arisen about approaches, techniques and policies as plans for the use of distributed networks for data sharing have evolved. Some questions are technical in nature, related to the actual use of the technology and characteristics of geographic information. Other issues are more a function of institutional environments and society in general.

A fundamental institutional question that has no answer, but is continuously debated, is 'what are the implications of cost-recovery policies on the evolution of the NSDI?' or more specifically, 'what are the differences in long-term availability and quality of spatial data when cost-recovery fees are imposed versus low-cost/ no-cost data access policies?' Other institutional issues include: What incentives (or constraints) are most effective in encouraging inter-agency coordination? How should copyright be applied to digital geospatial data? What are (or should be) the data-collection roles and responsibilities of federal, state and local agencies, as well as the private sector over any given piece of geography? What mechanism is required, if any, to 'oversee' the NSDI?

Many institutional/social questions exist in the area of standards. Exactly what standards are likely to best promote data sharing? Who best creates standards, for example, federal government, private industry? Who should be responsible for maintaining them? What are the incentives or mechanisms that will encourage the use of standards? Are 'certified' data sets needed and if so, who certifies?

Technical issues include basic questions about the ability to use different data sets from different computers across a distributed network. How is interoperability best promoted, for example, transfer standards or data structure specification? What role does the private sector play in promoting interoperability and what incentives exist to encourage their participation? How can multiresolution data sets collected by different organisations best be integrated and used? Who defines features and on what basis? What is the best way to represent metadata, textually, or graphically and how can metadata development be encouraged?

Some of the questions, particularly the technical ones, may have answers that can be found through academic research. Other issues, including many of the institutional ones, are only likely to be resolved as current actions are played out. For many aspects of the NSDI, a 'try it and we'll see what happens' approach may be the only course of action.

2.6 Conclusion

The use of geospatial data-processing technologies has spread rapidly throughout the USA in the last 20 years. These technologies play an increasingly vital role in helping society meet current and future environmental, economic and social challenges. As the use of these technologies has spread, so has the realisation that there are extensive requirements for data, but there is little data coordination, and limited ability to share data collection and maintenance efforts. These factors inhibit organisations' abilities to realise the full benefits of these technologies. Concurrent with this spread of geospatial data-processing technologies is the explosive growth in capabilities of and access to telecommunications technologies. The interest in improving the efficiency of the spatial data community and improved abilities to share data within the community provide the opportunity to develop a National Spatial Data Infrastructure that promotes data sharing.

Initial efforts to develop the NSDI, such as the clearinghouse and metadata standard, have focused on the ability to answer relatively simple questions. Who has what data? How can I access a set of data? How suitable are a set of data for my application? The success of future efforts, such as the development of standards for themes of data and a collaborative approach to creating and maintaining sets of framework data, will rely on the community's ability to agree on many institutional, policy, financial, as well as technical issues. Interests of organisations to create partnerships, to place responsibility for making decisions at lower levels and to collaborate to resolve issues that affect a geographic region, provide a force for overcoming these challenges and developing the NSDI.

References

EXECUTIVE OFFICE OF THE PRESIDENT (1994). *Coordinating Geographic Data Acquisition and Access: The National Spatial Data Infrastructure* (Executive Order 12906). Federal Register, **59**(71), 17671–4.

FEDERAL GEOGRAPHIC DATA COMMITTEE (1994a). *The 1994 Plan for the National Spatial Data Infrastructure: Building the Foundation of an Information-Based Society*, Washington, DC: Federal Geographic Data Committee.

FEDERAL GEOGRAPHIC DATA COMMITTEE (1994b). *Content Standards for Digital Geospatial Metadata*, Washington, DC: Federal Geographic Data Committee.

FEDERAL GEOGRAPHIC DATA COMMITTEE (1994c). *The National Geospatial Data Clearinghouse: A Report on Federal Agency Activities within the First Six Months of Executive Order 12906*, Washington, DC: Federal Geographic Data Committee.

FEDERAL GEOGRAPHIC DATA COMMITTEE (1995). *Development of a National Digital Geospatial Data Framework*, Washington, DC: Federal Geographic Data Committee.

GORE, A. (1993). *From Red Tape to Results: Creating a Government that Works Better and Costs Less*. Report of the National Performance Review. Washington, DC: US Government Printing Office, also on http://www.npr.gov/homepage/267a.htul

OFFICE OF MANAGEMENT AND BUDGET (1990). *Coordination of Surveying, Mapping and Related Spatial Data Activities*, Circular A-16, Washington, DC: Office of Management and Budget.

CHAPTER THREE

The European Geographic Information Infrastructure Debate

IAN MASSER and FRANÇOIS SALGÉ

3.1 Introduction

During the last year a number of important steps have been taken towards the creation of a European Geographic Information Infrastructure. In the process a considerable momentum has already been built up which has strong political support from the key stakeholders in the European geographic information community. Even though many of these developments are still at an early stage and there is as yet no formal political commitment to the establishment of a European geographic information infrastructure, it is felt that they are of such potential significance that a systematic review is justified. With this in mind, this chapter reviews the background to the present proposals, describes the main features of the proposals themselves and evaluates them both in general terms and also with respect to the US National Spatial Data Infrastructure. It also considers the implications for the European geographic information research community and identifies a number of key issues for future research.

3.2 Background

3.2.1 Context

The starting point for much of the current discussion of information infrastructures is the vision for Europe that was presented to the European Council in Brussels in December 1993 by the then President Jacques Delors in a White Paper entitled 'Growth, competitiveness and employment: the challenges and ways forward into the 21st century' (Commission of the European Communities, 1993). An important component of this vision is the development of the information society especially within the triad of the European Union, the USA and Japan. One result of this initiative was the formation of a high-level group of senior representatives from the industries involved under the chairmanship of Commissioner Martin Bangemann. The action plan for 'Europe and the global information society' prepared by this group was presented to the European Council at the Corfu summit in June

1994 (Bangemann, 1994). The group argued that recent developments in information and communications technology represent a new industrial revolution that is likely to have profound implications for European society. To take advantage of these developments it will be necessary to complete the liberalisation of the telecommunications sector and create the information superhighways that are needed for this purpose. With this in mind the group proposed 10 specific initiatives. These include far-reaching proposals for the application of the new technology in fields such as road traffic management, trans-European public administration networks and city information highways. These proposals were subsequently largely incorporated into the Commission's own action plan 'Europe's way to the information society' which was published in July 1994 (Commission of the European Communities, 1994a).

3.2.2 Geographic information

Parallel to these developments a number of important steps have been taken towards the creation of a European geographic information infrastructure. In April 1994, a meeting of the heads of national geographical institutes was held in Luxembourg which concluded that the time was right to begin discussions on the creation and supply of harmonised topographic data across Europe (Commission of the European Communities, 1994b). This view was reinforced by the letter sent to President Delors by the French Minister M. Bosson which urged the Commission to set up a coordinated approach to geographic information in Europe and the correspondence on this topic between the German and Spanish ministers and Commissioner Bangemann during the summer of 1994.

   As a result of these representations, a meeting of key people representing geographic information interests in each of the Member States was held in Luxembourg in February 1995. The basic objective of this meeting was to discuss a draft document entitled 'GI 2000: towards a European geographic information infrastructure' (Commission of the European Communities, 1995a) and identify what further actions should be taken in this respect. The main conclusion from this meeting was that 'it is clear from the debate that the Commission has succeeded in identifying and bringing together the necessary national representative departments that can play a role in developing a Community Action Plan in Geographic Information' (Commission of the European Communities, 1995b, p. 12). With this in mind it was agreed that DGXIII should initiate and support a widespread consultation process within the European geographic information (GI) community with a view to the preparation of a policy document for the Council of Ministers in late 1995 (de Bruïne, 1995). DGXIII's responsibilities include telecommunications, the information market and the exploitation of research.

3.3 Towards a European Geographic Information Infrastructure (EGII)

3.3.1 Introduction

This description of the proposals that have been put forward for a European Geographic Information Infrastructure (EGII) has been taken from the policy document dated 20 September 1995 which has been circulated to the main stakeholders in the European geographic information industry (Commission of the European Communities, 1995c). The current draft is the sixth to be circulated since February 1995 and has already been substantially modified as a result of the consultation process.

3.3.2 The European geographic information infrastructure

The policy document presents the case for a European Geographic Information Infrastructure and considers what steps must be taken to bring it into being. The case for a EGII is summarised in the following terms.

> The lack of a European mandate on GI is retarding development of joint GI strategies and is causing unnecessary costs in data acquisition and data conversion, retarding the development of new goods and services and reducing competitiveness. This situation can be improved through a European geographic information infrastructure that is set up at European government level and operated by and for the GI community. (p. 14)

It is envisaged that 'the EGII would be a stable, European set of agreed rules, standards and procedures for creating, collecting, exchanging and using GI' (p. 9). In addition, it is anticipated that the EGII would 'ensure that European wide base datasets are readily available' (p. 9).

The policy document goes on to point out that many elements of the EGII already exist in the growing number of digital geographic databases held by local, national and European providers and users in both the public and private sectors. However, to convert these elements into an effective infrastructure for Europe as a whole requires the following political decisions:

- Agreement of member states to set up a common approach to create European base data and to make this generally available at affordable rates.

- A joint decision to set up and adopt general data exchange standards and to use them.

- A joint decision to extend the mandates of NMAs and similar organisations to enable European level actions.

- Agreement to initiate European wide legal studies to address issues such as copyright, liability, misuse of information specifically with respect to GI (p. 12).

Once these matters are resolved it will be necessary to take the following steps to bring the EGII into being:

- Create conditions for the emergence of a plentiful, rich and differentiated supply of European GI, that is easily identifiable, easily accessible and competitively priced.

- Improve the possibilities for locating existing information via metadata services and for sharing such data across different applications.

- Ensure that the market for GI develops in a healthy and transparent way by stimulating the expansion of an electronic market place for GI.

- Encourage the creation of public/private partnerships so the wealth of public sector GI can be better exploited by the private sector for use in new and useful business and public sector applications.

- Support stronger co-operation between member state agencies (NMAs, National Statistical Institutes, Census Bureaux, Environmental Agencies, River and Coastal Authorities and so on) for creating geographic base data which is seamless across Europe.

- Research, review and support harmonisation of pricing criteria of national agencies providing GI, whether for base, thematic or statistical data, and to facilitate data acquisition particularly for simultaneous purchases from several agencies.

- Ensure that legal and regulatory initiatives at European level dealing with information law, such as copyright, privacy and liability of information providers, take into account the special characteristics of GI.

- Increase the awareness of information providers and potential users of the benefits of understanding the spatial aspects of their data, and the need to have certain skills.

- Ensure that European solutions are globally compatible.

- Explore the possibilities for promoting European GI research and development in areas relating to, for example, integration and visualisation tools, and spatial and temporal analysis techniques. (pp. 12–13)

3.3.3 Evaluation

The policy document is the product of an extended consultation process with the leading stakeholders in the European GI industry. As such it gives a clear indication of current thinking in Europe at the present time. Inevitably, given that the primary objective of the document is to make the case for political actions at the European Community level, a great deal of space is devoted to describing the potential benefits that will arise out of a more coordinated approach to geographic information provision at the European level, and there is relatively little detail regarding how key elements such as the development of metadata services or the creation of core data sets might be handled.

Despite these limitations, an important feature of the proposals is the extent to which they build upon the main strengths of Europe in the field of geographic information at the present time. These include some very good topographic data with a long history of informal cooperation between National Mapping Agencies as in CERCO (Comité Européen des Responsables de la Cartographie Officielle). This is evident in Pan-European initiatives such as the Multipurpose European Ground Related Information System (MEGRIN) which seeks to simplify cross-border trade and develop the market for European geographic information (Salgé, 1994). The proposals also draw upon the experience that has been built up at the European level by a number of agencies such as the Statistical Office of the European Communities (EUROSTAT) which is responsible for the coordination and harmonisation of the statistical data provided by the National Statistical Agencies. EUROSTAT has also developed a GIS system (GISCO) to assist the various directorates of the Commission in the development and monitoring of EU policies in fields such as agriculture, transport and environmental protection.

There are many obvious similarities and some interesting differences between these proposals and those that were put forward for the US National Spatial Data Infrastructure (NSDI) last year (Federal Geographic Data Committee, 1994; Tosta and Domaratz, see Chapter 2, this volume). For example, the goals of the EGII have much in common with the vision of the NSDI that 'current and accurate geospatial data will be readily available to contribute locally, nationally and globally to economic growth, environmental quality and stability and social progress' (Federal Geographic Data Committee, 1994, p. 1). There are also clear similarities between key elements of the EGII such as the need for metadata and the concept of core datasets and the National Geospatial Data Clearing House and the National Digital Geospatial Data Framework in the National Spatial Data Infrastructure.

Both initiatives claim that their objectives can be met largely by the refocusing of existing public expenditure. In the USA it has been estimated that Federal agencies alone spend $3 billion annually to collect and manage domestic geospatial data and that an additional $1 billion is also spent every year on global geospatial data (Federal Geographic Data Committee, 1994, p. 2). Similarly, it is estimated that about 0.1 per cent of Gross National Product every year or 6 billion ECU (about $7.5 billion) in the European Union is already spent on topographic data alone (Commission of the European Communities, 1995b).

There are also some important differences between the two initiatives which reflect the institutional contexts in which they are set. For example, there are a number of important issues relating to pricing and ownership of geographic information which will have to be resolved in the context of Europe before the EGII can be fully implemented which do not exist to anything like the same extent in the USA as a result of its Freedom of Information legislation. There is also a very different relationship in Europe between the Member States and the European Union to that which exists between the state and Federal governments in the USA. This reflects not only matters relating to national sovereignty, but also very different historical, cultural and legal traditions within Europe. By comparison with the US Federal government, European agencies have relatively limited powers, especially with respect to data collection. For this reason the main task of the Statistical Office of the European Communities is to coordinate the data collected by the National Statistical Agencies rather than to collect the data itself. Consequently there is no European level equivalent of the US Geological Survey (USGS) and none of the EC Directorates is involved in primary data collection. The task of topographic data collection is therefore left to the National Mapping Agencies of the Member States who have developed their own practices over a long period of time to meet national rather than European needs. These differences are reflected not only in the organisational arrangements that have come into being for the provision of topographic data but also in the content and overall availability of that data (Lievesley and Masser, 1993/4).

3.4 Implications for geographic information-based research

3.4.1 Introduction

This section considers the implications of the proposals that are being developed for the establishment of European Geographic Information Infrastructure for geographic information-based research from two different standpoints. First, the extent to which they are likely to meet the needs of the European geographic information research community is assessed with reference to the position statement on this topic prepared for the GISDATA scientific programme (Masser and Salgé, 1995). Second, some areas for research that will need to be given high priority in the future as a result of these proposals are identified.

3.4.2 The needs of the European geographic information research community

Throughout both the social and environmental sciences it has been argued that the lack of data in an appropriate form at the European level is a major deterrent to European-wide and transnational research. In the context of the social sciences, for example, a report on 'The social sciences in the context of the European Communities (European Science Foundation/Economic and Social Research Council, 1991) identifies the current gap between the quality of most national data and what is available at the European level as a major barrier to the effective utilisation of European social science research. For these reasons, one of the main starting points for the European Science Foundation's GISDATA programme was the perceived need to build up the European geographic information resource base by promoting greater data integration and the position statement sets out an agenda for further discussions from the standpoint of the European geographic information research community (Masser and Salgé, 1995). This statement argues that the basic objective of building up the European data resource base is 'to deliver harmonised data of high quality to researchers in the most efficient and cost effective way'. To achieve this it will be necessary to concentrate efforts on five main strategic tasks:

1   Identifying the key strategic data required for decision making at the European level.
2   Acquiring and preserving key strategic data.
3   Creating, documenting and integrating databases to make them as accessible and informative as possible.
4   Establishing network catalogues to help users find the information they need.
5   Facilitating wider and more informed use.

There are many parallels between these sentiments and those contained in the EGII proposals. The notion of key data sets has much in common with the concept of core data sets contained in the EGII. Similarly, the notion of metadata contained in the EGII corresponds with the idea of establishing networked catalogues. Both documents also stress the need to raise awareness in order to facilitate wider and more informed use of the data. Finally, the need for a high level of political commitment to achieve these objectives is also recognised in both proposals.

However, there are also some interesting differences between the two state-ments even at this stage of their development. The GISDATA statement, for example, extends the notion of the resource base to ensure that the historic record is preserved for future researchers. Similarly matters relating to access and availability are given special priority in the GISDATA statement. For this reason it is argued that investment will be required to develop existing facilities such as the national social science data archives to disseminate data and provide training for researchers. At the European level it is felt that this may require the creation of some large-scale facilities for this purpose.

3.4.3 Priorities for future research

The GISDATA statement also identifies four key issues for research that will need to be considered in the development of a European geographic information resource base. These issues are not in themselves new, but it is likely that they will become increasingly important in future debates about the EGII given the diversity of national, historical and cultural circumstances within Europe. These issues are as follows.

- *Co-ordination.* Some useful steps towards the harmonisation of key geographic data sets have been taken at the European level by EUROSTAT but there are still many technical and institutional problems which need further investigation. For example, a great deal of work is still required to agree common standards for the formatting and transfer of topographic data despite the efforts of CEN/TC 287 (European Committee on Standards Technical Committee for Geographic Information) over the last few years and the establishment of the MEGRIN network.

  At the European level there are also some important questions to be resolved regarding European Union policy towards data acquisition and access. For example, whereas the 1990 Directive on freedom of access to environmental data requires official agencies in member countries to make data available at reasonable cost, the 1995 Directive on personal data protection severely limits the collection and use of such data.

- *Copyright.* The development and dissemination of high-technology products and information services is outpacing the legal structures that regulate them. The use of GIS to integrate large amounts of spatially referenced data drawn from a variety of sources into new databases is a major innovation. There is a need for research to determine the extent to which the intellectual property rights of those involved in the creation of such databases should be protected.

  There are also considerable differences between European countries with respect to the nature and enforcement of existing copyright laws which require further investigation. Consequently, national data providers are likely to be differentially affected by the draft EU Directive on the legal protection of databases. This could have a detrimental effect on data resources in countries such as the UK and Spain where there are currently very restrictive copyright laws, while at the same time it may offer some measure of protection to those concerned with data provision in countries such as Italy and Greece.

- *Commodification.* Computer technologies such as GIS make it possible to provide data in a wide variety of formats, at differing spatial scales and to specifications determined by users rather than data providers. However, the additional costs of meeting these new demands are considerable and raise questions regarding the extent to which data providers should pass on some or all of these costs to the users of such data. These developments, together with increased pressures on government statistical agencies to recover some or all of the costs of providing data, mean that data is increasingly treated as a commodity which can be traded in its own right.

The extent to which data is treated as a tradable commodity varies considerably between different European countries and demands systematic study. Cost-recovery strategies for topographic data are particularly well developed in the UK, for example, where the National Mapping Agency, the Ordnance Survey already recovers over 70 per cent of the costs of data provision.

- *Confidentiality*. The growing demand for geographic information at the small area level conflicts with the legal requirement for data providers to protect personal privacy. Although a wide range of technical procedures can be used to reduce the risk of disclosure or personal information, it has to be recognised that public perceptions of confidentiality are as important as the actual risk.

  At present there are considerable differences between European countries with respect to both legislation on confidentiality and its enforcement which need further investigation. In some countries access to data is allowed only to privileged categories of users and the absence of open access restricts the potential for some geographic information-based research (see also Chapter 4, this volume).

In addition to these topics two other questions are also likely to require more sustained research effort than has hitherto been the case as a result of the EGII proposals. These are the need to evaluate the impacts of European-wide policies and directives on national initiatives and the need to monitor the development of geographic information services and products in the European context.

With respect to the former, it is important to utilise the experience developed in environmental initiatives such as the Coordinated Information on the Environment project (CORINE). Within the environmental field the impacts of the 1990 Directive (Directive 90/313 – Access to Environmental Information) on freedom of access to environmental information in different countries should be evaluated. There is also a strong case for policy analytic studies of other European initiatives not only stemming from the Commission, but also organisations such as CERCO and the European Umbrella Organisation for Geographic Information (EUROGI).

Insofar as the latter is concerned, it should be borne in mind that an important part of the argument for the development of a European geographic information infrastructure is the extent to which it will stimulate the growth of geographic information services and products, thereby enhancing the economic competitiveness of Europe as a whole. Consequently, it will be a matter of considerable importance to monitor the nature of the development of the European geographic information services industry in Europe and to study its geographic distribution and internal structure. This research area is not well developed at the present time and may require special initiatives to stimulate appropriate investigations of this important topic.

3.5 Conclusions

This chapter has reviewed recent developments in the European geographic information infrastructure debate and considered their implications for the European geographic information research community. As a result of this analysis a number of core research areas have been identified for future research. The findings of this analysis must be treated as provisional given the recent nature of most of these developments and the lack of detailed proposals for their implementation. Nevertheless, it is felt that the issues they raise are of such importance that they merit the most serious attention from the geographic information research community even at this early stage in their development.

References

BANGEMANN, M. (1994). *Europe and the Global Information Society: Recommendations to the European Council*, Brussels: Commission of the European Communities.

COMMISSION OF THE EUROPEAN COMMUNITIES (1993). *Growth, Competitiveness and Employment: The Challenges and Ways Forward into the 21st Century*, Brussels: Commission of the European Communities.

COMMISSION OF THE EUROPEAN COMMUNITIES (1994a). *Europe's Way to the Information Society: An Action Plan*, COM(94) 347 Final, Brussels: Commission of the European Communities.

COMMISSION OF THE EUROPEAN COMMUNITIES (1994b). *Heads of National Geographic Institutes: Report on Meeting held on 8 April 1994*, Luxembourg: Commission of the European Communities, DGXIII.

COMMISSION OF THE EUROPEAN COMMUNITIES (1995a). *GI 2000: Towards a European Geographic Information Infrastructure*, Luxembourg: Commission of the European Communities, DGXIII.

COMMISSION OF THE EUROPEAN COMMUNITIES (1995b). *Minutes of the GI 2000 Meeting 8 February 1995*, Luxembourg: Commission of the European Communities, DGXIII.

COMMISSION OF THE EUROPEAN COMMUNITIES (1995c). *GI 2000: Towards a European Geographic Information Infrastructure*, The EGII Policy Document, Luxembourg: Commission of the European Communities, DGXIII.

DE BRUÏNE, R.F. (1995). *A European strategy for geographic information*, Keynote address, Joint European Conference and Exhibition on Geographic Information, The Hague, 27–31 March 1995.

EUROPEAN SCIENCE FOUNDATION/ECONOMIC AND SOCIAL RESEARCH COUNCIL (1991). *The Social Sciences in the Context of the European Communities*, Strasbourg: European Science Foundation.

FEDERAL GEOGRAPHIC DATA COMMITTEE (1994). *The 1994 Plan for the National Spatial Data Infrastructure: Building the Foundation of an Information-Based Society*, Reston, VA: Federal Geographic Data Committee.

LIEVESLEY, D. and MASSER, I. (1993/4). An overview of geographic information in Europe, *Mapping Awareness*, **7**(10), 9–12, **8**(1), 33–6 and **8**(2), 35–7.

MASSER, I. and SALGÉ, F. (1995). Building the European social science resource base: the geographic dimension, *GISDATA Newsletter*, **5**, 9–13.

SALGÉ, F. (1994). The MEGRIN group, *Proceedings of EGIS '94 Conference, Paris 29 March–1 April*, pp. 1350–7, Utrecht: EGIS Foundation.

CHAPTER FOUR

Spatial Data as a Test Bed for National Information Policy

XAVIER R. LOPEZ

4.1 Introduction

How does the development of a national geographic information infrastructure affect scientific, technical, commercial and governmental processes? Will the availability of geographic information provide new areas of opportunity for scientists, engineers, data brokers and consumers of spatial data sets? Will impediments to public access reduce value-added processes?

Numerous studies have shown that increased diffusion of scientific and technical information will lead to greater innovation processes in a wide variety of applications ranging from basic science to the provision of commercial services (Malchup, 1984; Ballard, 1989; Pinelli *et al.*, 1990). The USA spends nearly US$65 billion each year on R&D infrastructure, and a major objective of this expenditure is to generate scientific and technical information (US Congress, 1990). Within the overall science and technology arena, the GIS sector is an example of one of the rare industries where relatively simple government action can directly fuel growth (International Trade Administration, 1994; Gartzen, 1995). Recent estimates show that the world-wide investment in GIS technologies by government and the private sector are approximately US$3.3 billion (Dataquest, 1995), with annual growth rates reaching nearly 30 per cent. A major portion of this investment has been in the production of spatial data sets.

Government agencies such as national mapping, statistical, and land registration offices are some of the largest producers and suppliers of spatially referenced data, and their dissemination policies can have a tremendous impact on the development of geographic information systems (GIS) industries. Indeed, government databases provide the raw material upon which innovative information industries add value to public information (Benjamin and Wigand, 1995). Government information is increasingly collected in digital formats which permits the data to be more easily used, shared and disseminated. Public, for-profit and not-for-profit organisations now seek access to spatial databases created by central and local governments. As a result, such data sets now have a wider audience, greater economic value, and increased political significance than previous analogue versions of the same works. Also, the information dissemination policies of government, through various legal and policy mechanisms to regulate public access and commercialisation of public information resources, can have a direct influence on the growth of the GIS scientific community, as well as contribute to the competitiveness of geographic information industry (de Bruïne, 1995).

Figure 4.1 highlights the legal and policy factors influencing the dissemination of spatial databases. This figure shows how the dissemination policies of various national agencies are determined by a complex legal, political and institutional milieu. An outcome of this setting are agency dissemination practices which will affect the various users and intermediaries differently. For example, pricing policy may be less of an issue to commercial value-added intermediaries than copyright and licensing controls. In contrast, academics and research scientists may be more sensitive to pricing impediments than copyright controls since they would have few intentions of reselling the information. By evaluating the end user and inter-mediary responses to comparable agency dissemination practices, one can begin to assess the benefits and drawbacks to each approach – thus forming the basis for test bed evaluation. The goal of such evaluation would be to inform policymakers of the benefits and drawbacks of pursuing alternative activities promoting public access and commercialisation of government information.

Most current research on data dissemination has been primarily concerned with information exchange at the intra- and inter-organisational level (Azad and Wiggins, 1995; Masser and Campbell, 1995; Pinto and Onsrud, 1995; Evans, Chapter 5 in this volume). Also, a significant number of activities in various countries have been concerned with particular aspects of data sharing. However, very little systematic knowledge exists concerning the dynamics of data exchanges between government and various intermediaries and end users within a nation as a whole. Furthermore, comparative evidence which sheds light on the similarities or differences among nations regarding data dissemination and value-added processes is absent. Therefore, the primary goal of the research outlined in this chapter is to extend the debate on current information dissemination practices by highlighting the various incentives and impediments promoting the use of government information for the maximum benefit of society. It is hoped that this work will help to guide further empirical and comparative studies on the impact of national and local policies on the information marketplace.

4.2 Importance of national mapping activities

National mapping agencies, statistical agencies and other large government suppliers of geographic information are key players in the development of a national spatial data infrastructure (NSDI) (see Tosta and Domaratz, Chapter 2 and Masser and Salgé, Chapter 3, both in this volume). It is therefore appropriate to examine how the information practices of these organisations meet changing national objectives. Although state and local tiers of government are critical institutions in the development of an NSDI, the attempt to compare alternative dissemination models requires that this analysis be limited to the national setting. Furthermore, since information law and policy (for example, public access, copyright, data protection, charging policies and so forth) is generally more explicit at the national than at more local levels, it was necessary to limit the analysis to

SUPPLIER   **END-USERS AND INTERMEDIARIES***

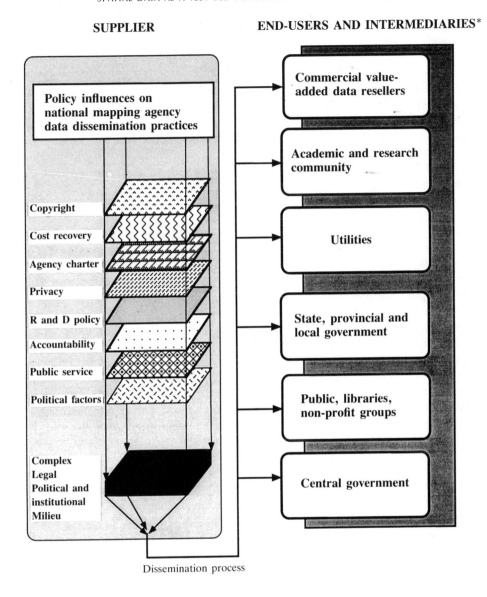

Dissemination process

* The user and intermediary groups above represent only a handful of
stakeholders that are directly and indirectly impacted by the data dissemination
policies of national mapping agencies.

Figure 4.1 Legal and policy factors influencing government information
dissemination practice.

national suppliers. However, the author does recognise that there are other
important government and commercial sources of geographic information.

Due to budget shortfalls, Central Treasury boards throughout many Western
countries have begun to pressure their line agencies to enforce copyright and data
licensing measures to support revenue-generating activities. As a result, the

creation of spatial data sets, which involves significant capital expenditures, is increasingly viewed as a lucrative revenue source to supplement ongoing data maintenance and system enhancements (Perritt, 1995; Rhind, 1991). These revenue generation developments place government mapping agencies in the difficult position of having to reconcile a number of competing interests: generating revenues from data sales; stimulating private sector innovation; and providing public access to government information resources. In light of ongoing budget cuts, national mapping agencies, like many public sectors agencies, are left to fend for their own institutional survival by emphasising revenue generation objectives. This is a logical outcome. However, it can also have the unintended consequence of undermining public access goals or national objectives aimed at improving downstream value-added information activities in the commercial and scientific community.

As one example, the UK's civilian mapping agency, the Ordnance Survey, is mandated by its Central Treasury to recover nearly 70 per cent of its total operating costs from the sale of data and services (Rhind, 1991, 1995). Stringent conditions of use for these databases are also clearly defined and vigorously defended. The French, Dutch, Swedish and Finnish national mapping organisations, along with the European Space Agency (ESA), also seek to achieve some level of cost recovery through the sale and license of publicly funded data sets. Even the World Meteorological Organisation (WMO), a United Nations organisation, is confronting this issue. In 1984, there was a proposal before the WMO Congress to replace the present principle of free and unrestricted exchange of meteorological and related information with a two-tiered data exchange system (World Meteorological Organisation, 1994). The outcome of this proposal would have led to restrictions on sharing important meteorological data among users internationally, and a potential decrease in derivative weather information products world-wide. This proposal was not adopted due to strong resistance from various countries, most notably the USA and developing countries (National Research Council, 1995). A major reason for the defeat of this proposal is that international programmes for global change research and environmental monitoring are based upon the principle of full and open exchange of government-funded scientific and technical information.

In situations where cost recovery has been utilised, the measured impact on the scientific community, on the public's access, and in the ability to stimulate commercial value-added opportunities is yet to be measured. However, there is a growing perception that certain applications of cost–recovery measures may have profoundly negative impacts on the innovation of the geographic information sector as a whole (Burkert, 1992; Litman, 1995; Maffini, 1990; National Research Council, 1995; Perritt, 1995). Hence, investigating alternative approaches and related outcomes from existing national information policies are called for. These test beds for dissemination and downstream innovation may provide an excellent opportunity to explore the casual linkages between information policy and national innovation in the sciences and industry.

Although the role of a mapping agency may be more important now then ever before, its role and contributions to national goals are shifting. The rapid uptake in the use of GIS, global positioning and remote-sensing technology by society at large is now beginning to outpace these agencies' abilities to keep up with the growing and increasingly diverse user demands for a wider range of specialised GIS databases. The commercial and non-profit sectors have seized upon this opportunity and are now becoming suppliers and intermediaries of geographic information. However, since many of these value-added activities generally require access to raw government data sets, the manner in which government leverages its information resources with other sectors of society can greatly influence the growth of this information supply chain.

4.3 Importance of intermediaries in the information marketplace

Information intermediaries also referred to as knowledge or information brokers are technological entrepreneurs which provide the necessary value-added processes to make public information resources useful to a wide end-user community (Pinelli *et al.*, 1990; Taylor, 1986). Commercial value-added resellers (VARs) use digitised attribute and map data from a number of sources and add value to produce standardised high quality products. National mapping data sets are commonly used due to their immediate availability. However, the demand for alternative sources of geographic information can be expected to soar. Intermediaries play an important role in augmenting the information dissemination goals of government by directly serving the end-user with enhanced information products and services. Traditionally, libraries and government depositories have provided the important link between government suppliers and end-users. However, in a networked society the concept of information intermediaries can be broadened to include commercial data vendors, scientists, academics, other government agencies and the general public (Benjamin and Wigand, 1995). The importance of commercial intermediaries in the value-added chain for geographic information is examined below.

The private sector is a major user and reseller of government information resources. This sector plays a significant role in developing a market for geographic information by providing spatial data products and services which cannot be met by government suppliers (Federal Geographic Data Committee, 1991). Some enhancements that VARs provide include combining data sets to meet end-user requirements, identifying new user markets, and providing informed interpretations of data usage.

Just as important, are VARs' ability to integrate various data sources, add value and inform users of the meaningful uses to which the data can be applied. As the public demand for specialised geographic information products grows with the growing adoption of GIS and digital mapping technology, commercial data vendors can be expected to perform a pivotal technology diffusion role. Over time, commercial data vendors are expected to provide the specialised end-user products and services which can support a variety of end-user applications such as in-vehicle navigation, desktop mapping and 3D environmental simulations (Openshaw, 1995).

Commercial data vendors compliment and enhance existing national dissemination activities by acting as an important information-transfer conduit. Some activities of private sector VAR activities, however, can raise significant concerns regarding personal information privacy, or data protection. This growing privacy issue needs to be considered more carefully as part of an overall national information policy strategy (Onsrud *et al.*, 1994).

Government policymakers must recognise the role of commercial data vendors if they are to promote responsible value-added processes (Branscomb, 1993). In other words, government must recognise VARs as partners in furthering the dissemination and diversification of public information and not as competitors. Currently, such recognition and cooperation is only beginning to evolve. There is considerable concern from the end-user and VAR communities that restrictive government dissemination practices continue to hinder the development of economically viable value-added information industries (Burkert, 1992; Gartzen, 1995; Perritt, 1994; Woodsford, 1991).

4.4 Effect of intellectual property on the public domain

A significant assumption underlying this discussion is the importance of promoting private sector intermediaries in the future evolution of an NSDI. The effect that alternative intellectual property approaches can have on stimulating these VARs is shown in Fig. 4.2. This figure identifies two alternative models describing the influence of government ownership control on the public domain of information and subsequent commercial value-added process. The two dissemination models are the Proprietary Spatial Data Dissemination Model and the Open Spatial Data Dissemination Model. The most significant difference between these two models is the fundamental role of the public domain in stimulating commercial value-added processes and in promoting government accountability. These models are described in detail below.

4.4.1 Proprietary Spatial Data Dissemination Model

The Proprietary Spatial Data Dissemination Model described on the top of Fig. 4.2 occurs where a single supplier, usually a national mapping agency, dominates the market for spatial data. Nearly all national mapping organisations would fall into this category. The defining feature of this model is the use of government copyright to enforce control of its information resources. In other words, by retaining the rights in its geographic information holdings, these resources are not available in the public domain. Spatial data are, of course, available as retail information products and are priced accordingly. Under this setting, geographic information enters the public domain only upon the expiration of copyright – ranging from 15 to 70 years after publication of a data set. Therefore, public data sets created in 1995 will not enter the public domain until the year 2010 at the earliest. Since the public domain in most countries is devoid of electronic databases and will be for the foreseeable future, end-users and intermediaries must resort to purchasing retail versions of the data from national providers, or seek alternatives elsewhere. If the impediments to acquiring the spatial data sets are too great, end-users and intermediaries will be forced to either go without the necessary information or they must recreate the same information content that has already been developed by government – usually at considerable cost. This situation may place commercial enterprises in the unfavourable situation of having to negotiate with their main competitor, which may be a government monopoly

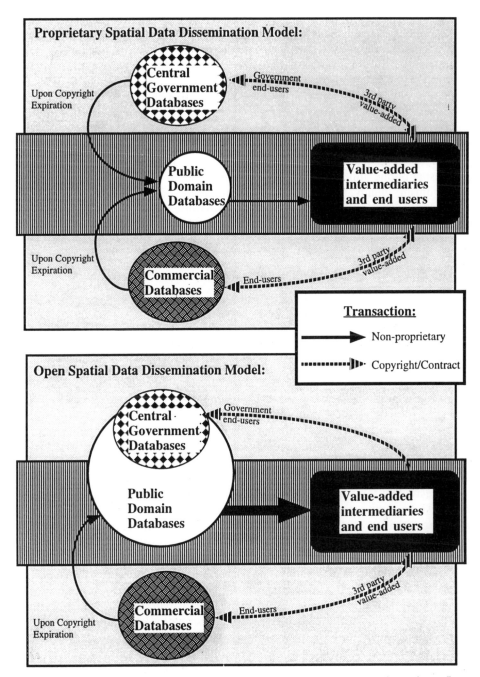

Figure 4.2 Effect of government copyright on the public domain and resulting flows of information.

supplier of spatial data. If commercial intermediaries do agree to purchase government information, these costs are simply passed on to end-users.

This approach of utilising government copyright to generate revenues potentially creates a drag on value-added activities, encumbers end-users with arbitrarily high pricing structures, and competes with emerging information-based enterprises (Commission of the European Communities, 1993, 1995a,b; Information, Market Observatory, 1995). Paradoxically, this proprietary approach, could lead to the formation of public information utilities which would then need to be regulated to ensure fair access and equitable pricing tariffs. Furthermore, the proprietary spatial data dissemination model may conflict with emerging GATT or NAFTA trade rules emphasising deregulation of government monopolies, or national policies to stimulate a national information industry. In an age of deregulating and privatising public utilities, promoting the development of such a utility may be an anachronism. Finally, although it has been heralded as a way to reduce the public's tax burden, proprietary dissemination can jeopardise growing public demands for more open and democratic government.

4.4.2 Open Spatial Data Dissemination Model

The Open Spatial Data Dissemination Model presented at the bottom of Fig. 4.2 demonstrates how placing government information resources into the public domain upon publication can stimulate diverse value-added activities, while fulfilling government accountability responsibilities. Inherent in this approach is the importance of government in contributing to a national information infrastructure upon which a commercial market in geographic information can flourish. This objective is met by increasing the flow of public domain databases to serve both government and commercial purposes.

As shown on the bottom graphic of Fig. 4.2, the vast amount of public domain data is immediately available to commercial, non-profit and other government sectors at the marginal cost of dissemination. Since the expansion of the GIS industry is tightly coupled to the availability of affordable spatial data, one could hypothesise that this approach allows the private and not-for-profit sectors to better capitalise on their national information resources (Samuels, 1993). Also, national mapping activities have been recognised as important national functions and have traditionally been supported through up-front appropriations. Although the functions of a national mapping organisation are under flux, this model recognises that the dissemination of spatial data is regarded as a process to enhance its primary mission. This approach attempts to stimulate the marketplace by using the geographic information assets which government has built up in its decades, if not centuries of mapping, census taking and other information collection processes and place them into the public domain. By making government resources available through the public domain, the likelihood of political control over information, a consequence of government monopoly, is eliminated. Likewise, all sectors of society are promised the same level of access to a low cost, non-proprietary geographic information infrastructure.

The principal drawback to the open-access dissemination approach is that it reduces the government's ability to commercialise its information resources. Another drawback is the inability to impose proprietary rights on the data, thereby eliminating any possibility for generating revenues above and beyond the cost of dissemination. Some also argue that the quality of data available in the public domain through an open-access approach is often worse than a comparable data set which is disseminated through cost recovery or proprietary partnership channels. Finally, an open-access approach impedes a promising mechanism for funding data conversion and maintenance activities. This leaves public agencies with no other option than to fund their data creation, maintenance activities from other means (for example, appropriations, savings from increased efficiency, fees on other services, or cross-subsidising data development).

4.5 National responses to information dissemination

Despite the much publicised importance of information in the overall national economy, policymakers in the past have shown great reluctance to explicitly target national information assets to stimulate economic and intellectual development. Rather, decisions concerning the dissemination of government information resources have been left to individual agencies to sort out, rarely with explicit policy statements. However, recent shifts in national priorities concerning the availability of geographic information to promote economic development, support an emerging information infrastructure, and monitor the global environment may require the development of explicit and coherent national information policies for the first time (Burkert, 1992; Sillince, 1994). As noted in the UK's Chorley Report, the importance of facilitating the development, dissemination and utilisation of geographic information throughout all aspects of society is becoming critical to the national interests (Department of Environment, 1987). In North America and Europe, this has been followed up with Presidential Orders and high-level policy papers dealing with access to government spatial data sets (de Bruïne, 1995; Commission of the European Communities, 1995a,b; Executive Office of the President, 1994).

4.5.1 US information policy setting

Since the founding of the US, the link between ready access to government information and a strong democracy has been significant. Four elements of US information policy are instrumental to the current setting: (1) the US Constitution; (2) the Freedom of Information Act (FOIA); (3) the US Copyright Act and (4) Federal information policy rules and guidelines.

Contemporary views of the role of government information are firmly grounded in constitutional principles, James Madison (1822) summed these up in stating that knowledge will forever govern ignorance, and that a popular government without popular information or the means to obtain it, is but a prologue to a farce or a tragedy or perhaps both. However, the people's right-to-know doctrine has traditionally been recognised through legislation rather than constitutional inter-pretation. The enactment of the US Freedom of Information Act (FOIA) in 1966, has served as an adjunct to constitutional elements of the US political system by ensuring an openness of government through information. Although the FOIA statute does not specifically identify digital data sets as a government record,

Federal and state courts have consistently held that computer records are public records for the purpose of FOIA. Until the enactment of FOIA, there were no statutory mandates compelling government to release information to the public (Branscomb, 1993). FOIA permits agencies to charge fees based only on the effort involved in fulfilling the request for information. There is no provision in the FOIA which would allow for fees based on commercial value of a record or data set.

To prevent Federal agencies from utilising intellectual property controls to circumvent their information dissemination responsibilities, Section 105 of the US Copyright Act precludes copyright protection for works of the Federal government. Unlike private entities, government presumably relies on justifications other than financial considerations for developing its works. As a result, Congress decided that it is in the public's best interest not to allow Federal agencies to copyright government information resources. The US Office of Management and Budget, an executive agency, has furthered the open-access principles outlined above by issuing Federal guidelines such as Circular A-130 and the Paperwork Reduction Act (PRA) which focus on improving the efficiency and effectiveness of Federal government record keeping (McClure and Hernon, 1989; Office of Management and Budget, 1992; Perritt, 1995). The importance of spatial data in improving the efficiency and effectiveness of US Federal service delivery and in boosting the performance of the domestic US geographic information industry was heightened in 1994 with a Presidential Executive Order requiring Federal agencies to collaborate with state and local agencies, the private sector and academia in the visionary goal of developing a National Spatial Data Infrastructure (NSDI), as doing so was critical to national objectives (Executive Office of the President, 1994).

4.5.2 European Union setting

It is difficult to describe the wide variety of responses by national suppliers of geographic information in Europe. Each member state in the European Union has unique policies which stem from a complex legal, cultural and political precedent (Lievesley and Masser, 1993). The dissemination of geographic information in various nation states data sets is determined by the collage of overlapping and sometimes conflicting policies and laws such as the privacy and anti-trust regulation, technology policy, and the competition law, national secrecy and security regulations, among others (Burkert, 1992; Lopez, 1995; Sillince, 1994). Until recently, there have been very few attempts to develop coherent national information strategies which would explicitly promote R&D advancement and commercialisation, while simultaneously attempting to harmonise differences within existing policies and regulations. This challenge is being confronted at the pan-European level with the proposed creation of a common European Geographic Information Infrastructure (Commission of the European Communities, 1995b).

Market studies, European Union technical reports and academic papers by GIS experts regularly suggest that European national data dissemination policies may be inadvertently holding back the competitiveness of the European Geographic Market Infrastructure (Blakemore and Singh, 1992; Gartzen, 1995; Hookham, 1994; Information Market Observatory, 1995; Rhind, 1991; Woodsford, 1991).

Although the linkage between public access to government data and the stimulation of the information sector has yet to be formally proven, a recent European Science Foundation (ESF) meeting of GIS experts concluded that the availability of geographic information across Europe is a significant problem which needs to be urgently addressed (Craglia, 1994).

In response to these concerns, the European Union (EU), has embarked on a number of strategic initiatives promoting the development of Community-wide geographic information markets. These initiatives are attempting to develop policies of access which exploit public sector information, link sources of European public sector information, and make better use of content resources in the public sector. Specific programmes include:

- *INFO 2000*. A Community programme to stimulate the development of a European multimedia content industry and encourage the use of multimedia content in the emerging information society. One of its principal action lines is to better exploit Europe's public sector information in order to stimulate interconnection throughout Europe and bolster the competitiveness of the European information industry (Commission of the European Communities, 1995a).

- *GI 2000*. This is a European Community policy document intended to raise the level of awareness regarding the development of a European Geographic Information Infrastructure (Commission of the European Communities, 1995b). The key issue addressed in this document is to provide a 'broad, readily available high-quality platform of base data within a uniform infrastructure across Europe so that every market niche is open to every entrepreneur, so that existing data can be combined to provide valuable information' (p. 3) (see also Chapter 3 in this volume).

- *PUBLAW*. This project is currently examining how European member states assert principles associated with a general right of access to government information. It is being undertaken by the European Union's DGXIII Legal Advisory Board (LAB). Preliminary findings from this project suggest that greater openness to government information is necessary to encourage the synergy between the public and private sectors in establishing a Community information market and to reduce possible distortions of competition in the European market for public sector information (Burkert, 1992).

A motivation for these EU and similar member state investigations is to identify dissemination policies which actively promote the greatest openness and innovation in the public and private information sectors. Likewise, it becomes important to identify restrictive policies which may result in reduced opportunities for the successful transfer of raw data sets to commercial data vendors and researchers. The scope of the following analysis is not to test this prediction, but to lay the theoretical foundation for future empirical research.

4.6 Alternative models for data dissemination

This section attempts to examine various models for disseminating government information, as well as their influence on value-added processes. Three approaches to disseminating government information – Cost-recovery, Public-Private Partnerships and Open Access – are described below.

4.6.1 Cost-recovery dissemination

Cost-recovery dissemination is motivated by public agency incentives to generate revenues from data sales, reduce the costs of dissemination and prevent commercial vendors from acquiring the data for virtually nothing and then selling derivative products for profit. This option relies on charging user fees to those who want to acquire a public data set. To minimise data leakage and prevent data piracy, the cost-recovery approach requires public information resources to be protected through intellectual property rights. Data is then licensed directly to public and commercial users. The licensing agreement stipulates allowable uses and regulates copying. Intellectual property control protects the public's trust by preventing free-riders from copying valuable public domain databases that may have cost hundreds of thousands of dollars and taken many years to develop. An example of this approach is that of the UK's national mapping organisation, the Ordnance Survey.

Data charges are based upon a formula which, in nearly all cases, results in fees and royalties above and beyond the actual cost of disseminating the data to the end-user. Although the justification is to undertake government dissemination activities in a more business-like manner, cost recovery becomes a means to generate public revenues necessary to fund sustained GIS database development. It is also a strategy for reducing the taxpayer's burden on developing spatial data. Under this scenario, government information is generally not considered a public good. Rather it is a scarce government commodity or service which is to be exchanged for a fee. Cost recovery attempts to utilise commercial market forces to redistribute what was once considered a public good or public information content.

Using a commercial sector approach to redistribute publicly funded factual information does raise important issues. As noted by the authors of *Reinventing Government,* user fees and intellectual property controls, which are intended to ration the dissemination of public goods and services, are not appropriate for collective goods such as government information. Such collective goods are intended to benefit the public as a whole and therefore demand should be encouraged (Osborne and Gaebler, 1993). In addition, basing annual revenue generation targets on the sale of GIS data raises the following challenges to a government supplier:

- Elasticity of demand is extremely difficult to determine.
- Users' database product requirements are unknown and will certainly vary.
- Users are price sensitive – ability to pay is not uniform.
- Revenues are likely to decrease over time as competitive data sources emerge.
- Potential user base is diverse, requiring product specialisation.
- Private sector may over time meet end-users' diverse market for database products.
- It is impractical to police illegal copying of public data sets.

Since the volume of public data entering the public domain is minimised through cost recovery, commercial value-added opportunities are also reduced. In addition, the cost-recovery approach has the effect of potentially silencing public interest challenges to policy formulation.

4.6.2 Proprietary partnership dissemination

For the purposes of this chapter, a proprietary partnership is narrowly defined as an agreement where the public agency grants a private or non-profit enterprise the exclusive right (or sole right) to disseminate publicly funded geographic information for a given time. There are many other forms of partnerships between government agencies and between government and the private sector where the public institution does not transfer ownership of the information and where the public domain supply of government information is not threatened. However, these alternatives are not considered in this chapter explicitly.

A major justification for many partnerships is to help finance costly government information database development activities. Without such public–private innovations, less information exists and the infrastructure to access it never gets built (Archer, 1995). Under the terms of a proprietary partnership, the agency turns over its geographic information to the private firm, which compiles, digitises and/or maintains the records. Copyright and licensing controls allow an agency to regulate access to the information, while providing a tradable asset (property rights to publicly financed data sets) to entice commercial partners.

Through the proprietary partnership approach, the exclusive vendor acquires all rights to disseminate the data. In more extreme examples, a public institution might relinquish all rights (ownership and dissemination rights) to the data. In return, the agency receives complete and updated copies of the compiled GIS data set for its own use or some percentage of the revenue resulting from the corporation's sale of data. This examination is limited to proprietary forms of partnerships whereby the rights to disseminate public information are transferred to a non-governmental organization.

The mechanics of this dissemination approach are straightforward. Users requesting access to the spatial data are referred to the commercial redistributor to obtain the data set. Since the public agency is not authorised under terms of the contract to provide access to the digital data, by wielding copyright and contract law and its position as exclusive distributor of valuable spatial data sets, the commercial vendor then controls the market for those data. In most cases, the public agency would attempt to ensure fair charging schedules to end-users.

One example of this approach is the Canadian Hydrographic Service which disseminates its digital raster navigation charts exclusively through a third party (Terry and Anderson, 1994). More recently, the US National Oceanic and Atmospheric Administration (NOAA) granted a consortium of commercial firms exclusive dissemination rights to its digitised raster navigation chart products. Both the Canadian Hydrographic Service and NOAA arrangements enable processing, customisation and dissemination of government data to be done by the private sector. Meanwhile, responsibility for the collection, management, and dissemination oversight is the responsibility of the government. This novel approach attempts to exclusively outsource a costly, but important function of government – the provision of public information.

The proprietary partnership is an innovative way reduce the public cost of disseminating government information while also generating revenues or in-kind data services (for example, data maintenance). However, developing exclusive channels for disseminating public information resources can raise significant problems. First, although the availability of high-quality electronic navigation chart products and services are welcomed by users, without some form of price regulation, geographic information may only be accessible to those who are able to afford it. Second, the inability to place nautical charts or other types of spatial data sets into the public domain, can lead to the development of government sanctioned information monopolies which would not necessarily be available to an open and diverse market. The development of an intermediary market is threatened by granting exclusive or sole licensing agreements. Similar to the cost-recovery approach, the outcome of the public–private partnership model is highly dependent on the potential market and the elasticity of demand by the user community. NOAA's agreement to grant exclusive distribution agreement does run counter to US Federal information policy. This has resulted in a recent lawsuit from a commercial vendor which has been refused access to NOAA digital navigation charts under the Freedom of Information (FOI) Act. The strong potential for litigation demonstrates the difficulty of establishing exclusive agreements with commercial vendors for the dissemination of government information in the USA.

Policymakers must be concerned with both the sustainability of existing dissemination functions and leveraging government information resources for national benefit. However, replacing a government monopoly in the distribution of information with a private one does not necessarily enhance downstream value-added market concerns. Since most monopolies lack the incentives to be responsive to changing external conditions, innovation in this important area could be curtailed needlessly (Perritt, 1995). Strategic partnerships aimed at outsourcing the dissemination of public information through sole distribution channels that then set arbitrary pricing and use conditions also threaten to undermine traditional economic imperatives for dissemination of public information – which is the promotion of a diversity of channels. The predicted outcome of exclusively licensing the dissemination of public information is a stagnant NSDI, whereby intellectual property controls and *ad hoc* pricing policies create a drag on value-added activities, encumber end-users with arbitrarily high pricing schedules and compete with emerging information-based enterprises.

4.6.3 Open access dissemination

In contrast to the previous two approaches, an open-access dissemination strategy requires public agencies to make their data available at the marginal cost of dissemination without enforcing copyright. In other words, government information resources automatically enter the public domain upon official publication. The objective of this non-proprietary approach is to comply with national legislation and/or policy aimed at ensuring the greatest public benefit by not attempting to artificially create scarcity of information resources. By reducing the charges for data, lowering the intellectual property impediments and involving key intermediaries in the dissemination process, government policy can significantly enhance the scale and scope of commercial value-added information processes, thus expanding the benefits of public information resources (McClure and Hernon, 1989).

Under this scenario, public data sets are priced at the marginal cost of dissemination, or at no direct cost to the user. Moreover, intellectual property controls such as copyright or contracts are discouraged (Office of Management and Budget, 1992). By allowing public databases to diffuse into the public domain, government fulfils its open records requirements while also providing a catalyst for commercial enterprise development. This approach reduces government's data dissemination burden by encouraging public agencies to focus on their mandated duties, rather than entering a commercial data market. In addition, technological advances such as electronic networks, such as the Internet, allow users to access government databases and metadata, while placing minimal burden on the agency. The principles underlying the Open Access Dissemination approach are consistent with most national information infrastructure (NII) policy goals.

The principal drawback to the open-access dissemination approach is that it reduces the government's ability to commercialise its information resources. This is a drawback, of course, only if the commercialisation of public information by government is an objective. Another drawback is the inability to impose proprietary rights to the data. This could result in an unfunded government service if databases development and maintenance efforts rely on data sales. Open-access policies also inhibit innovative proprietary partnerships for disseminating information resources that may not otherwise be available in either the public or private sectors. These activities would either be illegal or strongly discouraged. This leaves public agencies with no other option than to fund their data creation, maintenance activities from other means (for example, appropriations, savings from increased efficiency, fees on information services, or cross-subsidising data development).

Although the open-access approach raises some difficulty for the government data supplier, from the perspective of end-users and intermediary stakeholders, it does seem to be the ideal approach for leveraging existing information resources for the greatest national benefit. Its strongest asset is that it does not attempt to constrain the flow of public information through legal artefact to generate revenue. It promises to stimulate commercial and R&D innovation in geographic information, while protecting the public's right to hold its government's activities accountable – an important feature of advanced democratic systems.

Some argue that the data available in the public domain through an open-access approach is of lesser quality than a comparable data set which is disseminated through cost recovery or proprietary partnership channels. The concern for data quality, accuracy and completeness is increasingly important to ensure public safety and as protection from liability claims. However, focusing on data quality at only one stage of the information chain is misleading, since a large proportion of high quality derivative data sets which are copyrighted, licensed and generally more costly can be found further down the value-added data supply chain. For example, one of the main enhancements placed on raw public domain data sets by commercial value-added intermediaries is data quality. The availability of the US Census Bureau's TIGER/Line file is but one example of how dissemination policy has spawned dynamic value-added market chains for high-quality derivative spatial data products (Lopez, 1993, 1995).

What is important in any discussions concerning data quality is the ability for the end-user to choose from various alternatives which suit their needs. It is very likely that end-users require spatial data which is performance-based rather than specifications-based. In other words, they are increasingly concerned that a particular data product meets the performance requirements of their end-user applications, rather than the accuracy and scale specifications determined for government activities. More importantly, users are price sensitive. As desktop mapping applications become commonplace, end-users may not be willing or able to afford mission-critical data sets which were developed for much more rigorous applications. Instead, users will demand a choice of data products or services which meet both their application requirements and their pocketbooks. A fixed or captured end-user market for a single source of geographic information may not be an ideal channel to serve a growing and diversifying geographic information market, even though it may produce the quality data required by government. At this point it is useful to hypothesise how each of the three dissemination approaches affect the evolution of a national spatial data infrastructure.

4.7 Hypothesised outcomes of alternative dissemination approaches

Despite the benefits and limitations of each of the three dissemination approaches described above, policymakers must recognise that growing national and inter-national trends for increased openness in telecommunication and information networks, government accountability, and transborder data flows must be balanced with the requirements to ensure data protection. Figure 4.3 compares the predicted impacts that each of the three data dissemination approaches may have on the overall development of an NSDI. The graphic looks at the possible long-term (15-year) impact of the three comparative models for disseminating public information. Each shaded bar represents one of three classes of spatial data ownership (government, public domain and commercial). The volume of geographic information available through each of these ownership channels will vary depending upon the dissemination model adopted (for example, cost recovery, proprietary ownership and open access).

Under a cost-recovery approach the major supplier of end-user products will be the government. Since government can copyright and sell its databases, as would a private sector firm, the government competes directly with the private sector. This is shown in the figure as government suppliers dominating the spatial data market. Moreover, geographic information is not allowed to migrate into the public domain where the private sector can add value. Under this scenario, Fig. 4.3 depicts the public domain with almost no spatial data content. The predicted result of this scenario is an NSDI dominated by large government suppliers, an anaemic public domain, and a private sector which must either compete directly with a tax funded competitor or must form a partnership with its competitor.

In the proprietary partnership approach, government-owned information resources are transferred over to the private sector, again with little or no information resources entering the public domain. Figure 4.3 shows how both the government and public domain availability for spatial data is quite sparse. By contrast, it is the commercial sector which becomes the largest holder of spatial

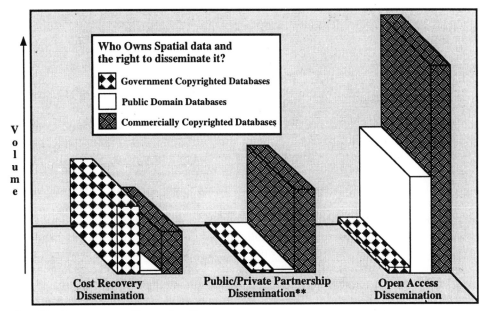

1. Cost recovery dissemination

Advantages:
- Reduced cost and increased revenues to agency
- Incentives for government data production
- Improved data products and services
- Consistency of government data and standards

Drawbacks:
- Curtailed public access to information
- Increasing cost value-added data to users
- Monopoly supply of primary data sources
- Creaming of lucrative data products
- Decreasing competitiveness of data market

2. Public private partnership dissemination

Advantages:
- Reduced cost and increased revenues to agency
- Incentives for government data production
- Improved data products and services
- Transfer of liability and system failure risks

Drawbacks:
- Curtailed public access to information
- Increasing cost of value-added data to users
- Monopoly supply of primary data sources
- Creaming of lucrative data products
- Loss of control over data dissemination

3. Open access dissemination

Advantages:
- Diverse channels for dissemination
- Decreased end-user costs for raw data
- Easy market entry for small inovators
- Rapid development of NSDI

Drawbacks:
- Limited supply of custom products and services
- Difficult to recover development costs
- Few fiscal incentives for data enhancements
- Unregulated re-dissemination of data

* The graphic above predicts the long-term impact of the three comparative models for disseminating public information. Each shaded bar represents one of three classes of spatial data ownership (government, public domain, and commercial). The volume of spatial data available through each of these channels will vary depending upon the dissemination model adopted (e.g., cost recovery, proprietary ownership, open access). These measures are comparative and are not to be considered as absolute values.

**The Public/Private Partnership model demonstrated here assumes that spatial datasets are either owned by a government agency, or were previously owned by the public agency and have since been licensed to an exclusive or sole distribution partner. As a result, distribution rights belong to the commercial firm which uses licensing and contract law to enforce copyright, pricing, third party uses and re-distribution.

Figure 4.3 Hypothesised impact of three dissemination approaches on the development of the NSDI.*

data sets – including those paid for by the taxpayers. This option has the benefit of allowing the private sector to use market forces to meet end-user requirements. However, if exclusive dissemination arrangements are pursued, the level of innovation and diversity of data products and services may not be fully optimised since VARs would be limited. Figure 4.3 shows that although the proportion of spatial data available from the commercial sector is large in comparison to government and the public domain, it is not as large as it could be, since competition is hindered through the use of exclusive dissemination conditions. It is likely that the cost for data will be high, since prices for data are not established by market forces, and there are strong tendencies to generate revenues from data sales.

The outcome of the third scenario, open-access dissemination, is very different from the previous two. Since government information is not controlled through copyright or license agreements it immediately flows into the public domain As a result, a low-cost avenue of raw public domain information resources becomes available to all – end-users, private sector, government users, value-added intermediaries – at the cost of dissemination. End-users and intermediaries can utilise the raw data resources placed into the public domain to serve for-profit or non-profit objectives. The figure depicts the appeal of the open-access approach by demonstrating how commercial value-added activities are stimulated by the availability of public domain information. Furthermore, the low barriers to market entry (that is low-cost data, no copyright), establish a very competitive commercial market for spatial data. Since government information is automatically placed into the public domain, there is very little which is available under government copyright. Although these diagrams are comparative and the outcomes are hypothetical, they provide a sketch for exploring how alternative approaches impact the development of a national spatial data infrastructure.

The first two approaches (cost recovery and proprietary partnerships) are excellent for exploiting the commercial value of a high quality geographic information, while the third approach fosters a process for adding value to raw government information to meet end-users' needs. The open-access approach invigorates private sector activities by reducing legal and pricing impediments and by not allowing government to compete unfairly in a geographic information market. In addition, there is very little theoretical justification or even anecdotal evidence demonstrating that proprietary methods for disseminating government information best serve national, scientific and commercial innovation. Although significant empirical work remains to be done, the open-access approach appears to provide the best solution for promoting the use of government information for the maximum benefit of society.

4.8 Conclusion

It is important to recognise that the suite of legal and policy factors which shape the development of national information policy are not static. They are a reflection of contemporary political, social economic and technological factors. National information policies are increasingly under pressure by global technological, economic and social forces. For example, the rapid globalisation of trade, the need for improved global telecommunications, and legal infrastructures which can facilitate the free flow of information across national boundaries are placing tremendous pressures on government agencies to adopt information dissemination policies which meet the needs of the emerging information society (Hlava, 1993).

These emerging techno-global factors have also heightened the need for public agencies both within and across national boundaries to cooperate in their data development and dissemination activities. This is particularly relevant with the rapid integration of the European Union (EU) and the ratification of GATT and NAFTA. The globalisation of trade, telecommunications and environmental monitoring compel national governments to cooperate, rather than compete, in leveraging their nation's information collection resources. The need for co-operation in the face of shrinking government budgets and growing regional environmental monitoring activities could lead to convergence of data develop-ment and dissemination objectives. At the same time, one must recognise that national power, considerations of national position, and the struggle among nations for comparative advantage will continue despite the growing calls for the globalisation of national information policies.

As geographic information becomes increasingly important in supporting public-sector decision making and national economic performance, information policy specialists must confront the issue of how government can best disseminate its resources to serve national objectives. Various approaches for disseminating information resources were described. These contrasting approaches can serve as national test beds for examining the effect of various information policies on public access, commercialisation and scientific advancement. The importance of utilising public information assets to serve society's needs is urgent. However, despite the lack of consensus on how geographic information should be disseminated, the global scope of trade, governing and environmental activities compound the urgent need to take into account the growing interdependence of national information collection activities. If North American and European efforts to develop a region-wide information infrastructure are to prevail, the issues raised in this chapter dealing with access, ownership and commercialisation of public informa-tion resources in promoting the use of government information for the maximum benefit of society must be resolved.

Acknowledgements

This chapter is based upon work partially supported by the National Center for Geographic Information and Analysis (NCGIA) under NSF grant No. SBR 88-10917. Any opinions, findings, and conclusions or recommendations expressed in this material are those of the author and do not necessarily reflect the views of the National Science Foundation. Portions of this chapter were previously published in: Lopez, X. R., 1995, The Impact of Scientific and Technical Information Policy on the Diffusion of Spatial Databases, *Proceedings of the Joint European Conference and Exhibition on Geographical Information*. 27–31 March 1995, The Hague, The Netherlands, 323–329.

References

ARCHER, H. (1995). Establishing a legal setting and organizational model for affordable access to government-owned information management technology, *Proceedings of Conference on Law and Information Policy for Spatial Databases*, pp. 13–24, University of Maine: NCGIA.

AZAD, B. and WIGGINS, L. (1995). Dynamics of inter-organizational geographic data sharing: A conceptual framework for research, in Onsrud, H. J. and Rushton, G. (Eds) *Sharing Geographic Information*, pp. 22–43, New Brunswick, NJ: Center for Urban Policy Research.

BALLARD, S. (1989). *Innovation Through Technical and Scientific Information: Government and Industry Cooperation*, New York: Quorum Books.

BENJAMIN, R. and WIGAND, R. (1995). Electronic markets and virtual value chains on the information superhighway, *Sloan Management Review*, **Winter**, 62–72.

BLAKEMORE, M. and SINGH, G. (1992). *Cost Recovery Charging for Government Information. A False Economy?*, London: Gurmukh Singh and Associates.

BRANSCOMB, L. M. (Ed.) (1993). *Empowering Technology*, Cambridge, MA: MIT Press.

DE BRUÏNE, R. F. (1995). *A European Strategy for Geographical Information*, Keynote address, Joint European Conference and Exhibition on Geographical Information, The Hague, 27–31 March 1995 (not in proceedings).

BURKERT, H. (1992. The legal framework of public sector information: recent legal policy developments in the EC, *Government Publications Review*, **19**(5), 483–96.

COMMISSION OF THE EUROPEAN COMMUNITIES (1993). *A Report to the Commission of the European Communities on an Evaluation of the Implementation of the Commission's Guidelines for Improving the Synergy between the Public and Private Sectors in the Information Market*. PUBLAW 2 Europe Final Report, Luxembourg: CEC Directorate General XIII.

COMMISSION OF THE EUROPEAN COMMUNITIES (l995a). *INFO 2000*, Council Directive. COM(95), Luxembourg: CEC Directorate General XIII.

COMMISSION OF THE EUROPEAN COMMUNITIES (l995b). *GI 2000: Towards a European Geographic Information Infrastructure (EGII)*, Discussion Document for Consultation with the European Geographic Information Community, Luxembourg: CEC Directorate General XIII.

CRAGLIA, M. (1994). Specialist meeting report. European data strategic review, Malgrate, Italy, *GISDATA Newsletter* **4**, 22–4.

DATAQUEST (1995). Market figures for the GIS industry, *Business Geographics*, **4**(1), 12.

DEPARTMENT OF ENVIRONMENT (1987) *Handling Geographic Information*, London: HMSO.

EXECUTIVE OFFICE OF THE PRESIDENT (1994). *Coordinating Geographic Data Acquisition and Access: The National Spatial Data Infrastructure* (Executive Order 12906). Federal Register, **59**(71), 17671–4.

FEDERAL GEOGRAPHIC DATA COMMITTEE (1991). *A National Geographic Information Resource. Annual Report to the Director, Office of Management and Budget*, Washington DC: Federal Geographic Data Committee.

GARTZEN, P. (1995). GIS in an ever-changing market environment, *GIS for Business: Discovering the Missing Piece in your Business Strategy*, pp. 63–6, GeoInformation International: Cambridge.

HLAVA, M. (1993). The Internationalisation of the information industry, *Bulletin of the American Society for Information Science*, **19**(3), 12–15.

HOOKHAM, C. (1994). The need for public-sector policies for information availability and pricing, *Proceedings of the 1994 Conference and Exhibition of the Association of Geographic Information*, **8**(2), 1–5.

INFORMATION MARKET OBSERVATORY (1995). *Geographic Information Systems in Europe: Problems and Potential, Information Market*, Observatory Working Report **95**/2, Luxembourg: CEC DG-XIII.

INTERNATIONAL TRADE ADMINISTRATION (1994). Computer Software and Networking, Chapter 27, in *US Industrial Outlook*, Washington DC: International Trade Administration.

LIEVESLEY, D. and MASSER, I. (1993). Geographic information in Europe: An overview, *Conference Proceedings of the Urban and Regional Information Systems Association*, pp. 153–65, Atlanta, GA: URISA.

LITMAN, J. (1995). Rights in government-generated data, *Proceedings of Conference on Law and Information Policy for Spatial Databases*, pp. 187–92, University of Maine: NCGIA.

LOPEZ, X. R. (1993). Database copyright issues in the European GIS community, *Government Information Quarterly*, **10**(3), 305–18.

LOPEZ, X. R. (1995). The impact of scientific and technical information policy on the diffusion of spatial databases, *Proceedings of the Joint European Conference and Exhibition on Geographical Information*, **2**, 323–9.

MADISON, J. (1822). Letter to W. T. Berry, 4 August 1822, cited in *Environmental Protection Agency v. Mink*, (1973), 410 US 73, pp. 110–11.

MAFFINI, G. (1990). The role of public domain databases in the growth and development of GIS, *Mapping Awareness*, **4**, 49–54.

MALCHUP, F. (1984). *Knowledge: Its Creation, Distribution, and Economic Significance*. Princeton, NJ: Princeton University Press.

MASSER, I. and CAMPBELL, H. (1995). Information sharing: The effects of GIS on British local government, in Onsrud, H. J. and Rushton, G. (Eds) *Sharing Geographic Information*, pp. 230–49, New Brunswick, NJ: Center for Urban Policy Research.

MCCLURE, C. and HERNON, P. (1989). *United States Scientific and Technical Information Policies: Views and Perspectives*, Norwood: Ablex Publishing.

NATIONAL RESEARCH COUNCIL (1995). *On the Full and Open Exchange of Scientific Data, Committee on Geophysical and Environmental Data*, Washington DC: National Research Council.

OFFICE OF MANAGEMENT AND BUDGET (1992). *OMB Circular A-130, Management of Federal Information Sources*, Washington DC: GPO.

ONSRUD, H. J. (1992). In support of cost recovery for publicly held geographic information. *GIS Law*, **1**(1), 3–6.

ONSRUD, H. J., JOHNSON, J. P. and LOPEZ, X. R. (1994). Protecting personal privacy in using geographic information systems, *Photogrammetric Engineering and Remote Sensing*, **60**(9), 1083–95.

OPENSHAW, S. (1995). A review of GIS in business applications, *Geographical Systems*, **2**, 153–68.

OSBORNE, D. and GAEBLER, T. (1993). *Reinventing Government: How the Entrepreneurial Spirit is Transforming the Public Sector*, Reading, MA: Addison-Wesley.

PERRITT, H. H. Jr. (1994). Commercialization of government information: comparisons between the European Union and United States, *Internet Research*, **4**(2), 7–23.

(1995). Should local governments sell local spatial databases through state monopolies? *Proceedings of Conference on Law and Information Policy for Spatial Databases*, pp. 52–72, University of Maine: NCGIA.

PINELLI, T. E., KENNEDY, J. M. and BARCLAY, R. (1990). *The Role of the Information Intermediary in the Diffusion of Aerospace Knowledge*, Paper Eight, NASA/DoD Aerospace Knowledge Diffusion Research Project, Springfield, VA: NTIS.

PINTO, J. K. and ONSRUD, H. J. (1995). Sharing geographic information across organizational boundaries: A research framework, in Onsrud, H. J. and Rushton, G.

(Eds) *Sharing Geographic Information*, pp. 44–64, New Brunswick, NJ: Center for Urban Policy Research.

RHIND, D. (1991). Data access, charging and copyright and their implications for GIS, *International Journal of Geographic Information Systems*, **6**(1), 13–30.

RHIND, D. (1995). Spatial data from government, *The AGI Sourcebook for Geographic Information Systems*, pp. 101–6, London: Association for Geographic Information (AGI).

SAMUELS, E. (1993). The public domain in copyright law, *Journal of the Copyright Society of the USA*, **42**(2), 137–82.

SILLINCE, J. A. A. (1994). Coherence of issues and coordination of instruments in European information policy, *Journal of Information Science*, **20**(4), 219–36.

TAYLOR, R. S. (1986). *Value-Added Processes in Information Systems*, Norwood: Ablex Publishing.

TERRY, B. F. and ANDERSON, N. M. (1994). 'A Canadian Public/Private Sector Initiative for the Production and Distribution of Electronic Charts', Unpublished paper, Ottawa: Canadian Hydrographic Service.

US CONGRESS (1990). *Helping America Compete. The Role of Federal Scientific and Technical Information*, United States Congress, Office of Technology Assessment, Washington DC: US Printing Office.

WOODSFORD, P. A. (1991). 'Demand for Geographic Data from the Private Sector in Europe', Unpublished paper presented at La Cartografia Española ante el Mercado Europeo, La Coruna, 4–6 November 1991.

WORLD METEOROLOGICAL ORGANISATION (1994). *Report of the Third Session of the WMO Executive Council Working Group on Commercialization of Meteorological and Hydrological Services*, Geneva: WMO.

CHAPTER FIVE

Infrastructures for Sharing Geographic Information: Lessons from the Great Lakes and the Columbia River

JOHN EVANS

5.1 Introduction

What does it take to share geographic information? It is unusual to see public agencies meaningfully sharing geographic information across organisational boundaries. Yet such sharing would seem especially fruitful in areas such as environmental management, where decision variables often cut across activity sectors, jurisdictions and regions, linked by physical pathways, adjacency, or other spatial relationships. Furthermore, the expansion of the Internet and increasingly advanced network software ought to make information sharing especially feasible, convenient and powerful.

Impediments to sharing seem to be partly technical (for example, adapting remote data to local purposes), and partly organisational (for example, adjusting to depend on outside sources). Incentives and impediments to sharing are both tied to the nature of geographic information: it is nearly always complex in structure and interpretation, rich in meaningful inter-relationships and difficult to understand or use without special tools. These aspects sharply distinguish geographic information from traditional alphanumeric data.

Information sharing often occurs in an *ad hoc* fashion, via methods that are devised anew with each interchange – single-use, single-purpose mechanisms. As the volume and frequency of information sharing grows, a more permanent mechanism often becomes beneficial, one that can be used repeatedly and for a variety of purposes. This multiuse, multipurpose mechanism (which may include a physical data network) is termed an information-sharing infrastructure.

The central question here is how to build and maintain such an infrastructure, and how to evaluate its success. Some useful background may be found in recent work on database interoperability, geographic information standards, networking technology and organisational behaviour. Using these building blocks, the research discussed in this chapter studies two examples of infrastructures to share geographic information: the Great Lakes Information Network and the Northwest Environmental Database in the Columbia River region. The study compares the principles set forth in the literature with the design, development and impacts of these existing systems, illustrating how these principles may be applied, adapted and extended to practical contexts.

5.2 Background

Several areas of recent research provide guidance for building and maintaining infrastructures for geographic information sharing. First, any such system provides some degree of connectivity between data systems. Connectivity in turn requires either setting standards or agreeing on a shared vocabulary: this is a new and growing area of geographic information research. As the infrastructure grows, Internet tools may help in network management and navigation. Finally, maintaining such an infrastructure depends on managing organisational aspects of collaboration. The following paragraphs describe each of these areas.

5.2.1 Levels of connectivity: physical, logical and semantic

In general, connectivity between data storage and retrieval systems may be of three types, or 'levels' (Wang and Madnick, 1989). Data networks provide *physical* connectivity – basic communication between computers. Sharing structured information between computer systems requires reconciling data models and query procedures, or *logical* connectivity. Finally, *semantic* connectivity bridges differences in data definitions (Siegel and Madnick, 1991), for accurate interpretation and use of shared information. Interoperable multidatabase systems, recently the focus of intense research (Hsu *et al.*, 1991; Litwin *et al.*, 1990; Sheth and Larson, 1990; Templeton *et al.*, 1987), offer connectivity at all three levels. Logical and semantic connectivity are the emphasis of data repositories (Jones, 1992), which extend data dictionaries (Narayan, 1988) through the use of enterprise models (Sen and Kerschberg, 1987). Most work on interoperability has been with alphanumeric data but some geographic extensions to data dictionaries (Marble, 1991) and repositories (Robinson and Sani, 1993) do exist. Geographic information, with its multidimensional data structures and relationships, may present quite different interoperability issues; yet multidatabase concepts and methods are useful for understanding and comparing geographic information sharing systems (Mackay and Robinson, 1992).

5.2.2 Standards, metadata and non-intrusive sharing

Data compatibility issues (for example, differing formats, definitions, or scales) often complicate the task of bringing together separate information resources for comparison, summary, or analysis. Traditional standards (US National Mapping Standards, Spatial Data Transfer Standard) are a valuable *lingua franca* for sharing geographic information; but sharing is often needed between autonomous organisations with different, yet well-established procedures or quality requirements. This need is addressed by newer, flexible standards based on *metadata* (structured data descriptions) and *queries* (data retrieval requests). These provide a shared vocabulary, a 'minimal standard' that defines the user's interaction with information, but does not intrude on the information itself. Noteworthy efforts include the US Federal Content Standard for Geospatial Metadata (Federal Geographic Data Committee, 1994) and Open Geodata Interoperability Services, based on 'object brokers' (Gardels, 1994). Metadata and queries promise a more elegant and efficient way to share geographic information than homogeneous standards; yet little is known about their practicality in real-life settings or their trade-offs *vis-à-vis* more intrusive methods.

5.2.3 Network tools and network management

As the Internet has come of age, so have tools for locating, retrieving, filtering, and using networked information – from the file transfer protocol (*ftp*), network file systems, and archive catalogues (*archie*) (Emtage and Deutsch, 1992), to Wide Area Information Servers (Kahle, 1991), Gopher, and especially the World-Wide Web (Schwartz *et al.*, 1992). Several tools have been developed specifically for locating and viewing networked geographic data (Menke *et al.*, 1991; Walker *et al.*, 1992). Frank (1994) sketches the role of data catalogues and navigational tools in several possible scenarios of a future spatial data infrastructure. These tools promise flexible and scalable ways to provide and use information across a growing network.

5.2.4 Organisational aspects of collaboration

Finally, the greatest obstacles to information sharing often seem to be behavioural, not technical (Croswell, 1989). Research suggests that geographic information sharing is difficult when organisations make quick technical choices to support limited internal purposes (Craig, 1995) or rely on inadequately trained staff (August, 1991). Inappropriate pricing (Rhind, 1992) or legal restrictions (Epstein, 1995) can also impede sharing. Conversely, geographic information sharing may be facilitated by negotiation (Obermeyer, 1995), a common, 'superordinate' goal (Pinto and Onsrud, 1995), a 'killer application' (Brodie, 1993), cost recovery (Taupier, 1995), and clear data ownership (Carter, 1992). While it may be difficult to change human factors solely for the purpose of information sharing, these findings help to identify the most fruitful organisational contexts and to steer clear of counterproductive organisational choices.

5.3 Case study overview

Principles from these four areas – connectivity, standards, networks, and organisational behaviour – are likely to help in building and maintaining infrastructures for sharing geographic information. The next sections illustrate, apply and adapt these principles in two real-world settings: the Great Lakes Information Network and the Northwest Environmental Database.

5.3.1 Study design

To create a test bed for the above principles, several coalitions of government agencies in the US were chosen for study based on their commitment to a distributed information infrastructure, their goal of scalable, non-intrusive information sharing and their focus on a shared natural resource. The intent was to spotlight cases where geographic information sharing seemed most likely to grow, to be sustained, to surmount organisational barriers, and to change tasks and processes of environmental management. To date, the study has focused on the Great Lakes Information Network (GLIN) and the Northwest Environmental Database (NED), both of which support the collaborative management of large interstate water resources in the US.

5.3.2 Great Lakes Information Network (GLIN)

In May 1993, the Great Lakes Commission, a partnership of eight US states, began experimenting with the Internet to enhance communication and coordination among the many groups concerned with the Great Lakes. The resulting Great Lakes Information Network (GLIN) links the Great Lakes states (Minnesota, Wisconsin, Illinois, Indiana, Michigan, Ohio, Pennsylvania and New York), the Canadian province of Ontario, and several Federal agencies in both the US and Canada (including the US Environmental Protection Agency, US Army Corps of Engineers and Environment Canada) as well as many other public and private groups concerned with the ecology or economy of the Great Lakes region (Ratza, 1996).

5.3.3 Northwest Environmental Database (NED)

In December 1980, the US Congress called on the Bonneville Power Administration (the Federal Energy Agency for the Pacific Northwest region of the US) and the Northwest Power Planning Council (representing the four states in the region) to mitigate fisheries and other environmental impacts of hydroelectric development on the Columbia River and its tributaries. They worked with the region's four states (Montana, Idaho, Washington and Oregon) and several Federal agencies and Indian tribes on a region-wide rivers assessment in 1984–6. Data from this river study formed the Northwest Environmental Database (NED), which was used to manage fisheries, wildlife and energy facilities in the region. NED was a geographic database on fisheries, wildlife, habitat and hydroelectric facilities, all located in the region; data come from a variety of sources and were shared without a network, via coordinators in each state's fish and wildlife agency (Bonneville Power Administration, 1996).

5.3.4 Dissimilar infrastructure choices

Information sharing on GLIN occurs via a decentralised data architecture based on the Gopher, World-Wide Web, electronic mail forums, and *ftp* servers. GLIN's focus has been on getting agencies connected to the Internet, either via dial-up to a central server or through local Internet providers, for decentralised growth of interrelated information on the Great Lakes.

In contrast, the Bonneville Power Administration and the Northwest Power Planning Council never set out to build a network, but to compile a consistent, region-wide set of data. Yet in order to make the database widely usable, and to allow for ongoing maintenance by state partners, they opted for a decentralised approach: they strengthened data management capacities in several agencies by purchasing software and hardware for geographic and other data management systems; they specified spatial locations using a powerful standard, the US Environmental Protection Agency's River Reach Files; and they built cross-reference tables to reconcile each state's data categories and stream identifiers to the common standards. To obtain a regional overview quickly, they chose a geographic scale of 1:250 000, later updated to 1:100 000. In addition, they also funded a full-time staff in each of the four states to coordinate data collection, structuring and distribution.

5.3.5 Comparable infrastructure impacts

Despite their different histories, technical choices, and modes of growth, these infrastructures have had comparable impacts on environmental management. They have enabled consistent region-wide strategies, provided some interstate collaboration opportunities, improved interdisciplinary 'ecosystem thinking', and enhanced public participation.

5.3.5.1 *Consistent, region-wide strategies*

Of the two infrastructures, NED presented the more dramatic impacts. In 1988, the Northwest Power Planning Council used NED to amend its long-term fish and wildlife plan declaring 44 000 miles of streams in the Columbia River region as 'Protected Areas': valuable fish and wildlife habitats precluded hydroelectric development. Participants in this amendment feel that such an achievement was unthinkable before the states pooled their information in a consistent form. NED was used by the National Marine Fisheries Service in applying the Endangered Species Act to salmon fisheries in the region. Regionally consistent data, as contained in NED, allowed the Service to assess and compare fisheries and their habitats by 'Evolutionarily Significant Unit', which requires far more data analysis than either surveying individual salmon runs or counting Pacific salmon as a whole.

Thus far, GLIN's tangible accomplishments are less dramatic; though it has enhanced communication, it is still largely an experiment that has not directly supported many official actions. However, in 1994–5, using GLIN's *ftp* and mail servers, a large, geographically dispersed team of scientists and software engineers jointly developed the Regional Air Pollutant Inventory and Development System (RAPIDS), used by the Environmental Protection Agency for emissions regulation and facility siting throughout the Great Lakes. Team members felt that their collaboration via GLIN allowed them to fashion a powerful product in record time.

Regional summaries and joint projects are nothing new; but these infrastructures have led to reusable strategies: in the Pacific Northwest, Protected Areas are reviewed every five years; also in the Columbia River region, Endangered Species listings are reviewed periodically; in the Great Lakes, RAPIDS will undergo modifications and extensions. These infrastructures also support multiple users and purposes: for instance, some state agencies in the Northwest have adopted NED's River Reach identifiers for their own habitat and natural resource data.

5.3.5.2 *Interstate collaboration*

The primary users and beneficiaries of these two infrastructures have been Federal agencies with a regional focus like the Environmental Protection Agency, the Bonneville Power Administration and the National Marine Fisheries Service. Yet both infrastructures could potentially help agencies in states, provinces, and tribes to collaborate and to coordinate their work. This has occurred, to a limited extent: air quality specialists in the Great Lakes states have used GLIN to collaborate on pollution assessment by developing RAPIDS; in the case of the Columbia River, NED was sometimes used to coordinate fishing regulations between upstream and downstream states (for example Idaho vs Washington) or for joint protection of boundary waters (for example, Washington vs Oregon). Certain fundamental aspects of these infrastructures have limited their wider use at the state level: Great Lakes' states have been slow to invest in networking technology; and several agencies in the Columbia River region found NED's geographic scale (1:100 000) and its qualitative assessments of habitat and resources ('good,' 'fair,' 'poor') inadequate for their site-specific work.

5.3.5.3 *Interdisciplinary 'ecosystem thinking'*

In both the Great Lakes and the Columbia River region, traditional planning and management approaches have tended to focus on one part of the ecosystem in isolation from other, related factors. In the Great Lakes, an exclusive focus on stemming effluents has overlooked species that depend on high levels of aquatic nutrients. In the Columbia River, overdependence on fish hatcheries has contributed to reducing the genetic diversity of fish stocks. The information-sharing infrastructures in both regions, however, have begun to facilitate communication and learning across disciplines and geographic areas. This cross-fertilisation has enabled the emergence of what some have called 'ecosystem thinking', which considers the health of the overall ecological system including human activities, at various spatial and temporal scales. Such cross-fertilisation is facilitated by GLIN's hypertext structure and NED's geographic layers and diverse data sources.

5.3.5.4 *Enhanced public participation*

An important benefit of these infrastructures has been better public access to information. In both cases, simply having information organised and accessible in one place has helped state and federal groups answer questions from the public. Open access to information is a high priority for the Great Lakes Commission, the Northwest Power Planning Council, the Environmental Protection Agency and others. In addition, GLIN's use of the Internet has provided channels for the public to communicate directly with federal and state agencies for questions or comments. Releasing detailed environmental and project data to the public does raise some concerns about misuse, staff burdens, and security; however, many (US federal agencies in particular) see it as either legally required or prudent in the light of recent court battles.

5.4 **Fitting theory to practice**

The following sections illustrate the groundwork elements outlined earlier with examples from the case study, highlighting challenges and lessons in each area.

5.4.1 Levels of connectivity: physical, logical and semantic

As described earlier, recent research outlines three kinds of connectivity in information sharing: physical, logical and semantic. These cases suggest that all three kinds of connectivity are important to an effective and lasting geographic information sharing infrastructure. For instance, GLIN's use of the Internet and the World-Wide Web provides physical connectivity among a large number of data sources. However, most of GLIN's information content was developed with little thought to logical or semantic connectivity: press releases, reports, meeting minutes, and the like. As participants have put these resources on-line, users have found it increasingly difficult to find available information on a given topic amidst hundreds of interrelated Web pages. Recently, a few GLIN participants have begun building a common keyword set and multisite search facilities, with a view to achieving some logical and semantic connectivity. Increased availability of structured data, both alphanumeric and geographic, may also facilitate meaningful data relationships and summaries across GLIN.

In contrast, with no network link between its data sources, NED had no physical connectivity; but its shared geographic data model gave it logical connectivity between its component databases; and its cross-reference tables, by reconciling data categories, offered some semantic connectivity. This enabled, at least in principle, a heterogeneous, decentralised, integrated information system, with each state partner in charge of data management details, yet able to provide comparable information to the regional level. In contrast with GLIN, which devoted a lot of effort to physical connectivity (simply getting its partners on-line), NED concentrated on data-centred issues of logical and semantic connectivity – ultimately reaching similar milestones in its impact on people's work. There are drawbacks. After the 1988 Protected Areas amendment, states tended to focus on their more detailed in-state data sets, at the expense of maintaining NED or keeping its cross-reference tables up to date. In the absence of a physical communications link among its participants, NED may have achieved fewer collaborative relationships among its participants, and thus, less reason to interchange and coordinate data across state lines.

For both of these infrastructures, a key challenge has been (for GLIN) to build or (for NED) to maintain a coordinated overall information resource despite differing priorities and procedures among participants. Their experience suggests that all three levels of connectivity (physical, logical and semantic) may be needed, at least to some degree, for effective, sustainable information-sharing infrastructures. However, despite the push in the literature towards fully interoperable solutions, these cases illustrate the benefits of partial connectivity at each level. Finally, GLIN's experience shows how a sharing infrastructure may grow in sophistication over time; and NED's experience suggests that various connectivity levels may usefully be tackled in any order.

5.4.2 Standards, metadata and non-intrusive information sharing

As detailed earlier, metadata and query languages are increasingly being added to traditional standards to enable non-intrusive information sharing. The two cases illustrate the feasibility of well-conceived, non-intrusive approaches; confirm the need for some kind of shared vocabulary; and highlight the challenges of keeping such systems up-to-date.

NED was quite forward looking in the late 1980s, with its goal of linking distributed data sources 'as is' without requiring them to conform to a single homogeneous standard. Its cross-reference tables, linking each state's stream identifiers with the River Reach numbers, are a simple, yet powerful example of metadata, and the River Reach system itself is a valuable shared vocabulary. Thus NED illustrates well the feasibility of non-intrusive data sharing.

However, while NED's small geographic scale (1:250 000-scale River Reach Files, later extended to 1:100 000 in most areas) enabled rapid data collection and compilation, several agencies felt that it omitted many ecologically significant stream reaches, and found it difficult to use for their site-level projects. Several would-be users chose to build their own separate databases at much larger scales (1:24 000) for site-specific work. Thus NED found its shared geographic reference difficult to maintain and adapt to changing needs. Furthermore, NED used mostly qualitative judgements ('good,' 'fair,' 'poor') to characterise habitats and natural resources along stream reaches. This choice enabled rapid data collection, and was sufficient for defining Protected Areas; but analytical and comparative uses often require quantitative data and a repeatable measurement method. For these and other reasons, NED is now being subsumed into an emerging aquatic resource information network, known as StreamNet, intended to serve the Pacific Northwest region with detailed, comprehensive information on fisheries and other river-related natural resources. (Bonneville Power Administration, 1996.)

GLIN has only recently begun to encounter questions of data compatibility. It began with a very hands-off philosophy, allowing approved participants to put on-line whatever they wished, in any form. As its sites have begun putting detailed maps and data on-line, and trying to make analytical use of these data, the need for some overall consistency has become more apparent. Although GLIN has no global standards for data format or quality, several partners have begun annotating each of their Web documents with an author and date 'signature', for a basic level of accountability. Some are adding keywords to this annotation to facilitate searching for information across the network. Yet GLIN is more a collection of documents, than a true distributed database: while keywords do help in locating relevant documents, GLIN's unstructured nature provides little logical or semantic connectivity, leaving to the user the task of extracting, comparing, or combining information across the various sites.

5.4.3 Network tools and network management

As detailed earlier, the recent proliferation of tools for handling networked information holds promise for building flexible, scalable infrastructures for geographic information sharing. GLIN, in particular, is finding that current network tools allow for scalable, accessible, navigable systems, but that managing their information content remains difficult.

GLIN's data architecture, based primarily on the World-Wide Web's inter-related screens of information, allows spontaneous, almost unlimited growth. Joining GLIN consists simply of putting data files on a Web or Gopher server and persuading one of GLIN's participants to include a hypertext pointer to the files. In addition, GLIN's use of Universal Resource Locators (URLs) provides a readily understandable, network-transparent interface, which allows new users to learn quickly. On the other hand, having invested heavily in networked connectivity, GLIN has found it difficult to initiate and manage project-related sharing of detailed statistical and geographic information without additional facilities for navigation and interpretation.

In the absence of a physical data network, NED was not conceived as a 'seed' for decentralised networked growth, but as a fixed set of databases and participants. It grew and evolved by traditional methods, however, as participants updated and expanded the data sets for which they were responsible.

5.4.4 Organisational behaviour

Finally, these cases illustrate how the organisational context can influence information sharing: especially superordinate goals, data access policies, and teamwork among sharing partners.

Shared, superordinate goals and 'killer applications' have affected the success of both of these infrastructures. For instance, declaring protected areas in the Pacific Northwest, rebuilding salmon stocks in the Columbia River basin, and controlling airborne pollutants in the Great Lakes are all important regional problems that no one agency could tackle on its own, and that require a lot of assessment and data analysis. Second, the cases show the need to articulate a shared goal from the outset, and to continue defining current and future applications of the infrastructure to maintain its relevance to specific needs. With its broadly framed objectives, GLIN as a whole has had trouble evoking greater commitments from state partners; and NED's guiding focus got diffused once the Protected Areas amendment was successfully passed. Third, an overly strong shared concern may actually impede collaboration: although the Northwest fisheries crisis has led many to support NED and related shared data systems, severe scarcity in some areas has sparked resource conflicts.

Pricing and access policies have also played an important role in both infrastructures. Most of their participants see the data products they have built as public goods, and freely give them to anyone who asks. Some agencies in the Great Lakes and Pacific Northwest do distribute data based on market-based or cost-recovery pricing; yet several of these have begun to relax their policies, allowing data recipients to 'pay' for data in a variety of ways such as enhancing a regional data layer by adding local detail, or token payments unrelated to the value of the data product.

Teamwork – a relationship which gives credit for accomplishments to the team rather than to any one participant – seems to have been the key to sustaining GLIN and NED. This relationship contrasts with organisational incentives that emphasise leadership over collaboration. These cases suggest that teamwork is built in part through frequent in-person meetings. In the Great Lakes, the RAPIDS team was able to 'build trust' through quarterly face-to-face meetings and weekly telephone conferences. In contrast, NED's coordination for several years consisted primarily of quarterly written reports: some participants have suggested that this 'maintenance mode' has contributed to its decline in recent years.

5.5 Conclusions

Sharing geographic information is an important but difficult challenge, in which multiuser, multipurpose infrastructures can play an important role. Several fields of research provide general principles for the design and maintenance of such infrastructures. The Great Lakes Information Network (GLIN) and the Northwest Environmental Database (NED) are useful test beds for illustrating, adapting and extending these principles. These two cases represent very different infrastructure choices: GLIN has focused chiefly on networking technology, NED on decentralised geographic information. Yet both have had comparable impacts on environmental management: consistent regional strategies, some interstate collaboration, interdisciplinary thinking and public participation. They have both faced challenges in broadening their interstate collaboration and in managing consistent decentralised information.

Together, these cases suggest the importance of partial data connectivity at multiple levels, physical, logical, and semantic; and they illustrate alternative growth paths over time. NED's experience confirms that non-intrusive strategies have been feasible for some time, but can be difficult to maintain; GLIN highlights the strengths and weaknesses of current networked information tools. Finally, both cases point out the influence of the organisational context (superordinate goals, open access policies, and teamwork) on the long-term effectiveness of such infrastructures.

These cases suggest several fruitful areas of practical, 'high-impact' tool development using current technology: (1) dynamic, networked base maps (for example, a River Reach File that could be directly used by all and extended to larger scales); (2) dynamic, networked directories and cross-reference tables linking multiple data sources; and (3) querying tools that would use these base maps and cross-reference tables to find, use and correlate diverse pieces of information. Emerging technologies, such as spatial query languages, object-oriented broker systems, or intelligent agents, may radically affect the choice of technical strategies; further inquiry is needed into their likely impacts and trade-offs.

This study also highlights opportunities for further research on the relationship between the technical and organisational design of such infrastructures and their actual impacts on environmental management. In this study, GLIN and NED were chosen based on the high 'quality' of their technical design — qualities such as connectivity, non-intrusiveness, scalability and their favourable organisational contexts. Studies of additional geographic information-sharing infrastructures (assessing impacts of alternative contexts and design choices) would clarify the relationships between their design and contexts and their strategic impacts.

Acknowledgement

This material is based upon work supported by the National Science Foundation under Grant No. SBR-9507271. Any opinions, findings, and conclusions or recommendations expressed in this material are those of the author and do not necessarily reflect the views of the National Science Foundation.

References

AUGUST, P. V. (1991). Use vs abuse of GIS data, *URISA Journal*, **3**(2), 99–101.

BONNEVILLE POWER ADMINISTRATION (1996). 'Streamnet: report on the status of salmon and steelhead in the Columbia River Basin, 1995', Portland: Bonneville Power Administration.

BRODIE, M. L. (1993). The promise of distributed computing and the challenges of legacy information systems, in Hsiao, D. K., Neuhold, E. J. and Sacks-Davis, R. (Eds) *Interoperable Database Systems*, DS-5, pp. 1–31, North-Holland: Elsevier.

CARTER, J. R. (1992). Perspectives on sharing data in geographic information systems, *Photogrammetric Engineering and Remote Sensing*, **58**(11), 1557–60.

CRAIG, W. J. (1995). Why we can't share data: institutional inertia, in Onsrud, H. J. and Rushton, G. (Eds). *Sharing Geographic Information*, New Jersey: Center for Urban Polcy Research, Rutgers University.

CROSWELL, P. L. (1989). Facing reality in GIS implementation: lessons learned and obstacles to be overcome, in Salling, M. J. and Gayk, W. F. (Eds). *Proceedings of the 1989 Annual Conference of the Urban and Regional Information Systems Association (URISA)*, pp. 15–35. **IV**, Washington, DC: URISA.

EMTAGE, A. and DEUTSCH, P. (1992). Archie: an electronic directory service for the Internet, in *Proceedings of the Winter 1992 USENIX Conference*, pp. 92–110, Berkeley, California: The USENIX Association.

EPSTEIN, E. F. (1995). Control of public information, in Onsrud, H. J. and Rushton, G. (Eds) *Sharing Geographic Information*, New Jersey: Center for Urban Policy Research, Rutgers University.

FEDERAL GEOGRAPHIC DATA COMMITTEE (1994). *Content Standards for Digital Geospatial Metadata*, Washington, DC: Federal Geographic Data Committee.

FRANK, S. (1994). The National Spatial Data Infrastructure: designing navigational strategies, *Journal of the Urban and Regional Information Systems Association*, **6**(1), 37–55.

GARDELS, K. (1994). Open GeoData Interoperability Services: an object-oriented framework for accessing distributed, heterogeneous geographic information. White Paper available from the OpenGIS Foundation, Cambridge, Massachusetts, USA.

HSU, C., BOUZIANE, M., RATTNER, L. and YEE, L. (1991). Information resources management in heterogeneous, distributed environments: a metadatabase approach, *Institute of Electrical and Electronics Engineers (IEEE) Transactions on Software Engineering*, **17**(6), 604–24.

JONES, M. R. (1992). Unveiling repository technology, *Database Programming and Design*, **5**(4), 28–35.

KAHLE, B. (1991). An information system for corporate users: Wide Area Information Servers, *Online Magazine*, **15**(5), 56–62.

LITWIN, W., MARK, L. and ROUSSOPOULOS, N. (1990). Interoperability of multiple autonomous databases, *Association for Computing Machinery (ACM) Computing Surveys*, **22**(3), 267–93.

MACKAY, D. S. and ROBINSON, V. B. (1992). Towards a heterogeneous information systems approach to geographic data interchange. Institute for Land Information Management Discussion Paper 92/1, Mississauga, Ontario: University of Toronto.

MARBLE, D. F. (1991). The extended data dictionary: a critical element in building spatial databases, in *Proceedings of the Eleventh Annual ESRI User Conference*, pp. 169–77. Redlands, CA: Environmental Systems Research Institute.

MENKE, W., FRIBERG, P., LERNER-LAM, A., SIMPSON, D., BOOKBINDER, R. and KERNER, G. (1991). A voluntary, public method for sharing earth science data over Internet using the Lamont view-server system, *Eos, American Geophysical Union Transactions*, **72**(8), 409–14.

NARAYAN, R. (1988). *Data Dictionary: Implementation, Use, and Maintenance*, Englewood Cliffs, NJ: Prentice-Hall.

OBERMEYER, N. J. (1995). Reducing inter-organizational conflict to facilitate sharing geographic information, in Onsrud, H. J. and Rushton, G. (Eds) *Sharing Geographic Information*, New Jersey, Center for Urban Policy Research, Rutgers University.

ONSRUD, H. J. and RUSHTON, G. (Eds) (1995). *Sharing Geographic Information*, New Brunswick, New Jersey: Center for Urban Policy Research, Rutgers University.

PINTO, J. K. and ONSRUD, H. J. (1995). Facilitators of organizational information sharing: a research framework, in Onsrud, H. J. and Rushton, G. (Eds) *Sharing Geographic Information*.

RATZA, C. A. (1996). The Great Lakes *Information Network*: the region's Internet information service in *Toxicology and Industrial Health*, **12**(3–4), 557.

RHIND, D. W. (1992). Data access, charging, and copyright and their implications for geographical information systems, *International Journal of Geographical Information Systems*, **6**(1), 13–30.

ROBINSON, V. B. and SANI, A. P. (1993). Modeling geographic information resources for airport technical data management using the Information Resources Dictionary System (IRDS) standard, *Computers, Environment and Urban Systems*, **17**, 111–27.

SCHWARTZ, M. F., EMTAGE, A., KAHLE, B. and NEUMAN, B. C. (1992). A comparison of Internet resource discovery approaches, *Computing Systems: the Journal of the USENIX Association*, **5**(4), 461–93.

SEN, A. and KERSCHBERG, L. (1987). Enterprise modeling for database specification and design, *Data and Knowledge Engineering*, **2**, 31–58.

SHETH, A. P. and LARSON, J. A. (1990). Federated database systems for managing distributed, heterogeneous, and autonomous databases, *Association for Computing Machinery (ACM) Computing Surveys*, **22**(3), 183–236.

SIEGEL, M. and MADNICK, S. E. (1991). A metadata approach to resolving semantic conflicts, Working Paper 3252-91-MSA, Cambridge, Massachusetts: Sloan School of Management, Massachusetts Institute of Technology.

TAUPIER, R. (1995). Comments on the economics of geographic information and data access in the Commonwealth of Massachusetts, in Onsrud, H. J. and Rushton, G. (Eds) *Sharing Geographic Information*.

TEMPLETON, M., BRILL, D., DAO, S. K., LUND, E., WARD, P., CHEN, A. L. P. and MACGREGOR, R. (1987). Mermaid: a front-end to distributed heterogeneous databases, *Proceedings of the Institute of Electrical and Electronics Engineers (IEEE)*, **75**(5), 695–708.

WALKER, D. R. F., NEWMAN, I. A., MEDYCKYJ-SCOTT, D. J. and RUGGLES, C. R. N. (1992). A system for identifying datasets for GIS users, *International Journal of Geographical Information Systems*, **6**(6), 511–27.

WANG, Y. R. and MADNICK, S. E. (1989). Facilitating connectivity in composite information systems. *Data Base*, Fall, 38–46.

CHAPTER SIX

European Micro-Data Availability: The Special Case of Britain

DANIEL DORLING

6.1 Introduction

This chapter presents the final results of a three-year study sponsored by the British Academy to collect and analyse the most detailed spatial data on UK human geography available, using new geographical information system techniques. The chapter discusses how the geographical information techniques developed to deal with this information could be used in a wider context.

This work was developed from the author's PhD thesis on visualising the social, economic and political spatial structure of UK geography. The author and his supervisor explored new methods of data compression, animation, smoothing, presentation and transformation in work published originally three years ago (Dorling and Openshaw, 1992). This methodological work has since been extended and is now being applied to many large data sets most of which have not before been analysed at these spatial scales (Dorling, 1993, 1995).

British spatial data presents a special case for a number of reasons, all connected with the pedigree of this information. First, it tends to be extremely complex in comparison with many other countries' data. The areal units have evolved differently in different parts of the country over time and boundary changes are common every 3 to 15 years. The UK was one of the first countries in the world to employ a Boundary Commission to change the boundaries of administrative units – which it appears to do as frequently as possible. Comparison over time is hence problematic. Second, the UK government is uniquely secretive about spatial data, having laws governing its use comparable with those of many military dictatorships. Third, there exists a huge wealth of official data which dates back further, and in more detail, than that of almost any other country.

Many of the problems of spatial complexity and of the sheer volume of information can be solved using Geographical Information Systems to create valuable data sets for research. The new UK data sets which have been drawn together and made spatially comparable for the first time by the author include the full British Censuses of 1971, 1981 and 1991 at ward level (10 000 spatial units by 15 000 variables); selected variables from earlier censuses by the highest resolution areal unit available, which can then be compared over time with the

Figure 6.1 An equal area ward map and an equal population ward cartogram of the UK (the keys show annotated county boundaries on the two projections).

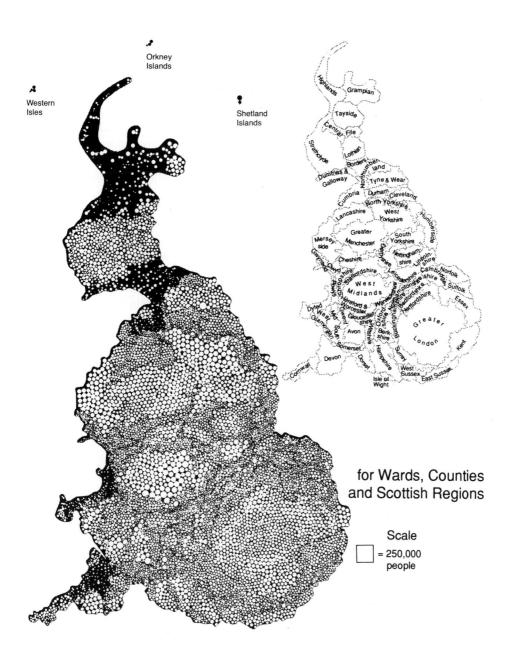

Figure 6.1 (continued)

same variables for later years; and the spatially coded results of all General Elections since 1955. The data set also includes mortality statistics recorded by postcode in the 1980s and made comparable with local authority series dating back to the 1930s; school exam results, house sales, information on wealth and on changing patterns of migration and commuting (at the full ward-to-ward level in both 1981 and 1991).

Visualisation techniques have been developed to analyse this information which is too vast to be studied by traditional statistical techniques. *A New Social Atlas of Britain* has been produced from these data sets which contains over 200 pages of novel colour maps (Dorling, 1995). Most of the maps use projections which distort space to highlight urban areas across the nation. The aim of this work has been to show how localities have changed over time, drawing patterns of social inclusion and exclusion across the many thousands of localities in the UK. In total over 2 000 000 individual variables are presented in this atlas. The preliminary drafts of prints taken from the atlas are used here to illustrate the extent and value of this geographical information.

This chapter argues that conventional quantitative statistical techniques mask the complexity of the spatial patterns in British society which computer visualisation has been able to illustrate and clarify. A persistent and particularly interesting problem has been that of how to visualise a spatial pattern evolving over time. Illustrations of various solutions used in the *New Social Atlas of Britain* are given in this chapter as well as alternative techniques and some of the lessons which are learned through trying to visualise so much of the social data which is available for a country, nationally, across so many localities.

The lessons learned from this work could be more widely applied to data from other countries and may challenge the assumptions that certain comparisons are not possible or that a given level of spatial resolution is acceptable. Geographical Information Systems are still rarely used to study data about social geography and when used for this purpose elementary mistakes are often made because the operators have been trained (and the systems are geared) to deal with information about the physical geography of places rather than their human geography. This situation is counter-intuitive because more money and effort is spent by government and private industry collecting spatial data about people than about any other subject. It is the marketplace which has nurtured Geographical Information Systems and yet these systems have not yet been thoughtfully applied to the problems that are of most interest to that market.

Finally, through illustrating how much material is available on the social geography of one country, the chapter questions the extent to which the authorities in the UK could be persuaded to allow some detailed geographically coded information to be made available to researchers from other countries. Currently the legal situation is that if a researcher from, say, Eire or the USA is working in collaboration with a researcher from the UK on mapping unemployment rates across the island of Ireland, they have to travel to the UK to undertake the work – despite the instant computerised links available. This is true even though information at the local community level, rather than at the individual or postcode level, is not commercially sensitive. International research into the problems of social geography is severely hindered by these restrictions, many of which only exist for political, military and other reasons which have long ceased to be relevant.

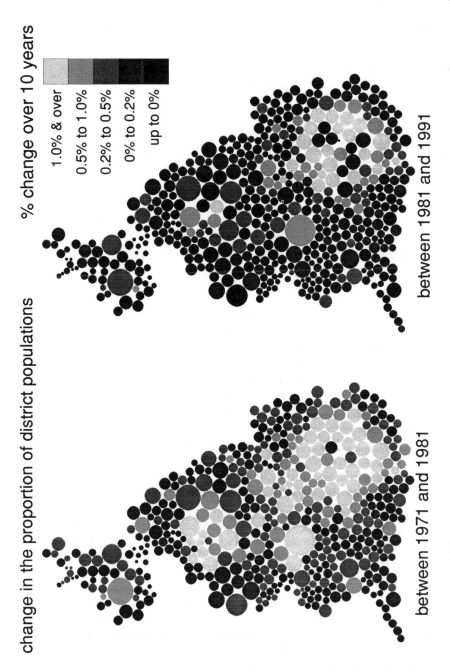

change in the proportion of district populations   % change over 10 years

1.0% & over
0.5% to 1.0%
0.2% to 0.5%
0% to 0.2%
up to 0%

between 1971 and 1981   between 1981 and 1991

Figure 6.2 Cartograms of the changing distribution of people resident in Britain between 1971 and 1991 who were born in the New Commonwealth.

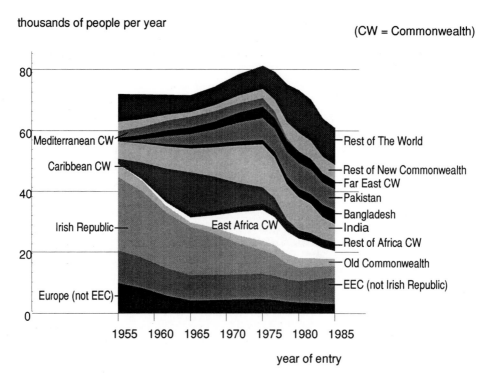

thousands of people per year

(CW = Commonwealth)

80

60 Mediterranean CW   Rest of The World

    Caribbean CW   Rest of New Commonwealth
                                                                         Far East CW
40   Pakistan
                                                                         Bangladesh
    Irish Republic   East Africa CW   India
                                                                         Rest of Africa CW
20   Old Commonwealth

                                                                         EEC (not Irish Republic)
    Europe (not EEC)

0
        1955 1960 1965 1970 1975 1980 1985

                                                year of entry

Figure 6.3   An equal population graph of the changing number of immigrants to the UK by their year of entry and their country of birth.

6.2 Drawing a new social geography of the UK

In many parts of the world information about very small localities stored on computers has become available for research. This information often contains hundreds of statistics about people living in thousands of localities. The work described in this chapter illustrates how a researcher can visualise these statistics to understand more of the localities of one country. An aim of this work is to show people facets of the society in which they are living which they would not otherwise recognise. The methods which are presented here could be applied to any part of the world for which the information is available, and the problems which are faced in dealing with this information are also universal.

The traditional means by which information about thousands of localities is presented is in equal-area choropleth map form. Unfortunately traditional maps distort this information by overemphasising statistics which relate to rural localities and under-representing the characteristics of the majority of the population who live in towns and cities. This situation applies to the UK, but an even stronger case could be made in many other parts of the world where rural areas are more sparsely populated and urban areas are more densely populated.

Figure 6.1 shows a traditional map of the 10 000 localities in Britain for which most local statistics are available: local government wards. The figure also shows an equal population cartogram in which each of these 10 000 localities is represented by a circle (the area of which is proportional to the number of people who live there). County and region boundaries are also included in the figure so that these places can be identified. On the cartogram the statistics relating to each person are given equal visual weight. Thus the majority of space on the cartogram represents localities within the largest cities in Britain.

Cartograms can also be used effectively when only a few hundred areas are being mapped. A key advantage of using cartograms in these cases is that the cartograms can be reduced to a small size while the areas in which most people live remain visible. This means that many cartograms can be placed side-by-side, allowing temporal trends in geographical patterns to be compared. Figure 6.2 shows two cartograms of Britain based on local authority districts. The first cartogram shows the changing proportion of the population of each area born in the New Commonwealth in the 1970s, the second shows this trend in the 1980s. The first cartogram demonstrates how the mainly black immigration of that time was confined to the cities of London, the Midlands, Manchester and Yorkshire. The second cartogram reflects the mortality of the largely white older New Commonwealth-born population. These inferences cannot be made from the cartograms themselves but are possible if enough related information is examined. To be able to present the quantity of information required to draw parallels such as these, between different temporal trends, the methods of graphical presentation need to be compact as well as being fair to the population presented.

6.3 Making area proportional to population

The principle of making area proportional to population can be extended from maps to diagrams of all kinds. This is often done intuitively. Figure 6.3 shows how the numbers of people immigrating to the UK from different countries changed each year from 1955. The vertical scale of the figure shows how many people were entering the country each year and the horizontal scale shows which year the information refers to. Thus, the area of the band of colour representing each area of origin is proportional to the total number of people who immigrated from that area over the period shown. This diagram contains a great deal of information presented in a compact and just way. Unfortunately, to be reproduced here all the illustrations presented have had to be converted into black and white, but even with this limitation it is possible to show a great deal of detail.

Traditional maps which draw places in proportion to land area can be useful, particularly when the population of interest lives in very remote areas. A good example in the UK is of people who can speak Gaelic or Welsh. Figure 6.4a presents maps and cartograms depicting their distribution in Scotland and Wales. Figure 6.4b shows how that pattern has changed over the last decade. The map of Gaelic speakers in Scotland in 1991 reflects how the proportion increases from less than 1 person in 50 speaking Gaelic in the South East of that country to over half the population being able to speak it in the North West. However, the equivalent cartogram (in its key) shows that 94 per cent of the population of Scotland lived in wards where less than 2 per cent of people could speak Gaelic. A similar, if less extreme pattern is seen in Wales. The cartogram of Wales shows the geography of change to be very different to that suggested by the map. Welsh speaking is not declining in most of Wales (by population) because most of the

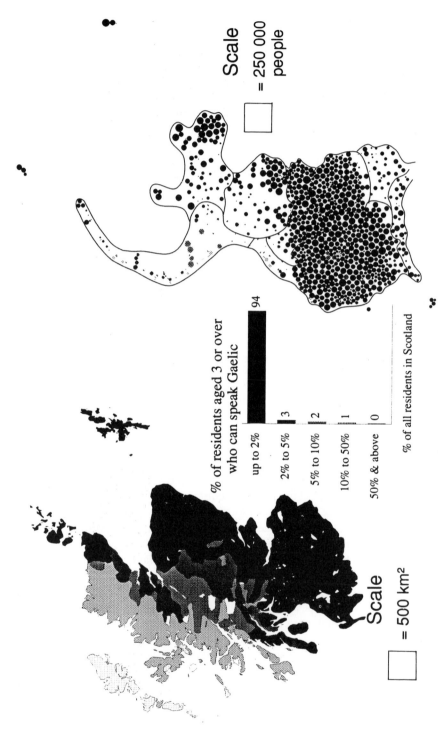

Figure 6.4a Maps and cartograms of the proportion of residents speaking Gaelic and Welsh in Scotland and Wales, respectively, in 1991.

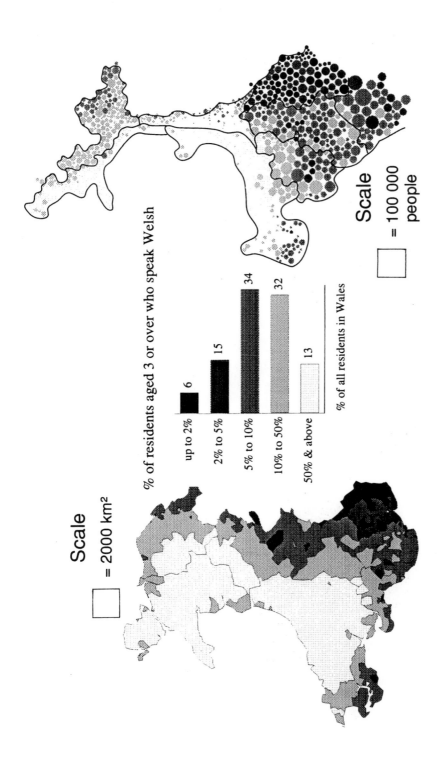

Figure 6.4a (*continued*)

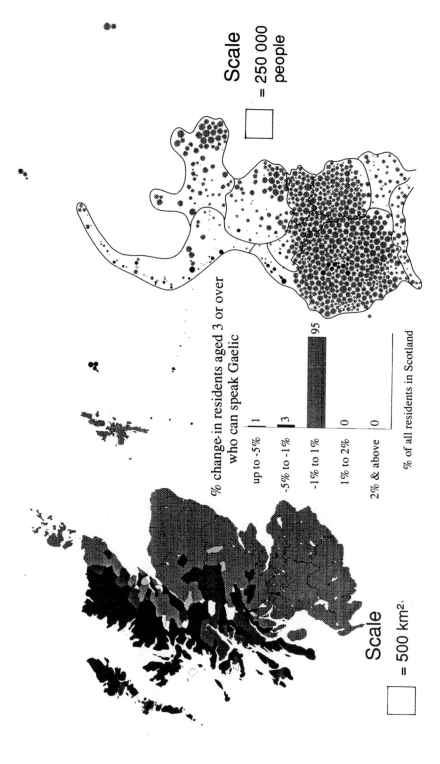

Figure 6.4b Changes to the maps and cartograms of Scotland and Wales in the proportion of residents speaking Gaelic or Welsh from 1981 to 1991.

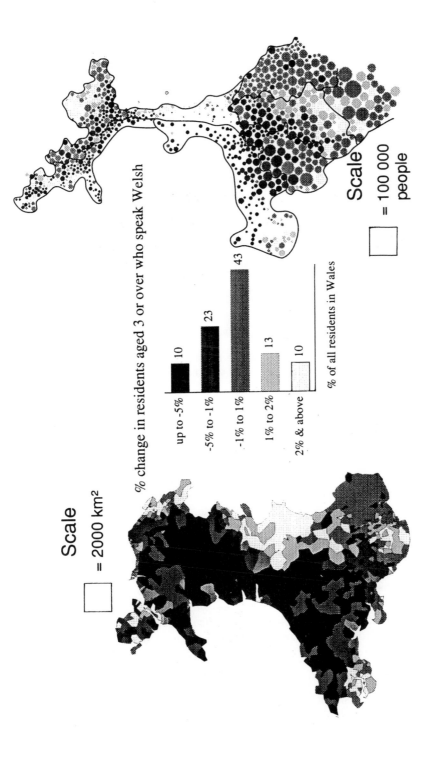

Figure 6.4b   (continued)

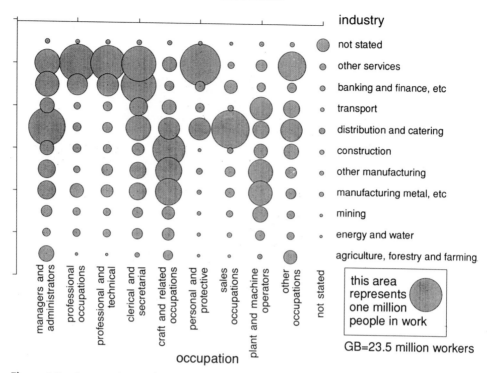

Figure 6.5   An equal population graph of the relationship between the stated occupation and the workplace industry of all workers in the UK in 1991.

Welsh live in the southern valleys where the effect of teaching Welsh in schools has had much influence.

The next two figures illustrate how the principle of making the components of graphics about people proportional to the numbers of people can be extended. Figure 6.5 shows how many workers in each industry were in various occupational groups in 1991. This type of graph is often given as an option on computer packages but is rarely used when it would be appropriate. The graphs shown in Fig. 6.6, in contrast, cannot be created easily with standard packages. In the main graph the vertical scale shows the proportion of all workers in each industry while the horizontal scale shows the proportion of those workers who are male or female, full-time, part-time or self-employed. Thus the total area of the bars is in proportion to the size of the workforce. For each industrial group a population pyramid is also drawn, again with its area in proportion to the total number of workers in that sector. These graphs can be difficult to read and they are certainly not simple to label. However, again they contain a great deal of information and can be argued to present it fairly.

6.4 Complexity and simplicity in visualisation

Occasionally very clear patterns are found in quite complex data. Often it is only after producing many graphics that these patterns are evident to the researcher. This is where the use of computers to visualise social data is most advantageous.

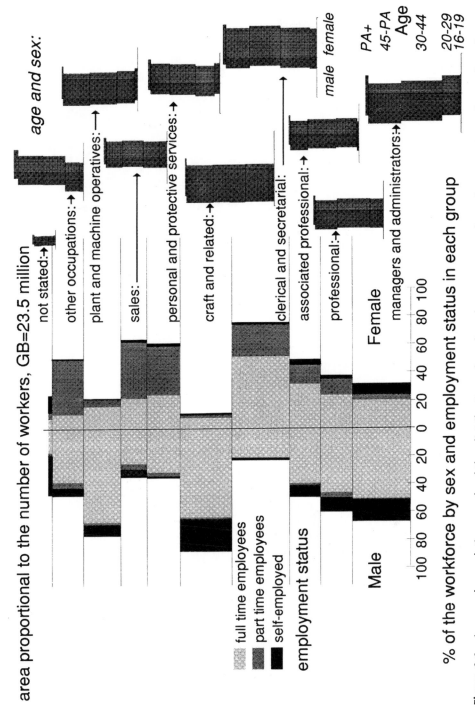

Figure 6.6 An equal population graph of the distribution of workers by age, sex and employment status for each major occupational group in the UK in 1991.

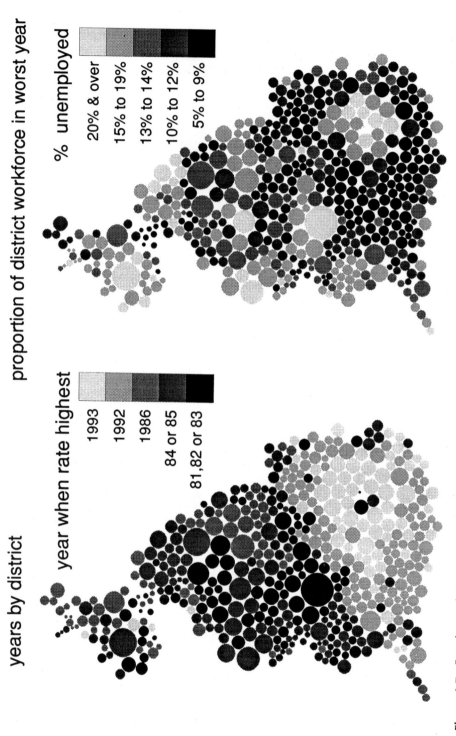

Figure 6.7 Equal population cartograms showing the year and rate of the highest levels of unemployment in the UK by district between 1981 and 1993.

Figure 6.7 presents another pair of district level cartograms. The cartogram on the left shows the year in which unemployment was highest in Britain between 1981 and 1993 in each district. The North/South divide can be seen to be abrupt. The cartogram on the right shows what that highest level of unemployment was in each district. Here the urban/rural divide can be seen to be most acute. Low levels of maximum unemployment circle the districts which make up the major cities (in which over a fifth of the workforce has been unemployed at one time or another). To understand these cartograms a new geography of Britain has to be learned. That appears to be harder for researchers who are most familiar with the traditional representation of this country. British cartographers, in particular, are often averse to these images, whereas people unfamiliar with Britain often expect to see London as the largest, rather than the smallest, place on the map.

When large quantities of data are being analysed standard types of graph often require modification. An example is the scatter-plot, in which overlapping dots are obscured. When there are many dots to plot, and many overlap, a misleading impression of the distribution can result. Figure 6.8 shows one method of dealing with this problem. In this figure a dot is plotted to show the relationship between the changes in house prices and unemployment in 459 districts each plotted 10 times to show the situation in each of 10 years. The dots are drawn with their areas proportional to the population affected. When two dots would overlap their populations are amalgamated and a single larger dot is drawn. The effect is reminiscent of the shading in old newsprint and the aim is similar: to darken certain parts of the paper more than others (in this case the parts which represent the experiences of more people). Thus, although a rough negative relationship can be discerned, it is apparent that most people in most years have experienced very little change in unemployment and have seen modestly rising housing prices. In the original version of this graphic the dots are coloured to show in which year these changes were strongest. That is not possible here in black and white.

It is possible to show the geographical distribution of average housing prices across 10 000 areas simultaneously. This is done in Fig. 6.9 where the average prices which buyers paid for houses in each ward over the 1980s are shown at 1991 prices. The prices are calculated from the mortgage book of a major building society. Every sale is linked to a ward through the postcode of the address of the property and a weighted average of the sale prices is made with the weights allowing for inflation up to 1991. Given that in some wards the size of this sample is quite low, it is remarkable how even the pattern is. This reflects how rigid local housing markets tend to be. The huge differences between the centre and the suburbs of London, or the West and East sides of Birmingham, are immediately apparent (Fig. 6.1 acts as a key to this figure). The pattern of low rates of unemployment in Fig. 6.7 can be seen reflected in the high levels of average housing prices in much of the South East. Through presenting graph after map after table after figure, given enough patience and space, readers can form their own views as to what facets of society appear to be interrelated most strongly and how. This is preferable to asking them to accept the results of statistical tests which disguise the prejudices and assumptions of their authors more than do a series of picture, each concentrating on a relatively simple subject.

It is important that the form of graphics used in a social atlas varies if the reader is to remain alert. Occasionally there is merit in using 'three-dimensional' charts. Figure 6.10 gives an example in which the almost exclusive rises which occurred

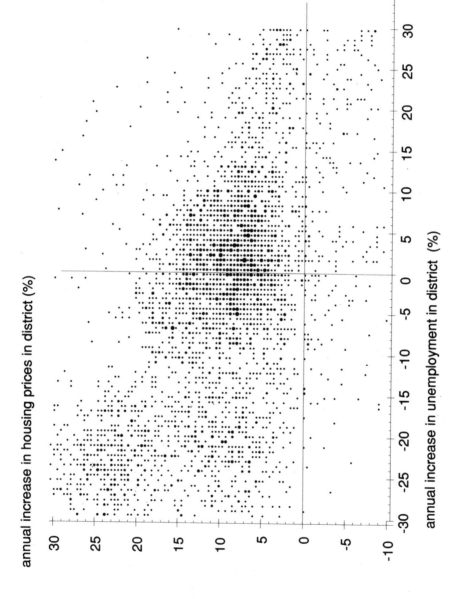

Figure 6.8 An equal population graph showing the relationship between changes in unemployment and house prices in the UK by district between 1981 and 1991.

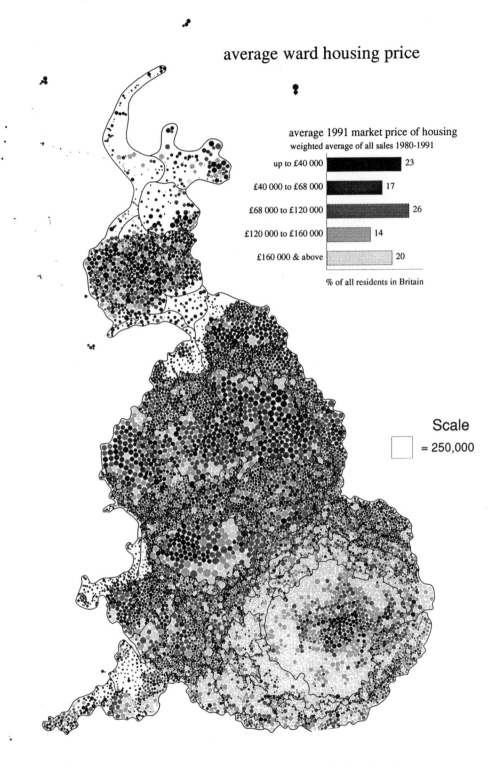

Figure 6.9 An equal population cartogram showing the distribution of average house prices in the UK by ward in 1991 (source unpublished Building Society data).

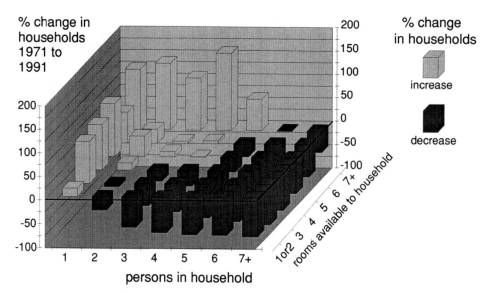

Figure 6.10 A graph showing the relationship between changing household size and housing size in the UK between 1971 and 1981 (not drawn using an equal population scale).

in one person or in seven-plus room households in Britain are emphasised. It is when a diagram or topic appears dull that pretty graphics have their most valuable uses. Figure 6.11 shows an opposing case. Here the chances of people dying by two causes of death by sex and single year of age are displayed, and how those chances are changing. A very simple form of graphic can be used in this example as the contents are of greater interest to many readers and because quite a lot of data is being portrayed which might only be confused by further embellishment.

Finally, Fig. 6.12 shows how a collage of cartograms can be used to depict a whole series of changes, in this case the results of all general elections in the UK since 1955. These cartograms do not actually show who won in each constituency, but instead show which party came second. Seven thousand results are included in this one graphic. In summary the graphic shows how the Liberal party has risen from the ashes of its post-war low to be seriously contesting a majority of seats in the south of England by the end of this century. The rise of nationalism in Scotland and the exit of main stream parties from Northern Ireland are also clear messages from the figure (clear at least in its original colour form). More importantly, by presenting this quantity of information the graphic can show to what extent these assertions are not universal. It is even possible to follow the fortunes of individual constituencies over time. A simple example is the Isle of Wight, the most southern constituency on the cartograms. There, the second placed party has changed from Labour to Conservative to Liberal. These maps can also incorporate the effects of boundary changes. If, for instance, you examine the figure closely you can see that the number of constituencies alters over time in each region. (A key is provided in Dorling *et al*. 1996.)

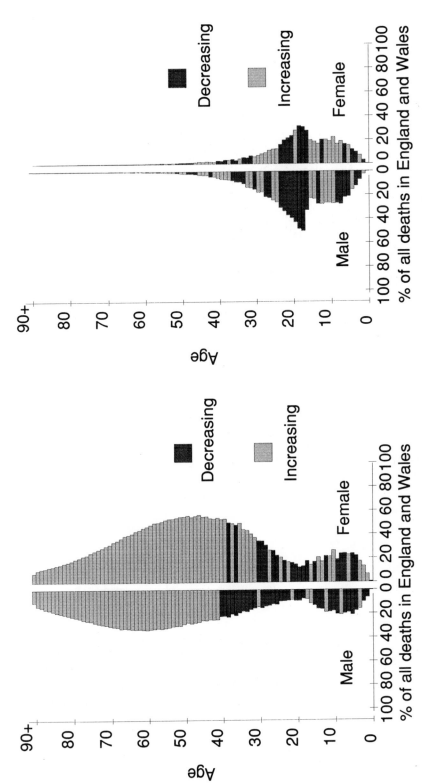

Figure 6.11a A graph showing the distribution of mortality from cancers in England and Wales by age and sex between 1981 and 1989 (not drawn using an equal population scale).

Figure 6.11b A graph showing the distribution of mortality from traffic accidents in England and Wales by age and sex between 1981 and 1989 (not drawn using an equal population scale).

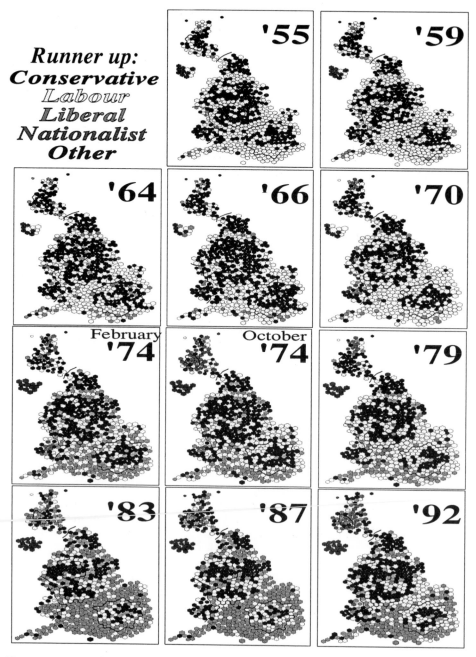

Figure 6.12 Equal electorate cartograms showing which political party came second in each constituency in every UK election since 1955.

6.5 Conclusion

This chapter has presented only 12 illustrations taken from a social atlas of the UK which contains over 200 maps and 200 graphs, most of which show patterns across 10 000 areas (Dorling, 1995). Hopefully, it has given a flavour of what can now be done with the wealth of information available and the ease with which that information can be manipulated. A home microcomputer was used to create all the maps and graphs shown here and to typeset the atlas. Apart from showing what is technically plausible this chapter argues that care is needed if the maps and diagrams which social scientists produce are to represent society fairly. One guiding principle is that equal numbers of people are equally represented. Another principle is to present as much information as possible when as complex an object as society is being studied. What we need to see cannot be predetermined. Simplification conceals the complexity of reality.

Acknowledgements

1 The data used in this chapter is Crown Copyright and is reproduced with permission of the Controller of Her Majesty's Stationery Office.

2 An earlier version of this chapter was included in the *Society of Cartographers Bulletin* (Dorling, 1994) and many of the figures are illustrative of the preliminary draft work for illustrations which appear in colour in *A New Social Atlas of Britain* (Dorling, 1995).

References

DORLING, D. (1993). Map design for census mapping, *The Cartographic Journal*, **30**(2), 167–83.

DORLING, D. (1994). Mapping and graphing the social structure of Britain, *Society of Cartographers Bulletin*, **28**(1), 7–18.

DORLING, D. (1995). *A New Social Atlas of Britain*, Chichester: John Wiley and Sons.

DORLING, D., JOHNSTON, R. J. and PATTIE, C. J. (1996) Using triangular graphs for representing, exploring and analysing electoral change, *Environment and Planning A*, **28**, 979–998.

DORLING, D. and OPENSHAW, S. (1992). Using computer animation to visualize spacetime patterns, *Environment and Planning B*, **19**, 639–50.

PART TWO

GIS Diffusion and Implementation

CHAPTER SEVEN

The Diffusion of GIS in Local Government in Europe

IAN MASSER and MASSIMO CRAGLIA

7.1 Introduction

Geographic information systems (GIS) are a multimillion dollar industry which is growing very rapidly at the present time. Estimates of its size vary according to the measures used. Payne (1993), for example, estimates that the European GIS market for hardware, software, and services was worth some 4 billion ECU in 1993. This estimate does not include the costs to government associated with the collection and management of geographic information. A recent report for the European Community (Commission of the European Communities, 1995) estimates that government spending on geographic information alone accounts for around 0.1 per cent of gross national product or about 6 billion ECU for the European Union as a whole. Similarly, in the USA the Office of Management and Budget has estimated that Federal agencies alone spent $4 billion annually to collect and manage domestic geospatial data (Federal Geographic Data Committee, 1994, p. 2).

The growth of the core GIS market shows little sign of slowing down. Dataquest (Gartzen, 1995), for example, estimate the revenue from GIS software worldwide is likely to grow at an average annual rate of 13 per cent up to 1998. At the present time, North America and Europe dominate the world GIS market.

The scale of the GIS industry makes it necessary to study the diffusion of this technology and its impact on society as a whole in some depth. This is because the technology has the potential to fundamentally change the use of geographic information in a wide range of circumstances and it must be recognised that not all of these changes will be beneficial (Wegener and Masser, 1996). The largest users of GIS technologies are central and local government agencies although the number of private sector applications has substantially increased in the last few years. In countries such as Germany and the UK, local government applications probably account for somewhere between a quarter and a third of the total market.

Given the size of the GIS industry and the extent to which it is dominated by a relatively small number of large users, it is surprising to find that relatively little systematic research has been carried out either on the development of the geographic information services industry itself or on the diffusion of geographic information technologies in key sectors such as local government (see for example, Masser and Onsrud, 1993). Local government is a particularly interesting field for diffusion research because it covers a wide range of applications in a great diversity of settings. There is also a strong national dimension to local government which is reflected in the expectations that people have with respect to local authorities, the tasks that they carry out and the professional cultures that have come into being to support them.

With these considerations in mind this chapter summarises the findings of research on the diffusion of GIS in local government in Europe which has been undertaken in the context of the European Science Foundation's GISDATA programme (Masser *et al.*, 1996). This includes nine national case studies of GIS adoption in Denmark, France, Germany, the UK, Greece, Italy, The Netherlands, Poland and Portugal as well as two more wide-ranging studies of organisational perspectives on GIS implementation and the development of contrasting scenarios which highlight some of the negative as well as the positive consequences of GIS adoption and implementation.

Of particular importance in this context is the extent to which the nine case studies, conducted by teams of national experts in each country, draw upon a common methodology based on the research on this topic that has been carried out at Sheffield over the last five years. This includes both survey and case study research on UK local government (see, for example, Campbell and Masser, 1992, 1995) as well as comparative research on GIS in the UK and other European countries (see, for example, Campbell and Craglia, 1992; Craglia, 1994). The methodology in the UK studies involved a comprehensive telephone survey of all 514 local authorities followed by detailed case studies, and was adapted in the other countries to allow for local circumstances. In some instances like Denmark, it was possible to carry out a complete survey of all local authorities, while in some others like Germany and Italy the surveys focused on significant subsets: large urban areas in Germany, regions and provinces in Italy. The other national studies focused more on local case studies backed up by secondary data to provide an overall picture of current developments.

Despite these differences, the findings of these case studies make it possible to carry out a more systematic cross-national comparative evaluation of GIS diffusion in one of the key stakeholders in the European GIS community in this chapter than has hitherto been the case. The next section of the chapter summarises the main features of the findings. As a result of this analysis a simple typology is developed in the third section of the chapter while the last section sets out an agenda for future research on both GIS adoption and implementation.

7.2 A comparative evaluation of European experience

The main features of GIS diffusion in local government in the nine countries investigated in the GISDATA project are summarised in Tables 7.1, 7.2 and 7.3. Table 7.1 compares the characteristics of these countries with respect to the institutional context within which GIS diffusion takes place in local government, the structure of local government and the extent of digital data availability.

7.2.1 Institutional context

From Table 7.1 it can be seen that there are important differences between the nine countries with respect to their political and economic stability. Recent political upheavals in Italy, for example, have profoundly affected the political culture that underlies both central and local government in that country. Poland is in the middle of a major structural transformation from a command to a market economy. Even relatively stable countries such as Germany are also undergoing important structural changes as a result of the incorporation of the East German Länder into the former West German state. In contrast, Denmark, France and The Netherlands have relatively high degrees of political and social stability which is also reflected in local government.

In several of these relatively stable countries, however, there have been moves to decentralise powers from central to local government over the last two decades. The impact of decentralisation is most pronounced in Denmark following the reforms of the 1970s, but similar trends can be found in France and Italy during the 1980s and more recently in Greece and Portugal. In the UK developments during the last decade and a half run counter to this general trend with more central government control over local government. Poland is also a country where central government control over local government is increasing once again after the decentralisation of power following the collapse of the former communist regime.

These changes have given rise to uncertainties about the responsibilities of the various layers of local government in several countries. In the UK, for example, this culminated in the abolition of metropolitan counties by the government in 1986. Similarly, the provinces in The Netherlands are struggling to find a role after changes in local government structure. In France, on the other hand, decentralisation has given new powers to the *départements* as it has to the provinces in Italy.

There are also important differences between countries with respect to the relationship between local government and the private sector. In Greece, Italy and Poland, for example, contracting out of data collection and IT tasks to the private sector is commonplace because of the limited resources at the disposal of local government. In the UK, local authorities are also being increasingly obliged by central government to put key tasks out to competitive tender to reduce costs.

In Portugal, on the other hand, central government is taking an increasingly proactive role to modernise local government. Like Greece, the diffusion of GIS in this country has been substantially accelerated by the availability of funds from the European Community.

7.2.2 Structure of local government

Table 7.1 shows only the tip of the iceberg in that it refers essentially to the number of authorities at various levels of local government in the nine countries and gives some indication of the size of these units in terms of population. As such, it does not deal explicitly with the great diversity that also exists with respect to the ways in which the various functions are discharged by the different tiers of local government in the nine countries.

Nevertheless, Table 7.1 highlights some important differences between the countries. Germany is the only country surveyed which has a Federal system of

Table 7.1 Key contextual factors in nine European countries.

	UK	Germany	Italy	Portugal	Denmark
Institutional context	Increasing pressures towards privatisation of local government tasks has effected GIS diffusion. Uncertain impacts of ongoing review of local government in terms of the number and functions of authorities	Stable and well-established local government culture. Incorporation of former East German *Länder* presents new opportunities	The 1990 reform of local government which extends powers and degree of autonomy of lower tiers has both constrained and created opportunities for GIS diffusion. Uncertain impacts of recent political upheavals on local government. Limited use of IT in local government	Pro-active Government measures to modernise local government. Establishment of a National Centre for Geographic Information to coordinate the implementation of a national geographic information system. Availability of EC funding to support modernisation	Widespread decentralisation of powers to local government since 1970. Extensive use of IT in local government since the 1970s
Structure of local government	Two tiers: 47 shire counties and 9 Scottish regions with 333 shire districts and 53 Scottish districts. Single tier of 69 metropolitan districts. Few districts have populations of less than 100 000	Three tiers: 16 *Länder*, 543 counties and large cities and 14 809 municipalities. Most of municipalities have less than 10 000 population	Three tiers: 20 regions, 103 provinces and 8100 communes. 87 per cent of communes have less than 10 000 population, 12 new Metropolitan Authorities	Three tiers: 7 regions (including the Azores and Madeira), 29 districts and 305 municipalities. Average size of municipality is 34 000	Two tiers: 14 counties and 275 municipalities. Half the municipalities have less than 10 000 population.
Digital data availability	Comprehensive digital topographic data service provided by Ordnance Survey. Service Level Agreement reached with local authorities in March 1993 boosted GIS diffusion	Strong surveying traditions and collaboration between local authorities with respect to the development and maintenance of digital topographic databases (for example ALK and MERKIS)	Slow progress made by National Mapping Agency and Cadastre. nearly half the regions developing their own digital topographic data-bases. Profusion of data sources	Diversify of sources. 65 per cent of country covered by 1:10 000 scale orthophoto maps from the National Mapping Agency. Some digital data at 1:25 000 scale in vector format from the Army Cartographic Services	Municipalities are the main provider of large-scale maps. They also co-manage the Cadastre with the Danish Survey and Cadastre and maintain a wide range of other register-based information systems

Table 7.1 (*continued*)

	Greece	France	Poland	The Netherlands
Institutional context	Limited financial and manpower resources at the disposal of local government leading to contracting out of IT functions to the private sector and University Laboratories. Role of EC funds in facilitating the diffusion of GIS	Long tradition of geographic information handling in large cities. Decentralisation measures since 1980 have given new powers to Départements and to new administrative structures for main urban areas	Very rapid change from centrally planned to market economy. Earlier attempts to decentralise power to local administration have recently been reversed, leaving some uncertainty over the responsibilities of different layers of government	Strong and stable central government with local authorities retaining a high degree of autonomy. Middle layer of provinces struggling to define its role
Structure of local government	Two tiers: 51 prefectures form upper tier, 361 cities and 5600 parishes form lower tier. Average population of parish is 1800	Three tiers: 22 regions, 96 Départements and 36 000 communes. Average population per commune is 1500. Variety of inter-communal structures especially in large urban agglomerations	Three tiers: 49 regions, 327 sub regions, and 2465 local authorities having an average population size of some 15 000	Two tiers: 12 provinces and local government divided between 650 municipalities and 100 polder boards in charge of water control. 50 per cent of municipalities have less than 12 000 people and 90 per cent less than 40 000
Digital data availability	Lack of digital data. 1:5000 scale digital contour data available only for some cities in the Athens and Thessaloniki areas. Proliferation of data produced by private sector databases	Small-scale digital data available from Institut Géographique National. Large-scale data in hands of Cadastre. Ongoing digitisation programme underway in conjunction with local authorities	Almost half of the country is covered in good quality cadastral information. The process of conversion to digital format has been undertaken but slow progress and organisational difficulties have led a number of local authorities to acquire their own data, often in partnership with private sector	Strong awareness of geographic information in The Netherlands and tradition of automated registers (population, etc). Slow progress of large-scale mapping project based on Cadastre has led to local authorities acquiring their own digital mapping individually or in partnership with utility companies and private sector

Table 7.2 Key features of GIS diffusion in the five countries where surveys were carried out using a similar methodology.

	UK	Germany	Italy	Portugal	Denmark
Coverage of findings	Comprehensive survey of all local authorities in England, Wales and Scotland in the second half of 1993	Survey of cities with 100 000 population in the first half of 1994	Comprehensive surveys of all regions and provinces in 1993/1994: some additional information for communes (1991) and Metropolitan areas (1993)	Survey of municipalities in late 1993/early 1994	Comprehensive survey of counties and municipalities in first half of 1993
Extent of diffusion	Almost universal in counties and Scottish regions. Half metropolitan districts. One in six shire/Scottish districts	Almost universal: 70/80 cities had GIS, 10 had firm plans	Two-thirds regions – one third provinces but another 1 in 3 with firm plans. Limited in medium/large cities	12 municipalities had GIS and a further 24 had AM/FM facilities	Over 80 per cent of all municipalities use register-based systems where georeferencing present. GIS/AM/FM almost universal in authorities with 50 000+. GIS in half counties
Geographical spread	North South divide: 32 per cent of Southern authorities had GIS. 24 per cent of Northern authorities had GIS	No significant geographic difference	Pronounced North/South divide among the regions (90 per cent N – 37 per cent S) and the provinces (50 per cent N – 20 per cent S)	Adoption levels highest in urban areas and in northern parts of Portugal	Urbanised regions generally have higher take up than less urbanised regions
Length of experience	70 per cent systems purchased since 1990	Most systems purchased since 1990	Half regional and 80 per cent provincial systems purchased since 1990	Some municipalities began GIS projects in 1990	Most GIS systems acquired since 1990

Table 7.2 (continued)

	UK	Germany	Italy	Portugal	Denmark
Main applications	Automated cartography and mapping for local planning and management	Surveying and topographic database management	Digital map production, strategic land-use planning. Environmental monitoring also strong in the provinces	Land use planning: automated mapping	Digital mapping. Also extensive use for utility management
Predominant software	Arc/Info in 25–30 per cent of county and metropolitan districts. Axis, Alper Records and G-GP in shire/Scottish/metropolitan districts	SICAD and ALK-GIAP (latter free to German local authorities)	Arc/Info for over half regional and provincial applications. Greater diversity at municipal level	Intergraph purchased by National Centre for Geographic Information facilities	Intergraph MGE and Autocad-based systems followed by Dangraf and GeoCAD
Perceived benefits	Improved information processing (60 per cent) especially improved data integration, better access to information and increased analytical and display facilities	Improved information processing: (65 per cent) faster information retrieval and increased analytical and display facilities	Improved information processing (51 per cent) especially automated map production and thematic mapping	Improved information processing, administrative reorganisation, data sharing with utilities	Improved information processing (66 per cent)
Perceived problems	Technical problems including lack of software and hardware compatibility. Organisational problems especially poor managerial structures and lack of skilled staff	Organisational problems especially lack of qualified staff and insufficient motivation of staff by management	Organisational problems especially lack of awareness, poor coordination and lack of skilled personnel	Bureaucratic inertia, lack of skilled personnel digital-data availability, lack of awareness, vendors, attitudes and limited follow-up support	Technical problems in small municipalities. Organisational problems in large municipalities

government. France, Italy and Poland have developed three-tier local government structures, whereas all the other countries have two-tier structures with the partial exception of the UK which has single-tier authorities in the main centres of population since the abolition of the metropolitan counties.

The UK also stands out from all the other countries with respect to the size of its lower tier authorities which, with relatively few exceptions, contain populations of over 100 000. Elsewhere in Europe, small is generally beautiful. Over 80 per cent of Italian communes have less than 10 000 population. In France the average population per commune is 1500 and the comparable figure for Greece is 1800. Most German authorities have populations of less than 10 000 and even in Denmark, where there was a major reorganisation of local government during the 1970s which resulted in a substantial reduction in the number of authorities, half the municipalities have populations of less than 10 000.

7.2.3 Digital data availability

Once again the UK stands out from all the other countries with respect to digital data availability given that its national mapping agency, the Ordnance Survey (OS), provides a comprehensive mapping service for both large- and small-scale maps. Digitisation of the former is nearly complete and local government access to this information has been substantially facilitated by the service-level agreement for the purchase of digital data reached between the local authority associations and the Ordnance Survey in March 1993. Through this agreement all authorities operate within the same framework and no longer need to enter into individual negotiations with the OS.

At the other end of the spectrum comes Denmark which also has a very high level of digital data availability but makes the municipalities themselves the providers of large-scale maps. These authorities also maintain the Cadastre together with the Land Registry as well as a large number of register-based systems.

In terms of digital-data availability Greece, Italy and Portugal in particular currently suffer from a proliferation of digital-data sources as a result of the limited progress made by their central government agencies with respect to the provision and coordination of digital data.

Elsewhere in France, Germany, The Netherlands and to some extent Poland, large-scale digital-data provision is closely linked to the maintenance of the Cadastre. In these countries the establishment of the data infrastructure required for municipal GIS is proceeding rather slowly despite the very strong surveying traditions in some of these countries.

It might be useful to point out that the focus of the research has been on the availability of digital mapping because the increasing availability and resolution of satellite data has not yet made a significant impact on local government in the countries analysed.

7.2.4 Survey findings

Table 7.2 summarises the main findings of the five national surveys. However, before comparing the findings of these surveys, attention must be drawn to a number of differences between them in terms of timing and coverage.

Table 7.3 Key features of GIS diffusion in the other four countries

	Greece	France	Poland	The Netherlands
Extent of diffusion	About 10 per cent of all cities have GIS facilities	Two-thirds of cities with more than 100 000 population have GIS. Widespread use in intercommunal agencies	30–40 mainly in small-medium sized towns (20 000–100 000 inhabitants). Few regional/sub-regional GIS	GIS is almost univeral among municipalities above 50 000 inhabitants and in approx. 50 per cent of those above 20 000
Geographical spread	No marked regional variations.	No discernible regional variations reported	No marked regional variations	No marked variations
Length of experience	Most systems purchased since 1990	Most systems purchased since 1990	Most systems purchased since 1990	Rapid take-up since late 1980s
Main applications	Surveying and topographic database management	Urban database management for surveying and planning	Topographic data, parcels data, cadastral information	Mainly topographic and thematic mapping
Predominant software	Arc/Info for over 80 per cent of local government applications	Arc/Info and Apic (French package)	Map Info, Autocad for small systems, Arc/Info for larger central government implementation	Intergraph MGE, Autocad and IGOS (Dutch package)

Both the UK and Danish surveys involved comprehensive telephone surveys of all local authorities which obtained a 100 per cent response rate. However, the Danish survey was carried out in the first half of 1993 between six months and a year before the other four surveys. This discrepancy in timing is particularly important given the rapid take-up of GIS in the smaller authorities.

In Portugal, a postal questionnaire was sent to all municipalities in late 1993 but not the regions. It achieved a response rate of 55 per cent which is good for a postal questionnaire but nevertheless the findings must be treated with some caution given that no information is available for non-responding authorities.

Because of the very large number of authorities in Germany, the German survey which took place in the first half of 1994, was restricted to the 86 cities with populations of over 100 000. A response rate of nearly 90 per cent was achieved in this case which is very high under the circumstances. Nevertheless, no information is available for either the higher levels of local government or authorities with less than 100 000 population.

In Italy, two comprehensive surveys of the regions and provinces were undertaken in late 1993 and early 1994 respectively, both achieving a 100 per cent response. However, the information on GIS in municipalities is limited to the findings of research carried out in 1991.

The main findings of the other four studies are summarised in a similar format to those of the five surveys in Table 7.3. It should be noted, however, that these studies are less comprehensive in coverage than the five surveys. In the case of Greece and Poland this presents relatively few problems given the low level of GIS diffusion in these countries. In the case of France and The Netherlands, it should be noted that much of the information is drawn from secondary sources which pre-date the findings of the five national surveys by several years.

7.2.5 Extent of GIS diffusion

Tables 7.2 and 7.3 show that the extent of GIS diffusion was lowest in Greece, Portugal and Poland. In these countries between 10 and 15 per cent of the cities had acquired GIS or AM/FM facilities. However, as the case of Portugal indicates, the situation is changing very rapidly as a result of proactive government measures and the extent of GIS diffusion is likely to increase dramatically in the immediate future.

Urban applications in medium and large cities predominate in Denmark, France and Germany. In Germany the use of GIS is almost universal in cities with populations of over 100 000, as it is in Denmark with reference to cities with over 50 000 population. Throughout these countries the extent of GIS diffusion is considerably lower in small towns and rural districts. In France and Denmark it is also lower at the département and county levels.

This relationship is completely reversed in the UK and Italy. In the UK the take-up of GIS in the counties is almost universal whereas only half the metropolitan districts had acquired facilities. In Italy, levels of diffusion at the regional and provincial levels are also very much higher than those for the cities. In both cases, however, the take-up in urban areas is generally higher than in rural areas.

In the UK, the obvious explanation for this difference is the extent to which the surveying and mapping functions that are attached to local government in most European countries are carried out, in this case by the Ordnance Survey. In addition there is no requirement in the UK to maintain a Cadastre as is the case in most other European countries. The case of Italy, however, is more difficult to explain and may reflect the new planning powers that have been given to the provinces.

7.2.6 Geographical spread

As noted above a general distinction can be made between urban, especially large urban, and rural areas with respect to the extent of GIS diffusion. In some countries, however, there is also a distinction between regions with respect to the diffusion of GIS. This is most pronounced in Italy with respect to the traditional divisions between the north, centre and south. At the regional and provincial level GIS adoption has reached nearly 100 per cent and 50 per cent respectively in the north as against less than 40 and 20 per cent for the equivalent authorities in the centre/south.

There are also clear differences in the UK between the north and Scotland and the south and eastern parts of the country. In this case, however, the ratio is reversed in favour of the wealthier south and eastern regions where the level of GIS adoption is 32 per cent as against 24 per cent in the north and Scotland.

Elsewhere, no pronounced regional variations have been reported for Denmark, France, Germany and The Netherlands. In Greece, Poland and Portugal the extent of diffusion so far is probably too low for any clear patterns to have emerged.

7.2.7 Length of experience

Both Tables 7.2 and 7.3 show that the vast majority of GIS systems in local government in virtually all countries have been purchased since 1990. This highlights the extent to which GIS in local government is very much a recent development in Europe.

7.2.8 Main applications

Tables 7.2 and 7.3 show that digital map production and digital mapping are the predominant local government applications in Denmark, France, Germany, The Netherlands and Poland. In most of them, these activities are closely linked to the maintenance of the Cadastre.

The main exceptions are Italy, Portugal and to some extent the UK. In Italy and Portugal land-use planning in some form appears to be the main GIS application. This is particularly the case in Portugal where the dissemination of GIS is closely linked to the preparation and approval of municipal land-use plans as well as automated mapping. In the UK, the position is more complex, but the emphasis, nevertheless, is on automated cartography linked to planning activities as well as the other technical services provided by local authorities.

7.2.9 Predominant software

Overall the findings summarised in Tables 7.2 and 7.3 point to the remarkable dominance of North American software, especially Arc/Info, in European local government. Arc/Info itself accounts for over 80 per cent of all GIS applications in Greece, half the regional and provincial applications in Italy and between 25 and 30 per cent of applications in UK counties and metropolitan districts. It also has a strong presence in France. On the other hand, in Portugal, Intergraph is the GIS market leader as a result of its acquisition as part of the implementation of the National Council for Geographic Information's national strategy.

In contrast, the German local authorities have remained loyal to their local software developers and the Danish authorities also make extensive use of local software packages. Many of these are custom-made for the specific tasks carried out in their local governments.

Despite the dominant market share of North American software in most countries there are also a large number of local tailor-made packages in use especially at the municipality level. Examples of these from the UK include G-GP, Axis and Alper Records (now Sysdeco). Widely used local packages in France include Apic and in The Netherlands IGOS.

7.2.10 Benefits and problems

Questions regarding the perceived benefits and problems to be derived from GIS were asked only in the five surveys. These revealed a high level of consensus of views throughout all the five countries involved. In the case of benefits there was general agreement that improved information processing was by far the most important benefit to be derived from GIS. The main reasons for this view were improved data integration and faster information retrieval together with increased analytical and display facilities. Irrespective of their nationalities, most respondents also identified similar organisational problems associated with recent developments in GIS. These included poor management structures and bureaucratic inertia together with lack of awareness and the shortage of skilled personnel.

7.3 Towards a typology of GIS diffusion in European local government

In overall terms the findings of the nine studies reveal a considerable measure of agreement regarding the perceived benefits and problems associated with GIS in local government in Europe. They also show that the length of local government experience in most cases is very similar between countries. It is useful at this stage, therefore, to concentrate on identifying some of the main differences between countries and exploring the extent to which a typology can be developed on the basis of these differences. From the evaluation above it would appear that two main dimensions can be identified which account for a large number of the differences observed between countries. This provides a starting point for further comparative evaluation.

The first dimension that emerges from the findings is related to the overall extent of diffusion and the level of digital-data availability in these countries. It measures essentially the links between data infrastructure and diffusion.

The second dimension is associated with the nature of the main GIS applications that are undertaken and the level of government at which they are carried out in the different countries. This dimension largely measures the professional cultures surrounding GIS applications.

When the nine countries are plotted in relation to these two dimensions in Fig. 7.1 it can be seen that all four quadrants of the diagram are occupied. The largest grouping consists of countries with high levels of diffusion, reasonable data availability where GIS is used essentially at the municipal level for surveying and mapping operations. Denmark is probably the best example of this category, but France, Germany and The Netherlands possess most of these characteristics.

The second largest group consists of countries with low levels of diffusion and restricted digital-data availability where GIS applications in local government are predominantly at the municipal level for surveying and mapping. Greece and Poland are the best examples of this category although Portugal shares some of these characteristics. However, there is an important difference between Portugal

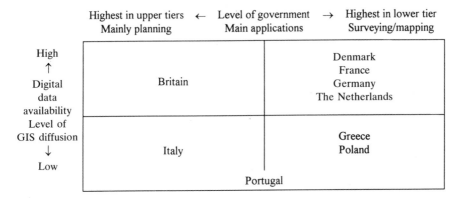

Figure 7.1 Typology of GIS diffusion in nine European countries.

and the other two countries with respect to the emphasis given to planning in Portugal. This places it somewhere between Greece and Italy in the table.

Each of the other two quadrants contains only one country. The UK stands out alone as a country with high levels of GIS diffusion and digital-data availability where GIS diffusion is greatest at the upper tier of local government and there is a strong planning tradition in local government. Similarly, the impact of GIS in Italy has been greatest at the upper tiers (regions and provinces) and mainly applied to planning even though the availability of digital data and the levels of GIS diffusion are lower than in the UK.

Two other matters are worth noting from Fig. 7.1. First, there is a broad correlation between relative stability and relative change in terms of the institutional context between the upper and lower halves of the diagram respectively. Given the changes that are already in progress in Portugal, for example, and to a lesser extent in Greece, Italy and Poland, the positions of these countries in the table may change considerably in absolute if not in relative terms over the next few years.

Second, there are also some interesting links between location in the diagram and the extent of the dependency on imported versus local software. The highest levels of dependency on North American software are generally to be found in the bottom half of the diagram in countries such as Greece and Poland where the overall level of GIS diffusion is still relatively low. In the top half of the diagram, on the other hand, levels of dependency are generally much lower and there are also a large number of local software products in use. This is particularly the case in Germany where these products are also the market leaders.

7.4 Conclusions

The findings of this comparative evaluation highlight some of the main similarities and differences between GIS diffusion in local government in nine European countries. They show that in all the countries surveyed GIS is a very recent phenomenon in local government and so far only a limited amount of operational experience has been built up to substantiate the claim that GIS implementation will lead to improved information processing. At this stage it is important to emphasise the distinction between adoption, the decision to acquire GIS facilities, and implementation, the use that is made of these facilities in practice.

7.4.1 GIS adoption

From the standpoint of GIS adoption it would appear from the analysis that two key factors account for most of the differences between the nine countries. The first of these measures the links between digital-data availability and GIS diffusion while the second reflects the professional cultures surrounding GIS applications in local government.

The question of digital-data availability is not simply a matter of the information rich versus the information poor. It is much more a question of central and local government attitudes towards the management of geographical information. For this reason those countries with relatively low levels of digital-data availability tend also to be countries where there has been a fragmentation of *ad hoc* data sources as a result of a lack of central government coordination. Conversely, those countries with relatively high levels of digital-data availability tend also to be countries where governments have created a framework in terms of responsibilities, resources, and standards for the collection and management of geographic information.

The question of professional cultures is closely linked to the responsibilities that have been given to local government in different European countries. In countries where local governments play an important role in maintaining land registers and/or cadastral systems, a highly organised surveying profession has come into being to carry out these tasks. Consequently, land information systems rather than geographic information systems tend to predominate in local government applications in these countries. Conversely, in countries such as the UK where land registration and digital topographic mapping are dealt with centrally, there is a greater emphasis on applications which are broadly linked to the various planning activities of local government. In this case the predominant professional cultures tend to be those of the land-use planner and, to a lesser extent, the transport engineer.

Within these two groups, however, it is also necessary to consider the nature of the responsibilities that have been given to local government and the resources placed at their disposal when comparing professional cultures. There are considerable differences, for example, as Craglia (1993) has shown elsewhere between the local planning functions and associated cultures that have developed in the UK and Italy. Similarly, there are also important differences between Germany and Greece with respect to the resources at the disposal of local government for land-management functions which affect the development of the surveying profession in each country.

As a result of this analysis it is felt that these two factors provide a useful starting point for the further analysis of the diffusion of GIS in local government in both Europe and elsewhere. What is now needed is further research which evaluates and refines the typology developed for the analysis in the context of local government in other countries. There is also a need for cross-national comparative studies of other key GIS application sectors. In this respect special priority should be given to the utilities sector because of its central position as the other key investor, alongside local government, in GIS systems technology.

The final question concerns the significance of the findings of the analysis with respect to diffusion theory as set out by Rogers (1993) and others. As noted above, the analysis has been primarily concerned with the adoption rather than the implementation of GIS. On this count it shows that there are considerable differences between countries with respect to their positions on the S-shaped curve. GIS diffusion in local government in the UK, Denmark and Germany in particular is already well past the critical point of take-off and is approaching saturation in relation to particular levels of government. Elsewhere, in Greece, Poland and Portugal in particular, GIS diffusion in local government is still at a relatively early stage on the curve, but, as the findings of the survey clearly indicate, the situation is changing very rapidly at the present time. Given the speed of change it will be necessary therefore to closely monitor events to keep track of the adoption process in these countries.

7.4.2 GIS implementation

It must be recognised that the primary focus of the above discussion has been on GIS adoption and that matters relating to GIS implementation and its consequences have only been discussed in general terms. Nevertheless, it is clear from this discussion that implementation is likely to become an area of growing significance in GIS diffusion research. GIS implementation research is concerned with the process whereby new technologies are adapted to meet the specific needs of organisations such as local authorities. The starting point for such research, as Campbell (1996) has demonstrated, is not classical diffusion theory but theories of organisational change. Given that the focus of this research is on the organisation itself rather than on the national circumstances within which that organisation operates, it may be expected that some striking similarities will be found between organisations in widely differing institutional contexts. In this way the findings of such research are likely to further highlight the diversity of responses to the adoption and implementation of new technologies such as GIS.

However, implementation research may also have a wider scope. As recognised by Wegener and Masser (1996) in their 'Brave new GIS worlds', the widespread introduction of GIS or GIS functions in manufacturing, services, retailing and leisure activities, logistics and travel, government and public and private sector planning may have consequences far beyond their direct impacts within their narrow field of application. GIS are already now indispensable for maintaining the efficiency of modern economies and the convenience of modern life and are expected to continue to open up new opportunities of economic and social activity. There are also fears that GIS may be used to undermine essential civil rights and societal values, create new patterns of dominance and dependency or impose on their users a certain view of the world which may be alien to their culture. These wider social and political consequences need to be investigated in more detailed social-science-based studies which take particular account of the negative as well as the positive consequences of these new technologies to counter the pro innovation bias of much diffusion research.

References

CAMPBELL, H. (1996). Organisational cultures and the diffusion of GIS technologies, in Masser, I., Campbell, H. and Craglia, M. (Eds) *GIS Diffusion: The Adoption and Use of GIS in Local Government in Europe*, pp. 23–45, London: Taylor & Francis.

CAMPBELL, H. and CRAGLIA, M. (1992). The diffusion and impact of GIS on local government in Europe: the need for a Europe wide research agenda, *Proceedings of the 15th European Urban Data Management Symposium*, pp. 133–55, Delft: Urban Data Management Society.

CAMPBELL, H. and MASSER, I. (1992). GIS in local government: some findings from Great Britain, *International Journal of Geographic Information Systems*, **6**, 529–46.

CAMPBELL, H. and MASSER, I. (1995). *GIS and Organisations*, London: Taylor & Francis.

COMMISSION OF THE EUROPEAN COMMUNITIES (1995). *GI 2000: Towards a European Geographic Information Infrastructure*, Luxembourg: DGXIII.

CRAGLIA, M. (1993). 'Geographic information systems in Italian Municipalities: a comparative analysis', Doctoral dissertation, University of Sheffield: Department of Town and Regional Planning.

CRAGLIA, M. (1994). Geographic information systems in Italian Municipalities, *Computers, Environment and Urban Systems*, **18**, 381–475.

FEDERAL GEOGRAPHIC DATA COMMITTEE (1994). *The 1994 Plan for the National Spatial Data Infrastructure: Building the Foundations of an Information-Based Society*, Reston: Federal Geographic Data Committee.

GARTZEN, P. (1995). GIS in an ever changing market environment, *Proceedings of the Joint European Conference and Exhibition on Geographic Information*, pp. 433–6, Basel: AKM Messen AG.

MASSER, I. and ONSRUD, H. J. (Eds) (1993). *Diffusion and Use of Geographic Information Technologies*, Dordrecht: Kluwer.

MASSER, I., CAMPBELL, H. and CRAGLIA, M. (Eds) (1996) *GIS Diffusion: The Adoption and Use of GIS in Local Government in Europe*, London: Taylor & Francis.

PAYNE, D. W. (1993). GIS markets in Europe, *GIS Europe*, **2**(10), 20–2.

ROGERS, E. M. (1993). The diffusion of innovations model, in Masser, I. and Onsrud, H. J. (Eds) *Diffusion and Use of Geographic Information Technologies*, pp. 9–24, Dordrecht: Kluwer.

WEGENER, M. and MASSER, I. (1996). 'Brave new GIS worlds', in Masser, I., Campbell, H. and Craglia, M. (Eds) Chapter 2, *GIS Diffusion: The Adoption and Use of GIS in Local Government in Europe*, pp. 9–22, London: Taylor & Francis.

CHAPTER EIGHT

GIS Diffusion in Greece: The Development of a Greek GIS Community

DIMITRIS ASSIMAKOPOULOS

8.1 Introduction

Geographic Information Systems (GIS) as a computer-based technology is still developing very rapidly mainly following the evolution of computer hardware and software industries (see Masser and Craglia, Chapter 7 in this volume). Because of the fast pace of current progress the present state of knowledge of GIS diffusion world-wide is fragmented and uneven both between countries and between different areas in the same country. As Masser (1992) argues it is difficult to see a European GIS community as a single entity because of the language, cultural and institutional differences, as well as the differences in technological thresholds. Greece is a small southern European country where it is possible to investigate the transfer and diffusion processes of GIS technology, through the development of the Greek GIS community, in ways that would not be possible in larger European countries such as the UK or Italy. In fact, Greece is a suitable laboratory to study GIS diffusion for at least two related reasons:

- Greece is a small country (population 10 million in 1991 census) where the common language and culture, coupled with the same system of government, allow the development of a relatively small GIS community on a national scale.
- GIS diffusion in Greece is a recent phenomenon of the last 5–6 years making it possible to study the development of the Greek GIS community at the early critical stages of its development.

The underlying assumption upon which this research is based is that GIS diffusion in Greece can be analysed through the study of the evolution of the Greek GIS community (Assimakopoulos, 1996). This implies that the concept of GIS diffusion can be investigated through the existing scientific and technological communities, as well as a number of other social actors and networks who participate in the construction of GIS innovations at local and national scales. Adopting this viewpoint it is possible to develop a theoretical model which is like a cognitive map of the diffusion journey plotted on the real network of key individuals, teams, or scientific and professional groupings who socially shape the adoption and implementation processes of GIS technology in Greece.

This chapter describes and discusses some of the main issues for GIS diffusion in Greece through the key notion of the Greek GIS community. It is divided into four main sections. The first describes the concepts of diffusion of innovations and technological communities. The second discusses the methods which were employed to investigate these key concepts. The third section presents some of the analytical findings in relation to the key concepts and the wider objectives of this research. Finally, the fourth section is an evaluation of the findings and a commentary on the value and the potential of the underlying theoretical concepts.

8.2 The main concepts

One way to present the two main concepts of GIS diffusion and GIS community is to adopt a technology-centred approach and start from GIS diffusion, then move to the informal social and communication networks which drive GIS diffusion over time, and finally study the formation process and structure of a newly emerging GIS community. An alternative way is to start from existing scientific and professional communities, or traditions of practice, and examine various individuals or organisations who adopt and implement GIS innovations, then again through emerging social and communication networks study GIS diffusion, as well as the evolution of a new GIS community. It is possible to start either from the new idea itself, or the social entities which shape the new idea through adoption and implementation processes. Tornatzky and Fleischer (1990) discuss the process of technological innovation in organisations, stating that the key point which differentiates diffusion research from, say, innovativeness research or other innovation/implementation process models is that diffusion takes as its starting point the innovation rather than the organization.

The study of GIS diffusion and implementation is becoming increasingly important as researchers and practitioners recognise the critical importance of the human, organisational, and institutional issues in the successful adoption and implementation of the technology. So far GIS diffusion and implementation research has mainly focused on either organisational or single administrative layer case studies in American or European local governments (Budić, Chapter 11 this volume; Campbell, 1994; Craglia, 1994; Innes and Simpson, 1993; Onsrud and Pinto, 1991). Existing scientific and technologial communities, as well as interpersonal social and communication GIS networks, are social entities which have not been used yet for the study of GIS diffusion and implementation processes. GIS like other computer-based innovations is a complex technology which crosses the boundaries of many well-established traditions of practice or scientific and technological communities. As a result, a broad range of social actors construct the meaning of GIS innovations at a national or international scale. People, no matter if they work in different departments or organisations, usually need to share experience and knowledge, or various information and material resources for effective GIS adoption and implementation processes. At the same time these social actors forge and maintain various GIS social and communication networks giving birth to an emerging GIS technological community of practitioners.

One could go so far as to argue that these people and organisations are responsible for the initiation of a GIS technological revolution. Constant (1980) argues that a technological revolution occurs not when the new system is operational, not when it is universally accepted, not even when it first works, but when it is accepted by even a significant minority of the relevant scientific and technological communities as the foundation for new normal practice. The outcome of this kind of GIS revolution is still an open question as the whole venture could lead to three possibilities. The most likely possibility is that the GIS revolution will merely be a stepping stone, as GIS facilitates in varying degrees all the existing scientific or technological communities which handle and analyse geographic information to modernise/innovate themselves. The GIS revolution perhaps could also lead to the establishment of a new and dedicated GIS scientific and technological community which would have as a mandate to design and update spatially referenced digital databases. Finally, the GIS revolution could lead to a combination of the two previous possibilities as a newly formed GIS community could work in parallel with existing scientific and technological communities which handle and analyse geographic information.

8.2.1 Diffusion of innovations

Diffusion of innovations has been studied by sociologists and communication theorists since the early 1940s. Rogers (1993) defines diffusion as the process by which an innovation is communicated through certain channels, over time, among the members of a social system. Consequently, GIS diffusion can be defined as the process by which information about GIS is communicated through certain media, or face-to-face, over time, among the members of different existing scientific and professional traditions. By this process a GIS community comes into being which is perceived as a collection of individuals and organisations who create and follow this new scientific and technological tradition of practice.

To put it in different words a GIS community refers to a collection of individuals and organisations that have a stake in technological development, such as data producers and suppliers, central and local government agencies, utilities, university departments and private sector firms. All these social actors or stakeholders participate with varying degrees in the GIS communities at a local, national or international scale sharing a common interest and performing complementary roles in the development of GIS applications. As a result, an emerging GIS community can be defined as a social and communication network of other separate networks who despite belonging to different geographical areas, professional groups or institutional settings share a common interest and function in complementary ways, in the development of GIS applications. This network of networks loosely connects individuals, teams and organisations who adopt and implement GIS in various settings facilitating the sharing of resources and the spread of information about GIS at an international, national and local scale.

Many researchers in the past have emphasised the social nature of diffusion. For example, Rogers and Kincaid (1981) argue that most individual members of organisations do not decide to adopt an innovation on the basis of their evaluation of the technical qualities and performance of the new idea. Instead, they depend on the subjective experience with the innovation of others like themselves, conveyed through peer networks, to give meaning to the new idea. Dearing (1993), also argues that people are more easily convinced of the personal value of an innovation when they learn information about an innovation from a person similar to themselves, while they stand to learn information which is of significant value from people who are dissimilar to themselves. Both the paradox of homophily/heterophily and the theory of 'weak network ties' can help us understand the creation process of a new technological community or the communication process among different professionals during the diffusion of a technological innovation like GIS.

Homophily is defined by Rogers (1983) as the degree to which people are alike in terms of certain attributes such as beliefs, values, education, social status and so forth and heterophily, the complementary concept, as the degree to which people are different in terms of the same set of attributes. In general, communication among homophilous people is easier, more precise, more rapid, and has more readily understandable (short-range) rewards, than communication with heterophilous people. The low-density/integration networks which are constructed by acquaintances ('weak ties') rather than close friends ('strong ties') often have a significant value in terms of greater information diversity. They stimulate the transfer of new ideas from, for example, the North American and northern European countries to countries on the periphery like Greece. However, weak ties that connect separate groups are more likely to convey new information about an innovation, and often involve heterophilous individuals.

The 'informational strength of weak network ties' (Granovetter, 1982) or the thesis that new ideas often flow between or among individuals who are distant in spatial, social, or communication terms is particularly important in the effective diffusion of technological innovations like GIS in that these weak ties can link separate individuals, groups, or organisations creating a whole new technological community. In fact, Granovetter (1982) argues that the macroscopic side of this communication argument is that social systems lacking in weak ties will be fragmented and incoherent. Individuals, for example with few weak ties, are deprived of information from distant parts of the social system and are confined to the provincial news and views of their close friends. The presence of both weak and strong network ties positions individuals within wider social and communication networks providing access to scarce resources, as well as facilitating the exchange of ideas and subsequently the cross-fertilisation of various technological traditions of practice.

In terms of GIS a number of recent publications (Azad, Chapter 10 this volume; Coppock and Rhind, 1991; Sinton, 1992; Steinitz, 1993; Tomlinson, 1988) have highlighted the role of key individuals (champions or opinion leaders) in the development of GIS since its early critical days. What is needed, however, to advance our understanding of GIS diffusion is an explicit discussion of the importance of various social and communication networks in the development of GIS. These kinds of social entities convey influence, support, or resources such as information and they can be viewed at the system level as the main routes which people and organisations use to socially construct the meaning of GIS innovations. There is a considerable gap in the literature concerning social and communication GIS networks which is maybe due to the fact that until recently social network analysis has not produced accessible theory and methods for non-sociologists or communication scholars (Scott, 1991; Wasserman and Faust, 1994).

8.2.2 Technological communities

The concept of technological community is another social entity which has not been used so far as a unit of analysis for GIS diffusion. A good indicator of the existence of technological communities are the different engineering societies, or bodies such as the Association for Geographic Information (AGI) in the UK. These kinds of professional groupings are well-institutionalised, highly specialised, and well-defined social entities which embody knowledge, development and innovation at the collective level. Such technological communities may be interchangeably composed of individual members, firms, or organisations. Various interest groups are usually also organised within their boundaries focusing on particular problems and technologies.

Technological traditions of practice are what binds communities of technological practitioners together. The analogy with Kuhn's (1970) 'paradigms' or 'exemplars' which scientists share in invisible colleges (Crane, 1972) is obvious. However, as Constant (1984) argues it is not clear that a technological tradition of practice comprises a set of specific exemplars in the same sense that a scientific paradigm, in its most narrow and precise usage does. A technological tradition has a knowledge dimension including hardware and software but it also has a sociocultural dimension as it includes social and communication stuctures as well as behavioural norms. In other words, a technological tradition of practice encompasses a wide array of different elements such as relevant scientific theory, specialised instrumentation, and accepted procedures and methods.

At a national scale, each of the existing technological communities (for example, geographers, surveying engineers, civil engineers and urban planners) related to geographic information has its own well-established technological tradition of practice which ensures social and economic status to its members. Within each technological tradition of practice a dominant technological framework has been developed composed of the concepts, methods and techniques employed by a scientific or technological community in its problem solving (Bijker, 1992). This technological framework in a sense creates a cognitive universe that does not allow the recognition of radical alternatives to normal practice. As a result each of the existing communities of practice inherently tends to be very conservative in order to preserve power positions or jobs for its members. This is the main inhibiting factor for the development of a new GIS discipline or science (Goodchild, 1992) as there is a key contradiction between the desire of existing scientific and technological communities to change incrementally while preserving the existing structures and the need for a new community of individuals and organisations which is only related to GIS.

8.3 Research methodology

To investigate the development of the Greek GIS community, a multisite and multistage approach was followed. The research has also adopted a network perspective which means that the unit of analysis is not the individual or the organisation, but an entity consisting of a collection of individuals and organisations and the set of linkages among them. According to Wasserman and Faust (1994), network methods focus on dyads (two actors and their ties), triads (three actors and their ties), or larger systems (subgroups or cliques consisting of individuals and organisations, as well as entire networks of different actors). As a result different sets of actors and network ties have to be analysed either manually or using specialised computer software. From a theoretical perspective a 'snowball sampling' approach (Rogers and Kincaid, 1981) is reflected in the organisation of the research which was carried out in three different stages between April 1992 and June 1994 using both qualitative and quantitative methods.

In the snowball sampling approach an original random sample of respondents ('starters') are asked to name their peers who then become the respondents in a second phase of data gathering, their contacts thus nominated become respondents in a third phase and so forth. In this way tracing and studying the chains of linkages is a process similar to that of a snowball rolling downhill as the sample grows slowly in the beginning and increasingly faster in later stages. The obvious advantage of the snowball sampling method is that the researcher does not arbitrarily impose the boundaries of the social system under study but he or she gradually uncovers them through the different respondents or participants in the research. Moreover, such a sampling method provides a significant advantage to reseachers who also want to use qualitative/ethnographic methods like observation and participant observation as it gradually allows identification and interaction with the respondents, following the network ties of the starters in a multistep sequence.

Qualitative methods are usually used to investigate the emerging nature of various social entities such as groups, organisations, or communities, and subsequently to formulate new theories and test new hypotheses in relation to these kinds of social actor. In this sense this research is mainly located in the qualitative, ethnographic or social anthropological tradition (Hammersley and Atkinson, 1983) as this kind of approach fits better with the problems and questions in hand for both theory development and testing. As a participant observer or merely observer, an ethnographer or social anthropologist can learn to interpret social phenomena in a similar way as the social actors that he or she studies. Moreover, the fusion of different kind of data collected by different methods can provide a more rounded general picture. Bryman (1988), for example, argues that researchers who locate their research in the qualitative tradition usually combine data derived from both quantitative and qualitative methods to produce a general picture.

More specifically, this research did not start from a random sample of Greek GIS experts but from key members of different networks which cover the whole spectrum of institutional, professional, disciplinary and organisational contexts in Greece. A number of qualitative and quantitative methods were also used to study the Greek GIS community in the early critical stages of its development. These include unobtrusive methods such as observation, and citations analysis, as well as unstructured discussions with many researchers and practitioners in various GIS events throughout the country. Semi-structured in-depth interviews were also carried out in three rounds in 1992, 1993 and 1994 based initially on an 8-page questionnaire and later on an extended version of this questionnaire. At the outset approximately 60 individuals working for almost 40 organisations all over Greece were interviewed in one or more occasions. Some 25 out of these people also provided quantitative sociometric/relational data about their personal GIS social and communication networks for the doctoral research.

A roster question, with a list of approximately 60 individuals who composed the Greek GIS community at the time, was also compiled from the list of contacts in early 1993. Subsequently, each interviewee was asked to answer how well he or she knew the GIS work and ideas of the rest of this group. As a result of this process 10 individuals were identified as the inner core of the Greek GIS community and they were studied in greater depth. Most of them (9 out of 10), had participated in the preliminary research. In 1993, all of them were reinterviewed with an open-ended questionnaire to develop their GIS profile.

It must be recognised that the 60 people who were identified and interviewed for the purposes of this research constitute the complete list of people who one might claim belonged to the Greek GIS community at the time. The precise definition of accurate boundaries for such an elastic social entity is not possible as there is an inherent fuzziness in the notion itself in terms of who is included and who is left out. To put it in a different way the concept of a GIS community is a social entity with indeterminate or fuzzy boundaries. Nonetheless, between 1992 and 1994 the approximately 60 Greek GIS experts who participated in this research represented, according to their peers, a large and substantial part of those responsible for constructing and managing the development of GIS applications in a wide range of settings throughout the country.

8.4 Findings of the research

Unlike the situation in the UK, where the government took the initiative almost a decade ago to study the impact of new technologies on the handling of geographic information (Department of the Environment, 1987), in Greece the absence of formal initiatives means that word of mouth and informal social and communication relationships are critical in linking individuals and small teams, as well as organisations. A lack of information resources coupled with a dominant political culture which puts a low priority on the use of information implies that individuals and small teams, rather than formal institutions and organisations, drive GIS diffusion. As a result the discussion below concentrates on individuals and small teams, rather than formal organisations that usually need to set up official procedures for the handling of geographic information.

The presentation of the research findings is organised in three subsections. The first presents a brief profile of the Greek GIS community based on individuals. The second presents a cognitive map of the Greek GIS community based on teams. It also discusses a number of issues related to GIS diffusion in Greece for both individuals and teams based on this ideal model. Finally, the third tests further the conclusions drawn in the previous analysis in respect to a particular issue: opinion leadership.

8.4.1 Individuals in the Greek GIS community

In 1993–4 it is estimated that the Greek GIS community was composed of approximately 60 individuals in some 40 organisations throughout the country

(Assimakopoulos, 1993). Half of these people are based in university departments. The rest are divided between public sector organisations (30 per cent), and private sectors firms (20 per cent). In terms of professional traditions, half work within the surveying engineering technological tradition of practice. Architects/planners account for about a quarter, and the remainder come from backgrounds such as civil engineering, information technology, geography, geology, and agriculture. In terms of academic qualifications 50 of the 60 interviewees have completed postgraduate degrees (PhD or Masters) either in Greece or abroad (North America, the UK and France). Forty have a PhD, with 10 having undertaken a GIS-related PhD in North American or Greek universities. A large section of the people interviewed have also participated in a number of events related to GIS technology in Greece (for example, Ursa-Net, ESRI user conference) or abroad (for example, European GIS conferences, Urban Data Management Symposia, GIS/LIS conferences or ESRI user conference).

The vast majority of the people who shape the Greek GIS community have a GIS in the department or organisation where they work. The remainder were still considering or had firm plans to buy a GIS in 1994. During the first phase of the GIS implementation process the computer hardware was usually an IBM-PC compatible with a 486 microprocessor. Later, when the applications had been developed, one or two workstations (SUN, HP-Apollo, DECstation) were usually used. There were also five mini-PRIME computers used by the Ministry of Defence. The Hellenic Military Geographic Service had four mini-PRIMEs in a token ring network and the Hydrographic Navy Service had a fifth. Overall in Greece there were approximately 400 GIS software licences in summer 1994. About 80 per cent of the software used for GIS applications Arc/Info for PC or workstation. Of the Arc/Info users half have only the PC version while the other half have both the PC and the UNIX versions. Ten people also used Erdas as a raster-based GIS package for satellite imagery and image processing. Intergraph controlled only 5 per cent of the GIS software market with approximately 15–20 licences around Greece at the time of the fieldwork.

In the early 1990s, most of the GIS applications in Greece had been undertaken by ministry agencies, utilities, and municipalities. The private sector had invested very little in the development of GIS applications. A major part of GIS investment in all the administrative settings comes from various European Union (EU) programmes. Through these funds, governmental organisations and utilities pull GIS technology in Greece by subcontracting the development of various applications. Private sector firms such as hardware and software vendors, and engineering consulting companies, together with universities and research institutes on the other hand push GIS innovations by developing a number of applications throughout the country. The great majority of the applications focus on cadastre, parcel management, tax assessment, urban, environmental and transportation planning oriented projects.

8.4.2 Social structure of the Greek GIS community

An important objective of the research was to identify the various GIS teams who constitute the Greek GIS community. Figure 8.1 provides a cognitive map of the Greek GIS community based on teams. The numbers on Fig. 8.1 represent GIS

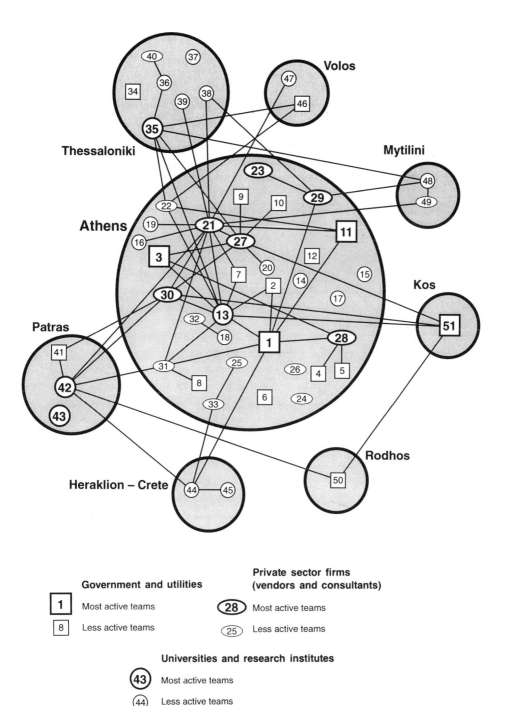

Figure 8.1 A cognitive map of the Greek GIS community in 1993–1994 based on teams.

teams who adopt and implement GIS innovations in a broad range of Greek settings. The different teams are grouped in three main categories (government and utilities, private sector firms including vendors and consultants, and universities and research institutes) according to their institutional position. Within each category the different teams are also divided into most and less active teams, in terms of the amount of GIS work that they have carried out throughout the country during the last 5 or 6 years, or the overall influence that they exercise for GIS diffusion.

Figure 8.1 shows that almost two-thirds of the total number of these Greek GIS teams are based in the Athens metropolitan area with only a third of them based in other cities (Thessaloniki, Patras, Heraklion, Volos, Mytilini, Kos, and Rodhos). Because the major GIS projects of the Greek central government and the utilites have been undertaken in Athens an even larger proportion of GIS investment is clustered in the Athens metropolitan area. As a result, in the early 1990s there were very few ties which directly connected people and teams in the periphery of the Greek GIS community. Most of the social and communications links need to be made through the centre. This has obviously considerable implications in terms of who is connected with whom, and how, throughout the country. In practical terms this means that the personal GIS networks of the individuals who shape the Greek GIS community overlap or are interconnected to a greater extent in Athens, where one finds all the main stakeholders who facilitate GIS diffusion in this city and, through it, to the rest of the country.

As can be also seen from Fig. 8.1 there are few direct links (for example, 35–46, 42–44) between teams on the periphery of the Greek GIS community. Undoubtedly these kind of network ties play a significant role for GIS diffusion as they increase the overall connectedness and integration of the social system under study. What is the nature of these network ties? How strong are they? And why do people forge and maintain this kind of GIS relationship? The 42–44 link connects two university-based teams which have a common interest in urban, regional and transportation planning. The 35–46 link connects a university-based team and a municipal team which also share a common interest in urban GIS applications and have a surveying engineering background. If we would like to focus on the former case, Graph 8.1 illustrates in detail the nature and strength of the link between teams 42 and 44 based on relational/sociometric data provided by two individuals from each team.

As can be seen from Graph 8.1 the dyads who come from the same team know each other's GIS work well. They have strong reciprocal ties with the highest possible score. Also, the triads have a remarkable symmetry in terms of the degree of reciprocality or the extent to which different individuals know the GIS work and ideas of their colleagues in a distant university. In terms of attribute data all four of the respondents are male, in their late 30s–early 40s, with postgraduate degrees, and three out of four share a civil engineering background as do most of the urban, regional and transportation planners in Greece. In both cases (42 and 44) the network links between and among the individual members of these GIS teams have a strong reciprocal nature based on a common disciplinary and professional background as well as a current interest in the development of particular types of GIS applications. As a result, an hypothesis which could be tested in future GIS diffusion research is that a high degree of homophily between and among the individual members of different GIS teams is a necessary and to

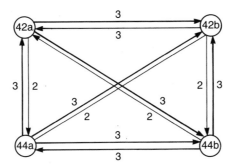

Graph 8.1 The social and communication GIS link between the two university teams 42 and 44 (see Fig. 8.1). The relational data derive from the personal GIS networks of two individuals from each team (3 means that the respondent knows the GIS work and ideas of the other person very well, 2 well, 1 hardly at all, and 0 not at all).

a large extent sufficient condition underlying the creation and maintainance processes of strong GIS network ties.

Moreover, if the most active of the teams of Fig. 8.1 are seen from a background discipline's perspective then the great majority of them (numbers 13, 27, 28, 29, 30 and 35) are members of the surveying engineering profession. This common academic and professional background provides the comfort of homophilous communication among surveying engineers in Greece, facilitating so far the sharing of GIS experience and ideas as well as creating the impetus for the further diffusion of GIS-technological innovation within the surveying engineering technological community. More importantly, as a result of this communication process the technological framework of surveying engineers is gradually becoming the dominant technological framework of the Greek GIS community. It is not a coincidence that a great deal of GIS applications focus on cadastre and parcel management, also emphasising issues of geometry, accuracy, and digital topographic data production.

However, as was pointed out previously, not only strong but also weak network ties can play a significant role for GIS diffusion as these kind of network ties integrate existing dense subsystems like small teams with other distant teams of the wider system, providing diversity and openess at the system level. Individuals who forge weak ties usually play the roles of bridges or liaisons. Bridges are individuals who belong to the local system and link separate groups or cliques at the local level. Liaisons are individuals who link cliques of the local system with distant cliques of the wider system. The local system in the context of this research is the Greek GIS community, while the distant or wider system is the international GIS community. A clique in network terms is a group in which each member is connected with all other members. GIS teams in Fig. 8.1 could be seen as cliques although these are not the only cliques within the Greek GIS community. To illustrate the concepts of bridge and liaison within the social system under study, two individual cases are presented from Fig. 8.1.

The first case is that of a surveying engineer who has recently set up his own GIS consulting firm (team 30). He directly connects team 42 with 30 as a bridge. He also indirectly connects 42 with 13. This individual completed a GIS-related

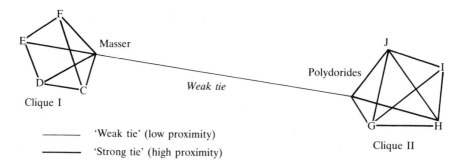

Graph 8.2   A weak tie between the teams of Professors Nikos Polydorides (University of Patras, Greece) and Ian Masser (University of Sheffield, UK).

PhD at the National Technical University of Athens (NTUA), thus he has strong overlapping ties with the surveying engineers of the team number 13 at the NTUA. As a result of this link he was invited to be a lecturer at the Ursa-Net meetings at Patras, and later he worked as a consultant for the development of an urban information system for the city of Patras together with the town planners of the team 42. His role so far is visibly constructive as he participated in various GIS events organised by planners throughout Greece, transferring ideas from the surveying engineering technological framework to the planning one and vice versa.

The second case illustrates a liaison between the American and European GIS communities through the head of team 42, Professor Polydorides. He is a former student of the Harvard University Graduate School of Design. Nowadays, as a prominent Greek Professor and secretary of Ursa-Net he maintains a number of weak ties with various European GIS experts providing openness to the whole of the Greek GIS community. As can be seen from Graph 8.2 which depicts the team of Professor Polydorides at the University of Patras and the team of Professor Masser at the University of Sheffield, there is no overlap between the personal communication networks of the two Professors so the possibility of information flowing from one system to another is dependent on that weak tie. If this weak informational tie collapses then the two teams become isolated and information exchange processes about GIS innovations are discontinued.

Apart from bridges and liaisons who usually function with varying degrees of inclusion in different technological frameworks there are also individuals who strictly comply with the dominant norms of their scientific discipline or professional tradition. Out of this set of individuals usually come opinion leaders, or in terms of GIS, people who play the role of GIS champion at the organisational or community level.

8.4.3 Opinion leadership

Apart from qualitative methods such as observation, or semi-structured interviews where respondents nominate opinion leaders, an alternative way to study opinion leadership is to analyse different personal networks and find within these social and communication networks who are the central individuals (Harkola and Greve, 1995; Valente, 1995). The Ucinet IV (Borgatti, Everett and Freeman, 1993) social network analysis computer software provides a number of centrality measures for identifying opinion leaders. Using a 25×25 GIS sociomatrix compiled from the personal networks of 25 individuals within the Greek GIS community, Freeman's (1979) degree centrality measures were calculated. The relationship which was measured is awareness of other's GIS work and ideas. This is a composite relationship consisting of various information exchange processes such as seeking advice or support in terms of GIS adoption and implementation, also taking into account other variables such as public image, trust or friendship. This measure can be also seen at the collective level as a good indicator for the existence (or not) of a GIS community since people who reciprocate these kind of GIS network ties can be viewed as actors who create and follow this new technological tradition of practice.

Table 8.1 shows the in- and out-degrees for each of these 25 GIS individual actors classified according to the three institutional categories which were used in Fig. 8.1. OutDegree is the sum of GIS ties which are initiated by a focal individual. InDegree is the sum of GIS ties which are received by a focal individual. These 25 individuals come from the whole spectrum of institutional, professional, disciplinary and organisational contexts within the Greek GIS community and belong to 20 different GIS teams as can be seen from Table 8.1 and Fig. 8.1. Table 8.2 divides into four categories these findings in terms of the in-degree reflecting centrality or opinion leadership at the individual level.

As can be seen from Table 8.1 the average in-degree of the Greek GIS vendors and consultants has also a much higher average than in- degrees of the other two categories as they are more visible within the Greek GIS community (see also the third column of Table 8.2). It is worth pointing out that the only individual in Table 8.2 who received more than 60 ties is the Greek ESRI vendor (see also team 21 in Fig. 8.1). Vendors' and consultants' GIS network ties also have a higher degree of reciprocality as the comparison of the differences between the average in-degrees and out-degrees of Table 8.1 indicates. Government and utilities GIS experts tend to initiate a few more GIS social and communication links than they receive because they are interested in what other people are doing with GIS innovations. On the other hand academics and researchers tend to receive a few more GIS ties than they initiate as they focus more on the development of particular applications. Within each institutional category the differences from the average in-degree are due to different scientific or professional backgrounds, spatial distance, or different types of GIS software as well as application areas.

The main feature of Table 8.2 is that the five people from all three institutional categories who receive between 40 and 60 network ties work within the surveying engineering technological tradition in the wider Athens area. As can be seen from Fig. 8.1 they belong to some of the most active teams (numbers 1, 13, 21, 27, 30) of the Greek GIS community. Most of them are GIS consultants and/or academics who were responsible for the development of a great deal of GIS applications in a variety of settings all over Greece during the last 5–6 years. Planners and other professionals who work in a different city or a different technological framework than the surveying engineering one are limited to the bottom half of Table 8.2. Individual prominence or centrality within various personal GIS networks generates a higher prestige and a wider reputation for surveying engineers in

Table 8.1 Freeman's degree centrality measures for 25 of the GIS personal social and communication networks of the Greek GIS community. The first number mirrors the team in which the individual partcipates in Figure 8.1

Government and utilities (I)			Vendors and consultants (II)			University and Research Institutes (III)					
No.	Out and in degrees		No.	Out and in degrees		No.	Out and in degrees		No.	Out and in degrees	
12	22	7	30	57	58	36	17	15	32	34	25
9	16	10	21a	72	61	19	13	21	13b	22	24
3	28	32	44b	31	28	16	34	19	42a	18	38
1a	39	36	21b	62	53	44a	26	22	42b	35	31
11	25	20	27	21	44	13a	33	41	38	4	16
2	22	26	31	26	33	18	3	12	48	23	21
1b	60	50									
Ave	30	26	Ave	45	46				Ave	22	24

Table 8.2 Four categories for opinion leadership based on Table 8.1. The criterion for the identification of opinion leaders is how many network GIS ties are received by each individual (Freeman's in-degree centrality measure)

Sum of network ties	(I)	(II)	(III)	Total number of individuals
Between 60 and 72	0	1	0	1
Between 40 and 60	1	3	1	5
Between 20 and 40	3	2	7	12
Less than 20	3	0	4	7
Total number of individuals	7	6	12	25

Greece. This in turn justifies a proportional allocation of financial and other rewards to this social group and its accompanying technological tradition.

As a result it seems that the emerging nature of a GIS community in Greece will be mainly dominated by the technological framework of surveying engineers. Moreover, if one considers that Arc/Info is the predominant GIS software representing almost 80 per cent of the Greek GIS market then the initial hypothesis about a centralised communication pattern of GIS diffusion in Greece is even clearer. Surveying engineers, Arc/Info users, and researchers and practitioners who are based in the wider Athens area drive to a large extent GIS diffusion, as well as the creation process of the Greek GIS community.

8.5 Evaluation

Because of the dominance of the surveying engineering scientific and technological community in the formation process of the Greek GIS community, it seems likely that GIS applications in Greece will continue to focus on cadastre, parcel management and tax assessment-oriented projects, with the main emphasis on the issues of geometry, accuracy and digital topographic data production. This situation is reflected in the structure and function of various GIS social and communication networks within the Greek GIS community. So far the key factor which sustains diffusion of GIS innovations in the wider Greek public sector is external to the local system. The availability of EU financial aid to modernise existing, or set up new, technological systems externally induces, or facilitates, GIS transfer and diffusion processes in central government agencies, utilities, and municipalities. EU funding can generally be obtained more easily if scientific or technological networks include as nodes not only social actors from the developed northern countries but also from the less-developed southern European countries like Greece. Adopting this point of view, many of the key members of the Greek GIS community have an outward perspective which gradually allows them to interconnect local nodes and networks with other nodes and networks at the European level.

This research investigated and mapped the big picture of GIS diffusion on a national scale by focusing mainly on the people who critically influence the current and future state of the art of this new technology in Greece. These people are aware of the GIS work and ideas of others within the system mainly because they need to share knowledge, and exchange experiences in relation to GIS innovations. Additionally, they tend to know each other because they not only collaborate but also compete for material or other resources. In the process these social actors gradually shape the embryo of a heterogeneous but increasingly dense GIS community at a national scale.

An important area of further research is to investigate the value and belief systems which underlie GIS diffusion processes including how the different social actors evaluate GIS innovations. It seems an irony that social actors must make their choice between the old and new technological tradition or system, before the old system is further refined or the new system well developed. It appears that the great majority of the social actors who participate in the construction of a GIS tradition in Greece must base their choices only on the perceived subjective promise, or the relative aesthetic appeal of the two systems. Such choices are exacerbated by the tendency of old and new systems to exhibit their strongest virtues along different dimensions. Therefore, the adoption and implementation of a new system often implies not only new hardware, software, data, skills, and organisational structures, but also a new value system, a new set of criteria by which technology is judged.

Apart from this area, GIS diffusion research needs to explore the issues surrounding social and communication networks as it is possible through these kinds of social entities to understand the nature and structure of an emerging GIS community. Some of the questions which need further investigation to advance our understanding of the social and communication structure of an emerging GIS community are the following.

- Is there a clear pattern in the arrangement of social and communication network ties for GIS diffusion at a national or international scale?

- Are the linkages between various GIS social actors of a homophilous or heterophilous nature?

- How strong/weak are the network GIS ties which connect individuals from different teams within a particular GIS community?

- Is it possible to identify various individual roles such as opinion leaders, bridges and liaisons through the different personal GIS social and communication networks?

- What can be learned from the relationships among individuals for existing scientific and professional communities which are related to a new GIS technological community?

Moreover a complementary set of questions could be related to the concept of technological community.

- Is there a dominant technological framework or tradition of practice in terms of GIS in a particular country (or a wider area such as English-speaking countries, central Europe, and so forth)?

- How are emerging GIS social and communication networks related to existing technological frameworks?

- Can we link the answers to the two previous questions to foresee the emerging nature of a GIS community at a national or cross-national scale? An understanding of these issues at the European level, for example, could be useful not only in satisfying pure scientific curiosity. It could help to evaluate the effect of European Union policies in terms of financial or other aid to its less favoured regions like Greece. It could also inform future policies for the allocation of resources as different social networks and existing technological communities alter an otherwise likely technological development trajectory having a significant impact on what solutions or technological systems can be put in place for such real problems as land use and transportation management and planning across Europe.

The answers to these questions are not easy in many cases as the large number of social actors involved in GIS diffusion impedes the collection of reliable network data, or the use of qualitative/ethnographic methods to study in depth the various social entities. The ability to fulfil both of these conditions make Greece a worthy case. In the near future, however, considerable advances in computer mediated communication technologies (for example, Internet, World-Wide Web) provide promising new research avenues to gather at least highly reliable GIS communication network data even for systems with hundreds of members which cross the boundaries of many countries (Rice, 1994). This kind of relational data could become the starting point for the study of GIS diffusion processes, or the formation process of a new GIS community, at a European or global scale.

Acknowledgements

This chapter has profited by careful readings and useful comments supplied by my supervisor Professor Ian Masser, and Dr Julia Harkola, Stanford University, as well as three anonymous referees. I would also like to express my gratitude to the 60 Greek GIS experts who participated in my research about the development of the Greek GIS community. For two years (October 1992–September 1994) this research was sponsored by an individual fellowship of the Human Capital and Mobility programme of the Commission of the European Communities (CEC, proposal number ERB4001GT921115). Any views expressed in this paper are those of the author and do not reflect in any way the views of the CEC.

References

ASSIMAKOPOULOS, D. G. (1993). The Greek GIS community, in Harts, J., Ottens, H. and Scholten, H. *Proceedings of EGIS '93, Genoa*, **1**, pp. 723–32, Utrecht: EGIS Foundation.

ASSIMAKOPOULOS, D. G. (1996). Greece: The development of a GIS community, in Masser, I., Campbell, H. and Craglia, M. (Eds) *GIS Diffusion: The Adoption and Use of GIS in Local Government in Europe*, pp. 147–162, London: Taylor & Francis.

BIJKER, W. E. (1992). The social construction of fluorescent lighting, or how an artifact was invented in its diffusion stage, in Bijker, W. E. and Law, J. (Eds) *Shaping Technology/ Building Society*, pp. 75–102, Cambridge, MA: MIT Press.

BORGATTI, S. P., EVERETT, M. G. and FREEMAN, L. C. (1993). *UCINET IV, Version 1.37*, Columbia, SC: Analytic Technologies.

BRYMAN, A. (1988). *Quantity and Quality in Social Research*, London: Unwin Hyman.

CAMPBELL, H. (1994). How effective are GIS in practice? A case study of British local government, *International Journal Geographical Information Systems*, **8**, 309–25.

CONSTANT, E. W. (1980). *The Origins of Turbojet Revolution*, Baltimore: Johns Hopkins University Press.

CONSTANT, E. W. (1984). Communities and hierarchies: Structure in the practice of science and technology, in Laudan, R. (Ed.) *The Nature of Technological Knowledge: Are Models of Scientific Change Relevant?*, pp. 27–46, Dordrecht: Reidel.

COPPOCK, T. and RHIND, D. W. (1991), The history of GIS, in Maguire, D. J., Rhind, D. W. and Goodchild, M. F. (Eds) *Geographical Information Systems: Principles and Applications*, **1**, pp. 21–43, London: Longman.

CRAGLIA, M. (1994). GIS in Italian municipalities, *Computers, Environment, and Urban Systems*, **18**, 381–475.

CRANE, D. (1972). *Invisible Colleges: Diffusion of Scientific Knowledge*, Chicago: Chicago University Press.

DEARING, J. W. (1993). Rethinking technology transfer, *International Journal Technology Management*, **8**, 478–85.

DEPARTMENT OF THE ENVIRONMENT (1987). *Handling Geographic Information, Report of the Committee of Enquiry Chaired by Lord Chorley*, London: Her Majesty's Stationery Office.

FREEMAN, L. C. (1979). Centrality in social networks: Conceptual clarification, *Social Networks*, **1**, 215–39.

GOODCHILD, M. F. (1992). Geographical information science, *International Journal Geographical Information Systems*, **6**, 31–45.

GRANOVETTER, M. (1982). The strength of weak ties: A network theory revisited, in Marsden, P. V. and Lin, N. (Eds) *Social Structure and Network Analysis*, pp. 105–30, Beverly Hills, CA: Sage.

HAMMERSLEY, M. and ATKINSON, P. (1983). *Ethnography: Principles in Practice*, London: Tavistock.

HARKOLA, J. and GREVE, A. (1995). 'The role of opinion leaders in the diffusion of a construction technology in a Japanese firm', Presentation at the Joint 15th Sunbelt and 4th European Social Networks Conference, London, July.

INNES, J. E. and SIMPSON, D. M. (1993). Implementing GIS for planning: Lessons from the history of technological innovation, *Journal of American Planning Association*, **59**, 230–6.

KUHN, T. (1970). *The Structure of Scientific Revolutions*, 2nd Edn, Chicago: Chicago University Press.

MASSER, I. (1992). We are beginning to see the creation of a European GIS community, *Geo Info Systems*, **2**, 19–27.

ONSRUD, H. J. and PINTO, J. K. (1991) Diffusion of geographic information innovation, *International Journal Geographical Information Systems*, **5**, 447–67.

RICE, R. E. (1994). Network analysis and computer mediated communication systems, in Wasserman, S. and Galaskiewicz, J. (Eds) *Advances in Social Network Analysis*, pp. 167–203, Thousand Oaks, CA: Sage.

ROGERS, E. M. (1983). *Diffusion of Innovations*, 3rd Edn, New York: The Free Press.

ROGERS, E. M. (1993). The diffusion of innovations model, in Masser, I. and Onsrud, H. J. (Eds) *Diffusion and Use of Geographic Information Technologies*, pp. 9–24, Dordrecht: Kluwer Academic Publishers.

ROGERS, E. M. and KINCAID, D. L. (1981). *Communication Networks: Toward a New Paradigm for Research*, pp. 109–10, New York: The Free Press.

SCOTT, J. (1991). *Social Network Analysis: A Handbook*, London: Sage.

SINTON, D. (1992). Reflections on 25 years of GIS, insert in *GIS World*, **5**.

STEINITZ, C. (1993). GIS: A personal historical perspective, *GIS Europe*, **2**(5–7), 19–22, 38–40, 42–5.

TOMLINSON, R. F. (1988). The impact of the transition from analogue to digital cartographic representation, *The American Cartographer*, **15**, 249–62.

TORANTZKY, L. G., and FLEISCHER, M. (1990). *The Process of Technological Innovation*, Lexington, MA: Lexington Books, D. C. Heath and Co.

VALENTE, T. W. (1995). *Network Models of the Diffusion of Innovations*, Creskill, NJ: Hampton Press.

WASSERMAN, S. and FAUST, K. (1994). *Social Network Analysis: Methods and Applications*, Cambridge: Cambridge University Press.

CHAPTER NINE

In Search of the Dependent Variable: Toward Synthesis in GIS Implementation Research

JEFFREY K. PINTO and HARLAN J. ONSRUD

> Success depends on three things: who says it, what he says, how he says it; and of these things, what he says is the least important.
>
> John Morley

9.1 Introduction

While within the GIS research community the problems of developing a generally agreed upon measure of implementation success (the dependent variable) are current and compelling, difficulties in attempting to determine system implementation success are actually long lived. For the past 25 years, researchers have been making attempts to develop a useful, generally accepted, and comprehensive model of system success within other areas of the Information Systems (IS) community (most notably, Management Information Systems, MIS): one that takes into account all relevant aspects of information systems and their operations. This research, far from creating agreement and synthesis, has not achieved the sort of widespread recognition that had been sought. Further, it has not made the obvious migration to the GIS literature. In fact, a casual glance at research into the implementation of geographic information systems in any of the major journals will quickly reveal that there are often as many different measures of system success as there are scientists studying the process. Unfortunately, the problems do not stop there. Another side-effect of our failure to achieve consensus is that even when two researchers agree on the same general metric of implementation success (for example, user satisfaction), they are often unable to agree on how that construct should be measured. What this phenomenon has created is a wide range of research studies, purportedly studying the same issues, and yet unable to agree as to what comprises 'success' or even how it should be measured.

This chapter intends to create a framework for the study of GIS implementation success by examining what past research in other information systems disciplines has taught us and how it can be incorporated into modern definitions of GIS implementation. The good news is that underlying much of the verbiage associated with all these differing models and outlooks, there are some general issues that can be brought forward, which will serve as a basis for establishing a more comprehensive model of successful information system implementation. Within this model, we will see that a number of divergent research streams can actually converge, and that often our charges that different types of system implementation produce different measures of success may not, in fact, be true. Research done on implementation in other fields (for example, MIS and management science) offers a great deal of utility and has much to teach us in our approach to better understanding successful GIS implementation.

9.2 The implementation 'problem'

Numerous studies have addressed the adoption and diffusion of technological innovations in other fields, including MIS, project management, and operations research and management science. An analysis of some of the relevant findings from these research streams can be an invaluable aid in helping refine the appropriate success measures for GIS implementation through an understanding of their historical context. It is important to note that the terms 'diffusion' and 'implementation' are used interchangeably within this section. Historically, the concepts of diffusion and implementation were independently derived from different sources. (Diffusion usually referred to the acceptance and use by some subset of the general population of scientific or technological innovations. The concept of implementation derived from the acceptance within organisations of new technical processes or models.) However, over the last several years, the two terms have come to engender the same idea (Onsrud and Pinto, 1991). Therefore, for the purposes of this chapter, diffusion and implementation are both intended to refer to the process through which an innovation is communicated through certain channels over time among the members of a social system (Rogers, 1983). Because implementation has been defined in terms of a change process that is accepted by organisational members, any definition of 'successful' implementation must incorporate these components. There are several definitions of implementation success, many of which share some common underlying properties. Early definitions of implementation success were usually predicated simply on use, or acceptance of the new innovation by its intended user (Lucas, 1975; Mitroff, 1975). The implementation process was generally seen as a bridge between system developer and user which, once having been crossed, was regarded as successful. Other definitions of implementation success have been more complex and thorough in their understanding of the nature of the factors relating to this success. Arnoff (1971), delineating conditions for implementation success in operations research models, states that:

> Unless our models result in decision rules and/or procedures, (a) which are implemented within the managerial decision making process, (b) which really work, and (c) which lead to cost effective benefits, our efforts are not successful. (p. 142)

Schultz and Slevin (1979) offered an important contribution to the discussion of implementation success through supplementing these decision rules by demonstrating a behavioural perspective of implementation. Their approach distinguished between three conditions by which any implementation should be judged. These criteria are technical validity, organisational validity, and organisational effectiveness. Technical validity and organisational validity are seen as necessary but not sufficient conditions for the implementation process to result in a positive change in organisational effectiveness.

9.3 Technical validity

The first criterion for implementation success, technical validity, is assessed in terms of the belief that the system to be implemented 'works'; that is, it is a correct and logical solution to the perceived problem. Obviously, should a new geographic information system package be fraught with difficulties and other system 'bugs,' it will not be used and hence should rightly be regarded as an implementation failure. Ashton-Tate Company lost a great deal of money and prestige in attempting to ship a new release of its flagship package DBase IV too early, before significant system problems had been debugged. Companies installing and attempting to implement the new database-management system would have been unsuccessful due to the lack of technical validity in the package.

9.4 Organisational validity

The second factor, organisational validity, is assessed as a measure of the congruence between the organisation and the system to be implemented. As such, it may be considered as the belief that the geographic information system is 'correct' for its users, that it will be accepted and used by the organisational members for whom it is intended. The key advance in system implementation theory that is suggested by the notion of organisational validity is the importance of including acceptance and actual use of the system in any equation of implementation success. A classic manifestation of organisational validity is the 'Not Invented Here' (NIH) syndrome found within many public and private organisations. NIH refers to the parochial attitude which suggests that unless a product or idea is created in-house, it is of no value. As a result, an externally derived idea will not be used by the company because it came from outside sources. Companies labouring under NIH frequently report serious difficulties implementing new systems or other innovations because, while being technically correct, they will not be used by the client groups for whom they are intended.

9.5 Organisational effectiveness

The 'bottom line' that any organisation seeks through the development and adoption of a new system or other innovation is some measure of improvement, either in productivity, efficiency, or environmental effectiveness. Early researchers recognised the importance of demonstrating some payoff by adopting a new information system. Based on the work of Schultz and Slevin (1979) and Schultz and Henry (1981), it was further determined that new definitions of implementa-

Table 9.1 Schultz and Slevin's (1979) model of implementation success

1	Technical validity	The assessment of whether or not the system's technology 'works'; that is, it is a technically accurate and appropriate solution to the problem at hand.
2	Organisational validity	Will the new system be accepted and used by its intended clients? Does the system alternative take into account organisation and behavioural considerations such as the organisation's culture, power and political realities, comfort level with computerisation, and so forth?
3	Organisational effectiveness	Does the new system influence the organisation's ability to perform its work more effectively, make its decisions more accurately, or lead to profitability? This factor seeks to assess the 'bottom line' worth of the new information system, not only into terms of cost/benefit analysis, but also in relation to its impact on organisational operations.

tion success need to be based on the perception of implementation as a process of organisational change. The basis for this view is the argument that it is not sufficient to regard an implementation as successful merely because the intervention of a new system causes organisational personnel to change their decision making processes; in effect, to make use of the GIS. Successful implementation requires that the results of the system lead to improved decision making. As a consequence of this definition, a new information system, such as a GIS, that is in place but is not improving decision making (that is, leading to enhanced organisational effectiveness) would not be considered successfully implemented.

9.6 New system success – modern definitions

In spite of earlier literature that has offered a fairly comprehensive look at the implementation of management science and operations research, the measurement of success within the context of information systems implementation has been somewhat less well developed. As recently as 1980, one of the leading researchers in the field of information systems identified the search for the dependent variable (system implementation success) as an issue that required rapid resolution if the field of management information systems was to position itself as an area in which viable research could be conducted (Keen, 1980). His point was that until such time as a generally agreed upon definition of successful system implementation was established, our attempts at researching and disseminating information pertaining to systems' implementation would be severely undercut by this failure to achieve a workable consensus. The importance of determining a suitable dependent measure for implementation success cannot be underestimated. As Keen (1980) stated:

> The evaluation of [information systems] practice, policies, and procedures requires an information system success measure against which various strategies can be tested. Without a well-defined dependent variable, much of information system research is purely speculative. (p. 61)

Since Keen issued this challenge, over 200 articles and books have been written on information systems and the quest to determine exactly what constitutes a successful system. While this research has predominantly examined the area of management information systems (MIS), there are sufficient parallels between the fields of MIS and GIS implementation to warrant their inclusion in this discussion of GIS implementation success.

One of the surprising aspects of the research that has been done to date is that it has been extremely difficult to build consensus on what constitutes a 'successful' information system. As a recent article pointed out, of the hundreds of studies that have been conducted on information systems' implementation and use, there are almost as many individualised measures of success as there are research studies (DeLone and McLean, 1992). Researchers and practitioners have rarely been able to agree on: (1) the important variables that should be considered when assessing success, and (2) how those variables should be measured; that is, what comprises an accurate measure of any specific variable deemed to be important. As you can see, the problem now becomes even more complex. Not only is there wide disagreement over the appropriate success variables, but even once they are employed by more than one researcher, there is often a widely divergent manner in which these constructs (as assessed by the variables) are measured.

The model that we propose for this chapter makes use of elements of success which have been gleaned from a variety of sources: some from original work on management science implementation as proposed by Schultz and Slevin (1975) and other research that is more contemporary, dealing with success within the context of management information systems. For example, DeLone and McLean's (1992) work in the area of MIS implementation success has been an invaluable resource. These measures of system success are equally applicable to the implementation of GIS as they are intrinsically tied to the management of a wide range of organisational information system applications.

9.7 GIS implementation success: towards a comprehensive model

Schultz and Slevin (1975) were correct in their early assessment of implementation success when they posited that any successful implementation effort is predicated on the technical efficacy of the system about to be introduced. In other words, the technical viability and workability of the geographic information system represents the first necessary but not sufficient condition for that system's successful adoption. Prospective users of any GIS must first be confident that the technology they are considering does, in fact, work; that is, that it accomplishes the tasks that its advocates claim it can perform. What this discussion is referring to is the notion of Information System Quality, as identified by DeLone and McLean (1992). Some of the more obvious measures of system quality, particularly within the context of GIS, would include system response time, ease of on-line use (user friendliness), and reliability of the computerised system (absence of consistent down-time). These all represent some of the more common and well-accepted determinants of a GIS's technical quality and should, rightfully, be addressed in assessing the chances for a successful introduction. Some technical characteristics are easily comparable across alternative GISs, such as built-in features, expandability, system speed, and so on. Other aspects of system quality (user friendliness) may be more qualitative and difficult to rate, let alone compare with GIS alternatives. Certainly, given the nervousness toward computerisation felt by many members of organisations, attempting to develop and introduce a non-user friendly system may be a difficult and ultimately futile process.

A second measure of technical viability of the GIS is the idea of information quality. The importance of information quality derives from the notion that any system is only as good as the information it delivers. In other words, rather than simply considering how 'good' an information system is (system quality), a better representation of quality would examine the outcomes of the GIS; that is, the accuracy of the data. In addition to data accuracy, some of the most important metrics for assessing the information quality delivered by a GIS are proposed data currency, turnaround time, completeness of the data produced, system flexibility, and ease of use among potential clients of the system. Additional elements of information quality are ease of interpretation, reliability, and convenience. These criteria are all examples of ways in which we can rate the quality of the information generated by the GIS and, as in the case of system quality, provide a context for the willing acceptance and use of the system by its intended target departments. It is important to note, however, that, as with any information storage and retrieval system, the quality of information produced lies in direct proportion to the quality of information inputted. Consequently, an additional implication of this model is the need to enforce quality control throughout the organisation, upstream of the GIS. A 'quality' GIS cannot obviate the work of poor quality inputting.

The third aspect of GIS success also relates to the earlier work of Schultz and Slevin's (1975) typology, as they identified their idea of organisational validity. Perhaps more than any other measure of the successful implementation of an information system is the importance of information use. Information use refers to the obvious point that information of any sort is only as good as it is accepted by organisational members and used in their decision-making processes; that is, the information is consumed by its recipients. Underscoring the difficulty of gaining system acceptance and information use is the problem of attempting to change employee behaviour. During a series of interviews the authors conducted with representatives from municipal governments in some Northeastern states, we were shown the town planning office's GIS, a PC-based version of a popular GIS product. The system was quite literally gathering dust as the members of the department continued to make use of paper maps and old charts for zoning and public works decisions. When we asked the planner why his subordinates were not performing these routine tasks with the town's GIS, he replied that the PC's monitor had broken down over 6 months ago, but that suited everyone just fine as they had never had much use for the GIS. Here was a clear-cut case of the problem with acceptance and use of a system. So disinterested were the planners in the GIS-created data that they had seized on the excuse of a broken PC as a basis for continuing with their old practices.

Beyond the obvious point that a successfully implemented GIS must be used is the more subtle issue of the different levels of use that may be found. For example, one study of information system implementation identified three different types of use:

1   Use of a system that results in management action.

2   Use that creates or leads to organisational change (different ways of performing standard tasks).

3   Recurring use of the system (Ginsberg, 1978).

Another study expanded the levels of use in a manner that is particularly relevant to GIS managers and users. In this research, four levels of use were explored:

1   use of the system for getting instructions;

2   use for recording data (for example, digitising parcel maps);

3   use for management and operational control; and

4   use of planning (Vanlommel and DeBrabander, 1975).

Finally, Masser and Campbell (1995), in their study of the diffusion and use of GIS in British local governments, noted that 'use' is often a problematic measure, depending upon the technological and organisational contexts within which a GIS is adopted. It is clear that when 'use' is broken down into its various components, it does not make sense for GIS managers to state that their systems are being used. The logical follow-up question to such a statement is asking how the system is being used. In other words, is the system being used to its fullest capabilities or is the organisation content to use the GIS to perform a few minor functions without really testing its capacity?

One of the most common (but difficult to assess) measures of information system success is user satisfaction. It is important to distinguish between system use and user satisfaction with the system. In many cases, a system may be used

Table 9.2   Toward a unified model of system implementation success

A System traits	
1 System quality	The system adheres to satisfactory standards in terms of its operational characteristics.
2 Information quality	The material provided by the system is reliable, accurate, timely, user-friendly, concise, and unique.
B Characteristics of data usage	
3 Use	The material provided by the GIS will be readily employed by our organisation in fulfilment of its operations.
4 User satisfaction	Clients making use of the system will be satisfied with the manner in which it influences their jobs, through the nature of the data provided.
C Impact assessment	
5 Individual impact	Members of the departments using the GIS will be satisfied with how the system helps them perform their jobs through positively impacting both efficiency and effectiveness.
6 Organisational impact	The organisation as a whole will perceive positive benefits from the GIS, through making better decisions and/or receiving cost reductions in operations.

to a marginal degree without generating much satisfaction by its users, particularly if there are no viable alternatives. In fact, a rule of thumb often suggests that in systems which are under-utilised, this state exists because they do not create much satisfaction from their users. User satisfaction, on the other hand, is an extremely relevant measure of implementation success in that it refers to the level of acceptance and positive feelings toward the GIS generated by using the system. It is important to note that user satisfaction has to be assessed after the fact; that is, the users must be in a position to evaluate the data generated, the ease with which they were able to create this data, whether or not the system meets its advertised goals, and so forth. In spite of these examples of issues that can influence user satisfaction, the concept remains very difficult to accurately assess. In effect, user satisfaction asks the manager to get inside the head of the users to see exactly what it is about the system that appeals to them. Even then, a second difficulty relates to determining relative levels of satisfaction; that is, does satisfaction mean the same thing to different users?

At this point, we need to consider other difficulties with the use of user satisfaction as a measure of implementation success. In particular, two additional questions arise.

1 Whose satisfaction should be measured?
2 How do we separate out individuals' general attitude toward computers from their satisfaction with this specific technology?

In other words, many groups or individuals within the organisation could potentially make use of GIS technology. An obvious problem arises when there is a wide discrepancy in terms of their relative levels of satisfaction with the GIS. Why is it working well for some departments and not for others? This question seeks to determine the reasons that the system's introduction is being handled better in some areas of the organisation than in others. Another difficulty with measuring satisfaction has to do with the potential for differential satisfaction at various organisational levels. Top management may be extremely pleased with the GIS while the lower, operational levels are only minimally applying the technology and doing everything possible to avoid using it. Obviously, the 'true' measure of user satisfaction in this case rests with the lower-level managers and staff who are ostensibly using a system with which they are quite dissatisfied.

The second concern regarding user satisfaction is one that has been well articulated in the information systems literature. This problem has to do with the obvious link between user satisfaction and overall attitude (positive or negative) about the use of computers. It is no secret that many individuals in most public and private organisations manifest a great deal of anxiety when confronted with computers. This 'computer anxiety' has been the cause of many voluntary withdrawals from companies as departments seek to computerise without spending adequate time retraining employees so that they will know what to expect. On the other hand, newer generations of American and European workers, growing up used to the presence and utility of computers, generally possess a much more positive attitude toward their capabilities. This positive attitude often reflects their general level of satisfaction with newly introduced computer systems such as GIS. The point that needs to be stressed for researchers is that prior to studying the implementation of a GIS, it is extremely important that adequate time is spent in assessing the general attitudes about computers on the part of impacted departments. If the perception is that these attitudes are not as positive as they should be, it can have a significant biasing effect on user satisfaction.

Organisations usually do not adopt new, often expensive technologies simply for their own sake. Rather, as a bottom line, they seek some form of return on their investment. In other words, what is the potential pay-off for using the GIS? This point underlines the final two assessments of implementation success: individual impact and organisational impact. Individual impact refers to the users' expectation that through use of the new GIS (that is, through the effect of enhanced information availability) they will derive positive benefits. In other words, using GIS technology in their operations will make them better employees, subordinates, governmental agents, or managers. In determining individual impact of a GIS there are a number of points that researchers need to bear in mind. For example, 'impact' could refer to an improvement in the subordinate's performance in that the GIS allows the employee to make better, more complete, or more accurate decisions. However, impact can also be assessed in other ways. For example, as has been noted:

> 'Impact' could be an indication that an information system has given the user a better understanding of the decision context, has improved his or her decision-making productivity, has produced a change in user activity, or has changed the decision-maker's perception of the importance or usefulness of the information system. (DeLone and McLean, 1992, p. 69)

You can see that, defined in this manner, the concept of user impact is truly multifaceted, comprising not just an assessment that the GIS makes better (more efficient and effective) employees, but that it creates a more professional, well-reasoned workforce, one that is capable of inculcating and using the GIS technology as a stepping stone to knowledge enhancement. Put in this light, GIS can have the power to teach as well as nurture better informed workers by allowing these employees to understand and grow with the system and its capabilities. As a result, when assessing individual impact, the manager may be at a disadvantage in knowing at what point to make that determination; that is, when will individual impact be realised to a degree that is noticeable and attributable to the influence of the GIS? We will discuss this issue of timing in impact assessment in more detail later in this chapter.

Another aspect of the bottom line assessment of a GIS's success lies in its relationship to organisational impact. As Schultz and Slevin (1979) have noted, any measure of system implementation success has to take into account some level of expected resultant increase in performance by adopting the technology. Top management must understandably seek some pay-off for their investment of time, monetary, and human resources in adopting GIS technology. Certainly, within the public sector, it is reasonable to expect municipal authorities to ask some tough questions regarding the expected gains from use of GIS prior to sanctioning its purchase. Campbell's (1993) study of GIS implementation in UK local government demonstrated that one of the most important perceived benefits of the GIS was its ability to improve information-processing facilities. Among the outcomes prized by her respondents was the GIS's ability to improve data integration, speed of data provision, access to information, and increased range of analytical and display facilities.

Unfortunately, the determination of organisational impact through the adoption of GIS technology may be highly uncertain and has been the source of enormous debate within the information systems field for years. In fact, many researchers are entirely uncomfortable with the idea that an information system must generate some concrete pay-off to its organisation, preferring to stress other ways in which the system has impacted on the organisation. For example, they argue, if the system has caused an organisation's employees to apply computerised technologies to new problem areas that had previously been ignored or thought inappropriate for its use, the system has demonstrated positive returns. Further, cost-reduction figures may be difficult to generate as they are often measured in terms of time saved through using the GIS over other, traditional work methods.

Assessing a system's impact on organisational effectiveness may further be problematic in that it is often hard to separate out the 'true' effects of the information system from other biasing or historical moderating effects. In other words, the organisation is not simply standing still while a new GIS is brought up and running. Other external events are affecting the organisation and its ability to function effectively. Consequently, it may be difficult to parse out the 'true' impact of a GIS from other activities and events that are influencing the operations of the organisation. These and similar issues have combined to make any accurate assessment of organisational impact difficult at best.

9.8 The assessment of success over time

One of the points that has been repeatedly stressed in this chapter is the necessity of developing an adequate programme in terms of knowing when to determine system implementation success. As previously mentioned, there are definite benefits involved in waiting until after the system has been put in place and is being used by its intended clients before assessing the success and impact of the system. On the other hand, we must be careful not to wait too long to determine system impact and implementation success so that the possibility exists of other organisational or external environmental factors influencing the organisation's operations to the point where we are unable to determine the relative impact of the GIS on operations.

Figure 9.1 illustrates the difficulty faced by the implementation team and researchers (Pinto and Slevin, 1988). This figure shows a simple time line, demonstrating the point at which various aspects of implementation success should be evaluated. Note that the time line as drawn has deliberately avoided any specific metrics; that is, we cannot posit the appropriate number of days, weeks, or even months that would necessarily have to elapse for each of these assessments. Rather, the time line simply illustrates the temporal nature of many of the dependent measure assessments. At the earlier stages in the implementation process, the typical assessments of success tend to revolve around issues such as system use and quality; that is, the GIS has been installed and is starting to be used by organisational members who begin to make preliminary evaluations of its quality. (Note from Table 9.2 how each of these issues is more comprehensively defined.) At this early stage, 'success' often rests with gaining the acceptance of organisational members to the new system and securing their willingness to actually use it in their activities. Note that it may be too early to make any accurate

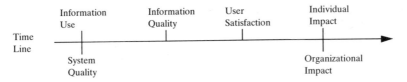

Figure 9.1 Assessment of system success over time.

determination of satisfaction or impact. Rather, at this stage, a number of projections regarding the system's success are being made. The GIS is examined in terms of its technical capabilities – have pilot project results been satisfactory? – including information use and system quality.

As the time line continues on to the right, additional aspects of system implementation success may be more accurately assessed. For example, once the system is up and running and is generally in use, it may now be possible to make some determinations about information 'quality'. In other words, as we become more knowledgeable about the system due to our continued use of its various features, we are in a better position to accurately gauge the quality of the information that it produces. An acceptably high level of information quality, coupled with the earlier measures of information use and system quality, can be a strong contributor to overall user satisfaction with the system – the state where the user of the GIS has begun to weigh the evidence of system value, and hence is more likely to have positive (or negative) feelings about the GIS. The important point to note here is that research assessing recently installed geographic systems may too quickly seek to acquire measures of user satisfaction, before the system's users can reasonably be expected to have enough data or exposure to the system to form a valid opinion.

Finally, at the end of the time line are issues of system impact on individual and organisational operations. As we have noted, 'impact' refers to the positive benefits that both individual users and the organisation as a whole derive from their use of the GIS. It seems clear that assessments of the benefits derived from a GIS are only apparent following widespread use of the system within an organisation. Consequently, as with the case of user satisfaction, we cannot reasonably expect to gain meaningful assessments of individual and organisational impact until enough time has transpired to allow the members of the organisation to arrive at informed opinions. Indeed, one should look with a degree of scepticism on research studies that purport to measure impact immediately following installation of a GIS. Such studies are far more likely to be measuring some initial organisational excitement regarding the capabilities of the GIS than they are to measure true impact.

The implications of this model for conducting implementation research are important because they suggest that our goal must be, where possible, to take into account temporal issues when assessing GIS implementation success. While it is beyond the scope of this chapter to posit the precise points in time when each of these various dependent measures should be determined, the time line in Fig. 9.1 does serve to illustrate the complexity involved in accurate implementation success measurement. Simply taking measures of these various items at one point in the implementation process, without allowing for the moderating effects of time, may

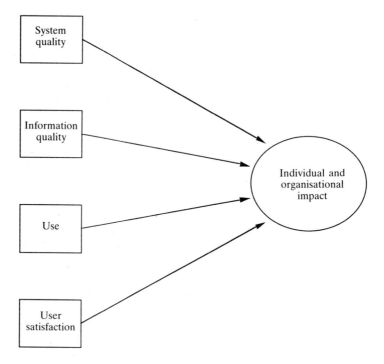

Figure 9.2 Modelling the determinants of system implementation success.

lead to misleading findings and inaccurate conclusions. A more conservative but likely, more meaningful, strategy is to allow for periodic assessments of various components of system implementation success over a time frame that will allow respondents to more accurately judge such qualitative issues such as information quality, user satisfaction, and individual and organisational impact.

Figure 9.1 suggests that while periodic assessments of the current state of the implemented system are important, an accurate determination of the ultimate success or failure of a GIS implementation is equally important. The difficulty lies in attempting to find a suitable reference point when the system has been transferred to the clients, is up and running, and is making some initial impact on organisational effectiveness. One of the important benefits of using such a 'post-installation' system assessment point is that it drives home the point to many implementation researchers and practitioners that the GIS implementation challenge does not end when the system is acquired and set up. In fact, in most instances, there is still a lot of hard work ahead. Implementation managers can foreshorten some of their time with post-installation user involvement depending upon the degree to which they and other relevant implementation team members consulted with clients at various earlier stages in the adoption process. However, it is important to point out that an accurate determination of the ultimate success or failure of an information system largely rests with those organisational factors, system use, user satisfaction, individual impact, and organisational impact. The common denominator underlying each of these factors has to do with the desire to fulfil the client's needs: matching the system to the client rather than attempting to alter the client's needs to fit the system.

9.9 Implications for research strategies

The discussion in this chapter has highlighted a number of salient issues that GIS researchers need to consider when attempting any sort of geographic system implementation assessment. These points are enumerated below.

1 *Installation is NOT implementation*.   One of the enduring difficulties with much of the research that is ostensibly done on GIS implementation is that it is essentially mislabelled: it could more properly be termed 'installation' research. The problem is that with many researchers, there is a basic difficulty in distinguishing between the activities necessary to acquire and install a GIS and those that pertain to its implementation within the organisation. A recent paper on GIS diffusion by Onsrud and Pinto (1993) attempted to make this point by distinguishing between a content model of GIS implementation (that is, those organisational and behavioural factors that contribute to implementation success) and a process model of the acquisition and installation process. A first, necessary condition of implementation research demands that a study's principal authors first make clear in their own minds the distinctions between those factors important to acquiring a GIS and those that relate to its implementation.

2 *Any GIS is only as good as it is used*.   What is the bottom line for any implementation effort? If we are willing to accept that 'successful' implementation can be measured on the basis of installing and bringing a system on-line, without any subsequent regard for its acceptance and use, we are misinformed. As Schultz and Slevin (1975) noted, the technical validity of a system is only the first step in the process of implementation. It is in successfully navigating the idea of organisational validity that we begin to understand the role of the client in determining success.

One of the most important points that past research and experience has taught us is that the client is the ultimate determinant of successful system implementation. This lesson, so fundamental to practising managers, is one that continually escapes the attention of researchers and information system theoreticians and must be continually relearned. We continue to deal with culture that is fascinated with the latest technological advances. One has only to pick up a newspaper or business magazine to see wide reference being made to the impending 'information superhighway', despite the fact that the average person may be over a decade away from realising many of the practical benefits of this family of technologies. In our drive to innovate, there is a very real danger that we will begin to pursue technology for technology's sake, rather than working to create systems that have practical and useful features. The oft-repeated statement, 'Of course it will be used – it's state of the art!', reveals a high level of naïveté about how innovations are received and perceived by the average person. It is vital that GIS implementation researchers understand the importance of client acceptance and use as a necessary condition of implementation success.

3 *Successful implementation may require system and client modification*.   A fundamental finding of Schultz and Slevin's earlier work referred to the concept of 'organisational validity'. What is important about this concept is the implication that any new GIS must be 'right' for the organisation toward which it is targeted. Further, the 'right' system refers to the importance of matching the needs and attitudes of the client organisation to the new technology. Many firms acquire geographic information systems that are under-utilised because they were inap-

propriate for the target organisation. However, it is important that the reader understands that an 'inappropriate' system is often not the direct result of technical difficulties or performance characteristic flaws. Many times a GIS will be perceived as inappropriate because it does not conform to the attitudes and value systems of the majority of organisational members. In other words, the cultural ambience of a firm must be taken into consideration when determining their technological needs.

We argue that the process of developing greater organisational validity for a GIS innovation involves a process of mutual adaptation between the system and the client organisation. Significant preliminary work is required from the project manager and team members as they scan the client and objectively assess attitudes and needs regarding an information system. If the team determines that it is not feasible to implement a GIS within the current organisational context, they need to begin formulating plans for how to create a more supportive environment. That process may require either modifying the GIS to suit the technological needs of the company, engaging in large-scale training programmes within the company to create an atmosphere of acceptance, or both. Unless project managers and their teams work to address potential problems of organisational validity, their highly developed and technically sophisticated systems are likely to fail without having been given a sufficient chance to succeed.

4 *When assessing 'success', give the system time to be incorporated into the client's operations.* In assessing the performance of an information system implementation, GIS researchers are faced with a significantly complex task. Because so much depends upon the acceptance and use of the system and its resultant impact on the client organisation's operations, it is often of very questionable utility to assess the system's impact before it can, in some sense, be determined. Too often, we have sought quick assessments of system impact as part of a general data-gathering process, whether 'impact' was yet a meaningful variable or not. A more appropriate methodology might be to sample an organisation's implementation effort at various points over some extended time frame following installation to gain better-informed data from the sample population being researched. Put another way, it may be possible to measure, with reasonable accuracy, such issues as system performance and client use at an early point following installation because they are more immediately derived success measures. On the other hand, any estimates of system impact made too early in the implementation process are likely to be misleading at best and utterly wrong at worst.

The obvious difficulty with making any sort of *post hoc* impact assessment is that of determining when best to make such an analysis. In other words, how long should a system be in place before it is analysed for its utility to organisational operations? Any new system will require a 'shake down' period while clients learn how to use it and adapt it to their activities. At the same time, they are also likely to be learning the various strong and weak points of the GIS and hence are usually unable to formulate an accurate assessment of the system for some time. On the other hand, the longer they take to assess implementation and performance success, the greater the likelihood that other intervening variables will interfere with their ability to give an honest appraisal of the GIS. For example, as new technological breakthroughs occur, any current GIS will begin to look old and increasingly cumbersome to its users, particularly when compared with the capabilities that new systems offer.

Clearly, a trade-off must be made between assessing implementation success too quickly and waiting too long. Whatever decision rule is adopted by GIS researchers, they need to make it with due regard to the various trade-offs that exist and the decision criteria must be applied consistently across all implementation efforts of a similar nature.

5 *Consider who stands to gain and lose from assessing GIS 'success' – remember the politics of the organisation.* An important complicating factor needs to be considered when seeking to determine GIS implementation success: the problem with attempts to develop a rational approach to assessing information system implementation success in the face of the irrationality that often accompanies organisations. This irrationality is usually manifested in examples such as the power and politics that accompany organisational activities, where one or more parties are intent upon furthering their own goals, even at the expense of the overall organisational good (Pinto and Azad, 1994). Normative models of how organisations ought to function are often notorious for failing to describe reality. Consequently, one facet of determining successful GIS implementation must be to consider the potentially self-serving effects of any party's willingness to label a GIS as either successful or failed. For example, from a power position, is it expedient for one organisation actor to dismiss a newly installed GIS because he perceives that it furthers the goals or power of another departmental manager? Within the context of geographic information sharing, Masser and Campbell (1995) made a similar observation in noting that 'organisational and political factors apparently offset in many instances the theoretical benefits to be obtained from structures that seek to promote information sharing' (p. 247). There is clearly no reason to suppose that those same factors will not affect attempts at posing overly rational methods for obtaining information on new system implementation success.

9.10 Conclusions

Much has been written and continues to be produced in efforts to develop a parsimonious model of geographic information system success. Any efforts toward a unifying model have to take into account aspects of the system itself: Does it work? Is it technically valid and usable? Further, unless the GIS is used in organisational activities and becomes an integral part of the organisation's operations, its impact is bound to be lessened. Finally, the 'bottom line' of any newly installed geographic information system lies in its impact on individual and organisational operations, in spite of the difficulties involved in attempting to develop an accurate and generalisable metric of information system impact.

This chapter has sought to offer a framework around which the various ideas of GIS success could be analysed. Just as there has been a profound increase in the number of studies that have begun to address issues in GIS implementation, there is a concomitantly compelling need to create a model of implementation success that can serve as a basis for future research. Without some generally accepted methodology for data collection and model for measuring implementation success, we in the research community face the very real danger of recreating, year after year, more research that is idiosyncratic and non-generalisable. Such a state is bound to ultimately damage our ability to accurately analyse geographic information system implementation and will simply marginalise our research.

What should be the appropriate response of the research community to this implicit threat? We suggest that one necessary avenue concerns working to create a comprehensive model of implementation success that may be applicable to all researchers engaged in implementation research. Once we have established a baseline set of dependent measures, it will be possible to generalise our findings across studies, nationalities and other limiting criteria; in effect, to make sure that all research is being performed on a similar basis that offers us the ability to compare and draw conclusions from as wide a research pool as possible. It is only through our concerted efforts to develop such a model of implementation success that these goals can be realised. As long as GIS researchers and scientists find it necessary to continually 'reinvent the wheel' in conducting their studies, we will continue to face the problems inherent in trying to draw a collective set of conclusions from implementation research.

It is necessary to note that managers and readers who are currently content with their system, regardless of its inability to measure up to several of the success determinants noted here, need to conduct a re-evaluation of their GIS and its true centrality to their department's mission and operations. Unless due care is taken to ensure that a GIS can be positively evaluated according to the criteria here established, it is unlikely that it is living up to its full potential – and neither, for that matter, is the department using it.

Acknowledgements

An earlier version of portions of this material was published in a chapter entitled 'Implementation Success and Failure' by Jeffrey K. Pinto in *Successful Information System Implementation: The Human Side* (Upper Darby, PA: Project Management Institute). Permission of the Project Management Institute to reprint this material is gratefully acknowledged.

References

ARNOFF, E. L. (1971). Successful models I have known, *Decision Sciences*, 2(2), 141–8.

CAMPBELL, H. J. (1993). GIS implementation in British local government, in Masser, I. and Onsrud, H. J. (Eds), *Diffusion and Use of Geographic Information Technologies*, pp. 117–46, Dordrecht: Kluwer.

DeLONE, W. H. and McLEAN, E. R. (1992). Information systems success: The quest for the dependent variable, *Information Systems Research*, 3, 60–95.

GINZBERG, M. J. (1978). Finding an adequate measure of OR/MS effectiveness, *Interfaces*, 8(4), 59–62.

KEEN, P. G. W. (1980). Reference disciplines and a cumulative tradition, *Proceedings of the First International Conference on Information Systems*, 9–18.

LUCAS, H. J. Jr. (1975). Behavioral factors in system implementation, in Schultz, R. L. and Slevin, D. P. (Eds), *Implementing Operations Research and Management Science*, pp. 203–16, New York: Elsevier.

MASSER, I. and CAMPBELL, H. J. (1995). Information sharing: The effects of GIS on British local government, in Onsrud, H. J. and Rushton, G. (Eds), *Sharing Geographic Information*, pp. 230–49, New Brunswick, NJ: Center for Urban Policy Research.

MITROFF, I. I. (1975). On mutual understanding and the implementation problem: A philosophical case study of the psychology of the Apollo moon scientists', in Schultz, R. L. and Slevin, D. P. (Eds), *Implementing Operations Research and Management Science*, pp. 237–52, New York: Elsevier.

ONSRUD, H. J. and PINTO, J. K. (1991). Diffusion of geographic information innovations, *The International Journal of Geographic Information Systems*, **5**(4), 447–67.

ONSRUD, H. J. and PINTO, J. K. (1993). Evaluating correlates of GIS adoption success and the decision process of GIS acquisition, *Journal of the Urban and Regional Information Systems Association (URISA)*, **5**(1), 18–39.

PINTO, J. K. and AZAD, B. (1994). The role of organizational politics in GIS implementation, *Journal of the Urban and Regional Information Systems Association (URISA)*, **6**(2), 35–61.

PINTO, J. K. and SLEVIN, D. P. (1988). Project success: Definitions and measurement techniques, *Project Management Journal*, **XIX**, No. 1, 67–72.

ROGERS, E. M. (1983). *Implementation of Innovations*, 3rd Edn, New York: Free Press.

SCHULTZ, R. L. and HENRY, M. D. (1981). Implementing decision models, in Schultz, R. L. and Zoltners, A. A. (Eds), *Marketing Decision Models*, pp. 123–34, New York: North-Holland.

SCHULTZ, R. L. and SLEVIN, D. P. (1975). Implementation and management innovation, in Schultz, R. L. and Slevin, D. P. (Eds). *Implementing Operations Research and Management Science*, pp. 3–20, New York: Elsevier.

SCHULTZ, R. L. and SLEVIN, D. P. (1979). Introduction: The implementation problem, in Doktor, R., Schultz, R. L. and Slevin, D. P. (Eds), *The Implementation of Management Science*, pp. 1–15, New York: North-Holland.

VANLOMMEL, E. and DEBRABANDER, B. (1975). The organization of electronic data processing (EDP) activities and computer use, *Journal of Business*, **48**(4), 391–410.

CHAPTER TEN

Leading and Managing the GIS Implementation Process: Beyond the Myth of the GIS Champion

BIJAN AZAD

10.1 Introduction: the importance of champions

A factor that has been strongly linked to the success of geographic information systems (GIS) implementation efforts is the presence of champions (Onsrud and Pinto, 1993). In general, this individual informally emerges usually in an organisation and makes a significant contribution to the success of GIS in the organisation (Montgomery and Schuck, 1993). Over 30 years ago in a seminal article, Schon (1963) identified the role of champions in the successful adoption of 25 military innovations. His conclusion was that in order to overcome the indifference and resistance that major technological change tend to bring about, a champion is needed to identify the idea as his or her own, to promote the idea actively and vigorously through informal networks, and to risk his or her position and reputation to ensure the innovation's success. Schon postulated that, '. . . the new idea either finds a champion or dies' (p. 84).

There is an abundance of studies in the field of innovations that have found strong support for Schon's position, that innovation success is closely linked with the presence of a champion (Maidique, 1980; Roberts, 1969; Schon, 1963, 1967, 1971). However, there are two significant criticisms that can be levelled against these studies. First, rigorous and systematic approach to the study of the championship process is rare in the innovation literature at large and GIS in particular. Second, even this existing knowledge and understanding of the championship process has not been adequately adapted to the GIS technology arena. This chapter attempts to address these shortcomings by recasting the GIS champion topic in terms of leading and managing the GIS technological change and implementation effort. In particular, the chapter has five goals: (1) to find out who the GIS champions are based on theoretically guided empirical research; (2) to discover first hand what are the central functions performed by champions; (3) to uncover any pattern to the set of activities they perform; (4) to suggest changes

– if any, based on the results of the case study – to the concept of innovation champion based on the evidence presented in this chapter; and (5) to make concrete recommendations regarding future research needs that bear on the important question of GIS champion.

The chapter focuses on the following aspects of the innovation championship in organisations: 'natural' emergent definition of champions by their peers; their central functions; and the structure of activities they engage in to see the GIS innovation through its implementation. Personality characteristics, leadership behaviours, and influence tactics are frequently discussed in the innovation (Doig and Hargrove, 1987) and champion literature (Kanter, 1988; Reich, 1987; Westley and Mintzberg, 1989). However, for the most part results of different studies have not included the central functions of and the activities of GIS champions as a generic process for comparison across situations. Consequently, little empirical evidence exists with respect to the championship process. What appears to be needed is the *development and testing of a model that integrates and puts forward an integrated process of championship functions and activities*.

The relationships posited in the model that is presented in this chapter will be inductively derived from the details of a 6-month extended case study with the aid of literature in the areas of entrepreneurship, organisational leadership, and influence (Bellman, 1992; Cohen and Bradford, 1991; Cornwall and Perlman, 1990; Pinchot, 1985; Tichy, 1983; Tichy and Devanna, 1986; Tichy and Ulrich, 1984).

10.1.1 GIS and champions

One cannot imagine a conversation about a GIS project that somehow does not touch upon the role of the GIS champion. Therefore, we can expect to see a multitude of references to the role of the GIS champion everywhere in the published literature. However, a cursory review of the literature on GIS implementation by this author revealed only a dozen or so cases that clearly point to the GIS champion. In fact, the only explicit definition of GIS champion functions was presented by Montgomery and Schuck (1993):

> Virtually all successful projects have a project champion. This individual, typically an executive in one of the organisations, is characterised as someone who has great personal interest in the implementation of a GIS and who is either familiar with the technology or is actively learning about it. He or she is often able to secure executive-level interest and support for a GIS project, based on technical practicality and economics, from one or more organisations. The champion identifies and recruits other potential participants. (p. 170)

Onsrud and Pinto (1993) identified the champion as one of the important correlates of GIS success in a large-scale survey of GIS users. The reference to the GIS champion appears in a variety of other works but only implicitly or indirectly (Antenucci *et al.*, 1991; Huxhold, 1991; Korte, 1992; Public Technology, Inc., 1991).

Upon close scrutiny, the keen observer discovers a lot of 'mythology' associated with the roles and tasks of the GIS champion. The observer may start thinking that there are no written guidelines on a set of activities that may be undertaken only that they exist, in the mind of the 'magic' GIS champion. Despite the varied interpretations of champions' functions, descriptions of what they actually do through the (GIS) implementation process has been presented only in rather general terms. It has been previously presented that, generally, champions:

- advocate new ideas and develop proposals (Rogers, 1983),
- define and re-frame problems (Doig and Hargrove, 1987; Rogers, 1983),
- specify innovative solutions (Rogers and Kim, 1985),
- broker the ideas among the many actors (Schon, 1967, 1971),
- mobilise opinion (Rogers, 1983); and
- help set the innovation decision-making agenda (Rogers and Kim, 1985).

The goal in this chapter is to build on this literature in order to develop and expand a more elaborate description of GIS champions' functions and activities based on empirical evidence and sound theoretical arguments. What specific functions do GIS champions serve and how? As central figures in the GIS implementation process, what are the scope and depth of their activities in that process? Do they work alone or with others? What strategies and tactics do they employ? In particular – to do with the case study in this chapter – is there a pattern in their functions and activities that is associated with enterprisewide GIS (organisation-wide) implementation?

10.1.2 Conceptual foundations

Before describing and analysing a particular case, it is useful to lay the conceptual foundation for this work. In this chapter the focus is on the government organisations deploying GIS technology, therefore we are concerned with 'public entrepreneurship' or 'public-sector championship' as a process of introducing GIS technology (innovation) to public-sector practice. In general, innovation represents a new combination of things that creates a disjuncture from standard operating procedures and the routine response of current systems. The newness of the idea is determined by its context, 'by the relevant unit of adoption' (Zaltman et al., 1973, p. 10). An idea is 'new' if those involved in developing and implementing it consider it to be new, although the idea may have been earlier developed elsewhere. Asking whether an idea is new in a particular context is different from asking for the source of the idea in that context. While both are important, it is the former that becomes a defining characteristic of an innovative idea (Kimberly and Evanisko, 1981; Rogers, 1983; Tornatzky et al., 1983; Zaltman et al., 1973).

The new idea is then translated or transformed into a more concrete reality, a *prototype*, which is a more developed and elaborate statement of the original idea. In GIS technology innovation, the prototype is usually called a 'pilot application' which is ultimately designed into a more robust system, an On-Line Transaction Processing/On-Line Analytical Processing (OLTP/OLAP) system. In this manner, if the pilot project is transformed into an OLTP/OLAP system then it is possible to state that the implementation process has started.

This evolution from innovative pilot to an OLTP/OLAP system suggests that various functional stages or requisites are necessary to produce and sustain an innovation – a GIS. Functional requisites, as a set of functions, goals or forms, assume that one knows what a 'developed' innovation looks like and then asks what was involved in getting it to that evolved state. Although some variation exists among the functional requisites or stages, four appear to be central.

- *Initiation* – the stage when the innovative idea develops and emerges.
- *Adoption* – the stage when the innovative idea evolves into a pilot application.
- *Implementation* – the stage when the innovative idea has been approved in theory and is now expressed as an OLTP/OLAP which is tested in operations.
- *Institutionalisation* – the stage when the innovative idea becomes established practice to the point where it is no longer considered an innovative idea.

Interpreting GIS championship in this manner enables definition of GIS champions generally as those individuals who participate in the first three stages: they develop a new application for GIS technology, translate it into a pilot project, and then help to implement it into public sector practice as an OLTP/OLAP system.

The literature also describes championship in many different ways: as technology champions, political champions, programme champions, bureaucratic champions, administrative and executive champions, and issue champions. Rarely are these terms defined, much less distinguished from one another. Interpretation becomes difficult when researchers use the terms interchangeably without clarifying either meanings or whether and how the terms can be distinguished from one another (Doig and Hargrove, 1987).

This diversity, while it provides rich examples of individuals engaging in interesting activities in the public sector, does not facilitate systematic study of and comparison among technology champions (and GIS in particular). This lack of conceptual clarity makes distinctions between champions and other actors involved in the GIS implementation process problematic. For example, how does one differentiate technology champions from project managers or others engaged in the GIS implementation process? Is there a difference in the first place?

In response to these conceptual problems in the literature and to the questions they raise, this chapter takes a 'broader' view of GIS championship. A GIS champion must have not only an innovative GIS application idea, but must also present a vision of pilot application design and translate that idea into a more robust application framework. Working with those who have the formal power and resource control (typically referred to as top management), the GIS champion seeks acceptance of the innovative pilot application into a more robust application and the eventual organisation-wide deployment and implementation of the application into an OLTP/OLAP. This distinguishes GIS champions from others in the GIS implementation process because the former are involved in three phases of the GIS implementation process – creation, design, and implementation – as they promote their technology applications innovations and ideas.

Decisions are not made independently of a context; they are strongly influenced by the positions and relative power of decision makers in their relationships to one another (Pfeffer, 1981). Therefore, a further analytic refinement of GIS champions distinguishes between champions on the basis of their locations throughout the institutional system of government. The assumption is that championship behaviour is likely to be associated with the champion's base of power. Knowing that an individual's power is contingent upon resources and that the position is one element in the calculus of resources (Pfeffer, 1981), it is reasonable to postulate that a champion's activities and behavioural pattern are likely to be related to position in the GIS implementation process.

To lay the groundwork for future comparisons between types of champions, especially in terms of their behavioural patterns, this chapter concentrates on GIS project champions. These are individuals who work from within the formal governmental system to introduce, translate, and implement innovative ideas into public sector practice as illustrated in a particular case – a State Department of Transportation (SDOT) with over 10 000 employees, responsible for 34 000 miles of roads and 8 district offices.

10.2 Methodology

The study of GIS championship in this chapter is approached in a manner approximating a pseudo-longitudinal design during an 8-month consulting assignment. The basic design of the research was along the lines of 'naturalistic inquiry' (Lincoln and Guba, 1985) and 'participatory action research' (PAR) (Argyris *et al.*, 1982) using multiple methods of data collection in multiple units of the same (but large) organisation, satisfying Yin (1989) recommended criteria for this type of case study research. (The limitations of this design are discussed in the conclusions.) It is also consistent with the proposed use of case studies by Benbasat *et al.* (1987) as far as information systems research is concerned: (a) phenomena is examined in a natural setting; (b) data are collected by multiple means; (c) one or few entities (person, group, or organisation) are examined; (d) complexity of entity is studied intensively; (e) exploration, classification, and hypothesis development are featured; (f) experimental controls and manipulation are not involved; (g) changes in site selection and data-collection methods may take place during the research; (h) 'why' and 'how' questions may be addressed through operational links traced over time; and (i) focus is on contemporary events.

In addition, lacking a guiding substantive theory and a viable operational model at the outset of this study made a flexible methodological strategy imperative – one that could facilitate the discovery of useful conceptual units and appropriate operationalisations. Such an approach needed to be open-ended, aimed at uncovering and exploring critical linkages in the social system. In the absence of knowing *a priori* what these might be, more formalised methods, such as survey research, were inappropriate. Consequently, a 'naturalistic inquiry' (Lincoln and Guba, 1985; Schatzman and Strauss, 1973) was pursued in order to lay the groundwork for the development of a 'grounded theory' (Glaser and Strauss, 1967) about GIS champions and GIS championship functions and activities.

10.2.1 Data collection

Data collection, beginning in the Autumn of 1994 and ending in the Summer of 1995, included archival research, in-depth interviews, and observations of key discussion groups, meetings, and GIS-related – as well as information technology – gatherings, such as presentations, joint application development (JAD) sessions, and committee meetings.

A total of 158 interviews (77 of which were with managers) were conducted with the GIS actors (including those later 'discovered to be' champions): the assistant secretaries and functional area managers both in the central and district offices and their staff, the representatives from other executive departments, some legislative staff members, and some members of local government.

Archival material consisted of official SDOT documents from the agency, oversight correspondence, internal memoranda, and position papers written by the GIS champions, presentations, agency minutes and handouts of group meetings, letters, and studies from various consulting organisations.

Direct observations occurred during committee meetings and presentations and in settings that ranged from JAD sessions, to a university campus, to the state capital. The researcher was a participant-observer and project manager for the consulting team that assisted the SDOT in their implementation plan development and execution (an average of two meetings were held with client actors on a monthly basis from December 1994 to June 1995). The GIS Technical Committee (GIS–TC), charged by the agency and convened by the SDOT project manager and head of the GIS–TC, was to develop an enterprise-wide GIS 'plan' for SDOT. All functional areas in the GIS debate participated in these meetings, giving the author a natural setting in which to observe their interactions with one another.

10.2.2 Sample

Identifying the GIS champions for this study was an emergent process. The author also drew on previous case study experience (Azad, 1990). As events unfolded, a few individuals – rather than the project manager who was the initial focus of the research – surfaced as critical actors in debates on a new technology idea (enterprisewide OLTP/OLAP GIS applications). The author eventually identified these individuals as 'GIS champions,' using an evolved definitional framework.[1]

The GIS champions who were the focus of this study, eight in total, came from various positions in the state government, not just from within the SDOT:

1 a policy analyst with the state oversight information technology agency (OITA);

2 a manager with the right-of-way function of the SDOT;

3 a GIS-planning specialist with the office of transportation planning;

4 the agency database administrator in the information systems (IS) office;

5 a GIS specialist attached to the IS office;

6 a district CAD manager;

7 a district transportation planning specialist; and

8 the GIS–TC manager and contract manager for the GIS project who was head of survey/Global Positioning System group within one of the SDOT district offices.

The GIS champions met the *a priori* criteria adopted by this research for GIS championship: 'person(s) responsible for stewardship of GIS projects within the agency'. The ideas of the GIS champions were a departure from the solutions traditionally proposed by the SDOT staff. In interviews, members of the SDOT and other major GIS groups in the state clearly characterised the GIS champions' ideas and recommended changes as 'radical'. In comparison, the IS office described their own orientation as an 'orderly change that occurs in sequential fashion, not bold change without study and research'. The GIS champions, on the other hand, referred to their pilot application ideas efforts as 're-engineering'. Their goal was not to 'tinker around the edges', nor to take the incremental path advocated by the IS office. They sought to 'transform' the agency rather than 'improve' it through technology.

Corroboration for this characterisation of the GIS champions' ideas as innovative applications also came from a larger audience beyond the SDOT community. In the GIS literature OLTP/OLAP applications in GIS are also described as 'radical change' in contrast to others undertaken so far (Antenucci, 1995).

10.2.3 Data analysis

Data analysis for the research reported in this study was inductive rather than deductive because of the emergent, case study and pseudo-longitudinal design (Lincoln, 1985). The approach to category development was informal. Nevertheless, categories were progressively identified in an interactive process as the data were collected and analysed, following the recommendations of Glaser and Strauss (1967) for the descriptive phase of theory-building research. As a next step, these categories were reviewed by some or all of the GIS champions and modifications were made where appropriate to include their observations.

10.3 Case summary

Rather than calling for a GIS application within just one function, as the prevalent form of implementation, the GIS champions studied in this case espoused the innovative idea of an enterprise-wide GIS. This idea, developed from 1992–4, advocated a total revamping of the present piecemeal approach to GIS, beginning with a re-examination of its underlying assumptions or 'givens'. Enterprise-wide GIS actually represented a cluster of ideas:

- Empower users on the desktop to have access to the information they need based on its spatial attributes.
- Initiate an enterprise-wide approach to application and database development for GIS based on principles of data sharing and application cooperation.
- Decentralise computing power to the local districts offices of SDOT and users through the use of client/server information technology.
- Increase economies for data/application acquisition and development based on the promulgation of standards for both data and applications.

The mechanism to drive this enterprise-wide GIS was the distributed client/server approach to data access. 'Client/server is the mechanism that drives all other changes . . . It provides the incentive to change,' said one GIS champion. It was anticipated that giving users the options of enterprise-wide distributed GIS versus *ad hoc* GIS would facilitate the adoption of the GIS standards. The *ad hoc* and stand-alone efforts of several workgroups making forays into the GIS arena was reaching its limits due to the lack of intra-organisational coordination and the reinventing of a multitude of applications. By some accounts three different SDOT district offices as well as the central office were developing an application suite that would assist in the assignment of project to different highway segments. Such news was not welcome as far as the IS office was concerned, trying constantly to satisfy different user groups' demands for 'tape extracts' of mainframe data on a weekly basis. The more important drawback was the 'reinventing' of the wheel in each office regarding this kind of application.

Initial efforts of the GIS champions to 'test' an enterprise-wide approach began in 1992–3 when they sought and won funding to create a SDOT-wide GIS-Technical Committee and a few pilot projects that encouraged experimentation with enterprise-wide GIS ideas. Pilot initiatives in 1992, although not originally identified with the enterprise-wide effort, were also incorporated. In retrospect, this laid the groundwork on which future GIS efforts would build, such as the project by a district to develop work-programme mapping. Although the application did not get to the users, it was considered a 'door opener' for future efforts.

In 1993, a major effort in which some of the GIS champions participated was the development of a topologically structured 1:24 000 scale map of the state highway system. This was a major turning point in the GIS history at SDOT because until then the only map that was state-wide in scope was a 'cartographic' product without any topological structure which was developed and maintained by the SDOT's survey mapping office.

Building on these initial efforts, in the spring of 1994 after intense lobbying, the GIS champions were able to convince the SDOT top management to launch an enterprise-wide GIS within SDOT.

10.4 Championship functions and activities

In this section a set of functions and activities for the champions are proposed based on the analysis of case study.

10.4.1 Functions

Our population of champions converged in their arguments (separately during the interview process) that all champions perform some or all of three functions. First, and foremost, they indicated that champions discover unfulfilled needs and select appropriate prescriptions – in this case enterprise-wide GIS – for how those needs may be met. That is, they are alert to opportunities. Second, as they seize these opportunities, champions bear the bulk of the reputational, emotional and frequently the career (read financial) risks involved in pursuing a course of action with uncertain consequences, that is, enterprise-wide GIS with, in this case, no track record of client/server integrated information systems within SDOT. Third, in pursuing these actions, champions must assemble and coordinate teams or networks of individuals and organisational units that have the talents and resources necessary to undertake the GIS innovation.

Discovering unfulfilled needs in the area GIS is not purely a technical activity and not necessarily difficult. Selecting the appropriate ways to satisfy those needs, however, often requires exceptional insight – and it is often the difference between success and failure. Champions indicated that they usually did not define what the unfulfilled needs were, but they were able to recognise the contextual nature of those needs and establish feasible approaches to meet them – in this case, duplication of efforts by different groups versus enterprise-wide GIS. That is, GIS needs were sometimes latent, and their nature was often not objective. Therefore, these needs existed primarily in terms of how they were defined and how they were conceived by the organisation and its institutional environment (funding approval agencies have been key in the SDOT case).

While our champions did not agree 100 per cent about the degree to which champions are defined by their willingness to engage in risky behaviour, it was clear that all champions, regardless of the locus of their activity within SDOT, bore the reputational and emotional risks involved in pursuing a technological course of action with uncertain consequences.

Besides discovering unfulfilled needs, selecting ways to meet them, and bearing the reputation and some of the career risks, all the champions agreed that they played a critical role in assembling the human, physical, and financial resources necessary to achieve the identified goals. Every champion is faced with the problem of assembling resources controlled by others, motivating members of a 'team' to work efficiently and minimising the tendencies to shirk. In particular, according to our champions, in the public sector the set of organisational problems that are faced are more severe than in the private sector. This is because the public sector champion must almost always create and maintain a *new* collective base to mount and sustain challenges to the existing organisational and institutional arrangement within and outside the agency.

10.4.2 Activities

GIS championship, like many other human activities, represents a complex, interrelated social process in which many activities occur concurrently and in which the conditions for action are only partly under the control of the actors being studied. This section describes – in the light of the case outlined above – the basic sets of GIS championship activities that emerged from the research. The activities presented here distil the essence of what these GIS champions actually did in this case. Due to space limitations, this section cannot do justice to the complexity and comprehensiveness of the GIS champions' actions. The following discussion of categories of activities does not follow an absolute chronology, nor should the activities be interpreted as a sequential set. Many activities were concurrent, and others were repeated over time. Instead, the activities are presented as logically and functionally required sets of actions that are part of the formulation and implementation phases of the GIS technology as they emerged in this case. The categories are as follows.

1 Idea generation activities

2 Problem-framing activities

3 Dissemination activities

4 Strategic activities

5 Demonstration project activities

6 Activities cultivating bureaucratic insiders and advocates

7 Collaborative activities with high-profile élite groups

8 Activities eliciting support from elected officials

9 Lobbying activities.

1 *Idea generation activities.* The GIS champions – as engineers, planners, policy analysts, researchers – traded in ideas, whether the development of wholly new applications or the enhancement of existing ones. For example, the ideas of the so-called New Mexico Pooled Fund study were key in the initiation process. This project involved the New Mexico DOT along with the Los Alamos Laboratory which together launched a large-scale pilot project to experiment with the development of GIS applications for transportation and to serve the requirements of the Intermodal Surface Transportation Efficiency Act of 1991 (ISTEA).

2 *Problem-framing activities.* Beyond idea generation, the GIS champions were compelled to engage in problem-framing and problem definition. In order to convince top management that their ideas deserved funding, the GIS champions needed to establish a clear link between the identified SDOT operational and policy problems and their proposed and preferred solutions – enterprise-wide GIS. There were two aspects to this. First, problem definition required identifying a current performance gap in various SDOT arenas. Second, the GIS champions had to convince top management that their solutions were preferable to others generated by the more traditional stand-alone approaches of the various workgroups.

3 *Dissemination activities.* In order to spread their ideas to as wide an audience as possible, the GIS champions employed a range of dissemination mechanisms. They wrote reports, position papers, articles and columns for newsletters. They taught workshops in state government environment. Interviewees cited telephone calls, personal contacts, articles and documents, and 'special meetings' that brought together GIS 'experts' with the state government leaders. In this regard a SDOT manager said of a GIS champion: 'He is phenomenal, the most effective organiser I've ever run into.'

4 *Strategic activities.* While not formalising their strategies into any written document, nor following any 'rational strategic planning process', the champions began to develop 'long-term strategies' to guide their overall efforts and 'short-term tactics' to cope with the changing political realities on a day-to-day basis. As one GIS champion commented:

> What we're doing — is both identifying short-term tactics and long-term strategies – or how we make this happen. Nothing like this happens that isn't orchestrated. This doesn't fall into place. You've got to decide where to push when you push, and what your position is.

5 *Demonstration project activities.* To gather evidence supportive of their ideas, two GIS champions – working with other change-oriented individuals – formed a group called GIS Applications Group (GAG). The group's purpose was 'to translate the momentum for GIS projects, into high potential application ideas for demonstration projects in order to test their efficacy and potential success' in the SDOT context (GAG Group Member, 1995). Results of the projects, if successful, would attract additional supporters and could then be used to argue for a more permanent status for the GAG and eventually for an agency-wide base map.

6 *Activities cultivating bureaucratic insiders and advocates.* The GIS champions were aware that strategically placed insiders within the bureaucracy would be important allies not only in the definition phase of the enterprise-wide GIS but also in the critical implementation phase. One example was particularly noteworthy in illustrating how the GIS champions developed a governmental network of like-minded allies. The GIS champions needed a change-oriented top manager to support an enterprise-wide GIS agenda. They were credited with 'working behind the scenes' with the governor's staff and key members of the Spatial Data Committee (SDC) to facilitate the appointment of the GIS–TC manager for SDOT, a self-proclaimed change agent.

7 *Collaborative activities with high-profile élite groups.* Elites, or 'thought leaders' as one GIS champion called them, were important to foster the innovation process. '. . . If you got the élites committed, the rest would fall into place' (GIS–TC member, 1995). Elite, high-profile participation was important for several reasons. First, it provided a source of prestige to support championship activities that needed testing in the SDOT GIS marketplace. Pilots and demonstration activities sponsored by the GIS champions were thus able to receive funding from the agency partially due to the credibility lent to their efforts by élite groups. Second, high-profile groups could contact those in control of resources to leverage the federal resources to fund pilots independent of SDOT funds, that would provide an 'objective assessment' of the enterprise-wide approach.

8 *Activities eliciting support from elected officials.* Cultivating a like-minded network within the governmental bureaucracy and among élite, high-profile groups outside of SDOT was vital but had to be complemented with support drawn from elected officials. Legislators and the governor were the vehicles through which enterprise-wide GIS funding requirements were translated into law. The opportunity came in the Spring of 1994. Responding to a comment from one of the governor's key policy advisers, who had mentioned that the governor's executive team was 'in the market for an overhaul of the information technology procurement', one GIS champion gathered together a group of other GIS champions to prepare a proposal. A meeting was scheduled for the next day with the governor's team. This two-to-three hour session, said one GIS champion, 'altered [the state's] GIS agenda'.

9 *Lobbying activities.* After the State Oversight Information Technology Agency adopted the enterprisewide approach to GIS, the championship efforts to persuade and sell the ideas of an enterprisewide GIS intensified and expanded in their scope. Lobbying activities began in earnest, focusing on the SDOT executives. One OITA analyst noted their impact:

This has been one of the few issues in all the years I've been here that I've seen develop that way . . . This developed almost in textbook fashion. The evidence in support of a comprehensive GIS approach was convincing! People came in support of that idea that you never would have expected.

10.5 Discussion and analysis

The discussion below addresses four related but distinct topics that were either targeted initially or emerged during the research process:

- Definition of championship functions
- The champions as a network of key stakeholders
- Activity structure for championship activities
- Congruence of results with innovation diffusion and organisational literature.

10.5.1 Championship functions

Based on the interview results and consultations with the eight identified champions, a number of functions were identified as core functions of championship. These functions were further refined after consultation with a number of sources on management and leadership (Tichy and Devanna, 1986; Westley and Mintzberg, 1989).

First and foremost, champions discover unfulfilled needs and select appropriate prescriptions for how those needs may be met – in many cases these are technological opportunities – that is, they must be alert to opportunities. Second, as they seize these opportunities, champions bear the reputational, emotional and, frequently, the career risk involved in pursuing a course of action with uncertain consequences. Third, champions must assemble and coordinate teams or networks of individuals and organisations that have the talents and/or resources necessary to undertake change.

Therefore, in summary the championship functions can be reduced to the following:

1 Alertness to opportunity.
2 Willingness to take risk in an organisational environment in pursuit of the opportunity.
3 Ability to coordinate the actions of other people to fulfil project goals.

10.5.2 Champion or network of champions

Other studies of information technology innovation and GIS diffusion emphasise the importance of one champion. Beath (1991) followed a similar and familiar path by focusing on one individual as an information technology champion in the organisation and proposed a series of hypotheses to enhance their mission in the organisation. The research discussed in this chapter, however, took a more open approach to the process of championship. In fact, the surprising result was that championship behaviour and/or functions are not limited to one individual. That is, it was discovered that the concept of champion as the 'domain' of one dynamic and forceful individual is a result of the heavily single biographical approach common to the study of championships and driven by a fixation on a 'great man' or 'great woman' approach to (technology-induced) change. Clearly, a small number of champions have truly transformed the GIS field by virtue of their actions (one can consult GIS magazines for witness). These are the likes of David Rhind, Larry Ayres, Nancy Tosta, Joel Morrison, Jack Dangermond, and so forth. There can be no doubt of the championship roles played by these individuals in their respective organisations and the industry.

However, this chapter adopted a more open approach not prejudging the existence of one 'key' champion. In fact, the research discovered that *champion-ship behaviour* is what is needed and that this is usually contained in more than one dynamic individual. In other words, there is a broad distribution of leadership talents and skills in the organisation and that GIS-related change can be propelled by actors other than the highly visible champion about whom the 'biographies' are written. The author was not able to determine the exact shape of distribution of leadership talents. However, if we assume that these skills are normally distributed across the large number of GIS implementation stakeholders within large organisations, then clearly the biographical approach to reporting championship behaviour has been 'guilty' of focusing our attention only on the tail end of the distribution. In this chapter, the champions' qualities are above average, but they are clearly not 'larger-than-life' individuals as is often reported. Campbell (1994) has reported similar scepticism of the notion of an 'omnipresent' GIS champion in successful implementations.

10.5.3 Activity structure of champions

The previous section categorised the depth and scope of the GIS champions' activities. On closer examination, the sets of activities can be grouped into logical categories or theoretical constructs (Glaser and Strauss, 1967) that capture in a more general way the range of championship actions. The four categories and the activities subsumed in each are outlined in Table 10.1.

Creative/intellectual defines the capability of the GIS champions to generate new ideas or to translate, apply and disseminate (broker) ideas from other domains into a new context. *Strategic* specifies the extent to which the GIS champions formulated strategies for action, both long term and short term, including heuristics for action. *Mobilisation/execution* marks the ability of GIS champions to take an innovative idea and move it through the governmental system (for example, legislative or administrative) and turn the agenda into action, which, in this case, culminated in legislative mandates. *Administrative/evaluative* indicates the level of the GIS champions' participation in the implementation process when pilot applications were organisationally tested and evaluated for their ultimate contribution and effectiveness.

The four categories of championship activities and the associated actions were *comprehensive and collective* in nature. Since their positions were in the lower rungs of the formal governmental system, the GIS champions had to be involved in a wide-ranging set of activities to empower themselves and their ideas. They could not rely on position power in government to press for funding of pilots. They

Table 10.1   Activity structure of GIS champions

Creative intellectual activities
1   Generate ideas
        Invent new GIS application ideas
        Apply models and application ideas from other GIS domains
2   Define problem and select solution
        Define performance gap
        Identify preferred solution alternative
3   Disseminate ideas

Strategic category activities
1   Formulate grand strategy and vision
2   Evolve political strategy
3   Develop heuristics for action

Mobilisation and execution activities
1   Establish demonstration/pilot projects
2   Cultivate bureaucratic insiders and advocates
3   Collaborate with high profile individuals/élite groups
4   Enlist elected officials
5   Form lobby groups and coordinate efforts

Administrative and evaluative activities
1   Facilitate application development administration
2   Participate in application development evaluation

---

had to search for funding sources and create novel but useful demonstration projects. They could not be guaranteed a place during official meetings where key policy discussions took place. Instead they had to cultivate bureaucratic insiders in order to keep themselves informed. Without automatically being included in the work-related networks among governmental officials, they had to forge their own contacts both in and out of SDOT.

One can also reason that without position or power in the formal governmental system, the GIS champions had to collectively work with and through others. While idea generation and problem definition can be carried out by a relatively small group of people, it was beyond the GIS champions' purview to sponsor legislation, bargain their way through OITA (governor's meetings), and ultimately pass, administer, and evaluate the laws. Under these circumstances, their influence could only be felt in working collectively with those who had the legal authority and responsibility to carry out such work, such as the OITA (which was the point of contact for the governor and legislators).

10.5.4   Congruence of results with innovation diffusion and organisational culture

There are several alternative explanations to account for the comprehensive and collective behaviour of the GIS champions. On the one hand, one could argue that these GIS champions provide an excellent example of what successful GIS champions do. In order for GIS champions to operate effectively, it would seem that they must be willing to engage in a comparable set of activities. These activities suggest a list of the necessary, although not sufficient, elements for successful GIS championship. Assuming this to be the case, one would expect to find a comparable set of activities for all GIS champions who are successful at introducing and implementing GIS innovative applications, even if operating in other domains and at other levels of government.

Another interpretation draws on the results of innovation-diffusion research to explain the comprehensive and collective character of entrepreneurial behaviour. In general, as Onsrud and Pinto (1993) explain, GIS innovations will be adopted more rapidly than others if they are perceived by receivers to have: greater relative advantage; greater compatibility; greater 'trialability'; greater observability and less complexity.

Although these are not the only qualities that affect adoption, research indicates they are the most important characteristics of innovations in explaining adoption rates (Rogers and Kim, 1985). It could be argued, therefore, that the radical innovation – enterprisewide GIS – espoused by the GIS champions was perceived by those within the SDOT to offer relatively little advantage, compatibility, 'trialability', and observability and be more complex than the practice it was intended to supersede. If these were the perceptions, one could have expected greater resistance to the introduction and implementation of such a radical idea. To counter this resistance, the GIS champions would have been compelled to engage in a comprehensive set of actions to ensure the continued viability of the innovative GIS applications approach. Thus, the greater the resistance to an innovative GIS application idea, the more comprehensive the GIS champions' activities need to be to overcome it. This hypothesised relationship can be summarised as the more radical the GIS innovative application, the higher the expected resistance to it and, therefore, the more comprehensive the activity structure of GIS champions needs to be to ensure viability of the GIS innovative application.

Given the described research design, it was not possible to ascertain which of the two interpretations is a more likely explanation of the GIS champions' behaviour in this particular case – the SDOT enterprise-wide GIS. However, the research documented what the GIS champions did and how they did it. Whether one, both or neither of these interpretations offers an accurate assessment of why the GIS champions acted as they did remains for future research to determine.

Others have proposed that the GIS championship behaviour is closely correlated with the 'organisational culture' (Campbell, 1994), and the importance of this 'culture' in information technology use (Barrett, 1992). In other words, it is the culture of an organisation which captures its ability to be adaptive and innovative and the existence of champions is just an indication of such capabilities. The organisational culture has developed into a strong subfield in the organisational behaviour since the publication of Peters and Waterman's *In Search of Excellence* (1982), and Kanter's *Change Masters* (1983) to name just two sources. The organisational culture may be an attractive concept for explaining such complex issues as leadership and management in the GIS implementation process. However, it is important to exercise care in using concepts that instead of being explanatory may perhaps be mystifying. In this regard, one of the seminal pieces in the field – Smircich (1983) – pointed out early on that the concept of culture has been borrowed from anthropology where there is no consensus on its meaning. Therefore, the elements and boundaries of organisational culture and its operational research significance should be delineated more clearly before it can be used as an explanatory tool. This is imperative if it is to serve as a basis for explaining GIS implementation success and failure as well as championship behaviour.

10.6 Conclusions

This case study examination of GIS championship identified three functions and a basic activity structure that potentially encompasses the work of eight GIS champions as they engaged in GIS technology innovation. The findings of this chapter are important insofar as they illustrate what may well be generic functions and activities associated with successful GIS technology innovation and championship. Using the methodology of 'naturalistic inquiry' (Lincoln, 1985) and 'participatory action research' (PAR) (Argyris et al., 1982) to lessen the problems associated with 'pure' and 'retrospective case histories', the research documented in detail what the main functions of GIS champions are and how they operated on a day-to-day basis to develop, translate and implement innovative GIS applications ideas into public sector practice.

From a research design perspective, however, this research is limited by analysis of a single innovation in a single setting. Given the in-depth, case study nature of this naturalistic inquiry, and PAR with its multimethod-research design, it was not possible to expand the research to include other sites and other GIS innovations. This limitation opens up the research to the criticism that these GIS champions were unique individuals operating in a very special context. As such, the study's findings would not be expected to be generalised to other GIS domains or other state, national or local settings.

One must remember, however, that naturalistic inquiry/PAR, 'Like experiments, are generalisable to theoretical propositions and not to populations or universes' (Yin, 1989). Comparable to laboratory research, this study does not represent a 'sample'. The goal has been an attempt to develop a model rather than to enumerate frequencies (that is, statistical generalisation). To establish the validity of this emerging model of GIS championship, one would need to design a series of studies to compare and contrast other GIS champions' functions and activity structures in other domains at different governmental levels. Consistency among findings would point to a fundamental set of functions and an activity structure characteristic of all GIS champions. As noted earlier, it is recommended that future research control the radicalness of the innovative idea to determine if the comprehensive nature of the activity structure is associated with the level of resistance to an innovation rather than associated with its success.

Follow-on research also should include comparisons between successful GIS champions and unsuccessful ones (those whose ideas do not survive the design and implementation phase of the innovation process). Are the functions and activities between the two groups different and, if so, how do they differ? A common set of functions and an activity structure for GIS champions moves the field closer to a theory of GIS championship that would enable specification of the necessary elements of successful technology policy innovation within organisations. In this context, GIS is only a technology enabler in the larger context of the re-engineering of the whole organisation which is far less technological and more managerial and organisational in nature. Future research on GIS championship should also focus attention on others involved in the innovation process.

The phenomenon observed here involved multiple actors in a complex transportation policy and operations environment using (GIS) technology to effect change. The emerging model of GIS championship suggests that champions can increase the probability of success by performing certain functions and developing an activity structure similar to the champions in this study. However, this model also recognises that these functions and the activity structure alone do not contain the necessary and sufficient elements for change. A perception of 'crisis', a growing recognition that a GIS application idea has merit, the existence of other skilful GIS actors to support the idea, these and other factors played a part in producing the GIS (innovation). Clearly, there are limitations to what GIS champions can do on their own.

In fact, the GIS champions in this research acknowledged their limits. They believed that the GIS innovation process 'could not be managed' because the results or outcomes were beyond the direct control of any single player or group of players. While *influencing the flow of events*, they did not control them. There were other important factors and players in the drama. This complexity suggests that future studies include the complete network of actors involved in the GIS technology innovation process. The research by Assimakopoulos (see Chapter 8 in this volume) which uses social network theory to study GIS diffusion in Greece is a step in that direction. These efforts would be expected to yield an even richer understanding of the GIS technology innovation process and the central part that GIS champions play in it.

Note

[1]All 77 managers (only) in the interviews were asked to identify one individual who in their mind had been responsible for the stewardship of GIS projects within the agency. The subsequent list was rank ordered and those mentioned eight times or more were included as champions. The number 8 was a 'natural' cut-off, since two of the champions were mentioned nine times and the other six each eight times. This accounted for 66 of the 77 responses. The rest were all individual name responses save one name that was mentioned twice.

References

ANTENUCCI, J. C. (1995). Tracking trends, *Geo InfoSystems*, 5(5), 52–4.

ANTENUCCI, J. C., BROWN, K., CROSWELL, P. L. and KEVANY, M. (1991). *Geographic Information Systems: A Guide to the Technology*, New York: Van Nostrand Reinhold.

ARGYRIS, C., PUTNAM, R. and SMITH, D. M. (1982). *Action Science*, San Francisco: Jossey-Bass Publishers.

AZAD, B. (1990). 'Implementation of GIS in Local Government: 12 Cases', unpublished dissertation proposal, Cambridge, MA: Massachusetts Institute of Technology.

BARRETT, S. (1992). 'Information Technology and Organizational Culture: Implementing Change', unpublished paper, Bristol: University of Bristol.

BEATH, C. M. (1991). Supporting the information technology champion, *MIS Quarterly*, **15**(3), 355–72.

BELLMAN, G. M. (1992). *Getting Things Done When You Are Not in Charge*, San Francsico: Berret-Hoehler.

BENBASAT, I., GOLDSTEIN, D. and MEAD, M. (1987). The case research strategy in studies of information systems, *MIS Quarterly*, **11**(3), 369–86.

CAMPBELL, H. (1994). How effective are GIS in practice? A case study of British Local Government, *International Journal of GIS*, **8**(3), 309–25.

COHEN, A. R. and BRADFORD, D. L. (1991). *Influence without Authority*, New York: John Wiley.

CORNWALL, J. R. and PERLMAN, B. (1990). *Organizational Entrepreneurship*, Reading, MA: Addison-Wesley.

DOIG, J. W. and HARGROVE, E. C. (Eds) (1987). *Leadership and Innovation*, Baltimore: Johns Hopkins University Press.

GLASER, B. and STRAUSS, A. (1967). *The Discovery of Grounded Theory: Strategies for Qualitative Research*, Chicago: Aldine.

HUXHOLD, W. E. (1991). *An Introduction to Urban Geographic Information Systems*, New York: Oxford University Press.

KANTER, R. M. (1983). *Change Masters*, New York: Simon & Schuster.

KANTER, R. M. (1988). Change-master skills: What it takes to be creative, in Kuhn, R. (Ed.), *Handbook for Creative and Innovative Managers*, New York: McGraw-Hill.

KIMBERLY, J. R. and EVANISKO, M. J (1981). Organizational innovation: the influence of individual, organizational, and contextual factors on hospital innovation technological and adminisrative innovations, *Academy of Management Review*, **24**(4), 689–713.

KORTE, G. B. (1992). *The GIS Book*, Santa Fe: OnWord Press.

LINCOLN, Y. S. (1985). The substance of the emergent paradigm: Implications for researchers, in Lincoln, Y. S. (Ed.) *Organizational Theory and Inquiry*, pp. 137–57, Newbury Park, CA: Sage.

LINCOLN, Y. S. and GUBA, E. G. (1985). *Naturalistic Inquiry*, Newbury Park, CA: Sage.

MAIDIQUE, M. (1980). Entrepreneurs, champions and technological innovation, *Sloan Management Review*, **21**(2), 72–83.

MERRIT, R. L. and MERRIT, A. J. (Eds) (1985). *Innovation in the Public Sector*, Newbury Park, CA: Sage.

MONTGOMERY, G. E. and SCHUCK, H. C. (1993). *GIS Data Conversion Handbook*, Fort Collins, CO: GIS World, Inc.

ONSRUD, H. and PINTO, J. (1993). Evaluating correlates of GIS adoption success and the decision process of GIS acquisition, *URISA Journal*, **5**(1), 18–39.

PETERS, T. and WATERMAN, R. (1982). *In Search of Excellence*, New York: Simon & Schuster.

PFEFFER, J. (1981). *Power in Organizations*, Boston: Pitman.

PINCHOT, G. (1985). *Intrapreneuring*, New York: Harper & Row.

PUBLIC TECHNOLOGY, INC. (1991). *Local Government Guide to Geographic Information Systems: Planning and Implementation*, Washington, DC: PTI.

REICH, R. (1987). Entrepreneurship reconsidered: The team as hero, *Harvard Business Review*, **73**, 77–83.

ROBERTS, E. B. (1969). Entrepreneurship and technology, in Gruber, W. H. and Marquis, D. G. (Eds) *The Factors in the Transfer of Technology*, Cambridge, MA: MIT Press.

ROGERS, E. M. (1983). *Diffusion of innovations*, 3rd Edn, New York: The Free Press.

ROGERS, E. M. and KIM, J. (1985). Diffusion of innovation in public organizations, in

Merrit, R. L. and Merrit, A. J. (Eds), *Innovation in the Public Sector*, Newbury Park, CA: Sage.

SCHATZMAN, L. and STRAUSS, A. (1973). *Field Research*, Englewood-Cliffs, NJ: Prentice-Hall.

SCHON, D. (1963). Champions for radical new inventions, *Harvard Business Review*, **69**, 23–32.

SCHON, D. (1967). *Technology and Change*, New York: Delacorte Press.

SCHON, D. (1971). *Beyond the Stable State*, New York: Random House.

SMIRCICH, L. (1983). Concepts of culture and organizational analysis, *Administrative Science Quarterly*, **28**(3), 339–58.

TICHY, N. M. (1983). *Managing Strategic Change: Technical, Political and Cultural Dynamics*, New York: Wiley-Interscience.

TICHY, N. M. and DEVANNA, M. A. (1986). *The Transformational Leader*, New York: John Wiley.

TICHY, N. M. and ULRICH, D. O. (1984). The Leadership challenge – a call for the transformational leader, *Sloan Management Review*, **26**, 56–64.

TORNATZKY, L. G., EVELAND, J. D., BOYLAN, M. G., HETZNER, W. A., JOHNSON, E. C., ROITMAN, D. and SCHNEIDER, J. (1983). *The Process of Technological Innovation: Reviewing the Literature*, Washington, DC: U.S. National Science Foundation.

WESTLEY, F. and MINTZBERG, H. (1989). Visionary leadership and strategic management, *Strategic Management Journal*, **10**(3), 17–32.

YIN, R. K. (1989). *Case Study Research: Design and Methods* (revised edition), Newbury Park, CA: Sage.

ZALTMAN, G., DUNCAN, R. and HOLBECK, J. (1973). *Innovation and Organizations*, New York: John Wiley.

CHAPTER ELEVEN

GIS Technology and Organisational Context: Interaction and Adaptation

ZORICA NEDOVIĆ-BUDIĆ

11.1 Introduction

The introduction of computerised geographic information systems (GIS) into organisational environments is a complex process of mutual adaptation. The new technology upsets the established organisational processes, while the organisational context and culture modify the technological set-up and use. Knowledge about the relationship between information system technology and organisations in which it is incorporated would facilitate GIS implementation and increase the likelihood of its success. This chapter summarises previous research on organisational change and diffusion of new technology and reviews the issue of technology-organisation fit. Results of a recent case study in four local government agencies are presented as evidence of an impact of organisational contextual factors on the process of GIS initiation and implementation. Socio-technical system design is the recommended GIS management approach alternative to traditional data processing.

11.2 Mutual impact and dependence: unanswered questions

The experience and research accumulated over the last few years show that the difficulties associated with introduction of geographic information systems (GIS) in local governments are primarily due to the complex and idiosyncratic nature of organisations as adopters of the new technology (Budić, 1993b; Campbell, 1991). Introducing GIS technology exerts change within an organisational system: organisational structure and procedures, flow of information and communication patterns may all be altered (Campbell, 1991); and social and political processes already taking place within organisations are intensified (Markus, 1983; Pinto and Azad, 1994) so that 'changing technology blows into the organisation like changing weather' (Kraemer *et al.*, 1989, p. 2). The activities of organisations adopting new technologies are not limited to choosing among hardware and software: what organisations actually make are 'important social choices contained within the organisation design undertaken to use the technology' (Davis and Taylor, 1979, p. 380). This is why the agencies that have already acquired GIS technology experience various degrees of success in implementing it (Budić, 1994; Campbell and Masser, 1991; Onsrud and Pinto, 1993). Paralleling the suggestion that implementation of an innovation (that is, GIS technology) may lead to organisational restructuring and new processes, are the claims that the process of technological adaptation to organisational internal environment, commonly referred to as 'reinvention', bears equal significance (Rice and Rogers, 1980; Rogers and Kim, 1985). The adopters change the innovation, particularly if it is more complex and irreversible. The reinvention provides a better fit with local circumstances and/or adaptation to changing conditions. The technology is customised.

Both the organisational redesign and the adjustment of technology to the organisational context aim at establishing an optimal fit between the organisation and the technology. The ultimate goal of the mutual adaptation is successful implementation of the technology, that is, full utilisation of GIS for performing organisational tasks and the internalisation of GIS into organisational structure and processes. However, the consideration of interaction between the technology and organisational context has been absent from GIS practice.

The need for 'organisational re-engineering' and the role of technology in the re-engineering have been recognised just recently (Strand, 1995), but only sporadically applied. Research, too, offers little on this important subject, particularly with respect to GIS technology in public organisations. Most of the recent research studied GIS diffusion (Budić, 1993a; Campbell, 1991; Campbell and Masser, 1991; French and Wiggins, 1990; Masser and Campbell, 1994; Wiggins, 1993) or analysed the impact of various GIS implementation factors (Azad, 1990; Brown and Brudney, 1993; Budić, 1993b; Budic and Godschalk, 1994; Nedović-Budić and Godschalk, forthcoming; and Onsrud and Pinto, 1993).

The singular focus of diffusion of innovation research on either the factors or the process resulted in two separate streams of research criticised for their narrow view (Downs and Mohr, 1976, 1979; Rogers, 1983). Combination of factor and process-oriented research in GIS, although still rare, promises better understanding of the interaction between organisations and technology and easier identification of potentials and constraints to successful diffusion of GIS technology (Azad, 1994; Onsrud and Pinto, 1993; Schultz *et al.*, 1987). The integrated approach challenges the traditional research boundaries between the system-rationalist, and institutionalist or interactionist perspectives of organisational behaviour (Craglia, 1994) but, by so doing, inherits the tension between the positivist and the phenomenological view of innovation diffusion.

The questions regarding the technology-organisation interaction abound: How does the introduction of GIS technology affect organisational structure and processes? How is the organisational change manifested? Do the efforts in matching technological and organisational aspects of GIS introduction result in more successful implementation of the technology? How do we determine whether the fit is good? Do particular technological and/or organisational configurations (that is, structure, processes) facilitate/impede successful implementation of GIS technology? Which organisational structures, processes or characteristics are conducive to different phases of GIS diffusion?

Through review of conceptual models and previous research on information systems (IS), management information systems (MIS) and organisations, this chapter looks at two areas that are important for understanding the organisational dynamics associated with the introduction of GIS technology and for devising effective GIS implementation strategies: (1) organisational change and diffusion of new technology; and (2) technology-organisation fit.

The prospective beneficiaries from the new practice and knowledge are numerous. Despite the intensified diffusion of GIS technology among local governments in the 1990s, there are still many organisations left to face the implementation process in future. The most recent survey conducted by the International City Management Association reported an average of about 30 per cent diffusion rate among US local governments (Sprecher, 1994). Masser and Campbell (1994) discovered a similar trend in the UK.

11.3 GIS diffusion and organisational change

11.3.1 Organisation and technology defined

Organisations can be defined as social systems consisting of individuals working cooperatively toward a common goal under authority and leadership (Heffron, 1989). Pfeffer (1981) and Danziger *et al.* (1982), however, see those individuals as driven by their own diverse interests, values and motivations. Organisations are structured and relatively stable organisms, often viewed as inflexible and driven by inertia, particularly those in the public sector (Rogers and Agarwala-Rogers, 1976; Scott, 1990). Despite the relative stability, organisations engage in adoption of innovation constantly (Rogers, 1983). They are dynamic systems vulnerable to failure of a new technology (Howard, 1985). Organisational internal context (which is the focus of this chapter) involves two major aspects (King and Kraemer, 1985).

1 Characteristics and change/stability of the organisation, including:
    (a) mission (functions, tasks, goals and objectives);
    (b) structure (complexity, formalisation and centralisation);
    (c) resources (size, budget, staff, technological, knowledge);
    (d) operations (size and scope of tasks and functions, structure of work, procedures and practices);
    (e) social relations (communication, coordination, coalitions, power basis and conflict); and
    (f) culture (shared philosophies, ideologies, myths, values, beliefs, assumptions and norms).

2 Motivation for incorporating GIS technology, with a recognised gap in organisational performance as the most common motivation behind organisational engagement in adoption of innovations (Feller and Menzel, 1977; Rogers, 1983; Zaltman *et al.*, 1973).

Technology, including computerisation, is defined by Heffron (1989) as the 'processes and methods that an organisation uses to accomplish its substantive goals' (p. 118). Because public organisations' work processes are based on knowledge rather than on machines and their raw material are people, paper and complex social, political and economic problems, the notion of technology is comprehensive. It encompasses knowledge, tools, techniques and action. Bretschneider (1990) found that the information processing and management reflects the differences between public and private organisations. Perrow (1967) categorises organisational technology into four groups depending on the analysability and variability of the raw materials: craft, non-routine, routine, and engineering.

11.3.2 Innovation process and organisational change

Two simultaneous processes occur during the incorporation of GIS technology in local governments: innovation process and the process of organisational change. The two are apparently interrelated and mutually dependent, but rarely observed or studied together. In addition, there is no practical coordination between them during the introduction of GIS technology.

The general diffusion of innovation paradigm views the innovation process as consisting of several steps with the innovation decision as the central element. Both Zaltman *et al.*'s (1973) and Rogers' (1983) model of innovation process in organisations distinguish two main phases: initiation and implementation, encompassing the steps that precede and the steps that follow the innovation decision (Table 11.1). Translated into the realm of GIS technology, the initiation involves activities related to gathering information about the technology, evaluating the information against the organisational needs and calculating costs and benefits associated with the technology. Implementation includes installment of GIS technology, database development, system maintenance and utilisation of the technology.

The process of organisational change parallels the innovation process. It consists of three steps: (a) unfreezing; (b) moving/change and (c) refreezing (Lewin, 1952). With initiation of an innovation the organisation experiences a 'disturbance' and leaves a temporarily achieved equilibrium; it is followed by the implementation process which is characterised by organisational response/reaction to the innovation; finally, after the needed change is achieved, the organisation settles into a routine and a state of new equilibrium. Both the innovation process and the organisational change process are described as consisting of differentiated and sequential phases. The reality, however, asks for a more flexible interpretation of the phases, as they overlap, change order, and/or do not necessarily happen.

Kwon and Zmud (1987) were among the few to relate the innovation process with the organisational change process and turn attention to the organisational dynamics corresponding to each phase of introducing the new technology (Fig. 11.1). Their framework correlates unfreezing to the initiation stage; change/moving to the adoption and adaptation; and refreezing means acceptance, use (performance, satisfaction) and incorporation of the technology. Mutual adjustment and adaptation between the technology and the organisational internal context occurs during the moving/change phase.

Table 11.1 Innovation process models: stages and substages*

Rogers (1983)	Zaltman, Duncan and Holbek (1973)
I Initiation	I Initiation
1 Agenda-setting	1 Knowledge-awareness
2 Matching	2 Formation of attitudes
'The decision to adopt'	3 Decision
II Implementation	II Implementation
3 Redefining/restructuring	1 Initial implementation
4 Clarifying	2 Continued-sustained implementation
5 Routinising	

*Several other authors modelled the innovation process with slight variations but essentially the same content: Hage and Aiken (1970) — evaluation, initiation, implementation, and routinisation; Nolan (1973) — initiation, contagion, control and integration; and Rice and Rogers (1980) agenda-setting, matching, redefining, structuring and interconnecting.

Unfreezing	Change	Refreezing
|	|	|
Initiation	Adoption	Acceptance
	Adaptation	Use/Performance/Satisfaction
		Incorporation

Figure 11.1 Organisational change process and innovation process (adapted from Kwon and Zmud, 1987).

The authors offer a comprehensive and thorough review of research findings from various fields of study to show the differential effect the various factors exert during the introduction of new technology, depending on the phase of the innovation process. Their review points to many unexplored areas, and raises awareness about a very important but neglected aspect of diffusion research. By examining each phase separately, they introduce a more precise and complex definition of the innovation process and thus partially address the problem that is claimed to have frustrated the building of the general theory of the diffusion of innovation (Downs and Mohr, 1976; Rogers, 1983; Zaltman *et al.*, 1973). Kwon's and Zmud's main points relevant for this study and the findings of other research are summarised in the following sections.

11.4 Research findings

11.4.1 General findings

Research on GIS diffusion provides evidence consistent to the findings of the research on other computerised technologies and information systems. A summary of the main factors held across the study areas as the best predictors of adoption follows.

1 Organisational size as it is associated with the availability of financial resources (budgets) and larger staff (Budić, 1993b; Campbell and Masser, 1991; Feller and Menzel, 1977; French and Wiggins, 1990; King and Kraemer, 1986).

2 Stability of organisational structure and small turnover in key personnel (Budić, 1993b; Campbell, 1990; Kanter, 1989).

3 Political stability of organisational environment, that is, the absence of organisational conflict (Budić, 1993b; Campbell, 1990; Croswell, 1991; Markus, 1983).

11.4.2 Stage-dependent findings

Previous research confirms the significant influence of the organisational context on the process of adoption of computerised technologies. Interestingly, it also points to differential and often bidirectional relationships, depending on the phase of innovation diffusion in which the effects are observed. Different characteristics of the organisational structure are conducive to initiation and different to the implementation of computerised technology (Heffron, 1989). Table 11.2 presents the summary of the research in this area.

The table shows the instability of findings in this poorly researched area of technological diffusion, but confirms that the characteristics of organisational structure and different phases of the innovation process are interdependent. The instability is often due to the inconsistency in definitions of various concepts addressed in this research. This is primarily the case with the concepts of adoption and implementation. The term adoption is often used in general diffusion research to identify acquisition of an innovation, an important initial action in the innovation process. Another way of defining adoption is as one of the final states of diffusion which signifies a successful implementation (acceptance, internalisation) of an innovation. More accurate and consistent information about the direction, strength and quality of the relationships between the organisational context and the process of technological incorporation would provide valuable pointers to current and future GIS developers.

11.4.3 County government case study (Budić, 1993b)

A multiple case study of four local government agencies, conducted in 1992, confirmed an association between outcomes of introduction of GIS technology and the context within which the process was taking place. The agencies studied were all located within the same county government in North Carolina and all had an opportunity to introduce GIS technology by 1988. Data were collected through interviews and observations in the following four agencies: School District's Transportation Department (TD), Tax Assessor's Mapping Section (MS), Planning Department's Community Assistance Section (CA) and Comprehensive Planning Section (CP). The organisational elements included: organisational function, structure, resources, operation, stability, social relations and motivation for incorporating GIS technology. Contextual elements differed among the four agencies and affected the GIS diffusion to a varying degree (Table 11.3). Evidence

Table 11.2 Organisational structure and innovation process (adapted from Kwon and Zmud, 1987)

	Complexity	Centralisation	Formalisation	Informal network
Kwon and Zmud (1987)	+	−	−	+
Initiation	+	+	+	+
Adoption	+	−	+/−	+
Adaptation	−	+	+	
Use	+	−	+	
Performance	−	−	−	
Satisfaction				
Incorporation				+
Pierce and Delbecq (1977)				
Initiation	++	− −	−	
Implementation	+	−	+	
Adoption	+	+	+	
Zaltman and Duncan (1977)				
Initiation	+	−	−	
Implementation	−	+	+	
King and Kraemer (1986)				
Adoption	−			

on each agency's behaviour with regard to the studied factors is provided in the Appendix to this chapter and in other published sources (Budic, 1993b; Budic and Godschalk, 1994; and Nedovi-Budic and Godschalk, 1996).

In summary, characteristics of the internal organisational context determined to a certain extent the success of both GIS initiation and implementation. Table 11.4 displays the general direction of the influence of particular factors by agency. The positive and negative signs indicate compatibility between the outcome of a phase and the conditions present. A positive sign means that a successful initiation or implementation took place in an environment which was already conducive to that particular process. A negative sign indicates conditions that were detrimental to either initiation or implementation of GIS technology and that actually impeded the diffusion of GIS technology. For instance, although most of the organisational factors in the Mapping Section were favourable for GIS implementation, it did not happen primarily due to an unstable and conflicting environment and lack of motivation for its application at the higher administration level. The positive outcome of GIS implementation in the Community Assistance Section may be attributed to available resources and motivation to implement GIS and conducive organisational structure, despite its instability and conflict. Initiation and implementation in the Transportation Department were undertaken under almost ideal internal organisational conditions.

The star and the box signs reveal a disparity between the outcome of GIS initiation or implementation and the organisational conditions. The star indicates a successful outcome in spite of unfavourable conditions; the box indicates a failure to initiate or implement a GIS in spite of favourable conditions. For instance, in

Table 11.3 Case study findings: internal organisational context of the four agencies

Factor	Transportation department (School system)	Mapping section (Tax office)	Community assistance section (Planning)	Comprehensive planning section (Planning)
Function	School bus routing	Property mapping	Information provision	Design, update and implementation
Structure	Large complex centralised	Large complex formalised centralised	Small fragmented decentralised informal	Small fragmented decentralised informal
Resources	Very good	Very good	Good	Inadequate
Operations	Large scale (about 45 000 students)	Large scale (about 100 000 parcels)	Small projects	Major project
Changes/stability	Some/relatively stable	Major/very unstable	Frequent/very unstable	Frequent/very unstable
Social relations	Cooperative	Conflicting	Segregated conflicting	Segregated conflicting
Motivation for GIS	High	Low	Medium	Low

Table 11.4 Internal organisational factors in relation to the outcomes of GIS initiation and implementation in the four agencies

Agency	TD	TD	MS	MS	CA	CA	CP	CP
Phase outcome	IN	IM	IN	IM	IN	IM	IN	IM
	Yes	Yes	Yes	No	Yes	Yes	No	No
Function	+	+	+	□	+	+	□	□
Structure	★	+	★	□	+	★	□	−
Resources	+	+	+	□	+	+	−	−
Operations	+	+	+	□	+	+	□	□
Change/stability	+	+	★	−	★	★	−	−
Social relations	+	+	★	−	★	★	−	−
Motivation for GIS	+	+	+	−	+	+	−	−

IN = Initiation   Yes = Positive outcome
IM = Implementation   No = Negative outcome
+ = Positive outcome in favourable environment
− = Negative outcome in unfavourable environment
★ = Positive outcome in unfavourable environment
□ = Negative outcome in favourable environment

the Transportation Department the initiation went smoothly regardless of the centralised organisational structure. In the Mapping Section, a GIS was also initiated despite the unfavourable organisational structure and its instability and conflict. While in the same agency the structure was adequate for implementing GIS technology, the process was suppressed by the negative influence of other factors. Finally, the Comprehensive Planning Section neither initiated nor implemented GIS technology. A few favourable elements of the section's in context could not offset the overwhelming negative organisational conditions.

The organisational contextual factors, clearly, influenced GIS initiation and implementation with different strength. Four out of seven analysed aspects of internal organisational context came out as very important for initiation and/or implementation of GIS technology: (a) motivation for GIS was crucial for both initiation and implementation; (b) availability of resources was also supportive for both initiation and implementation (financial, staff, time); (c) organisational instability was an obstacle to successful implementation; and (d) presence of organisational conflict was very detrimental for implementation of GIS technology. Characteristics of organisational structure were also influential, but did not show as clearly as the first four factors. The final outcome, however, depended on a trade-off among the many factors and resulted from complex and non-linear organisational dynamics. Analysis of social networks and major actors (or GIS champions), as presented by Assimakopoulos (see Chapter 8, this volume) and Azad (see Chapter 10), respectively, would contribute valuable additional information for a better understanding of the diffusion process.

Finally, the relationship between organisational function/operations and GIS technology was not explicitly considered in this study, because all four agencies had a clear GIS application, although for tasks of different nature. The functions ranged from more structured and routine activities, such as bus routing and property mapping, to more complex and non-structured tasks, such as those carried out by the Planning Department sections. While the nature of the functions performed by the agencies could have impacted on the success of GIS implementation, the actual GIS adoption level in each of them was measured only relative to the optimal use of GIS technology for that particular agency's functions. Also, the fact that the Transportation Department and the Mapping Section had similarly straightforward and focused tasks, but implemented GIS technology with varying degrees of success, indicated that organisational function was not a significant factor in the comparative study of GIS adoption in the four cases.

The importance of organisational function, however, is manifested at the organisational unit level and in fact appears crucial for selecting and customising the technological innovation (in this case GIS). The following section discusses the ways in which organisational functions and structure match the technology.

11.5 Technology-organisation fit

11.5.1 The ongoing debate

The claim that the introduction of new technology induces changes is undisputed. Organisational mission, structure, operations, learning and knowledge, social relations, resources and constraints may all change as result of computing dynamics

(King and Kraemer, 1986). The purpose of these changes is the mutual adjustment of the technology and the organisation. Leavitt (1965) maintains that task, people and structure change in order to 'neutralise' the impact of the newly introduced technology and achieve organisational homeostasis. Markus (1984) confirms the need to change organisational structure, job, communication patterns and inter-organisational relationships, in case of mismatch between the system features and the existing organisational design. Moreover, the fit is declared important for organisational effectiveness and better predictor of organisational performance than technology, structure or technology and structure combined (Alexander and Randolph, 1985). Federico *et al.* (1980), on the contrary, are alert to the limited evidence of organisational changes taking place because of IT and the lack of knowledge about appropriateness, impact and the cost-effectiveness of particular organisational models with regard to the effective use of information systems.

The source of primary impetus, that is, whether the structure adjusts to technology or the technology is customised to meet the organisational context, remains to be determined. The empirical evidence is mixed, although the technological determinism is mostly rejected (Heffron, 1989; Scott, 1990). Through a case study of computing in local governments, Kraemer *et al.* (1989) demonstrated that computing activities reflect the organisational structure. They provided a striking example of a municipal government's attempt to develop a centralised system for financial management. The system was inhibited by an already existing decentralised organisational structure and achieved successful implementation only after complying with it. The authors interpreted the findings from the 'reinforcement politics' perspective (the use of information systems for the maintenance of existing patterns of power) and placed the ultimate force in management activity. The 'political formula' is an appealing and plausible explanation, given the value of information as a source of organisational power (Pfeffer, 1981) and the role of organisational structure in determining the flow of information (Duncan, 1983). In addition, Kraemer *et al.*'s (1989) example illustrates very well the intersection and usefulness of both the rationalist and the interactionist approaches to understanding organisational behaviour with respect to the introduction of new technology.

11.5.2 Conceptual models

Unlike the scarce empirical evidence, the theoretical propositions and conceptual models of organisation-technology interaction are more available. From the rich body of literature on information systems two approaches are isolated to help understand the match between the technology and the organisation: (a) using the extent of routinisation of organisational tasks and functions as a guidance; and (b) adjusting the technological system and the organisation according to the criterion of function.

Routinisation continuum. Alexander and Randolph (1985) developed a simple measure of the organisational-technology fit that involves scaling organisational structures with respect to how mechanistic they are and scaling technologies with respect to the routine involved in their application. Their research suggests matching mechanistic structures with routine technologies and organic (flexible) structures with non-routine applications. Government agencies as people-process-

Table 11.5 Markus' conceptual framework on system and organisational features (adapted from Markus, 1984)

System type	System function	Key design features	Related organisational features
Type 1 Operational	To structure work	Work rationalisation; work routinisation	Workforce composition, job design, organisational structure, workflow coordination, organisational culture
Type 2 Monitoring and control	To evaluate performance and motivate people	Standards, measures, evaluation, feedback, reward	Job design, organisational culture
Type 3 Planning and decision	To support intellectual processes	Models, data manipulation	Workforce composition, job design, organisational structure, organisational culture, centralisation versus decentralisation
Type 4 Communication	To augment human communication	Communication procedures, communication, mediation	Spatial and temporal factors; communication channels and networks
Type 5 Interorganisational	To facilitate interorganisational transactions	Procedures for inter-organisational transactions, mediation of inter-organisational transactions	Related with customers or suppliers relation with competitors

ing organisations and organisations with an insufficient knowledge base should adopt a non-routine technology and its matching organisational structure (Heffron, 1989). Aldrich (1972) and Dewar and Hage (1978) add that routine, automated technology increases organisational formalisation, standardisation and centralised control.

Another way of viewing the extent of routinisation is through a decision-making perspective. Gorry and Morton (1989) categorise decisions as structured, semi-structured and unstructured. Those decisions differ in their use for management (operation, control and strategic planning, respectively) and in their information requirements. Consequently, they require different system designs too.

Function criterion. Markus (1984) takes an interactionist perspective to information system development. This perspective relates four organisational features: technology, structure, culture and politics. In her book *Systems in Organisations – Bugs and Features*, Markus offers a thorough and comprehensive framework for understanding the interaction between the information systems technology and organisational context. Her starting point is the classification of system types based on corresponding organisational functions: (a) operations, (b) monitoring and control, (c) planning and decision, (d) communication and (e) interorganisational activity. Each system/function type has a corresponding set of features including key design features and related organisational (operational) features (Table 11.5).

In addition, the impact of each operational feature and the nature of the impact is evaluated. The concepts from the initial framework are taken a step further toward operationalisation. For example, in the case of operational systems, the impact on the organisational structure is manifested as organisational integration (versus differentiation) of the work flow and horizontal structure; in the case of planning and decision systems, the impact on the organisational structure is through increased centralisation (versus decentralisation) of decision making. These examples illustrate the applicability of the schema developed by Markus for an informed guidance and control of technological implementation. Due caution, however, needs to be exercised in employing the schema across various organisational contexts and different technologies. Also, the non-linear relationship and interaction between the factors warns against a direct and mechanistic use of the model.

11.6 Conclusion

Organisational theory and research on IS/MIS implementation is an excellent starting point for understanding the interaction between local government organisational context and GIS innovation. This heritage provides still limited, but valuable information potentially applicable to the GIS field. First, GIS developers need to be aware of the variable quality, direction and strength of relationship with organisational structural elements depending on the phase of the innovation process. Second, they need to be prepared to cope with both technological change (reinvention) and organisational change associated with development of a GIS.

Continuous deliberate action can turn a haphazard or random introduction of GIS technology into a planned activity to adjust organisational operation and make the best use of the technology. The information acquired through various streams of research provides useful guidance for introducing GIS technology into the public organisations. Careful design and management of the GIS implementation process is essential for ensuring desired outcomes of GIS technology in the local governments (Nevodić-Budić and Godschalk, 1994; Campbell and Masser, 1991; Kraemer et al., 1989). Controlling the change process and directing the mutual adjustment of technology and organisation should be the most prominent GIS management activity. Although all organisational processes, factors and structures are not manipulable, they can be at least considered as either facilitators or obstacles to GIS diffusion. Creating the culture of change alone can also substantially contribute to the successful incorporation of GIS technology. GIS implementation strategies and organisational [re]design promises to be more effective if they utilise this new knowledge.

Recognition of the dynamic relationship between the organisational technology and its social and political systems requires application of GIS development approaches that are alternatives to traditional data processing and structured information system design. From a wide range of available methods that have evolved over the past several decades (Hirschheim and Klein, 1992), socio-technical system design seems the most suitable to handle the organisational dynamics associated with the introduction of GIS technology (Scott, 1990; Eason, 1988; Markus, 1984). The socio-technical approach assumes that both the technical system and the social system must be designed and that they must be designed interdependently. Contrary to the traditional system building which is fully prespecified, emphasises technical aspects and involves professional control, the socio-technical approach is more open, flexible and sensitive to organisational features and context. It involves both the system designers and the system users and allows for communication and integration of their different views. In summary, the socio-technical approach considers simultaneously the technology and the organisational structure and processes and matches them in order to achieve the system objectives.

Clearly, there is sufficient advice available for our use, but the gaps in knowledge are substantial too. In the GIS area, the question of organisational change and adaptation associated with the introduction of the technology is entirely unexplored. Insight into the manner in which the technology-organisation match occurs and isolation of organisational configurations that support various GIS applications will contribute practical information for more effective initiation and implementation of the technology. It will provide the tools for GIS managers and advance the general knowledge about computing in local governments.

Appendix A1.1 **Transportation Department, City/County School System**

Organisational structure. The Transportation Department was part of a large, complex and centralised organisation. As the previous research suggests, this kind of environment could obstruct both initiation and implementation of GIS technology. Both phases, however, proceeded smoothly partially because the Transportation Information Management System (TIMS, a GIS-based bus routing tool) was incorporated in only one organisational unit.

Change/stability; social relations. Relative organisational stability and a some-what competitive, but generally cooperative environment were conducive to both GIS initiation and implementation. A merger between the city and the county schools in 1985 did not affect the implementation. The merger happened before the initiation of TIMS was contemplated.

Resources. The resources for TIMS initiation and implementation were adequate. The Department as well as the School System had a history of successful automation. Although the GIS-related expertise and equipment were not available in-house, there was a sufficient level of confidence and experience with com-puterised technology to launch a new undertaking in computerisation.

Financial resources for developing and using TIMS were provided internally and externally. Internal financial resources matched a state grant. Outside sources continued to be the major funding base for the TIMS.

Motivation. The Transportation Department had a clear motivation to acquire and implement TIMS. The initiative came from a higher administrative level (Operations Director and Transportation Director). The motivation for incor-porating GIS technology could clearly be tied to the nature and type of operations performed by the agency. Computerisation of school bus routing was seen as an opportunity to switch from a manual to an automated system of operation management. Considering the scale of the departmental tasks (transporting 60 per cent of about 45 000 students and operating over 400 buses), TIMS was expected to substantially increase efficiency. There was a strong belief at the higher administrative level about tangible benefits in terms of time and energy savings, reduction in expenses (computed at about $9000 per bus annually) and easier assignment of children to buses. In addition, the use of TIMS was to be mandated for all school districts in North Carolina.

Appendix A1.2 Mapping Section, Tax Assessor's Office

Social relations/conflict/stability/change. The major obstacle to GIS initiation and implementation in the Mapping Section was a political intra-organisational conflict and related organisational restructuring that occurred during 1988 and 1989. The power struggle primarily concerned the issue of controlling the County mapping function. From the early 1980s there was a disagreement about the purpose and activities of the Land Records Department. The Department perceived itself as servicing various county government units with mapping products. The Tax Assessor's Office perceived it as a single function agency focusing on land parcel mapping in support of property assessment and taxation.

Additional threats were perceived as coming from the increased independence of the Land Records Department. Because of a proactive leadership and managerial style of its head, it was becoming a strong entity. The head was one of the leaders of the county computerisation efforts including development of the county On-line Assessment Information System (OASIS). The head's in-depth knowledge and usage of the system added another layer to the perceived threat (this time technological) from the Land Records Department.

A few interviewees interpreted the conflict as a personality clash between the head of the Land Records Department and the director of the Tax Collection Department. The power struggle was complemented by a number of activities: building alliances and coalitions, initiation of a committee to determine the necessity to remerge Land Records under Tax Office and a lot of 'back stage' performances. In November 1989, several days before dismantling the Land Records Department and after a year of political battling and pressuring, the head of the Land Records Department resigned. The Land Records Department was then merged under the Tax Assessor's Office. It was renamed the Mapping Section and had its staff reduced from 21 to only 8 mappers. The mappers who did not stay within the section were reassigned to duties in other sections of the Tax Assessor's Office.

Although the political conflict was not explicitly related to GIS technology, the fact that the head of the former Land Records Department was the major proponent of the technology affected the progress of GIS implementation. GIS technology was partially sacrificed due to general political resentment toward the Land Records Department head's personality and ideas. Despite the unfavourable conditions for GIS initiation, in 1988 the head of the former Land Records Department decided to acquire Arc/Info and a PC from the departmental slack funds. The act was approved by the County Manager.

Organisational instability and conflict, however, strongly affected the implementation of GIS technology. After the merger and resignation of the head of the former Land Records Department, the implementation of GIS technology in the Mapping Section of the Tax Assessor's Office was halted. Occasionally, specific actions were taken to ensure that the implementation did not continue. For instance, contacts between the main GIS user in the Mapping Section with other GIS users within the County government were controlled and participation of the Mapping Section staff members in GIS related matters within the government and in the Arc/Info user group was suppressed; and in 1991 a decision was made by the Board of Commissioners not to allocate funds to match a possible state grant for GIS development.

Although the size of jurisdiction (about 100 000 parcels) called for the automation of parcel mapping, digitising of the parcels was never initiated. While the head of the former Land Records Department believed that there was sufficient capacity to develop a GIS database in-house, there was a concern in the Tax Assessor's Office about a large up-front investment needed to develop and maintain the county GIS. A move to revive GIS implementation in 1992 by officially assigning a staff member responsible for GIS development, employed a strategy of political bypass. Similar to the approval of the Arc/Info purchase, the decision came from the County Manager's Office. This formal charge with GIS technology facilitated coordination on GIS-related matters within the County and with the City. The GIS staff member from the Mapping Section participated in the Joint City/County GIS Committee, but had not vested authority to pass recommendations or engage in decision-making. Except for the guidance and overview tasks, creation of this new position within the County government was not accompanied with any other concrete measure to advance the implementation of GIS technology.

Motivation. The head of the former Land Records Department had a clear motivation to incorporate the technology, with the goal of enhancing organisational efficiency and effectiveness. The large scale of the tax-mapping operation justified an effort in automation. The head also pursued an organisational prestige agenda, with high expectations about governmental performance. The motivation for GIS technology disappeared with the head's resignation. In addition, the fact that one rather expensive information system (OASIS) was recently developed and successfully utilised, was a counterargument for the introduction of GIS. The system was excellent and served very well the main organisational purpose – tax records and property management.

Organisational structure. Centralised and formalised organisational structure was not favourable for initiation of GIS technology, but was conducive to its implementation. The large size of the organisation and its complexity were conducive to the successful acquisition of the technology, but may have prevented its implementation. Once the technology was acquired, the conditions for its implementation would have been good, had not the conflict and restructuring happened.

Resources. Finally, resources were adequate for implementing GIS. The Tax Office had a history of successful automation. Two departments, the Data Processing and the Land Records Department, were candidates to actively help with GIS implementation.

Financially, the County government did not seem to have experienced problems, although measures were continuously taken to reduce spending. That philosophy, as expressed by a couple of interviewees, was a double-edged sword and eventually affected the quality of government services provided to citizens and reduced the competitiveness of the government in its ability to attract and keep highly skilled employees.

Appendix A1.3 Community Assistance Section, City/County Planning Department

Motivation. The head of the Planning Department's Transportation Section, who initiated the purchase and use of GIS, was strongly motivated to incorporate GIS technology for departmental activities. He believed in the utility of computerised technology for increasing efficiency in task performance. Benefits from the technology were primarily seen in freeing up planners' time from data collection to do more creative activities. Due to its orientation toward data-handling activities, the Community Assistance Section was also a logical beneficiary of the technology. It became the departmental GIS home after the main GIS users moved to that section.

Resources. The computer-related experience of two GIS users (the main GIS user and the Head of the Transportation Section who promoted GIS) was adequate to build on. The Head of the Transportation Section led the departmental computerisation efforts. He initiated, installed and helped maintain a local area network (LAN) in the Planning Department.

Financial resources were secured through state and federal funding for transportation planning. It was sufficient for the small-scale GIS implementation that was targeted by the Planning Department. More substantial funding was never sought.

Organisational structure. Organisational structure considerably influenced both GIS initiation and implementation. Although the small size of the organisation was not too conducive to effective initiation of GIS technology, its complexity, lack of normalisation and decentralised nature were favourable. A fragmented organisational structure and informality in functioning were not supportive for GIS implementation and, in fact, strongly impacted on the diffusion of the GIS toward individual employees as well as toward other sections, especially toward the Comprehensive Planning.

Change/stability. Unstable and conflicting organisational environment was probably the major factor that contributed to the modest diffusion of the technology. The Planning Department had frequent reorganisations and reassignment of positions. Initially it comprised only the Operations Section and the Study Section. Major reorganisation occurred in 1988. Reassignment of employees from one section to another created interpersonal conflicts. The most recent reorganisation in 1992 resulted in the creation of the Street Naming and Addressing Section. The new section involved employees previously staffing a number of other sections. Almost all interviewed employees either were at some point part of some other section of the Planning Department or worked for another unit of the County government.

Social Relations. Finally, the Planning Department was organisationally very fragmented, with employees perceiving their units and themselves as rather separated, particularly with regard to work-related communication. While these conditions did not seem to have strongly influenced the initiation of GIS technology in the Transportation Section and the Community Assistance Section, they might have hampered the implementation of the technology, particularly its diffusion toward individuals and other organisational units within the Planning Department.

Appendix A1.4 Comprehensive Planning Section, City/County Planning Department

The prevailing organisational environment of the Planning Department partially determined the Comprehensive Section's response to GIS technology and the resulting reluctance to use it in the process of updating the Comprehensive Plan. Even though the data-collection phase was an ideal opportunity to actually engage in computerising a database, the Head of the Comprehensive Planning Section and his staff did not find GIS technology to be a suitable tool for the project. Rather, the planning tasks themselves were so overwhelming that no additions to the project were considered.

Motivation. Motivation to introduce GIS technology in the comprehensive planning process was clearly missing. In another words, organisational benefits from using the technology were not seen as worth the trouble of switching to a new technology. At that time the technology was not perceived as an opportunity for better and more efficient management of large amounts of data. There was no action to initiate development of a GIS. Also, utilisation of GIS-related services available in the Community Assistance Section and the Mapping Section were limited. The traditional, manual mapping was seen as more appropriate for the task. This was particularly true with the generation of presentation maps, for which GIS technology was perceived as offering inferior quality, clearly reflecting the drafting (craft) culture of the section employees.

Resources. In addition to the unfavourable organisational circumstances and lack of motivation for employing GIS technology, the Comprehensive Planning Section did not have adequate resources for either GIS initiation or implementation. Even though the section did not have major financial problems and had sufficient funding for its ongoing activities, there were no additional funds provided. The section staff did not receive any aid in human resource to help the comprehensive plan updating process. No funds were committed for training or hiring additional personnel to operate a GIS for projects undertaken in the Comprehensive Planning Section.

Finally, unlike in the Community Assistance Section, the technological infrastructure was not a strong aspect of the Comprehensive Planning Section's history and environment. Even the software available through the departmental LAN was under-utilised. The local area network was used only occasionally and generated a lot of frustration. None of the section employees was an intensive computer user. The knowledge base was insufficient to build on.

References

ALDRICH, H. E. (1972). Technology and organization structure: A re-examination of the findings of the Aston Group. *Administrative Science Quarterly*, **17**(1), 26–43.

ALEXANDER, J. W. and RANDOLPH, A. W. (1985). The fit between technology and structure as a predictor of performance in nursing subunits. *Academy of Management Journal*, **28**(4), 844–59.

AZAD, B. (1990). 'Implementation of geographic information systems', PhD proposal, Department of Urban Studies and Planning, Cambridge, Massachusetts: Massachusetts Institute of Technology.

AZAD, B. (1994). *Phased/incremental GIS implementation: dynamics of implementation phases and management/organizational challenges*, Paper presented at URISA '94, Milwaukee, Wisconsin, 7–11 August, (not in proceedings).

BRETSCHNEIDER, S. (1990). Management information systems in public and private organizations: an Empirical Test, *Public Administration Review*, **50**(5), 536–45.

BROWN, M. M. and BRUDNEY, J. L. (1993). *Modes of Geographic Information System Adoption in public organisations: Examining the effects of different implementation structures*, Paper presented at the annual meeting of the American Society for Public Administration, San Francisco, CA, 17–21 July (not in proceedings).

BUDIĆ, Z. D. (1993a). GIS Use Among South-eastern Local Governments — 1990/1991 Mail Survey Results, *URISA Journal*, **5**(1), 4–17.

BUDIĆ, Z. D. (1993b). Human and Institutional Factors in GIS Implementation by Local Governments, Doctoral dissertation, University of North Carolina at Chapel Hill.

BUDIĆ, Z. D. (1994). Effectiveness of Geographic Information Systems in local planning, *Journal of the American Planning Association*, **60**(2), 244–63.

BUDIĆ, Z. D. and GODSCHALK, D. R. (1994). Implementation and management effectiveness in adoption of GIS technology in local governments, *Computers, Environment and Urban Systems*, **18**(5), 285–304.

CAMPBELL, H. (1990). The organisational implications of Geographic Information Systems for British Local Government, in Harts, J., Ottens, H. and Scholten, H. (Eds). *Proceedings of EGIS '90 Conference*, **1**, pp 145–57. Utrecht: EGIS Foundation,

CAMPBELL, H. (1991). *Organisational Issues and the Utilisation of Geographic Information Systems,* Regional Research Laboratory Initiative, Discussion Paper Number 9, Sheffield, UK: Economic and Social Research Council, Regional Research Laboratories.

CAMPBELL, H. J. and MASSER, I. (1991). The impact of GIS on Local Government in Great Britain, in *Proceedings of Association for Geographic Information Conference*, Birmingham, pp. 2.5.1–2.5.6, London: AGI.

CRAGLIA, M. (1994). Geographical Information Systems in Italian Municipalities, *Computers, Environment and Urban Systems*, **18**(6), 381–475.

CROSWELL, P. L. (1991). Obstacles to GIS implementation and guidelines to increase the opportunities for success, *Journal of the Urban and Regional Information Systems Association*, **3**(1), 43–56.

DANZIGER, J. N., DUTTON, W. H., KLING, R. and KRAEMER, K. L. (1982). *Computers and Politics: High Technology in American Local Governments*, New York: Columbia University Press.

DAVIS, L. E. and TAYLOR, J. C. (1976). Technology organisation and job structure in DUBIN, R. (Ed) *Handbook of Work, Organisation and Society*, pp. 379–419, Stokie, IL: Rand-McNally.

DEWAR, R. and HAGE, J. (1978). Size, technology, complexity and structural differentiation, *Administrative Science Quarterly*, **72**(5), 328–46.

DOWNS, G. W. Jr. and MOHR, L. B. (1976). Conceptual issues in the study of innovation. *Administrative Science Quarterly*, **21**(4), 700–14.

DOWNS, G. W. Jr. and MOHR, L. B. (1979). Toward a theory of innovation, *Administration and Society*, **10**(4), 379–408.

DUNCAN, J. W. (1983). *Management: Progressive Responsibility in Administration*, New York: Random House Business Division.

EASON, K. (1988). *Information Technology and Organisational Change*, London: Taylor & Francis.

FEDERICO, P., BRUN, K. E. and MCCALLA, D. B. (1980). *Management Information Systems and Organizational Behaviour*, New York, NY: Praeger Publishers.

FELLER, I. and MENZEL, D. C. (1977). Diffusion milieus as a focus of research on innovation in the public sector, *Policy Sciences*, **8**(1), 49–68.

FRENCH, S. P. and WIGGINS, L. L. (1990). California planning Agency Experiences with automated mapping and Geographic Information Systems, *Environment and Planning B: Planning and Design*, **17**(4), 441–50.

GORRY, G. A. and MORTON, M. S. S. (1989). A framework for management information systems, *Sloan Management Review*, **31**(1), 49–61.

HAGE, J. and AIKEN, M. (1970). *Social Change in Complex Organizations*, New York: Random House.

HEFFRON, F. (1989). *Organization Theory and Public Organizations*, Second Edn, Englewood Cliffs, NJ: Prentice Hall.

HIRSCHHEIM, R. and KLEIN, H. K. (1992). Paradigmatic influences on information systems development methodologies: Evolution and conceptual advances, *Advances in Computers*, **34**, 293–392.

HOWARD, E. W. (1985). New technology and organisational change in Local Authorities, *Computers, Environment and Urban Systems*, **10**(1), 9–18.

KANTER, R. M. (1989). Mastering the dilemmas for innovation, *California Management Review*, **31**(4), 38–69.

KING, J. L. and KRAEMER, K. L. (1985). *The Dynamics of Computing*, New York: Columbia University Press.

KING, J. L. and KRAEMER, K. L. (1986). The dynamics of change in computing use: A theoretical framework, *Computers, Environment and Urban Systems*, **10**(1/2), 5–25.

KRAEMER, K. L., KING, J. L., DUNKLE, D. E. and LANE, J. P. (1989). *Managing Information Systems*, San Francisco, CA: Jossey-Bass.

KWON, T. H. and ZMUD, R. W. (1987). Unifying the fragmented models of information systems implementation, in Boland, J. Jr. and Hirschheim, R. A. (Eds) *Critical Issues in Information Systems Research*, pp. 227–51, New York: John Wiley.

LEAVITT, H. J. (1965). Applied organisational change in industry: Structural, technological and humanistic approaches, in March, J. G. (Ed.) *Handbook of Organizations*, pp. 1144–1170, Chicago, IL: Rand-McNally.

LEWIN, K. (1952). Group decision and social change, in Newcomb, T. M., Hartley, E. L. and Swanson, G. E. (Eds), *Readings in Social Psychology*, pp. 459–73, New York: Henry Holt.

MARKUS, L. M. (1983). Power, politics and MIS implementation, *Communications of the ACM*, **26**(6), 430–44.

MARKUS, L. M. (1984). *Systems in Organizations – Bugs and Features*, Marshfield, MA: Pitman.

MASSER, I. and CAMPBELL, H. (1994). Monitoring the take-up of GIS in British Local Government, in *Proceedings of the Annual Conference of Urban and Regional Information Systems Association (URISA)*, pp. 745–54, Milwaukee, WI.

NEDOVIĆ-BUDIĆ, Z. and GODSCHALK, D. R, (forthcoming). Human factor in adoption of Geographic Information Systems (GIS): A Local Government case study, *Public Administration Review*.

NOLAN, R. (1973). Managing the computer resource: Stage hypothesis, *Communications of the ACM*, **16**(7), 399–405.

ONSRUD, H. J. and PINTO, J. K. (1993). Evaluating correlates of GIS adoption success and the decision process of GIS acquisition, *URISA Journal,* **5**(1), 18–39.

PERROW, C. (1967). A framework for the comparative organizational analysis, *American Sociological Review*, **32**(2), 194–208.

PFEFFER, J. (1981). *Power in Organizations*, Marshfield, Massachusetts: Pitman.

PIERCE, J. L. and DELBECQ, A. L. (1977). Organizational structure, individual attitudes and innovation, *The Academy of Management Review*, **2**(1), 27–37.

PINTO, J. K. and AZAD, B. (1994). The role of organizational politics in GIS implementation. *URISA Journal*, **6**(2), 35–61.

RICE, R. E. and ROGERS, E. M. (1980). Reinvention in the innovation process, *Knowledge: Creation, Diffusion, Utilization*, **1**, 499–514.

ROGERS, E. M. (1983). *Diffusion of Innovation*, Third Edn, New York: The Free Press.

ROGERS, E. M. and AGARWALA-ROGERS, R. (1976). *Communication in Organizations*, New York: The Free Press.

ROGERS, E. M. and KIM, J. (1985). Diffusion of innovations in public organizations, in Merritt, R. L. and Merritt, A. J. (Eds), *Innovation in the Public Sector*, pp. 85–108, Beverly Hills, CA: Sage.

SCHULTZ, R. L., SLEVIN, D. P. and PINTO, J. K. (1987). Strategy and tactics in a process model of project implementation, *Interfaces*, **17**(3), 34–46.

SCOTT, W. R. (1990). Technology and structure: An organizational-level perspective, in Goodman, P. S., Sproull, L. S. and Associates (Eds), *Technology and Organizations*, pp. 99–143, San Francisco: Jossey-Bass.

SPRECHER, M. H. (1994). ICMA survey dissects Local Government IT use, *URISA Journal*, **6**(2), 92–4.

STRAND, E. J. (1995). GIS plays a role in business process re-engineering, *GIS World*, **8**(4), 34–6.

WIGGINS, L. L. (1993). Diffusion and use of Geographic Information Systems in public sector agencies in the United States, in Masser, I. and Onsrud, H. J. (Eds), *Diffusion and Use of Geographic Information Technologies*, pp. 147–63, Dordrecht: Kluwer.

ZALTMAN, G. and DUNCAN, R. (1977). *Strategies for Planned Change,* New York: John Wiley.

ZALTMAN, G., DUNCAN, R. and HOLBEK, J. (1973). *Innovations and Organizations*, New York: John Wiley.

PART THREE

Generalisation

CHAPTER TWELVE

Generalisation: Where Are We? Where Should We Go?

JEAN-PHILIPPE LAGRANGE

12.1 Introduction

12.1.1 Generalisation: a definition

Many definitions have been proposed for generalisation (see for example McMaster and Shea, 1992). If we paraphrase the definition given by them for cartographic generalisation we may state that 'generalisation is the process of deriving, from a data source, a (possibly cartographic) data set through the application of spatial and attribute transformations, so as to reduce in scope the amount, type, and (if cartographic) cartographic portrayal of encoded data consistent with the targeted product and intended audience'.

Actually it is possible to distinguish between object generalisation (or model-oriented generalisation, or data generalisation), which is intended to derive a geographical database, and cartographic generalisation, which aims at deriving a map or a cartographic data set. This point is further discussed elsewhere (for example Grünreich, 1995; Müller *et al.*, 1995). However, these two process categories differ mainly in their respective objectives, and only to a lesser extent in their approaches, quality criteria or tools. Only issues strictly related to aesthetic quality in cartographic generalisation are specific to cartographic generalisation, for the rest both generalisation categories primarily aim at preserving as much information as possible when reducing the data.

Other differences may be considered, based on the technical context of the generalisation process. When producing cartographic products a mapping agency will look for best quality first, whereas for visualisation purposes in a GIS context the criterion which comes first is automation or speed of computation. This clearly shows that various solutions need to be devised which meet different quality requirements.

12.1.2 Geographic databases and maps

Geographic databases are often said to be almost completely different from conventional maps, and indeed computer-based technologies clearly offer a greater flexibility. However, even if we leave aside the fact that most of the existing geographic databases merely result from map digitising, there are many similarities between maps and geographic databases.

12.1.2.1 *Information and the meaning of data*

First, it is worth noting that the existing data structures (vector and raster data structures) are used to implement 'well-defined' objects – or entities, that is to say objects which are given a meaning and a boundary. Therefore, the outputs of such databases look quite close to standard or choropleth maps. The representation constructs are basically the same: points or point symbols, lines and polygons.

In fact, it is true that theoretically we may consider objects which have fuzzy boundaries, or objects for which the spatial representation is a continuous field (Burrough and Frank, 1995). However, as atomic objects are easier to handle, conceptually and from a computer-science perspective, our geographic databases remain based on a decomposition of the real world into abstract or logical units as stated in Goodchild (1990). Another reason for this modelling choice is that, like maps which are basically communication means and rely on a cartographic language intended for communication (Freytag, 1993), geographic databases must answer the information communication requirements. As a matter of fact, information, as opposed to rough data, may be defined as being abstracted and synthesised data.

It has been often argued that maps are intended to communicate an overall message by graphics' means (and therefore do not render actual features but graphic substitutes which suggest the message, possibly a set of houses), while GIS provides a representation of actual features, through some interpretative filter (see Morehouse, 1995). In fact, it is worth noting that the initial interpretative process, for example during stereoplotting, involves some kind of generalisation, so that the result is not always a representation of 'true features' but may be, as well as in the map case, a representation of an area. In this respect, the discussion and example provided by Spiess (1995), are interesting to compare with the Swiss example given by Morehouse (1995). Whichever the activity, GIS or cartography, the issue of modelling comes first (Mark, 1991; Grünreich, 1995), and geographic databases most often rely on the map metaphor.

Then, we are often led to the question: What do these objects mean? In other words we get a digital (or cartographic) representation of the real world, and we want to make use of this representation in order to perform some reasoning task about the real world. This obviously implies that we are aware of the intended meaning of the entities used for abstracting and communicating information in this real world. A special case of such activities is the generalisation of geographic information in order to derive more abstracted information, the latter being intended to be better suited for a given family of reasoning tasks. Consequently, for a successful generalisation to be carried out, we need to consider both the 'meaning' of initial data and the family of applications targeted for the derived product. In the cartographic context, these constraints are often phrased in terms of knowledge of the geographic space (semantic and structural knowledge[1]), and knowledge of the objective of the map (also related to structural knowledge).

Obviously, geographic databases may carry more information than maps. While cartography is restricted to use the well-known seven graphic variables defined by Bertin (1973), and in fact we are often limited to only three attributes (for example, the case of roads: usually classification, width and number of lanes), geographic databases allow us to store as many attributes as desired. In some sense *a usual geographic database may be seen as being equivalent to a collection of maps*, corresponding to the various thematic selections that are allowed. Again, the communication of such information will dictate a reduction to one or a few maps, as the graphic display is the most efficient and convenient communication means for spatial information (Jacob, 1992). In fact this is even true of such simple data as tables of numbers (Bertin, 1977). From a pragmatic point of view, most GIS tools are equipped with functionalities for graphic screen display and cartographic output.

12.1.2.2 *Resolution and scale*

Another difference which is often put forward is that a geographic database is not as constrained by scale considerations as a map is. Further, the idea of 'scaleless' databases is very common. It is clear that, owing to the considerations expressed above, we should be cautious about such an idea. Indeed, the currently available techniques allow us to constitute only multiscale, or multiresolution, databases.[2]

In fact, and roughly speaking, a map, whose scale is e, and which is printed with a best resolution r allows us to represent features which are not smaller than $\frac{r}{e}$ or features which are separated by a distance greater than or equal to this ratio. In other words, $\frac{r}{e}$ corresponds to the ground resolution of the map. If we consider this question from a closer point of view, it is well known that the map resolution r depends on the type of symbol under concern (shape, colour, and so forth.), but we can still define ground resolutions for the represented geographic phenomena. This ratio illustrates the influence of scale on map content and the applied thematic selection, but also the close relationship between the map and the geographic database (which, as part of the data capture specifications, a ground resolution – geometric and semantic – is defined too).

12.1.3 Generalisation in Geographic Information Systems

The above discussion intends to recall that, when dealing with geographic databases, we are generally faced with problems which are common to the cartographic activity. To be more specific, if we consider the main GIS activities, using a classical schema (Fig. 12.1).

We may see that when we want to integrate or analyse data (upper left activities), we need to know the 'original' meaning of the initial data in order to handle them adequately. This meaning is often transmitted by means of a scale associated with the database, but in fact this information is far from being sufficient. More metadata are needed beside the conceptual schema.

Similarly, when we want to visualise data – or information – or derive new data,

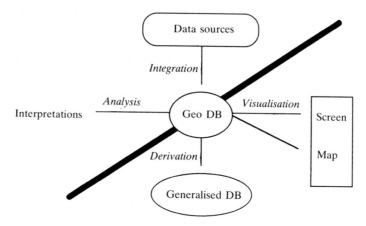

Figure 12.1 GIS activities.

we cannot merely rely on syntactic class definitions, or on class names. Actually, class names provide us with more than a lexical anchor, as we may have some common-sense knowledge on the words used, but these words may also be misleading when coming from one language to another one (Kuhn, 1994).

This weakness has given rise to many studies on geographic metadata or meta information, and on geographic information standardisation, especially in the context of geographic data exchange and digital map libraries. Again such meta information is fairly related with generalisation levels, at least when considering a database which results from the digitising of existing maps, or which is derived from another database by means of generalisation.

When a cartographer generalises an area, he or she can take advantage of his or her knowledge of the geographic reality and cartography as a communication activity in order to achieve the best result. On the other hand the computer process has to rely only on the geographic objects provided, that is to say something pretty similar to the cartographic items the cartographer starts with. Unfortunately, the computer is not skilled in interpreting such objects in terms of geographic reality but a sensible transformation process needs to consider more than the given graphic items.

From a technical perspective, if we consider visualisation or derivation as being some kind of computation of a view of a data set ('view' being used here in the database sense of the word), then what are the possible forms of such a view? First, it may be computed by using the usual (for example, relational) database operators: selection, projection, join and so forth. In such a case we merely get a restructured extract of the data set. We may, in addition, use a grouping operator and some aggregation functions (sum, average and so on) to further restructure our query result. In such a case, and if our grouping applies only to non-geometric attributes, we get some elementary kind of statistical generalisation of data. Going one step further, we look for spatial and thematic generalisation. This is nothing but foreseeable: just as a database view is a 'mechanism' for deriving a new representation on, or view of, the real world, generalisation is a special case of this activity which consists in deriving a 'simplified' representation from an initial one, when considering geographical data. If we come back to technical considerations, this means that in the context of GIS tools, a true view mechanism should integrate generalisation functionalities (Claramunt and Mainguenaud, 1994; Rigaux and Scholl, 1994).

So far, we may conclude that we need only functionalities for the so-called 'statistical generalisation' (Brassel and Weibel, 1988) or 'model-generalisation' (Müller *et al.*, 1995). If, however, we now consider the needs for graphic (screen or paper) display, the 'cartographic generalisation' also becomes important.

12.2 A quick overview of past research on automated generalisation

This is not the place for a lengthy and detailed history of past research, and besides the main trends are rather well known, but it is still interesting to consider briefly the current achievements and where we come from. More details and complementary elements may be found in McMaster and Shea (1989), Buttenfield and McMaster (1991) and Müller *et al.* (1995). Three main phases may be schematically considered.[3]

First, during the 1960s and the early 1970s, the focus was put on geometric simplification, quite often with the goal of compacting the data. The famous Douglas' algorithm was thus introduced under the title 'Algorithms for reduction of the number of points required to represent a digitised line or its caricature' (Douglas and Peucker, 1973). The context of these research efforts was the progressive digitising of geographical data, with most of the algorithms dealing with line simplification (although some tools for building outline simplification were also developed, for example, in Germany), and aiming at simplifying the output of line digitising.

Then, for about a decade, research has evolved from this starting point, while still often keeping a focus on geometry. As it is impossible to devise any general algorithm for simplification, several studies have been conducted, with the goal of assessing the applicability of algorithms, either by quantifying the effects of the algorithms (McMaster, 1983, 1986, 1987a and b; Müller, 1987), or by characterising the geometry of features (Beard, 1991; Buttenfield, 1986, 1987, 1989, 1991; Freeman, 1978; Jasinski, 1990; McMaster, 1995; Müller, 1986; Thapa, 1988). Consequently, some classifications of the algorithms and algorithm sequencings have been proposed (McMaster, 1989). Some algorithms for other generalisation operations have also been proposed, for example conflict resolution (Christ, 1979; Jäger, 1990; Monmonier, 1987) and land-cover generalisation (Monmonier, 1983), but significantly less efforts have been devoted to such issues. In fact, the research activities of this period may be considered to be more cartography oriented than before (instead of being purely geometry oriented), but dealing mostly with independent generalisation,[4] which means that the tools developed allow us to generalise one feature at a time and independently of other features. Each feature is considered as being isolated from any context, and only its geometry is taken into account. As a result, these tools basically allow for a one theme at a time (or one layer at a time) generalisation: one operation is applied successively to every feature of a theme, and then another theme may be processed, or another operation considered.

Currently, and since the end of the 1980s, we may say that research is characterised by attempts to go beyond the independent generalisation of geometry.[5] Very briefly, the focus is now often put on the global generalisation process (generalisation operations formalising, generalisation process understanding and formalising and so forth), and on considering the spatial and semantic context of features and the geographical phenomena involved in the generalisation. Significantly, several workshops or specialist meetings have been devoted to the issue of generalisation explicitly under the heading of 'knowledge' (for example, NCGIA meetings in 1989 in Syracuse, reported in Buttenfield and McMaster, 1991, and in 1993 in Buffalo, and the 1993 ESF GISDATA meeting in Compiègne reported in Müller *et al.*, 1995. Finally, it is now admitted that generalisation cannot be considered only from the cartographic point of view but also from the point of view of deriving new (generalised) databases from existing ones, and that such derivations are even quite often a prerequisite for map production. These questions are discussed further below.

12.3 Some fundamental issues for research

Among the key issues identified during the GISDATA Specialist Meeting, in Compiègne, we may highlight 'semantic object definition, geographical analysis and knowledge formalising' as necessary prerequisites for graphical conflict resolution, and 'data quality', while the Initiative 8 Specialist Meeting in Buffalo (1993) was devoted to knowledge formalising (see also Chapter 13, this volume). All in all, knowledge or information are the words which come first. If we have a brief look at the broad categories of knowledge involved in generalisation (Fig. 12.2), it turns out that the digital representation (geometrical arcs and descriptive attributes) is not sufficient for making good generalisation decisions. Indeed, these decisions should depend upon the targeted product and on what we know of the represented features and their importance in depicting reality. Besides, they rely upon knowledge of the generalisation activities and operations (in order to choose actions and then trigger algorithms) and quality assessment (to control the process). Again, it is worth noting that most of these knowledge categories are also of interest when analysing geographical data.

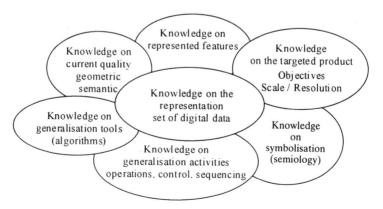

Figure 12.2   A view on knowledge categories for generalisation.

12.3.1 Knowledge on intrinsic properties of objects

Three broad knowledge categories have been identified, (for example, Armstrong, 1991): structural, procedural and geometric knowledge, where structural knowledge refers to expertise in the scientific fields (geology, hydrology and so on, hence knowledge on the actual features) and in mapping, procedural knowledge is expertise in choosing the appropriate generalisation operators, and geometric knowledge is expertise in identifying the appropriate geometrical representation.

Knowledge on the features to be generalised is obviously of prime importance (see Mark, 1989), bearing in mind that a cartographer is not only a graphic artist but also someone who has some skills in geology, geomorphology and topography. Such knowledge includes some structural knowledge and semantic knowledge, that is to say the meaning of data, which is rarely provided in the existing geographic databases, even when considering the sometimes existing metadata. However, it is clear that, for example, streams and roads are different classes of features, which carry different semantics and the appearance of which is (usually) different. Further, subcategories may be considered, which are never, or almost never, considered in databases. For example, a meandering river is something different from a large and hardly sinuous river, a sinuous mountain road cannot be compared with a straight road, even if their sizes are comparable, which implies that these characteristics must be maintained during the generalisation process.

This knowledge also needs to be related to the targeted product objectives, which means that, when merely considering the selection of features, their importance as determined by such knowledge will guide the process. As a consequence, a great deal of interest is being given to acquiring such knowledge, or using and learning such knowledge, but since it is most often not available there are more efforts on extracting it from the given representations.

Since there is still a lot to do before we are just able to simplify (in the broadest sense of the word) geometry, several attempts have been or are being made which aim at analysing feature geometry, so far mainly for linear features. These analyses may consist in classifying on a 'shape basis', such as in McMaster, 1995, or in Buttenfield, 1991, or in attempts to both decompose the features while classifying the resulting parts, for example, Buttenfield, 1987, Plazanet, 1995 (and in Chapter 17 of this volume).

It is worth noting that such research involves the use of geometric measurements which may be useful not only for classifying or decomposing shapes, but also for validating a generalisation transformation. Indeed, shape maintenance, or controlled transformation, is one of the main quality criteria for generalisation, at least for cartographic generalisation and when considering only geometry changes, since shapes must be in accordance with the nature of the feature (see, for example, Spagnuolo, Chapter 16, this volume). Subsequently, such research paves the way for developing tools which allow us to assess a change at different levels, from the elementary level of a partial quantitative measurement up to the level of a qualitative validation of shape.

Besides, such decompositions and qualifications, once conveniently coupled with knowledge on generalisation operators and algorithms, should allow us to apply in an opportunistic way the best tools and tunings to the various parts of the features to be generalised (Plazanet et al., 1995).

So far, there have been rather few attempts in this direction, and the current achievements do not allow us to consider that the problem is solved. Only a few feature classes have been considered, with promising but still not definite results. Such research needs to be tackled for various types of features and shapes, and therefore a lot still remains to be done.

Another type of knowledge is the importance of features for the targeted map or derived product. Such knowledge is needed to automate the selection process, but also, when extended to comparative relationships between features, for taking decisions in conflict resolution. If we first consider the intrinsic part of feature importance, that is to say, how important a feature or feature class is (or merely, in a binary form, should it be selected or not), some observations should be made:

- In most practical cases this knowledge is provided by the cartographer, who will in an interactive role be even more practical. A cartographer will interactively define the selections to apply, and later edit manually the data to delete further items.

- These operations relate to the map, or product, design activity, and we would like ideally that the system comprises enough structural and semantic knowledge, so that the cartographer just needs to define the true objective but not all the intermediate conditions which must be satisfied to reach the goal.

- However, relatively few research efforts have been devoted to such issues (for example, Hutzler and Spiess, 1993; Müller et al., 1993; Richardson, 1994), and kept closely to what might be termed 'surface knowledge', either by the use of some type of learning activity (Müller et al., 1993; Richardson, 1994) to relate directly feature classes with map classes, or by the intensive use of human/computer interaction.

It is true that going further leads quickly to taking into account the relationships which exist between features, as the representation of the geographic space is actually given not only by the logical entities but also by these relationships, and that much remains to be done in analysing our modelling techniques and their weaknesses (what is actually modelled, what is missing).

12.3.2 Knowledge on interactions between objects

If we go a step further we are now faced with the problems of spatial conflicts and the choice of actions in non-independent generalisation. While intrinsic knowledge may be sufficient for independent generalisation we then need knowledge on interactions between objects. At first, and rather sketchily, we may say that conflict resolution is a generalisation operation (just as simplification), while the choice of actions should be considered at the higher level of process management. This will be discussed further below.

The management of interactions also entails quality issues which so far have not been much considered. An example of such a quality factor could be the topological quality of a result, but also the homogeneity in density or the relative densities of feature classes or the relative arrangement of objects.

12.3.2.1 *Spatial conflicts*

Spatial conflict resolution is not the only problem in generalisation, but it is clearly crucial for the automation of generalisation. As a matter of fact, competition for map space has even been considered as a paradigm for map design (Mark, 1990). This problem is indeed at the heart of generalisation automation.

- Although we have tools for other operations that we can run systematically (for example Douglas' algorithm) we do not have any tool for solving conflicts in a similar way, but only some partial solutions for special cases.
- Some techniques exist which allow us to detect conflicts, but which do not give an account of the nature of the conflict. Only in rather simple situations are we able to deduce automatically the proper action to be undertaken.
- Conflict resolution has strong relationships with generalisation process scheduling. When an operation happens to give rise to a conflict, should the conflict be solved immediately, or should this action be delayed and then until which state? Semi-manual techniques often consist in applying simplification tools, and then processing the spatial conflicts, sometimes by merely undoing locally simplification modifications. However, a more satisfactory and automated strategy could integrate more closely conflict resolution and other generalisation actions.

It is worth noting that, as opposed to operations such as caricature, enhancement or typification, conflict resolution is not only needed for mapping (as often said) but also for resolution change in a 'pure' GIS context. Indeed, if we consider a raster organisation of data (straightforward representation of the discrete space, in which we actually operate in a digital environment) a mere resampling when resolution becomes larger is likely to introduce conflicts, that is to say false topologic relationships between objects. Therefore, this problem of conflict resolution cannot be considered as being only a matter of visualisation.

However, relatively few research efforts have been devoted to conflict resolution or conflict understanding. From Christ (1979) until now there have been probably less than a dozen studies on the various facets of this issue. At least three subareas still deserve further research.

1 Some analysis and formal classification of conflict types have been proposed (for example, Mackaness, 1994; McMaster and Shea, 1992), but related techniques for the detection and qualification of conflicts are still largely missing. We may detect proximity, but in a mixed situation is the problem first of congestion or coalescence?

2 Techniques for 'adaptative', or general, conflict resolution (*how* to generalise) are also still to be developed. Although various techniques for displacement have been proposed (Christ, 1979; Jäger, 1990; Monmonier, 1987), there is at present no general technique or strategy; depending on the context, conflict resolution may involve more than displacement (Müller and Wang, 1991), but also typification and aggregation.

3 Integration of conflict resolution in the generalisation process: interactions with other actions, scheduling (*when* and *why*). As soon as conflict resolution

involves different operations, and is triggered by various actions (instead of representing an isolated step in a linear process) we need to investigate how this task fits in the overall process. In fact, we may think that the process might be organised by the conflict-resolution steps as in Ruas (1995).

Although spatial conflicts may appear in limited numbers, this may be enough to prevent the application of a fully automated process. Indeed, costly and lengthy controls need to be performed, which may interrupt the process from time to time and (quite often) require manual interventions. This in turn may represent one or two days of work on a large map sheet. Besides, conflict resolution may be at the heart of the generalisation process, at least in some cases.

12.3.2.2 *Contextual generalisation*

Although conflict resolution is a prominent and fairly common example of a generalisation task which involves relationships between objects, there are many other situations in which the cartographer must think of an area as a whole, hence departing from the strict decomposition into individual features. As a matter of fact, such situations allow him or her to express his or her cartographic art, which makes it possible to render the geographic reality, or the geographic phenomena. He or she evaluates a situation rather than considers an object, and takes decisions based on spatial relationships between objects (orientation, relative location and proximity) and on the meaning of the initial abstraction ('points, lines and polygons') that he or she starts with.

There are many such situations, which explains the huge number of particular rules sometimes identified. In part, this may explain why some decisions will remain difficult to automate or, put in different words, why cartography may remain some kind of art. This applies mainly, if not only, to high-quality cartography, while the need for the ability to take such decisions may be less stringent for more 'ordinary' cartography, especially GIS displays.

However, there are special cases which may exhibit some generality. First, spatial and thematic structures. Such structures include buildings and streets in urban areas which, at least in old European towns, form patterns that characterise different urban areas (old, possibly medieval, town centre, nineteenth-century areas and new residential suburbs). If conveniently portrayed, such structures are often easy to identify visually.

It is clear that for the generalised version to convey the same meaning, the apparent structures need to be identified and considered when generalising. For example, building generalisation has so far mainly been tackled as being polygon (building outline) simplification tailored for this special class of polygons. As a result, if the building outline simplification works well, aggregation often fails to render the urban texture: streets and arrangements of buildings are not actually taken into account.

All in all, the only domain for which a significant part of the generalisation process as implemented in digital environments involves contextual knowledge may be land cover generalisation. Indeed, some rules are often used which are contextual in nature (for example, Schylberg, 1992, Richardson, 1994). This, however, remains an isolated case except for more specific studies such as hydrography generalisation in Müller *et al.* (1993).

12.3.3 Generalisation process – operations and scheduling

Several generalisation models and classifications of operations have been proposed. Although the operations proposed are rather close to each other (Beard, 1987; Beard and Mackaness, 1991; Buttenfield, 1985; McMaster and Shea, 1992 and others), there are more variations in the proposed models, from very abstract models focusing on states to models which focus on planning[6] (Brassel and Weibel, 1988) or put a significant emphasis on control, in a broad sense of the word, (McMaster and Shea, 1992).

So far we have missed several important building blocks or assessments. First, there is still some gap between the operations (rather low level, close to the algorithms) and the abstract tasks (decide when, why and how, elaborate a sequencing). We know what are the upper level tasks, but they remain difficult to translate into sequences of operations. Just as in planning activities, several intermediate analysis levels of increasing detail may help in constructing a solution (for example, decomposition of the working area into subunits, such as in Ruas, 1995). Further, we are missing tools needed for assessing a situation (which in some cases may be used as well for control activity). For these conceptual issues, the effort may now go on the intermediate levels, and on the techniques needed to implement them.

From a complementary point of view, approaches such as 'amplified intelligence' (Lee, 1993; Weibel, 1991; Weibel et al., 1995) and associated machine learning techniques may prove to be useful, by allowing us to extract experimentally well-suited sequencings. Briefly, the motivation for such an approach lies in the fact that it is extremely difficult to acquire knowledge from a cartographer (lack of formalisation, non-cooperative attitude and so forth) while it is possible to record his or her choices when using the tools intended for automated generalisation. But again, for such techniques to be useful at a deep level, we need to be able to recognise and qualify a situation so that a local choice may be related to such a situation.

At this stage, a lot remains to be done in this area, and ideas borrowed from the Artificial Intelligence field of planning may be useful, once the relationships between the higher levels and the operations are formalised (and quite likely, some intermediate levels identified). So far and from a practical point of view, approximated solutions are devised by cartographers using a 'try and check' approach. Such solutions tend to be difficult and costly (many attempts are needed, each of them entailing a lot of checking) while most often lead to many manual editing interventions.

12.3.4 Quality control and assurance – assessment of generalisation results

Quality is – or should be – a major factor in any production activity. In the research context quality assessment is needed for validating the applicability of some developed tool or method. This should be enough for motivating research in this direction for the digital cartography world. However, this is a subject to which attention has not been paid until quite recently, with a lack of assessment techniques as a result (Fisher, 1993), while large-scale experimental studies only began in 1994 within the framework of the OEEPE[7] Working Group on generalisation.

Further, the importance of such studies goes beyond the usual philosophical motivations stated above. In fact from a research perspective, tools developed for quality control are also needed for choosing the right actions and tuning the tools used, hence for progressing in the automation of generalisation. From a practical point of view, the advantage of disposing of quality assurance and quality control procedures and tools is twofold.

- First, this would guarantee a good homogeneity of results, even if the process remains partly interactive and is run by two different operators. In turn, this would guarantee that the resulting products are compatible, and may be assembled if needed. This is not the case with conventional manual cartography as the artistic freedom of cartographers tends sometimes to lead to striking variations.

- Second, as digital data and GIS tools diffuse in governmental agencies and private companies, map making is no longer a privilege of skilled cartographers. However, the currently available tools provide little automation and guidance, which often results in poor-quality maps. This is not a purely theoretical consideration as these maps may then be used in decision-making with poor consequences.

12.3.4.1 *Objective and quantitative criteria*

While qualitative criteria are difficult to formalise, several studies have been devoted to inherently quantitative or objective criteria and especially to geometric measures and simple semantic or topologic criteria. Besides, related work may be found in the fields of cartometry (Maling, 1989) and GIS quality (accuracy, semantic exhaustivity and so on).

However, surprisingly, no one system or generalisation product has already been evaluated using a complete range of such measures. Only partial objective assessments are available, while global assessments rely on human expertise and visual control. More experimental studies are needed for a better understanding of the scope of the available measures.

Moreover, geometric quality in the sense of accuracy maintenance is not enough. Again, the information conveyed by digital representation encompasses more than locational information. We often identify features (or subparts of features) on the basis of shapes (meanders or a series of hairpin bends). A cartographer who wants to assess the quality of a map will check not only the accuracy but the way the reality is portrayed and, at the local level, how shapes are rendered. Techniques for identifying and characterising shapes are therefore not only useful for maintaining them when needed, but also for checking how well they are maintained.

If we go to the global level, we then have to assess the global information content of the data set or map. Although some principles were proposed long ago (for example, the radical law, Topfler and Pillewizer, 1966, or information theory, Knöpfli, 1982) we are far from being able to assess the global content of a product, typical cartographic considerations (homogeneity in density and so forth) being left aside. Such an issue is fairly related with some ongoing studies on geostatistical analysis, which aim at characterising and/or segmenting GIS data sets, especially for quality assessment purposes.

12.3.4.2 *Subjective and non-quantitative criteria*

The measurements discussed in the previous section may be (partly) considered as related to the subjective appreciation of the cartographer or end-user, and this is most likely true of information density. As a subjective criterion becomes quantified, this quantification tends to be objective, although it may miss part of the necessary checking. Little has been done in the field of aesthetic quality, or clarity, and on deeper assessments of the information content which tend to be subjective. This by no means implies that such criteria are meaningless. Besides, some inputs may be found in the cartographic tradition and in the field of graphic semiology.

12.3.5 **The GIS context**

So far GIS provides only fairly simple displays of digital data (even cartographic symbolisation is still being improved because of initial limits). Several research studies have considered various kinds of 'intelligent zooming' techniques intended to enhance the current zoom in/zoom out usual functionality of GISs, either by relying on some kind of sampling (Merret and Shang, 1994), or prestored multiple representations (see Timpf, Chapter 14, this volume), or precomputed generalisation levels (van Oosterom and Schlenkelaars, 1993) but restricted to the use of Douglas' algorithm.

Such approaches, while providing an improvement in display capabilities, are still restricted in scope, either by their inability to provide a meaningful derived display (but a mere filtering) or the restriction to prestored representations (which in turn suffer from data volume and redundancy). It is true that interpolation techniques (Monmonier, 1991) may partly overcome the limitation to prestored representations, but if some changes need to happen in-between two levels (features which vanish, typification of a set of features or caricature) the system needs either more prestored levels (at the expense of a greater data volume) or significant generalisation capabilities.

We ideally would like on-the-fly generalisation functionalities to be available. Unfortunately this is not the case, and there are at least two impediments for such a capacity.

1   We do not have at hand a complete set of automated generalisation operations (meaning that we do not even have at least one algorithm for every needed operation).

2   We are not able to control, from an information point of view, the effect of most of our algorithms, nor to assess the information consistency of the global result.

Besides, it is well known that the performances of most of the current generalisation algorithms, using existing hardware technology, are far from being compatible with 'real-time' computation needs.

However, it is worth noting that in the GIS context and as opposed to conventional mapping, the speed of computation becomes relatively more important than high cartographic quality. Therefore, the integration of generalisation functions in these display facilities should deserve more attention, and pave the way for more flexible displays. Moreover, the derivation of data sets by means of generalisation is becoming more and more important in the production context (because of customers' demand), while requiring a lot of costly manual editing. Again, the integration of every available generalisation and quality control function in GIS tools should be considered as a priority. In the longer term, such functions would then be available for devising new and enhanced view mechanisms – for allowing a user to derive the representation he or she needs by means of database queries.

12.4 Conclusion

So far we dispose of isolated tools or tool-boxes which do not allow for a significant degree of automation except for rather small-scale reduction ratio and for large-scale ranges (for example, change achieves around 50 per cent of automation for 1:10 000 down to 1:25 000, and similarly MGE/MG – which, however, provides classical tools for small-scale generalisation – achieves a good automation degree for the same type of scale range and scale reduction ratio). In other cases (first of all, smaller scale ranges) we do not go beyond about 20 per cent of automation on average. However, these are very important scale ranges from a market point of view.

An exception may be raster map generalisation (for example, land cover maps). For this type of map satisfactory automated processes may be devised, even for small scales (1:100 000). Perhaps because these maps are in essence statistical maps, the portrayed 'features' are themselves more abstract, mainly at the small scales, whether in vector maps we get abstract representations of common life features.

Such a level of automation remains unsatisfactory for most applications: National Mapping Agencies could achieve significant gains if the processes become more automated; customers of geographic data sets who need to plot cartographic products (such as local authorities) are still faced with the choice of either investing a lot in such production (by paying a cartographic agency for producing the needed maps, for example) or accepting a lesser quality, and finally GIS users have to work with limited display capabilities.

In order to further automate generalisation processes we need to progress the analysis of features and of feature shapes. This is even needed for really automating such 'simple' tasks as simplification. Besides, we are still currently missing some tools such as caricature, or true typification. However, these tools also require a good understanding of shapes and feature natures.

Going one step further, progress should be made in the analysis and discovery of patterns of structures of objects/features. This would be a first step toward contextual generalisation, one could almost say true generalisation as opposed to independent generalisation.

The integration of tools is another essential research area (sequencing, process scheduling and planning), for which an important case study is conflict resolution. Progress in this direction is needed to design good generalisation process models, which will alleviate the need for a human operator to control and monitor the process.

Last but not least, is quality assessment and control. As stated above, progress in this direction is not only needed from a research point of view but from a very practical point of view. Many studies have been devoted to devising new algorithms for already partially solved tasks, but this is an area of research which still deserves more attention.

Finally, generalisation issues have strong interactions with GIS issues and cannot be thought of being of interest only for cartography:

- GIS data often, if not always, result from some generalisation process, which has implications for managing and using such data.

- Several research directions for generalisation share significant common problems with GIS research issues.

- Generalisation is a needed functionality for several GIS activities (data visualisation, cartographic output, and derivation of data sets) which are quite common or are becoming quite common.

- In both domains understanding and taking into account the *information content* (or intended meaning) of geographical digital data are deep and crucial issues.

It is often argued that not only is the automation of generalisation not feasible, but it is undesirable, because of the artistic content of mapping. Nevertheless some progress has been made in this direction, and is desirable even if this will lead to less art in standard map series. The key point is that digital generalisation will ultimately allow for a better assessment of the products: greater stability, homogeneity, objective and systematic assessment. We should not forget that in traditional cartography the assessment usually relies more on the confidence put into the cartographer, or on partial evaluation, than on exhaustive checking of the product. As a result there are sometimes striking variations in aspect and quality but also gross errors not discovered at control time but only later when the product is being used.

In other words, changing the art of generalisation into a technique of generalisation would have important benefits beyond the expected economic gains.

Notes

1   See Section 12.3, p. 192.

2   Or, once on-the-fly generalisation is available, we will be able to manage master databases while disposing of facilities for deriving information at the needed abstraction level.

3   We use here an analysis which has been used formerly several times.

4   We use here the wording 'independent generalisation' to denote actions which consist in transforming, or omitting, a feature without any regard to the surrounding features, in other words by considering only this feature independently of others.

5   Which by no means should imply that research on geometry generalisation is, or should be, given up.

6   Process identification, scheduling and so forth.

7   Organisation Européenne pour les Etudes Photogrammétriques Expérimentales.

References

ARMSTRONG, M. P. (1991). Knowledge classification and organization, in Buttenfield, B. P. and McMaster, R. B. (Eds) *Map Generalization: Making Rules for Knowledge Representation*, pp. 86–102, New York: Longman Scientific and Technical.

BEARD, M. K. (1987). How to survive on a single detailed data base, in Chrisman, N. (Ed.), *Proceedings of Autocarto 8*, pp. 211–20, Baltimore: ASPRS/ACSM.

BEARD, M. K. (1991). Constraints on rule formation, in Buttenfield, B. P. and McMaster, R, B. (Eds), *Map Generalization: Making Rules for Knowledge Representation*, pp. 121–35, New York: Longman Scientific and Technical.

BEARD, M. K. and MACKANESS, W. (1991). Generalization operations and supporting structures, *Proceedings of Autocarto 10*, pp. 29–45, Baltimore: ASPRS/ACSM.

BERTIN, J. (1973). *Sémiologie Graphique*, Paris: Mouton, Gauthier-Villars.

BERTIN, J. (1977). *La Graphique et le Traitement de L'information*, Paris: Flammarion.

BRASSEL, K. and WEIBEL, R. (1988). A review and conceptual framework of automated map generalization, *International Journal of GIS*, **2**(3), 229–44.

BURROUGH, P. A. and FRANK, A. U. (1995). Concepts and languages in spatial and temporal analysis: Mapping out the GIS hyperspace, *International Journal of GIS*, **9**(2), 101–16.

BUTTENFIELD, B. P. (1985). Treatment of the cartographic line, *Cartographica*, **22**(2), 1–26.

BUTTENFIELD, B. P. (1986). Digital definitions of scale-dependent line structures, *Proceedings of Autocarto London*, pp. 497–506, London: Autocarto London.

BUTTENFIELD, B. P. (1987). Automating the identification of cartographic lines, *The American Cartographer*, **14**(1), 7–20.

BUTTENFIELD, B. P. (1989). Scale-dependence and self-similarity in cartographic lines, *Cartographica*, **26**(1), 79–100.

BUTTENFIELD, B. P. (1991). A rule for describing line feature geometry, in Buttenfield and McMaster (1991) pp. 150–71.

BUTTENFIELD, B. P. and MCMASTER, R. B. (Eds) (1991). *Map Generalization: Making Rules for Knowledge Representation*, New York: Longman Scientific and Technical.

CHRIST, F. (1979). Ein Programm zur vollautomatischen Verdrängung von Punkt- und Linienobjekten bei der kartographischen Generalisierung, in *Internationales Jahrbuch für Kartographie*, pp. 41–64, Bonn: Kirschbaum Verlag.

CLARAMUNT, C. and MAINGUENAUD, M. (1994). Identification of a definition formalism for a spatial view, in Schraeder, F. H. (Ed.), *Proceedings of the International Workshop on Advanced Geographic Data Modelling (AGDM'94)*, Delft, The Netherlands: Netherland Geodetic Commission.

DOUGLAS, D. H. and PEUCKER, T. (1973). Algorithms for reduction of the number of points required to represent a digitized line or its caricature, *The Canadian Cartographer*, **10**(2), 112–22.

FISHER, P. (1993). 'Formalizing the evaluation of cartographic knowledge', Unpublished, position paper for the NCGIA Initiative 8 Specialist meeting on 'Formalizing cartographic knowledge'.

FREEMAN, H. (1978). Shape description via the use of critical points, *Pattern Recognition*, **10**, 2–14.

FREYTAG, U. (1993). Map functions, *Cartographica*, **30**(4), 1–6.

GOODCHILD, M. (1990). *Geographical Data Modelling*, NCGIA Technical Report 90–11, Santa Barbara: NCGIA.

GRÜNREICH, D. (1995). Development of computer-assisted generalization on the basis of cartographic model theory, in Müller *et al.* (1995) pp. 47–55.

HUTZLER, E. and SPIESS, E. (1993). A knowledge-based thematic mapping system — the other way around, in von Peter Mesenburg, H. (Ed.), *Proceedings of ICC'93*, **1**, 329–40.

JACOB, C. (1992). *L'empire des Cartes, Approche Théorique de la Cartographie à Travers l'histoire*, Paris: Albin Michel.

JÄGER, E. (1990). 'Untersuchungen zur kartographischen Symbolisierung und Verdrängung im Rasterdatenformat', PhD dissertation, Hannover: Institut für Kartographie, Universität Hannover, Nr 167.

JASINSKI, M. (1990). 'The comparison of complexity measures for cartographic line', NCGIA Technical Report, January 1990, Santa Barbara: NCGIA.

KNÖPFLI, R. (1982). Generalization – A means to transmit reliable messages through unreliable channels, in *Internationales Jahrbuch für Kartographie*, pp. 83–91, Bonn: Kirschbaum Verlag.

KUHN, W. (1994). Defining semantics for spatial data transfers, in Waugh, T. C. and Healey, R.G. (Eds), *Proceedings of the 6th International Symposium on Spatial Data Handling*, pp. 973–87, London: Taylor & Francis.

LEE, D. (1993). From master databases to multiple cartographic representations, in von Peter Mesenburg, H. (Ed.), *Proceedings of the International Cartographic Conference*, pp. 1075–85, Cologne: German Cartographic Society.

MACKANESS, W. (1994). Issues in resolving visual spatial conflicts in automated map design, in Waugh, T. C. and Healey, R. G. (Eds), *Proceedings of the 6th International Symposium on Spatial Data Handling*, pp. 325–40, London: Taylor & Francis.

MALING, D. H. (1989). *Measurements from Maps – Principles and Theory of Cartometry*, Oxford: Pergamon Press.

MARK, D. (1989). Conceptual basis for geographic line generalization, *Autocarto 9*, pp. 68–77, Baltimore: ASPRS/ACSM.

MARK, D. (1990). Competition for map space as a paradigm for automated map design, *Proceedings of GIS/LIS*, pp. 96–106, Anaheim: American Society of Photogrammetry and Remote Sensing.

MARK, D. (1991). Object modelling and phenomenon-based generalization, in Buttenfield and McMaster (1991) pp. 103–18.

McMASTER, R. B. (1983). A mathematical evaluation of simplification algorithms, *Proceedings of Autocarto 6*, pp. 267–76, Ottawa: Autocarto.

McMASTER, R. B. (1986). A statistical analysis of mathematical measures for linear simplification, *The American Cartographer*, **13**(2), 103–17.

McMASTER, R. B. (1987a). Automated line generalization, *Cartographica*, **24**(2), 74–111.

McMASTER, R. B. (1987b). The geometric properties of numerical generalization, *Geographical Analysis*, **19**(4), 330–46.

McMASTER, R. B. (1989). The integration of simplification and smoothing algorithms in line generalization, *Cartographica*, **26**, 101–21.

McMASTER, R. B. (1995). Knowledge acquisition for cartographic generalization: experimental methods, in Müller *et al.* (1995) pp. 161–79.

McMASTER, R. B. and SHEA, K. S. (1992). *Generalization in Digital Cartography*, Association of American Geographers Resource Publications in Geography.

MERRET, T. H. and SHANG, H. (1994). Zoom tries: a file structure to support spatial zooming, in Waugh, T. C. and Healey, R. G. (Eds) *Proceedings of the 6th International Symposium on Spatial Data Handling*, vol. 2, pp. 792–804, London: Taylor & Francis.

MONMONIER, M. (1983). Raster mode area generalization for land use and land cover maps, *Cartographica*, **20**(4), 65–91.

MONMONIER, M. (1987). Displacement in vector and raster mode graphics, *Cartographica*, **24**(4), 25–36.

MONMONIER, M. (1991). Role of interpolation in feature displacement, in Buttenfield and

McMaster (1991) pp. 189–203.

MOREHOUSE, S. (1995). GIS-based map compilation and generalization, in Müller *et al.* (Eds) (1995) pp. 21–30.

MÜLLER, J. C. (1986). Fractal dimension and inconsistencies in cartographic line representation, *The Cartographic Journal*, **23**, 123–30.

MÜLLER, J. C. (1987). Optimum point density and compaction rates for the representation of geographic lines, in Chrisman, N. (Ed.), *Proceedings of Autocarto 8*, pp. 221–30, Baltimore: ASPRS/ACSM.

MÜLLER, J. C. and WANG, Z. (1991). *Area-Patch Generalisation: A Competitive Approach*, ITC Technical Report, Eindhoven: ITC.

MÜLLER, J. C., LAGRANGE, J. P. and WEIBEL, R. (Eds) (1995). *GIS and Generalization — Methodology and Practice*, London: Taylor & Francis.

MÜLLER, J. C., PENG, W. and WANG, Z. (1993). Procedural, logical and neural net tools for map generalization, in von Peter Mesenburg, H. (Ed.), *Proceedings of ICC'93*, vol. 1, pp. 192–202, Cologne: German Cartographic Society.

VAN OOSTEROM, P. (1995). The GAP-tree, an approach to 'on-the-fly' map generalization of an area partitioning, in Müller *et al.* (1995) pp. 120–32.

VAN OOSTEROM, P. and SCHLENKELAARS, V. (1993). The design and implementation of a multi-scale GIS, *Proceedings of EGIS'93*, pp. 712–22, Utrecht: EGIS Foundation.

PLAZANET, C. (1995). Measurement, characterization and classification for automated line feature generalization, in Peuguet, D. J. (Ed.), *Proceedings of Autocarto 12*, vol. **4**, pp. 59–68, Anaheim, CA: American Society of Photogrammetry and Remote Sensing.

PLAZANET, C., AFFHOLDER, J. G. and FRITSCH, E. (1995). The importance of geometric modelling in linear feature generalization, in Weibel, R. (Ed.), *Cartography and GIS Special issue on Map Generalization*, **22**(9), pp. 291–305.

RICHARDSON, D. E. (1994). Generalization of spatial and thematic data using inheritance and classification and aggregation hierarchies, in Waugh, T. C. and Healey, R. G. (Eds), *Proceedings of the 6th International Symposium on Spatial Data Handling*, vol. 2, pp. 957–72, London: Taylor & Francis.

RIGAUX, PH. and SCHOLL, M. (1994). Multiple representation modelling and querying in Nievergelt, in Roos, J., Schek, T., Widmayer, H.-J and Widmayer, P. (Eds) *Proceedings of IGIS'94*, International Workshop on Advanced Information Systems, pp. 59–69, Berlin: Springer Verlag.

RUAS, A. (1995) Formalismes pour l'automatisation de la généralisation: métrique, topologie, partitionnement hiérarchique et triangulation locale, *Proceedings of the International Cartographic Conference*, Barcelona, forthcoming.

SCHYLBERG, L. (1992). 'Rule-based area generalization of digital topographic map', PhD dissertation, Lantmateriet, Sweden.

SPIESS, E. (1995). The need for generalization in a GIS environment, in Müller, J. C., Lagrange, J. P. and Weibel, R. (Eds), *GIS and Generalization-Methodology and Practice*, pp. 31–47, London: Taylor & Francis.

THAPA, K. (1988). Automatic line generalization using zero-crossings, *Photogrammetric Engineering and Remote Sensing*, **54**(4), 511–17.

TOPFLER, F. and PILLEWIZER, W. (1966). The principles of selection, *The Cartographic Journal*, **3**(1) pp. 10–16.

WEIBEL, R. (1991) Amplified intelligence and knowledge-based systems, in Buttenfield and McMaster (Eds) *Map Generalization: Making Rules for Knowledge Representation*, pp. 172–186, London: Taylor & Francis.

WEIBEL, R., KELLER, S. and REICHENBACHER, T. (1995). Overcoming the knowledge acquisition bottleneck in map generalization: the role of interactive systems and computational intelligence, in *Proceedings of COSIT '95*, pp. 139–56, Berlin: Springer.

CHAPTER THIRTEEN

Formalising Cartographic Knowledge

ROBERT McMASTER and BARBARA BUTTENFIELD

13.1 Introduction

It is not surprising, given the now almost complete metamorphosis of cartography into the digital realm, that researchers have turned to artificial intelligence in general, and expert systems specifically, to solve some of the more difficult problems of cartographic design. By necessity, then, research in automated cartography has increasingly relied on the potential for knowledge bases and expert systems. This has resulted in a series of prototype systems being developed, with varying degrees of success, in the areas of type placement, generalisation, projection selection, and lineage data management; other areas, however, even where strong conceptual frameworks exist, have been for the most part ignored. Thus, even with the existence of several well-developed conceptual frameworks in the area of cartographic symbolisation, such as those by Dent (1993), Hsu (1979), and MacEachren and DiBiase (1991), there has been little effort to develop a fully operational expert system to assist users selecting symbols. Given recent developments in knowledge acquisition methodologies and representational frameworks, it may now be possible to build knowledge bases for problems previously considered too difficult, such as map generalisation. The cartographic research community has put increased emphasis on building cartographic knowledge bases, designing fundamental knowledge acquisition techniques, and attempting to implement more advanced acquisition techniques, such as the application of neural networks.

13.2 Cartographic knowledge formalisation

This chapter will review current efforts in the acquisition and representation of cartographic data for the potential of developing expert systems. For the purposes of this chapter, knowledge formalisation will be identified as knowledge acquisition plus knowledge representation. Acquisition involves the process of extracting, through a variety of techniques and sources, primary cartographic information. This may involve sources such as textbooks, government documents and specifications, and from processes such as logging user activity. Representation involves the structuring of the acquired cartographic information, as in the design of the rule base or semantic net. The potential for cartographic expert systems relies, in the end, on the ability of cartographers to acquire and represent complex types data, such as those needed for automated generalisation.

In a previous attempt to comprehensively address the potential for expert systems in cartographic design, Buttenfield and Mark (1991), provided the following insights:

- Automated mapping requires a large degree of human interaction in order to produce products of acceptable quality.

- Many of the aspects of the cartographic process require human judgement, and the application of aesthetic principles.

- The question of expert systems for the design of more abstract maps, such as statistical maps, has been addressed very little in the literature.

- A major part of the research agenda for developing a cartographic expert system is to extract the rules from the programmes and articles discussed above, to put them into a standard form, and to enter them into an expert-system shell for evaluation and eventually integration with other expert system modules.

This study included a graphic that depicted the potential role of an expert systems approach for each of the major components of map design. In this illustration (Fig. 13.1) the length of the bar estimates the applicability (low, medium or high) of an expert-systems approach to various components of map design. The shaded area of each bar indicates the current progress in each. Note, for instance, that whereas there is much potential for an expert-systems approach in feature selection, little work has actually been completed. One excellent project in the identification of appropriate features, however, was carried out by Richardson and Müller (1991) for small-scale Canadian maps. Alternatively, the bar for label placement under PRODUCTION indicates that there has been much progress made. After a general discussion of both expert systems and techniques for cartographic knowledge acquisition, several examples of successes in formalising cartographic knowledge will be detailed. It should be noted that work by Mackaness *et al.* (1986) and Fisher and Mackaness (1987) has also addressed the possibility of an expert-systems approach for cartographic design.

13.3 Expert systems

One can find multiple definitions of expert systems in the literature. Mishkoff (1985) defines expert systems, or knowledge-based systems, as computer programmes that contain both declarative knowledge (facts about objects, events and situations) and procedural knowledge (information about courses of action) to emulate the reasoning processes of human experts in a particular domain or area of expertise. Brownston *et al.* (1985) explain that an expert system is a computer program, often written in a production-system language, that has expertise in a narrow domain. They assert that the difference between using traditional programming like C or FORTRAN, and expert-system rules, is that rules are easier to change than low-level code. Rules, they state, can also be given in a more descriptive way than traditional programming. According to Weiss and Kulikowski (1984) expert systems are computer programs that manipulate symbolic knowledge and heuristics to simulate human experts in solving real-world problems. Shea

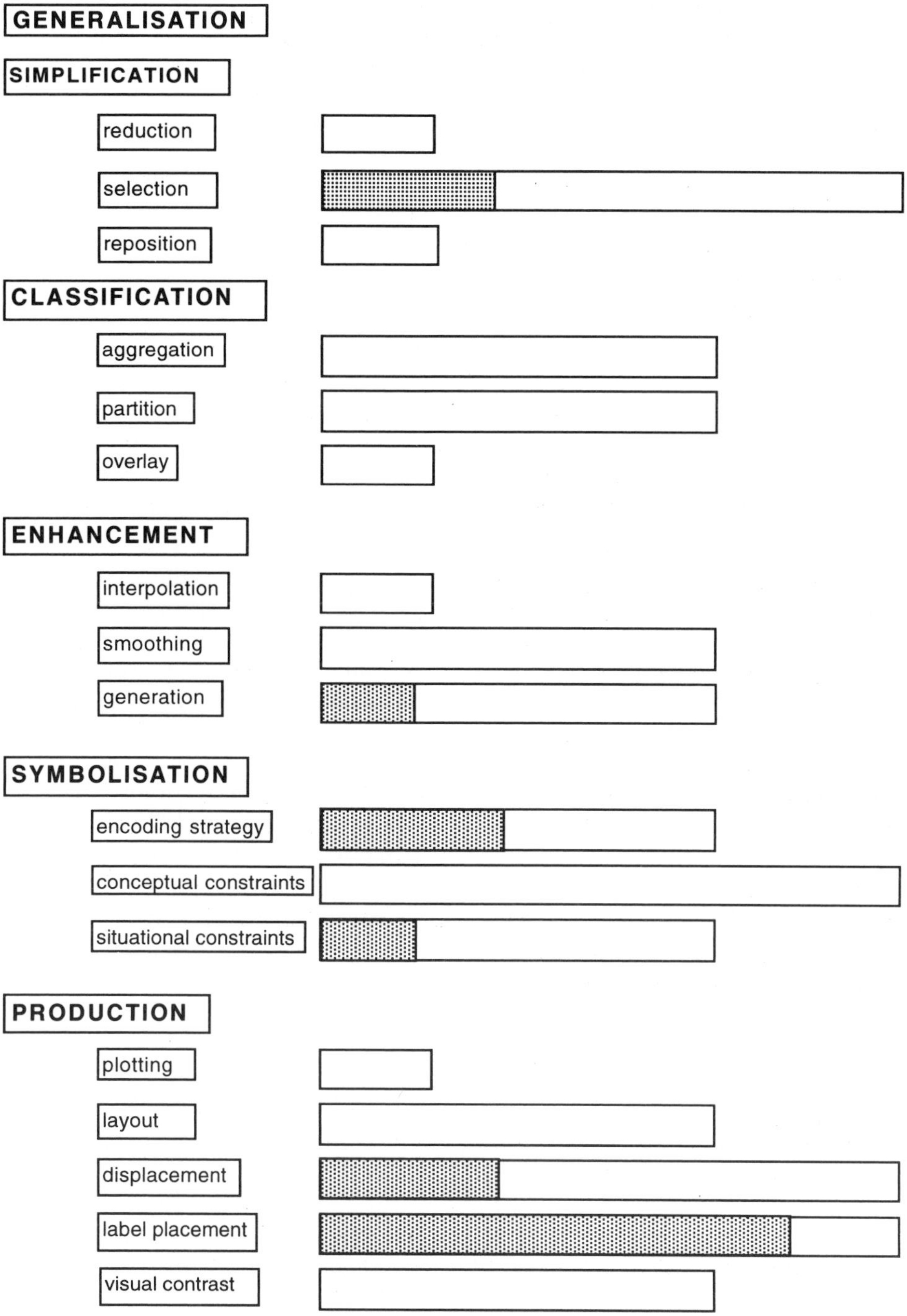

Figure 13.1 Buttenfield and Mark's (1991) role of expert systems in map design.

(1991) proposes 'Expert systems not only embody expert knowledge, they also have the ability to recount the steps taken to solve a problem, as well as gain proficiency at a particular task' (p. 6). Before proceeding, a few crucial definitions related to expertise, knowledge, and rules are provided from the work of Brownston *et al.* (1985).

> *Expertise*: Proficiency in a specialised domain. An expert system is said to have expertise in its domain if its performance is comparable to that of a human with five to ten years of training and experience in the domain (p. 447)
>
> *Knowledge*: Any information that can be represented as either *declarative knowledge* or *procedural knowledge*, for example, in the form of rules, entries in data memory or another database, or control strategies. Knowledge may be specific to a task domain or general enough to be independent of all domains (p. 448)
>
> *Rules*: A unit of representation that specifies a relationship between situation and action. Rules are ordered pairs that consist of left-hand side and a right-hand side. . . . In production systems, rules are the units of production memory and are used to encode procedural knowledge. A rule is also called a production (pp. 453–4)

Typically, an expert system consists of three components: the actual knowledge base; the interpretative inference engine; and the user interface (Fig. 13.2). The knowledge base contains the declarative (factual) and procedural (heuristic) knowledge necessary for the particular domain – for example, cartographic design. In the context of formalising cartographic knowledge, populating the knowledge base remains the major impediment. How can the appropriate cartographic knowledge be acquired and subsequently represented in existing rule bases? The inference engine is made up of rules that are used to control how the rules in the knowledge base are used or processed. Finally, the user interface allows communication or interaction between the expert system and an end-user. Several studies, such as those by McMaster (1995), McMaster and Chang (1993), McMaster and Mark (1991) and Weibel (1995), have suggested that the user interface itself may be used to acquire cartographic knowledge.

All expert systems may be typified by a typical life cycle (Fig. 13.3). This production flow, designed by McGraw and Harbison-Briggs (1989), identifies the steps of (1) problem statement and requirements, (2) feasibility study, (3) refinement of requirements, (4) design decisions, (5) prototype version, (6) delivery system and (7) maintenance. As highlighted in Fig. 13.3, and detailed shortly, most of current effort in building expert systems for cartography is now in stages 3, 4 and 5; up to this point in time, very few completed delivery systems have been documented. However, several interesting prototype systems have been developed, including those for type placement, projection selection and, most recently, cartographic generalisation and a few systems have actually reached production, as in the case of the Institute Géographique Nationale's (IGN) type placement procedure.

In building such systems, two types of individuals normally participate: knowledge engineers and domain experts. The knowledge engineer is the designer of the expert system, while the domain expert provides the context-specific declarative and procedural knowledge. The difficulty in the process, of course, is the problem of knowledge acquisition – the transferral of the context-specific knowledge from the domain expert to the knowledge engineer, and the con-comitant representation (structuring) of the knowledge, normally in the form of

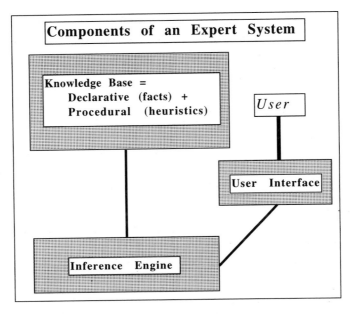

Figure 13.2 Components of an expert system.

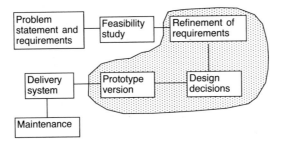

Figure 13.3 Typical life cycle for an expert system (modified from McGraw and Harbison-Briggs, 1989).

rules, in the expert system. McGraw and Harbison-Briggs (1989), in providing a more detailed description of the knowledge acquisition (KA) process, divide the KA process into three phases, including the basic steps of problem definition, initial KA sessions, and knowledge implementation. For instance, in conducting the initial KA sessions, the possible techniques include developing sample problems, using interviewing techniques, and using protocol analysis. However, such generic techniques are often not appropriate for acquiring cartographic knowledge. The next section outlines the different strategies for classifying cartographic knowledge and, based in part on this classification, suggests a set of proposed techniques for acquisition.

13.4 Knowledge classification

At the most basic level, two types of knowledge may be identified: declarative and procedural. Declarative knowledge provides the textbook knowledge of a field, including facts, concepts and descriptive information about a particular domain. Brownston *et al.* (1985) contend that such declarative knowledge can be retrieved and stored, but cannot be immediately executed; to be effective, it must be interpreted by procedural knowledge. Examples of declarative knowledge in cartography include type placement and style, colour selection for certain maps such as hypsometric charts, and many minimal feature sizes for government charts. Procedural knowledge, on the other hand, represents the how-to aspects of a field, including procedures, strategies, or heuristics that are employed to solve problems or achieve goals. Brownston *et al.* (1985) asserts that procedural knowledge can be immediately executed using declarative knowledge.   In the domain of cartography, examples of procedural knowledge include the selection and sequencing of generalisation operators, the identification of appropriate interpolation routines, and the selection of an error statistic for a given planimetric position on a map. In the context of cartographic generalisation, Armstrong (1991) identified three types of knowledge, including: (1) geometrical, or what cartographers recognise as a set of feature descriptions encompassing absolute and relative locations; (2) structural, or the intrinsic expertise of the scientist, such as the demographer, meteorologist, geomorphologist, and (3) procedural, or the operations and sequencing of operations necessary for generalisation.

From a different perspective – computer science – five classes of knowledge have been proposed (McGraw and Harbison-Briggs, 1989). These may be identified as declarative, procedural, two classes of semantic, and episodic (Table 13.1).

Note that two levels of semantic knowledge are identified: that which identifies the major concepts and vocabulary and that which identifies the decision-making procedures and heuristics.   In cartography, the concept of detailed semantic knowledge has become the basis for 'richer' more meaningful databases, such as the USGS's DLG-E files (Buttenfield, 1995). A classification better suited for the needs of cartographers might expand on the declarative/procedural types to include both semantic and structural. It also appears that, as proposed by several researchers in the discipline, knowledge can be both deep (database) and surface (map) (Table 13.2). The bottom half of this table details the types of features/processes that might be included in each of the four basic categories.

Weibel (1995) has identified a series of potential methods/uses for capturing cartographic knowledge (Table 13.3). Early research efforts put emphasis on the potential for interviews and the analysis of (mostly) government documents. However, due to the loosely structured nature of much of this information, the extraction of definite rules proved difficult if not impossible. More current activity appears to emphasise the potential for machine learning and amplified intelligence strategies, although both are now at an incipient stage.

13.5 Formalising cartographic knowledge

Several of the terms used need further definition. Reverse engineering involves a detailed study of maps – of the same geographical area but of different scale – in order to ascertain the types of features present, how these features change with scale and the procedures that might have been used in, for instance, the generalisation of the features. One can then trace individual features as they change through scale, and identify the 'procedures' that describe the geometric

Table 13.1 McGraw and Harbison-Briggs (1989) types of knowledge

Knowledge	Activity	Suggested technique
Declarative	Identifying general (conscious heuristics)	Interviews
Procedural	Identifying routine procedure/tasks	Structured interview Process tracing Simulations
Semantic	Identifying major concepts/vocabulary	Repertory grid Concept sorting
Semantic	Identifying decision-making procedures/heuristics	Task analysis
Episodic	Identifying analogical problem-solving heuristics	Simulations Process tracing

Table 13.2 Classification of cartographic knowledge

	Surficial (low?)	Deep (high?)
Declarative (geometric) Procedural (algorithms) Structural Semantic		

Surficial structural	= Ridge and valley lines
Deep structural	= Knowledge of generating hydrologic process (Radial drainage based on geology) Knowledge of urban transportation growth
Surficial geometric	= x–y coordinate pairs, pixels
Deep geometric	= Relationships of proximity, neighbourliness, built into topological data structures
Surficial procedural	= Tolerance values for algorithms
Deep procedural	= Sequencing and iteration of algorithms Differential generalisation
Surficial semantic	= Basic attributes (road class)
Deep semantic	= Density of buildings along road Quality of vegetation w/in 100 m

changes. Amplified intelligence, on the other hand, provides the user with some control of a given process, such as symbol selection, but allows the computer – using an expert system, for instance – to make most of the decisions. The application of amplified intelligence has also been suggested as a possible means for knowledge acquisition, as in the direct-manipulation user interface discussed above.

So what has been accomplished to date? The following section will report on four efforts, considered reasonably successful at formalising cartographic knowledge for either a fully automated or expert-systems strategy. These include

Table 13.3 Weibel's (1995) classification of knowledge acquisition techniques

Method	Potential use
Conventional KE interviews	Establish initial framework
	Background for other methods
Analysis of documents	Establish initial knowledge base (procedural knowledge)
	Significant source
Reverse engineering	Formalise rules for selection
Map analysis	Acquisition of procedural knowledge
Machine learning	Interpretation of large numbers of facts from reverse engineering
	Refinement of initial rules
Neural networks	Replacement of operators by more holistic approaches
	Classification (structure recognition)
Amplified intelligence	KA through interactive logging
	Acquisition of procedural knowledge
	Evaluation of operators

efforts in map projection selection, type placement and vector/raster generalisation. For each, the raw cartographic information is provided, followed by an example of the formalisation. It should also be noted that, while the examples may appear somewhat dated, the emphasis is on how original cartographic knowledge may be formalised.

13.5.1 Map projections

Basic rules for the selection of map projections have existed for many years. Conventional wisdom – and textbook knowledge – tells users, for instance, that for a given application (for example, surveying), only a conformal projection should be selected. Over many centuries, specific projections have been designed for most cartographic applications and regions of the globe. An excellent example of a framework for the selection process was provided by Snyder (1987), in which a detailed decision tree was designed (Fig. 13.4). Using this conceptual structure for map-projection selection knowledge, combined with map-projection classification knowledge, an expert system – MaPKBS – based on production rules and frames was developed by Nyerges and Jankowski (1989). As described by Jankowki and Nyerges (1989), 'The main purpose of knowledge specification is the formalisation of the domain knowledge' (p. 86) and, based on Snyder's decision tree, includes:

- categories of the size of geographic area to be mapped – world, hemisphere, continent, ocean, sea, country, region;
- geographic area attributes – location, directional extent;
- geometric attributes describing distortion of the area on a map projection – shape, size, direction, distance, scale;
- trade-offs and interdependencies between geographic areas and their attributes – geometric attributes describing distortion, map functions and different map projections.

Figure 13.4 Map-projection selection decision tree. (*The American Cartographer*, vol. 16, No. 1, p. 36, used with permission of the American Congress on Surveying and Mapping.)

Figure 13.5 depicts the representational structure of the knowledge base in MaPKBS, detailing how the control of the reasoning process is carried out using production rules and frames.

13.5.2 Type placement

Any respectable cartographer can recite many of the rules provided by Edward Imhof in his seminal work on name placement. The paper most accessible in

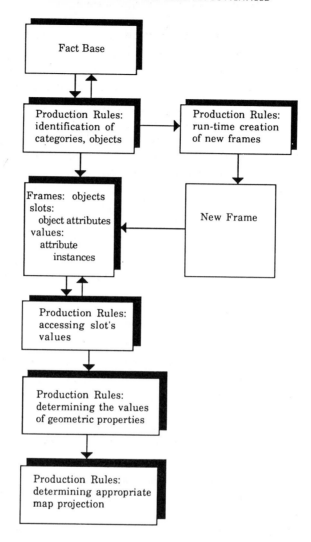

Figure 13.5 Representational structure of the knowledge base in MaPKBS. (*The American Cartographer*, vol. 16, No. 2, p. 90, used with permission of the American Congress on Surveying and Mapping.)

English (by Imhof), translated by George McCleary in 1975 and published in *The American Cartographer*, provides 'general principles and requirements' for the cartographer. Presented here are six principles (rules) to guide type placement:

1 The names should in spite of their incorporation into the dense graphics of the map, be easily read, easily discriminated, and easily and quickly located.

2 The name and the object to which it belongs should be easily recognised.

3 Names should disturb other map contents as little as possible. Avoid covering, overlapping, and concealment.

4 Names should assist directly in revealing spatial situation, territorial extent, connections, importance, and differentiation of objects.

5 Type arrangement should reflect the classification and hierarchy of objects on the map; variation style and size help to do this.

6 Names should not be evenly dispersed over the map, nor should names be densely clustered. Here name selection and name arrangement are important.

Figure 13.6 shows Imhof's (1975) graphical solutions for several punctiform type problems. Mower (1993), using these and other guidelines, utilised a parallel computer for solving the type placement problem. In building such rule-based type placement programs, as explained by Mower; 'A general-purpose, name-placement procedure must be able to map arbitrary regions over a broad range of scales, extract features in the region and rank their importance, and resolve competitions for map space in favour of features of greater importance. At smaller map scales, insufficient space will exist to place all features and their labels in an acceptable manner.' Mower's application of SIMD (single-instruction, multiple data) architecture to a GNIS (Geographic Names Information System) for Central New York State results in Fig. 13.7. Most professional cartographers would find this result quite acceptable.

13.5.3 Raster-based generalisation

In the problem domain of cartographic generalisation, which represents one of the significant unsolved problems in automated cartography, there have been several research efforts to formalise knowledge and build expert systems. Using maps from the National Land Survey of Sweden, Schylberg (1993) completed a detailed study of the potential for a rule-based approach to raster-format land use/land cover mapping for the generalisation processes of amalgamation, simplification and smoothing. Some of the rules extracted from the National Land Survey Handbooks include:

- Lakes are included at the scale of 1:50 000, if they have a minimum size of $25 \times 25\,m^2$ and at the scale of 1:100 000 if they have a minimum size of $50 \times 50\,m^2$.

- If the width of a river over a stretch longer than 500 m is greater than 10 m on a map of scale 1:50 000 and 20 m on a map of scale 1:100 000 it is represented with double shore lines, otherwise as a single line.

Regarding land cover, the handbook states the following.

- Open fields should be included if the area is greater than $50 \times 50\,m^2$.
- Forests are included if the area is greater than $100 \times 100\,m^2$.
- Wetlands smaller than $50 \times 50\,m^2$ at a map scale of 1:50 000 and smaller than $100 \times 100\,m^2$ at a map scale of 1:00 000 are not included.

Guidelines for representing area features on the 1:25 000 map of Sweden include:

Cover	Smallest area (in m^2)	Number of 5 m pixels
Hydrology	175×175	1 237
Wetland	600×600	14 400
Forest	245×245	2 401
Built-up areas	500×500	10 000

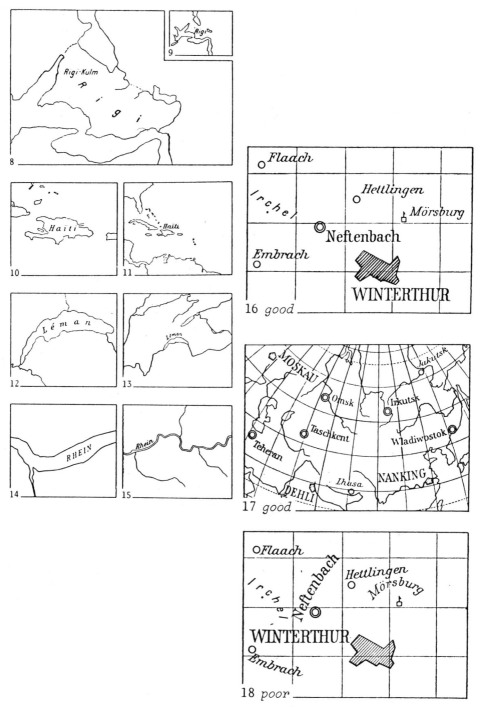

Figure 13.6 Imhof's (1975) name placement rules for punctiform features. (*The American Cartographer*, vol. 2, No. 2, p. 131, used with permission of the American Congress on Surveying and Mapping.)

Figure 13.7 Mower's (1993) parallel architecture automated name placement strategy. (*Cartography and Geographic Information Systems*, vol. 20, No. 2, p. 78, used with permission of the American Congress on Surveying and Mapping.)

Using such rules, Schylberg (1993) designed an approach for simplifying a raster database through deletion (Fig. 13.8), as well as amalgamation, simplification and smoothing. One can see from this illustration how the process actions (inference engine) acts on the rules to eliminate features. Figure 13.9 depicts the cartographic result of the application of one set of rules to the feature classes of water, wetland, forest, open fields and built-up areas.

13.5.4 Vector-based generalisation

Another study, completed by Nickerson (1988), applied a rule-based approach for the generalisation of vector-format maps. The research was based on a model of the generalisation process designed specifically for expert systems and consists of five tasks: (1) the four distinct feature modification operations; (2) symbol scaling; (3) feature relocation and symbol placement; (4) scale reduction; and (5) name placement. As described by Nickerson and Freeman (1986), the generalisation process is one of effectively converting the *source map*, of scale *1:s*, symbol size *a*, and area $w \times h$ to an *intermediate* scale map, of scale *1:s*, symbol size *ka*, and area $w \times h$, before final conversion to the *target* map, of scale *1:ks*, symbol size *a* and area $w/k \times h/k$. It is at the intermediate map scale that features are relocated and symbols replaced due to overlap and interference. The production rules used by Nickerson (1988) (Fig. 13.10) were coded in FORTRAN in the MapEx system. For example, rule 7.1 states:

> IF   new_scale is greater than or equal to 100 000, and the feature has a route number specified and the feature is not a divided highway,
> THEN   simplify the feature, and write the simplified feature.

The experimental MapEx program, through the application of complex rules, could delete, simplify, combine, and displace linear features.

13.6 Consistency in expert terminology

A different – nonalgorithmic – approach to formalising knowledge for cartographic generalisation was reported by Rieger and Coulson (1993) in order to answer the question: 'what is the present level of understanding and agreement among cartographers of the process of generalisation?' Specifically, the study was designed to capture the 'generalisation knowledge' of a small group of expert cartographers with the following objectives:

1 determine the type and variety of procedures used for generalisation;
2 the definitions of those procedures;
3 the characteristics of the procedures; and
4 the conceptual system of each expert.

Their methodology asked 23 'expert' cartographers about their knowledge of specific operations of generalisation, such as simplification, smoothing, amalgamation and displacement. The group included academically and technically trained cartographers employed in universities, government map-producing departments, and private map-producing companies. Partial results of this study are provided in Table 13.4.

Their findings included the following.

1 Of the 11 terms for the procedures which were taken from the cartographic literature, the terms were not defined in the same way by all the experts and a few were not understood at all.
2 Simplification, aggregation, and classification had the most varied definitions.
3 Smoothing, symbolisation, selection and elimination were similarly defined by most of the subjects.
4 Displacement and exaggeration were placed in the middle.
5 Typification and induction were not understood by an overwhelming majority of the subjects and therefore no definitions could be given.
6 If it is accepted that expert systems should be based on the knowledge that the community of experts holds in consensus, then the domain of cartographic generalisation is currently not suited to expert-systems technology.

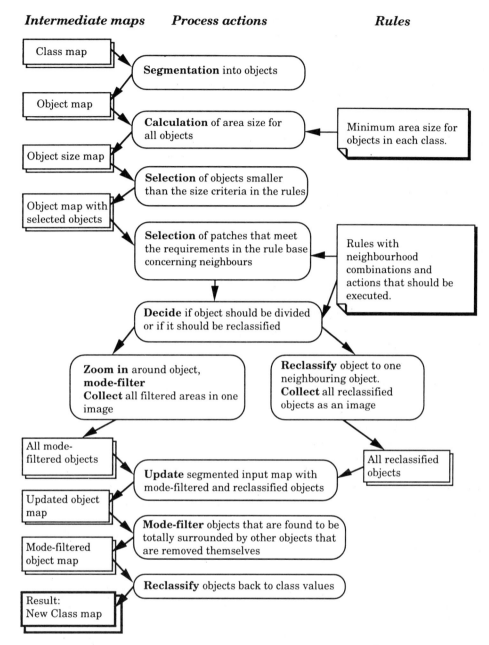

Figure 13.8 Schylberg's (1993) procedure to simplify a raster database. (With kind permission of Lars Schylberg.)

This study indicates that it may be necessary to confirm through user testing and surveys, before the development of a rule base for a given problem domain and that there exists enough consistency within the professional community to warrant the application of an expert-systems approach. For certain applications, including type placement and projection selection, such consistency exists. For others, such

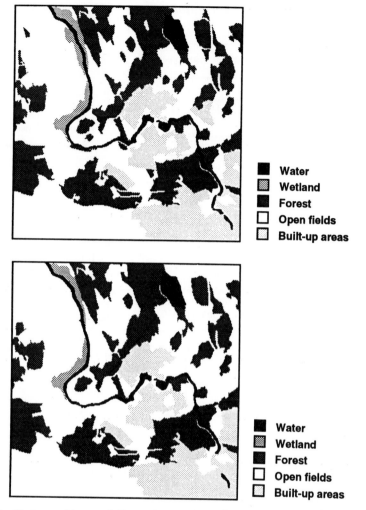

Figure 13.9   Cartographic result from the application of Schylberg's (1993) rule base. (With kind permission of Lars Schylberg.)

as generalisation, more basic research on identifying and classifying the unique operations is needed.

13.7 Conclusions

It appears that fully automated solutions to cartographic design will rely, to some degree, on an expert-systems strategy that will necessitate improved methods for knowledge formalisation — the acquisition and representation of cartographic knowledge. Although one cannot point to widespread acceptance of an expert-systems approach to cartographic design, the research community, based on recent activities, can be optimistic about their potential. In addition to work on the classification of cartographic knowledge, researchers – in several disciplines – have now developed a series of techniques for knowledge acquisition, including reverse

Ruleset for roads begin

Rule 7.1

IF   new_scale is greater than or equal to 100 000, and the feature has a route number specified, and the feature is not a divided highway,

THEN   simplify the feature, and write the simplified feature

Rule 7.2

IF   new_scale is greater than or equal to 100 000, and the feature is a divided highway,

THEN   get the entire divided highway, and combine all pieces into a single feature, and simplify the feature, and write the simplified feature

Rule 7.3

IF   new_scale is greater than or equal to 250 000, and the feature is an interchange,

THEN   get all other pieces of the interchange, and replace them with the symbolic interchange symbol, and write the symbolic interchange symbol

Figure 13.10 Production rules used by Nickerson (1988).

engineering, structured interviews, extracting textbook knowledge and amplified intelligence. Implementation of these techniques can now be found in multiple areas of design such as projection selection, type placement and cartographic design. In the area of type placement, the formalisation of knowledge has led to acceptable solutions based on manual standards. Experiments in the development of knowledge bases for cartographic generalisation have led to several promising prototype systems, but no widespread production systems. In other areas, such as in projection selection and cartographic symbolisation, there has been little progress beyond conceptual frameworks. It is unlikely that a comprehensive expert system for all aspects of the cartographic process will be developed, yet domain-specific modules can be successfully implemented. Work is now progressing in multiple directions, including projects involving reverse engineering, amplified intelligence and the development of rich semantic databases, which may be considered, in effect, an expert-systems approach. Perhaps the most active area in the application of expert systems is in cartographic generalisation where, for this particular problem domain, research in both North America and Europe is continuing the attempt to apply expert systems. However, most researchers would now admit that, while the progress in formalising knowledge on the generalisation

Table 13.4 Number of experts using given generalisation terms in the Reiger and Coulson (1993) study

Term	No. of experts using term ($n = 23$)
Simplification	12
Classification	21
Displacement	20
Selection	5
Elimination	18
Exaggeration	21
Aggregation	17
Symbolisation	23
Smoothing	20
Induction	1
Typification	5

process has been positive, a totally automated system now, or perhaps ever, is unlikely. Thus, it is promising that cartographers are slowly coming to the conclusion that, indeed, many of the decisions on design are perhaps best left in the hands of the user.

References

ARMSTRONG, M. (1991). Knowledge classification and organization, in Buttenfield, B. P. and McMaster, R. B. (Eds), *Map Generalization: Making Rules for Knowledge Representation*, pp. 86–102, London: Longman.

BROWNSTON, L., FARRELL, R., KANT, E. and MARTIN, N. (1985). *Programming Expert Systems in OPS5: An Introduction to Rule-Based Programming*, Reading, MA: Addison-Wesley.

BUTTENFIELD, B. P. (1995). Object-oriented map generalization: modelling and cartographic considerations, in Muller, J. C., Lagrange, J.-P. and Weibel, R. (Eds) *GIS and Generalization: Methodology and Practice*, pp. 91–105, London: Taylor & Francis.

BUTTENFIELD, B. P. and MARK, D. M. (1991). Expert systems in cartographic design, in Taylor, D. R. F (Ed.) *Geographic Information Systems: The Computer and Contemporary Cartography*, pp. 129–50, Oxford: Pergamon Press.

BUTTENFIELD, B. P. and McMASTER, R. B. (Eds) (1991). *Map Generalization: Making Rules for Knowledge Representation*, pp. 3–20, London: Longman.

DENT, B. (1993). *Cartography: Thematic Map Design*, Dubuque, Iowa: Wm. C. Brown.

FISHER, P. and MACKANESS, W. A. (1987). Are cartographic expert systems possible?, in Chrisman, N. S. (Ed.) *Proceedings, AUTO-CARTO-8, Eighth International Symposium on Computer-Assisted Cartography*, pp. 530–40. Falls Church, VA: American Society of Photogrammetry and American Congress on Surveying and Mapping.

HSU, M.-L. (1979) The Cartographer's Conceptual Process and Thematic Symbolisation, *Cartography and Geographic Information Systems*, **6**(2), 117–127.

IMHOF, E. (1975). Positioning names on maps, *The American Cartographer*, **2**(2), 128–4.

JANKOWSKI, P. and NYERGES, T. (1989). Design considerations for MaPKBS-map projection knowledge-based system, *The American Cartographer*, **16**(2), 85–95.

MACKANESS, W. A., FISHER, P. and WILKINSON, G. G. (1986). Towards a cartographic expert system, *Proceedings AUTO-CARTO LONDON, International Symposium on Computer-Assisted Cartography*, pp. 578–87, London, UK: RICS.

MACEACHREN, A. M. and DiBIASE, D. (1991). Animated Maps of Aggregate Data: Conceptual and Practical Problems, *Contemporary and Geographic Information Systems*, **18**(4), 221–229.

MCGRAW, K. L. and HARBISON-BRIGGS, K. (1989). *Knowledge Acquisition: Principles and Guidelines*, Englewood Cliffs, NJ: Prentice Hall.

MCMASTER, R. B. (1995). Knowledge acquisition for cartographic generalization: experimental methods, in Müller, J. C, Lagrange, J.-P. and Weibel, R. (Eds), *GIS and Generalization: Methodology and Practice*, pp. 161–79, London: Taylor & Francis.

MCMASTER, R. B. and CHANG, H. (1993). Interface design and knowledge acquisition for cartographic generalization, *Proceedings, Eleventh International Symposium on Computer-Assisted Cartography*, pp. 187–96, Minneapolis, MN.

MCMASTER, R. B. and MARK, D. M. (1991). The design of a graphical user interface for knowledge acquisition in cartographic generalization, in *Proceedings GIS/LIS'91*, pp. 311–20, Atlanta: GA.

MISHKOFF, H. (1985). *Understanding Artificial Intelligence*, Dallas TX: Texas Instruments Inc.

MOWER, J. E. (1993). Automated feature and name placement on parallel computers, *Cartography and Geographic Information Systems*, **20**(2), 69–82.

NICKERSON, B. (1988). Automated cartographic generalization for linear features, *Cartographica*, **25**(3), 15–66.

NICKERSON, B. and FREEMAN, H. (1986). Development of a rule-based system for automated map generalization, Proceedings, *Second International Symposium on Spatial Data Handling*, pp. 537–56, Seattle, Washington.

NYERGES, T. and JANKOWSKI, P. (1989). A knowledge base for map projection selection, *The American Cartographer*, **16**(1), 29–38.

RICHARDSON, D. E. and MÜLLER, J. C. (1991). Rule selection for small-scale map generalization, in Buttenfield, B. P. and McMaster, R. B. (Eds), *Map Generalization: Making Rules for Knowledge Representation*, pp. 136–49, London: Longman.

RIEGER, M. and COULSON, M. R. C. (1993). Consensus or confusion: cartographers' knowledge of generalization, *Cartographica*, **30**(2),(3), 69–80.

SCHYLBERG, L. (1993). Computational methods for generalization of cartographic data in a raster environment, Doctoral thesis, Royal Institute of Technology, Department of Geodesy and Photogrammetry, Stockholm, Sweden.

SHEA, K. S. (1991). Design considerations for an artificially intelligent system, in Buttenfield, B. P. and McMaster, R. B. (Eds), *Map Generalization: Making Rules for Knowledge Representation*, pp. 3–20, London: Longman.

SNYDER, J. P. (1987). Map projections – a working manual. US Geological Survey Professional Paper 1395. US Geological Survey, Reston: US Government Printing Office.

WEIBEL, R. (1995). Three essential building blocks for automated generalization, in Müller, J. C., Lagrange, J.-P. and Weibel, R. (Eds), *GIS and Generalization: Methodology and Practice*, pp. 56–69, London: Taylor & Francis.

WEISS, S. M. and KULIKOWSKI, C. A. (1984). *A Practical Guide to Designing Expert Systems*, Totowa, New Jersey: Rowman & Allenheld.

CHAPTER FOURTEEN

Cartographic Objects in a Multiscale Data Structure

SABINE TIMPF

14.1 Introduction

Geographic Information Systems manage data with respect to spatial location and those data are presented graphically as a map or sketch. There are a number of similar graphics applications, where a database of entities with some geometric properties is used to render these entities graphically for different tasks. Typically, these tasks require graphical representations at different levels of detail, ranging from overview screens to detailed views (Herot *et al.*, 1980). A function to draw cartographic sketches quickly and in arbitrary scales is needed as to date most applications only allow changes to twice or half the original scale before major distortions occur. This calls for further progress in map generalisation, a notoriously difficult problem. Efforts to achieve automated cartographic generalisation have been successful for specific aspects (Freeman and Ahn, 1987; Powitz, 1993; Staufenbiel, 1973), but no complete solution is known, nor are any to be expected within the immediate future as discussed by Lagrange, and McMaster and Buttenfield in Chapters 12 and 13 in this volume.

14.1.1 Motivation

Displays should be more adaptable to users' needs and their tasks instead of creating and presenting a static view of the data (Lindholm and Sarjakoski, 1994). In current approaches, objects to render are selected for each map and then transformed into cartographic objects to construct a map. This conforms to the traditional view of the cartographic process (Fig. 14.1). Generalisation is done each time a new map is constructed and each time a new selection of the entities to display is performed. The term 'entities' refers to the things in the world, whereas the term 'objects' refers to things in the database.

The simplification we propose is to transform the entities just once into cartographic objects. We assume that geographic objects have a slow change rate and that the database is updated at appropriate intervals. The cartographic objects

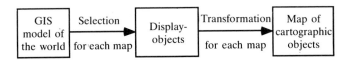

Figure 14.1 Traditional view of the cartographic process.

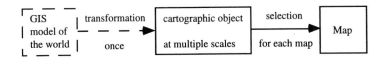

Figure 14.2 Selection process.

are stored at multiple scales in a single database (Beard, 1987) from which they are selected for each map output (Fig. 14.2).

This database will not be much larger than the most detailed database assumed in current proposals (assume that every generalised representation is a quarter of the size of the previous one, then the total storage requires only one-third more capacity than the most detailed data set). Generalised representations can be collected from existing maps or for some cases produced automatically. A similar approach has already been presented by van Oosterom (1989), who proposed to use an R-tree for storage of the generalised objects. Our approach is broader, in that we model the underlying conceptual organisation of cartographic objects in a map and are not yet concerned with implementation methods.

The data structure chosen is a multiscale cartographic forest, where renderings for objects are stored at different levels of detail. A forest is a collection of trees and therefore essentially a directed acyclic graph (DAG). This extends ideas of hierarchies or pyramids and is related to quad-trees (Samet, 1989b) and strip-trees (Ballard, 1981). The structure of the multiscale forest is based on a trade-off between storage and computation, replacing all steps which are difficult to automate by redundant storage. The resulting forest structure is more complex than the hierarchical structures proposed in the literature so far (Samet, 1989a and b). Objects may change their appearance considerably, for example, they change their spatial dimension, or change from a single object to a group of objects. Special attention is required in the case of new objects appearing when we zoom in, or disappearing when we zoom out. In this chapter we will consider the operation zooming as a new perspective on the problem of multiscale representations. We are convinced that the results will have an impact on the problem of generalisation. The multiscale structure will also support multilevel modelling and combinations of objects in different scales in one map as introduced by Bullen in Chapter 27, this volume.

From the database, the map output is constructed as a top–down selection of pre-generalised cartographic objects, until a sufficient level of detail is achieved. Pre-generalised cartographic objects can be taken from existing map series. The dominant operation is 'zoom', intelligently replacing the current graphical representation with a more detailed one that is appropriate for the selected new scale. Methods to select objects for rendering are based on the principle of equal information density, which can be derived from Töpfer's radix law (Töpfer and

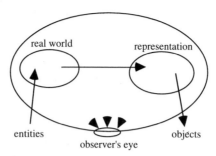

Figure 14.3 Entities versus objects.

Pillewitzer, 1966) and has been used by cartographers (Beck, 1971). We propose a relatively simple method to achieve equal information density, namely measuring 'ink'.

14.1.2 Structures of geographic objects

There are several types of geographic objects in the world. Couclelis (1996) made the distinction between objects and entities (Fig. 14.3). Entities are things in the real world that we can perceive. In our representation (our mental model) entities become objects: representations of things in the real world.

Most geographic objects do not have directly visible and sharp boundaries (geological layers, soils); others do not have visible boundaries but boundary zones or undetermined boundaries (for example, forest, sand dunes) (Burrough and Frank, 1995).

There are two views (or models) of the world which have impact on the data structure chosen. The field view sees the world as a continuum and therefore implies that there are no real boundaries unless we define them. The object view divides the world into objects with different properties and sharp boundaries. Both views have their advantages and drawbacks. We will try to incorporate both views in our data structure, although traditional paper maps adhere to the object view of the world. Pantazis proposes (Chapter 23 in this volume) a conceptual framework in which both views coexist. Unfortunately the classes in his framework cannot be used for a classification of graphical objects to be used for our approach.

14.2 Multiscale structure

One approach for structuring cartographic data is to consider cartography as a language with its own syntax and vocabulary (Palmer and Frank, 1988; Robinson and Petchenik, 1976; Youngmann, 1977). These objects are combined from a graphical vocabulary, which provides the atoms for graphical communication (Bertin, 1967; Head, 1991; Mackinlay, 1986; Schlichtmann, 1985). Highly simplified, cartographic objects can be differentiated by dimension (points, lines, areas) and the cartographic variations (object drawn as symbol, object representing a scaled representation, a feature associated with text, and text without a delimited graphical feature). This results in a dozen geometric categories (Morrison, 1988).

The approach selected here is to construct a multiscale directed acyclic graph (DAG) for each semantic class that exists in a map and to specify rules for their interaction. Semantic classes are water bodies, railroads, roads, settlements, labels, and symbols (Hake, 1975; Staufenbiel, 1973). They represent the first stage in a characterisation of object features which roughly corresponds to first level conceptual groups humans have. The second stage will be realised through the 12 categories mentioned above. We will have to examine the feasibility of this approach.

14.2.1 Data structure

The idea is that every entity is represented at multiple scales in a forest (Buttenfield and Delotto, 1989), that there are multiple graphical renderings for every cartographic object, organised in increasing graphical detail and pre-generalised. This includes that an object may split in subobjects, each with its own graphical rendering (Fig. 14.4). Generalised representations can be collected from existing map series or, for some cases, produced automatically. This circumvents cartographic generalisation at the expense of storage.

In Fig. 14.4, a multiscale DAG for houses is shown. At a very high level (meaning small-scale) the graphical object 'house' is not rendered at all, at a lower level it is rendered as a symbol, then as a generalised geometrical representation, and as a geometrical description. Between each of these renderings a jump in the representation method is made.

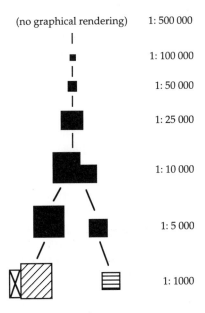

Figure 14.4 Example of a multiscale tree structure.

Table 14.1   Types of changes of object representations for smaller to larger scale

Continuous changes	Discrete changes		
	Slight change	Complete change	New object
1 No change in appearance 2 Increase of scale	3 Change in symbol 4 Increase in detail 5 Appearance of label	6 Change in dimension 7 Shift to geometric form 8 Split into several objects	9 Appears

In the next lower level, the geometry is depicted more clearly and it is shown that the object is, in fact, an aggregated object, and the representation method remains the same. The most detailed rendering is again made possible through a jump in the representation method. These jumps correspond to special operations of the zooming process, for example specialisation and disaggregation.

Considering existing map series, where the same objects are mapped at different scales, a strategy for creating hierarchies follows (Table 14.1). The list demonstrates that objects may change their spatial dimension in the generalisation hierarchy. A particular problem is posed by objects which are not represented at small scale and seem to appear as one zooms in. This appearance of objects can be accounted for in the indexing data structure which will again be a graph. Most of the change types (1–4, 6, 7 in Table 14.1) are not reflected in the data structure (Fig. 14.5a).

However, change types 5, 8, and 9 affect the data structure. The change types 'appearance of a label' and 'object appears' require a new link to come into the existing DAG from the outside (Fig. 14.5b). The change type 'split into several objects' requires that more than one link is leading from a single node (Fig. 14.5c). Of these two structural change types, namely new link and several links from one node, the first is the more important and also the most difficult to handle.

14.2.2 Why a directed acyclic graph?

A directed acyclic graph (DAG) is a well-known and documented structure in graph theory (Perl, 1981) and has many applications in the database area (Güting, 1994). The logical data structure suited for zooming would be one that links object representations of one level or scale with object representations of another level or scale in a hierarchical fashion. In our structure the nodes of the DAG contain the object representations and therefore the information necessary to render the objects. The directed links determine the direction of the zooming-in process.

In graph theory a forest is a collection of trees while trees are acyclic graphs (Ahuja *et al.*, 1993; Perl, 1981). In our data structure each object group (water bodies, railroads, roads, settlements, labels, and symbols) is represented by a forest (Fig. 14.6), for example, a number of houses at a certain location on the map can build a single tree because of their spatial proximity. Houses in a different area of the map build a second tree, and so forth. All 'house-trees' combined thus

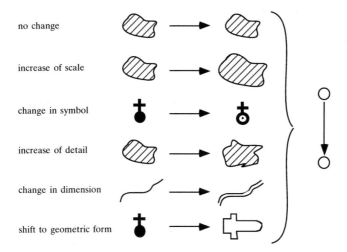

Figure 14.5a Examples for the changes that do not affect the data structure.

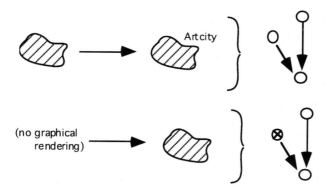

Figure 14.5b Examples for the change types 'appearance of label' and 'object appears'.

Figure 14.5c Example for the change type 'split into several objects'.

build a forest. This is true for all object groups. The whole collection of forests is again a forest. The forest contains the information for rendering in the nodes and the information for selection on the links.

When implementing the structure we use an approach similar to the Reactive Tree (van Oosterom, 1989). The difference is that we first classify the graphical

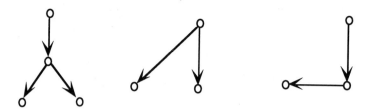

Figure 14.6   An example for a forest.

objects and construct several trees that are spatially intertwined. They may then be represented in a reactive tree. There is still research to be done to define the relationships between the different trees and forests in order to preserve the topology between objects from different groups.

14.2.3 Moving about in the structure

Zooming is a concept that originates from the metaphor of the sound of an airplane flying towards the earth. This means that as we get nearer to the object of interest, we see more detail. In computer graphics this has been partly realised as getting nearer to the focus of the window of interest while enlarging the information contained in the window (Fig. 14.7a). Volta (1992) studied a content zoom in which the categories of the window of interest are shown in more detail (Fig. 14.7b). For example, three major soil classes are differentiated into a detailed schema of several dozen classes. We identified the need for an intelligent zoom (Frank and Timpf, 1995), that realises both requirements for zooming (Fig. 14.7c).

By intelligent zoom we understand a zoom operation, which respects the known principle of equal information density (Beck, 1971; Töpfer and Pillewitzer, 1966). It implies that more detail about objects becomes visible as the field of vision is restricted and the scale is increased. This leads immediately to a hierarchical data structure where objects are gradually subdivided in more details. This hierarchical structure is applied to all geometric objects, not only to lines as in strip trees (Ballard, 1981) or in the model of Plazanet (Chapter 17 in this volume), to a pixel representation of an area as in a quadtree (Samet, 1989a and b) or the pyramid structures used in image processing (Rosenfeld *et al.*, 1982). When continuously zooming, the jumps in representations could be smoothed by using a morphing algorithm. Literature on 2D-morphing of geometric features is abundant and well researched (see for example Sederberg and Greenwood, 1992).

14.3 Selection criterion

The problem to address is the selection of the objects in the forest which must be rendered. Two aspects can be separated, namely, the selection of objects which geometrically extend into a window and the selection of objects to achieve constant information density. The selection of objects which extend into the window is based on a minimal bounding rectangle for each object and a refined decision that

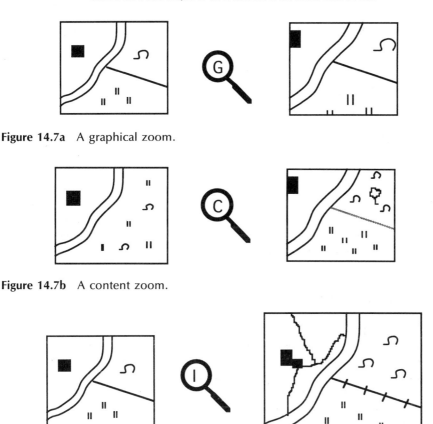

Figure 14.7a A graphical zoom.

Figure 14.7b A content zoom.

Figure 14.7c An intelligent zoom.

can be made based on object geometry. In order to ensure fast processing in the multiscale forest, the minimal bounding rectangles must be associated with the forest and forest branching, such that complete subforests can be excluded based on window limits. This is well known and the base for all data structures which support fast spatial access (Samet, 1989a and b).

The interesting question is how the depth of descent into the forest is controlled to achieve equal information density. In data structures for spatial access, an 'importance' characteristic has been proposed (van Oosterom, 1989). It places objects which are statistically assessed as important higher in the forest so that they are found more quickly. The method relies on an assessment of the 'importance' of each object, which is done once when the object is entered into the cartographic database. When a cartographic sketch is desired, the most important objects are selected for rendering from this ordered list. The usability of this idea is being studied for a particular case, namely the selection of human dwellings (cities) for inclusion in a map (Flewelling and Egenhofer, 1993). A method based on an ordering of objects is not sufficient for the general case.

The cartographic process is fundamentally constrained by the limit that one piece of paper can carry only one graphical message. The cartographic selection process is mostly dealing with the management of the resource map space and how it is allocated. The graphical density of the displays should be preserved over several scales. A relatively simple practical method for uniform graphical density is to measure 'ink', that is, pixels which are black. One assumes then that there is a given ratio of ink to paper. This ratio must be experimentally determined, measuring manually produced good maps (for example, 1 inked cell per 10 cells of paper). The expansion of the forest involves progressing from the top to the bottom, accumulating ink content and stopping when the preset value for graphical density is reached. The ink content should not be measured for the full window, but the window should be subdivided and ink for each subdivision optimised. We have not yet found a suitable method for subdividing the window and optimising the ink content. This selection principle does not avoid that two objects can be rendered at the same location. It requires afterwards a placement process (similar in kind to the known name-placement algorithm) to assign a position to each object (Herot *et al.*, 1980). It also requires a method for displacement of objects as proposed in for example, Bundy *et al.* (1994).

14.4 Conclusions

A number of applications require graphical presentations of varying scale, from overview sketches to detailed drawings. A simple scale change is not sufficient to produce drawings which humans can easily understand. The problem is most visible in cartographic mapping, where map scales vary from 1:1000 to 1:100 000 000, covering a range of 10^5. Over the last five centuries, cartography has developed useful methods to produce overview maps from more detailed ones. Several complex filtering rules are used to delete what is less important, simplify the objects retained and so forth. Unfortunately these rules have not yet been formalised to produce a fully automated system.

The proposed multiscale forest is a method to produce maps of different scales from a single database. This avoids the difficulty of cartographic generalisation at the expense of storage. Objects are stored at different levels of generalisation, assuming that – at least for the difficult cases – the generalisation is done by humans, but only once. Building a multiscale forest is probably a semi-automated process where automated processes are directed by a human cartographer. All operations where valuable human time is necessary are done only once and the results are stored.

The concept is based on a trade-off between computation and storage, replacing all steps which are difficult to automate with storage. These steps are performed initially, while the remaining steps, which can be easily automated, are performed each time a query asks for graphical output.

The resulting forest structure is more complex than spatial hierarchical structures proposed in the literature so far, as objects may change their geometric appearance considerably. For example, they change their spatial dimension, or change from a single object to a group of objects. Special attention requires the case that objects seemingly appear as we zoom in. The proposed forest is related to quad-tree or strip-tree structures, but generalised: objects of all dimensions can be stored and the dimension of an object between steps of generalisation can change.

Acknowledgements

I thank Werner Kuhn and Andrew Frank for their inspiring discussions on this topic. I also thank all participants of the summer institute for their good comments on my presentation. Finally, I thank the anonymous reviewers for their detailed comments.

References

AHUJA, R. K., MAGNANTI, T. L. and ORLIN, J. B. (1993). *Network Flows: Theory, Algorithms and Applications*, Englewood Cliffs, NJ: Prentice Hall.

BALLARD, D. H. (1981). Strip trees: A hierarchical representation for curves, *ACM Communications*, **24**(5), 310–21.

BEARD, K. (1987). How to survive on a single detailed database, in Chrisman, N. R. (Ed.) *Proceedings of Auto-Carto 8*, pp. 211–20, Baltimore, MD: ASPRS and ACSM.

BECK, W. (1971). Generalisierung und automatische Kartenherstellung, *Allgemeine Vermessungs-Nachrichten*, **78**(6), 197–209.

BERTIN, J. (1967). Sémiologie Graphique, Paris: Gauthier-Villars.

BUNDY, G. Ll., JONES, C. B. and FURSE, E. (1994). A topological structure for the generalization of large scale cartographic data, in Fisher, P. (Ed.), *Proceedings of GISRUK '93 Conference*, pp. 87–96, Leicester: University of Leicester.

BURROUGH, P. and FRANK, A. U. (1995). Concepts and paradigms in spatial information, are current Geographic Information Systems truly generic?, *International Journal of GIS*, **9**(2), 101–16.

BUTTENFIELD, B. P. and DELOTTO, J. (1989). *Multiple Representations: Report on the Specialist Meeting*, Report 89-3, Santa Barbara, CA: NCGIA.

COUCLELIS, H. (1996). Towards an operational typology of geographic entities with ill-defined boundaries, Chapter 3, in Burrough, P. and Frank, A. (Eds), *Natural Objects with Indetermined Boundaries*, pp. 45–56, London: Taylor & Francis.

FLEWELLING, D. M. and EGENHOFER, M. J. (1993). Formalizing importance: Parameters for settlement selection, in *11th International Conference on Automated Cartography*, pp. 167–175, Falls Church, VA: American Society of Photogrammetry and American Congress on Surveying and Mapping.

FRANK, A. U. and TIMPF, S. (1995). Multiple representations for cartographic objects in a multi-scale tree – An intelligent graphical zoom, *Computers and Graphics Special issue: Modelling and Visualisation of Spatial Data in Geographical Information Systems*, **18**(6), 823–83.

FREEMAN, H. and AHN, J. (1987). On the problem of placing names in a geographic map, *International Journal of Pattern Recognition and Artificial Intelligence*, **1**(1), 121–40.

GÜTING, R. H. (1994). *GraphDB: A Data Model and Query Language for Graphs in Databases*. p. 155, Hague: FernUniversität Hagen, Informatik-Bericht.

HAKE, G. (1975). *Kartographie*. Berlin: Göschen/de Gruyter.

HEAD, C. G. (1991). Mapping as language or semiotic system: review and comment, in Mark, D. M. and Frank, A. U. (Eds), *Cognitive and Linguistic Aspects of Space*, pp. 237–62, Dordrecht: Kluwer Academic Publishers.

HEROT, C. F. *et al.* (1980). A prototype spatial data management system, *Computer Graphics*, **14**(3), 63.

LINDHOLM, M. and SARJAKOSKI, T. (1994). Designing a visualization user interface, in MacEachren, A. and Taylor, D. R. F. (Eds), *Visualization in Modern Cartography*, pp. 167–84, Oxford: Pergamon, Elsevier Science Ltd.

MACKINLAY, J. (1986). Automating the design of graphical presentations of relational information, *Transactions on Graphics*, **5**(2), 110–41.

MORRISON, J. (1988). The proposed standard for digital cartographic data, *American Cartographer*, **15**(1), 9–140.

VAN OOSTEROM, P. (1989). A reactive data structure for geographic information systems, in Anderson, E. (Ed.), *Proceedings of Auto-Carto 9*, pp. 665–74, Baltimore, MA: ASPRS and ACSM.

PALMER, B. and FRANK, A. (1988). Spatial languages, in: Marble, D. (Ed.), *Third International Symposium on Spatial Data Handling*, pp. 201–10, Sydney, Australia.

PERL, J. (1981). *Graph Theory*, (in German). Wiesbaden: Akademische Verlagsgesellschaft.

POWITZ, B. M. (1993). Zur Automatisierung der kartographischen Generalisierung topographischer Daten in Geo-Informationssystemen. PhD-thesis, University of Hannover.

ROBINSON, A. H. and PETCHENIK, B. B. (1976). *The Nature of Maps: Essays toward Understanding Maps and Mapping*, Chicago: The University of Chicago Press.

ROSENFELD, A., SAMET, H., SHAFFER, C. and WEBER, R. E. (1982). *Applications of Hierarchical Data Structures to Geographical Information Systems*, University of Maryland: Computer Vision Laboratory, Computer Science Center.

SAMET, H. (1989a). *Applications of Spatial Data Structures: Computer Graphics, Image Processing and GIS*, Reading, MA: Addison-Wesley.

SAMET, H. (1989b). *The Design and Analysis of Spatial Data Structures*, Reading, MA: Addison-Wesley.

SCHLICHTMANN, H. (1985). Characteristic traits of the semiotic system 'map symbolism', *Cartographic Journal*, 23–30.

SEDERBERG, T. W. and GREENWOOD, E. (1992). A physically shaped approach to 2-D shape blending. *Proceedings of Computer Graphics SIGGRAPH'92*, pp. 25–34, Reading, MA: Addison-Wesley.

STAUFENBIEL, W. (1973). *Zur Automation der Generalisierung topographischer Karten mit besonderer Berücksichtigung grossmasstäbiger Gebäudedarstellungen*, Hannover: Wissenschaftliche Arbeiten WissArbUH 51, Kartographisches Institut Hannover.

TÖPFER, F. and PILLEWITZER, W. (1966). The principes of selection, *Cartographic Journal*, **3**, 10–6.

VOLTA, G. (1992). Interaction with attribute data in Geographic Information Systems: A model for categorical coverages, Orono: University of Maine, Master of Science.

YOUNGMANN, C. (1977). A linguistic approach to map description, in Dutton, G. (Ed.), *First International Advanced Study Symposium on Topological Data Structures for Geographic Information Systems*, pp. 1–17, Cambridge, MA: Laboratory for Computer Graphics and Spatial Analysis, Harvard University.

CHAPTER FIFTEEN

Efficient Handling of Large Spatial Data Sets by Generalised Delaunay Networks

TERJE MIDTBØ

15.1 Introduction

Modern technology makes it possible to sample huge amounts of data that describe spatial phenomena. The handling of the data requires efficient methods and data structures. New and powerful computers, sophisticated algorithms and appropriate data structures open the possibilities for solving advanced and complicated problems that earlier were impossible to tackle. However, while better methods for spatial modelling lead to more efficient analysis of large data sets, better methods for data acquisition will make the amount of data even larger. Consequently, it is still desirable to choose a representative selection of the data for presentation or further analysis.

A set of scattered points in three-dimensional space is the basis for the modelling. The scope of this chapter is to describe efficient topological models for the point sets, so that it is possible to handle various operations and analysis based on the sampled points. Various types of mathematical functions can be used to make models of a surface through the point set. However, for surfaces of a random variation, like terrain surfaces, representation by small plane elements has turned out to be most efficient. When a surface is divided into small planes, each plane can be represented by three points, each containing x, y and z-coordinates. Connecting these points will make a triangle. Together the triangles form a triangular irregular network (TIN). One of the great advantages of using triangles in surface modelling, is the possibility of adapting the triangle size to fit variation in the terrain surface.

Delaunay triangulation is definitely the most used triangulation technique for triangular modelling of terrain surfaces. This method is based on achieving a well-proportionate triangular geometry between the horizontal projections of the data points. The geometric conditions focus on nearness between the neighbouring nodes in the network. There are several algorithms for the generation of Delaunay networks. This chapter is based on the incremental insertion of points and the swapping of invalid triangle edges. However, in a program system that handles triangular meshes some other basis features are necessary as well. Midtbø (1994) shows how points are deleted from the triangulation in a dynamic way by using the swapping approach from the insertion process. Another important feature in the system is the procedure for merging together adjacent networks. These routines make it possible to handle a single part of the model and subsequently merge it to other parts. Working on smaller parts of a large model is more surveyable and easier to handle. This chapter shows how the TIN can be stored in a blockwise manner to maintain easily the topological information in a surface model. The chapter also gives a practical example on how the blockwise storage of the TIN can be utilised for efficient analysis on certain parts of the TIN.

By using an incremental algorithm for the Delaunay triangulation it is possible to make a *qualified selection of points*. During the triangulation all the *candidate points* (points which are not yet included in the network) are evaluated, and the points that make the most significant contribution to the model are chosen for the triangulation. The chapter shows how this qualified selection of points is used to make models of changeable resolution. It is possible to zoom the model in and out, and the precision of the triangular model can be chosen to meet our needs.

Finally, the chapter draws some parallels to three-dimensional Delaunay networks. While using an incremental algorithm during the construction of a tetrahedral network it is possible to use the methods for qualified selection of points, also for three dimensions.

15.2 Basic algorithms

15.2.1 Incremental method

Delaunay triangulation is a widely used method to achieve a well-proportionate triangular network between measured data points. The research described in this chapter is based on the incremental method for Delaunay triangulation (Guibas and Stolfi, 1985; Lee and Schachter, 1980) where the triangulation stays as a Delaunay network all the way during the triangulation process. The 'initial value' for the triangulation is a valid Delaunay network (at least one triangle). Figure 15.1 shows the repeated steps of the algorithm. A point is included in the network by connecting it to the vertices of the enclosing triangle by three new edges. Finally, edges affected by the insertion will be rearranged using the swapping approach proposed by Lawson (1972, 1977), until each triangle meets the circle criterion. The affected edges are found by the *maximum angle-sum criterion* described below.

15.2.2 The maximum angle-sum criterion

Lawson (1977) shows that by using the max-min angle test we will achieve a Delaunay triangulation and Midtbø (1993b) has used the maximum angle-sum criterion to determine if the diagonal of a quadrilateral is valid in a Delaunay triangulation. The maximum angle-sum criterion is closely related to the InCirc test in Guibas and Stolfi (1985) and can be formulated in Definition 1.

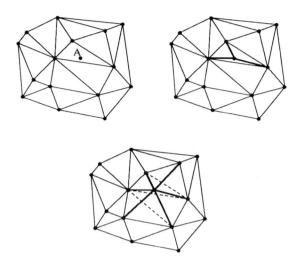

Figure 15.1 Insertion of a point into a Delaunay triangulation.

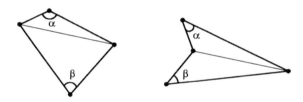

Figure 15.2 The angles are calculated in the 'diagonal free' corners.

Definition 1 In a quadrilateral formed by two adjacent triangles in a Delaunay triangulation, the diagonal goes between two opposite vertices where the sum of the interior angles of the quadrilateral is greater or equal to π.

It is useful to observe that, in the quadrilaterals, only the angles of the vertices of the interior diagonal can exceed π. This can be utilised to form an algorithm where only the cosine of the inside angles of the quadrilateral determines whether the diagonal is to be swapped.

Figure 15.2 shows two quadrilaterals. For these and other quadrilaterals we know that

$$0 \leqslant \alpha \leqslant \pi$$

$$0 \leqslant \beta \leqslant \pi$$

By investigating the nature of cosine for these angles, it can be concluded that $\cos \alpha + \cos \beta < 0$ when a diagonal has to be swapped.

15.2.3 Removing points from a Delaunay triangulation

The insertion of points into TINs has been well studied, while methods for removing points from TINs have been given less attention in literature. Midtbø

Figure 15.3 The edges are swapped with no respect to the circle criterion. (The three last interior edges are deleted.)

(1994) describes some algorithms where the edge-swapping process from the point insertion is reversed. The removal of points should be an important feature of a program system that handles triangular meshes. Nevertheless, many existing triangulation programs do not support this feature. Midtbø (1994) concludes that in most cases a very simple algorithm is the most efficient. This algorithm swaps all the edges away from the point to remove (Fig. 15.3). We only need to be sure that no edges adjacent to a concave angle are swapped. When only three edges remain inside the polygon, the edges are deleted. After this initial step is completed, the maximum angle-sum test is used for the quadrilaterals that lie inside the original polygon.

The running time for this algorithm is $O(n^2)$, although Midtbø (1994) presents more sophisticated algorithms running $O(n\log n)$. However, the number of edges connected to a point in a Delaunay triangulation for a random data set is about six (average). Because of less testing and calculations an $O(n^2)$ algorithm may be better suited than an $O(n\log n)$ algorithm for a small n (Aho *et al.*, 1987). Some practical examples in Midtbø (1994) further support this finding.

15.3 Qualified selection of points

A triangle formed by three points represents an exact description of a plane in space. When candidate points that lie inside the horizontal projection of the triangle fall in the triangle plane, these points will make no contribution to the model. It is possible to extend this algorithm to ignore 'almost redundant' points as well. This means that a point may not be included in the triangular mesh if it lies close to one of the triangles. Consequently, it is possible to establish surface models of various precisions. The precision is set by a predefined threshold value for point inclusion.

When a generalised surface model is required, it is an advantage that the selected points give the best possible description of the surface. During the insertion process in the incremental algorithm, a *split point* is chosen from the points that are enclosed by the present triangle. Usually the most distant point in each triangle makes the most significant contribution to the model, and is consequently chosen as the split point (Fig. 15.4).

After the triangle is split and the surrounding edges are reorganised, every triangle that is influenced by the reorganisation will have a new split point. When all the points lie closer to the triangle than the predefined threshold, λ, no more

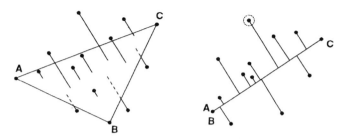

Figure 15.4 The most distant point from the triangular plane becomes the split point.

points will be included inside that triangle. The triangulation for the entity area is terminated when every point, that is not included in the network, is closer to a triangle than λ.

The triangles to split can be chosen in random succession. This is simple and efficient; experiments show that the results are quite good. However, the surface models do not necessarily become identical when one data set is processed twice with different point succession. The variation is small, however, and the precision of the surface model will still be the same. It is possible to avoid this 'problem' by selecting the triangle to split in a more sophisticated way. During the triangulation we can, for example, always split the triangle which has the most distant point. Then variations caused by different point succession only occur when we have to choose between two (or more) triangles with equal distance to the most distant point.

15.3.1 Data structures

To handle the qualified selection of points *and* the swapping operation in the incremental algorithm it is necessary to use a well-arranged and efficient data structure, as for example the *dual-edge* structure (Heller, 1990). In this structure edges are the basic elements. Each edge has pointers to an endpoint, the next edge in the triangle and the geometric identical edge in the neighbouring triangle. In addition it is important to use an efficient point-handling structure during the construction of the TIN. The running time of the triangulation algorithms depends to a large degree on the point-handling structure. Quad-tree and grid representations are proposed by Bjørke (1988), Lee and Schachter (1980) and McCullagh and Ross (1980). In the dual-edge structure the triangular network itself represents the storage structure for the points in the data set. For each triangle there will be a pointer to a list consisting of all points enclosed by the triangle. When points that lie inside a triangle are stored as attributes to the triangle, no extra time is spent to localise those points.

15.3.2 Threshold for point inclusion

The threshold for point inclusion depends on the problem to be solved and will vary in type and size. Figure 15.5 shows a typical threshold, λ_p. Every point that is closer to their enclosing triangle than λ_p will not be included in the mesh.

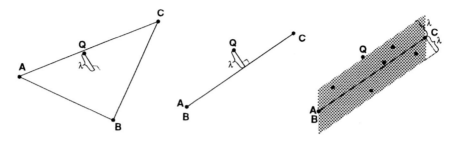

Figure 15.5 (a) Threshold value, λ_p, for the perpendicular distance; (b) a cut through the triangle; (c) region inside the threshold value.

The threshold in Figure 15.5 is designed for the perpendicular distance from point $Q(x_Q, y_Q, z_Q)$ to the triangular plane, $\triangle ABC$, calculated by Eq. (15.1)

$$dist_p = \frac{ax_Q + by_Q + cz_Q + d}{\sqrt{a^2 + b^2 + c^2}} \tag{15.1}$$

where a, b, c and d are given by the vertices of the triangle. When we are searching for the most distant point from a triangle, it is less time consuming to only compare the numerators of Eq. (15.1). The denominator is equal for all points enclosed by a triangle. Consequently the split point is determined by Eq. (15.2)

$$s_p = \max |ax_i + by_i + cz_i + d|, \quad (i = 1, 2, \ldots n) \tag{15.2}$$

n is the number of points enclosed by the horizontal projection of $\triangle ABC$.

For certain tasks it may be advantageous to keep separate threshold values for each side of the triangle. The threshold value can also depend on where the point lies in the model. Some areas may be more important than others and we will have:

$$\lambda = f(x, y, z) \tag{15.3}$$

Example. For maritime charts it is important to include all peaks that can be dangerous for navigation. Consequently, the threshold value above the triangles are very important and should be rather small, perhaps even zero. Below the triangles, the size of λ is of minor importance and can be much greater. For certain areas, for example through a fairway, it might be interesting to maintain a higher level of precision. In that case the threshold is set by Eq. (15.3). Further, maritime charts have to keep a higher level of detail in shallow waters. The threshold is consequently set by the z parameter in Eq. (15.3) where λ grows when the depth is deeper.

15.3.3 Models of changeable resolution

Together, the insertion by a qualified selection of points and the deletion of points paves the way for models of changeable resolutions. Here the chapter outlines a possible strategy for the handling of such models. During the triangulation of certain areas, the incremental triangulation can be stopped at various levels of precision. In the beginning of the triangulation λ is set very large, such that a few

Figure 15.6 Zooming the model.

Table 15.1   Classes of contribution

Class 1	$dist_p > 50$ m
Class 2	50 m $\geq dist_p >$ m
Class 3	20 m $\geq dist_p >$ m
Class 4	10 m $\geq dist_p >$ m
Class 5	$dist_p \leq 1$ m

triangles describe the area. From this rough model we can localise certain areas where a higher level of precision is desired. For these areas a smaller threshold is given and the triangulation continues. In this way we can *zoom* in the terrain model. If we need to zoom out again, it is convenient to store the distances from the included points to their triangles. This distance can be stored as an attribute to the point. The distance tells us what contribution a point makes to the model. When we are zooming out, the points are removed from the network according to the distance, and no extra calculations are required to localise the points to delete. Without any 'history' from the triangulation, zooming out will be much more time consuming. This is because it is necessary to localise the enclosing Delaunay triangle of the point.

Figure 15.6 shows an example of the zooming operations in a TIN. A 100 m $\times$ 100 m height data base is the basis for this zooming. A contour map of the model is drawn to visualise the zoom. A heavily generalised model covering Voss is generated and shown in scale 1:1 000 000. From this rough model it is possible to select certain areas that we want to have a closer look at. A new λ and scale is chosen, and a new more detailed model can be constructed. The resolution of the data points that represents the surface is decisive for how much the model can be zoomed in.

For large surfaces of high resolution, the amount of data will grow very large. When the same data set is to be used for triangulation more than once, it will be convenient to classify the points after the contribution the points give to the model. The first time the area is triangulated the points will be submitted to their respective classes.

Example. The points are sorted depending on how close the points that are included lie with regard to the enclosing triangle. Table 15.1 shows an example of how the data set can be divided into classes of contribution.

For a new triangulation of the same area, the points to be included can be selected by their class. For a triangulation with a precision of Class 1, none of the points in higher classes have to be considered. In the same way it is easy to zoom out by removing every point from one or more classes.

A model of low resolution is required when the operations on a large model includes heavy calculations and simulations. For more accurate calculations in certain areas, the corresponding TIN can be refined for these areas. Such methods can be used, for example, in the calculation of the extension of the area covered by a mobile telephone transmitter. In a low-resolution model advanced simulations of the trace and reflection of the signal can be calculated in an efficient way.

Similar methods can be used in the calculation of the dispersion and reflection of a navigation system for an airfield. In this case the threshold value increases with the distance from the transmitter.

15.4 Block-based management of the surface model

Particularly for terrain surfaces, data sets for surface modelling very often consist of a huge number of points. When the surface model is used for a practical purpose, only parts of the set may be necessary for computer processing. If there are proper routines for merging triangular meshes, this means that we can handle data sets with an endless number of points. When the requested calculations are completed in certain areas, this part of the network is removed from the computer memory. New areas can then be merged in the network and the surface model will behave like a seamless model. It also means that large data sets can be handled by small computers, where memory is a serious limitation. Working in smaller areas separately is also much faster than handling large, heavy models in one operation.

The generalisation feature depends on an efficient data structure for handling additional points. When these points are stored in lists that are subordinated to the triangles, no extra calculations have to be executed to locate the points that lie inside the triangles. This makes the incremental algorithm run slower when a huge number of points lie inside each triangle. It is slower especially in the beginning, when the area is covered by large triangles. If the area is partitioned into smaller blocks, the triangulation is much faster despite the blocks having to be merged at the end of the process (Midtbø, 1993b).

Further, when there are large areas to triangulate, it is necessary to 'swap' computer memory to the disk. Thus it is advantageous to partition the area into blocks, where the data for each block is localised together in memory. The 'partitioning method' is also useful for triangulation on parallel computers where the data blocks can be triangulated simultaneously in different processors.

15.4.1 Dividing the data set into blocks

Before doing the block-based triangulation, the data set has to be divided into smaller subsets. The data set can be divided into a predefined block pattern, or the size and shape of the blocks can be determined from the point scattering. Figure 15.7 shows various methods for the division of the data set. In Fig. 15.7a the data set is divided into predefined blocks. The partitioning is fast because the boundaries between the blocks are parallel to the coordinate axis. The method is well suited for uniformly distributed data sets, but when the point scattering differs over the area, some of the blocks might be empty. Figure 15.7b shows a block configuration that is better adapted to this situation. The blocks are split by a point quad-tree method (Samet, 1984). If there are too many points inside a block, this is split into four new ones. Figure 15.7c shows how the data set can be divided into blocks where the corner points are members of the data set. It is advantageous to construct the block boundaries along *break lines* when information concerning sudden variation in the surface is given from the data set. No reorganisation of edges is required when two adjacent blocks that share this 'natural boundary' are merged. The determination of corner points, and especially the sorting of points over lines that are not parallel to the coordinate axis, will be more time consuming for this kind of block partitioning.

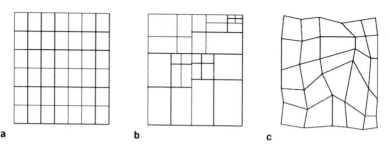

Figure 15.7   Methods for subdivision of the data set.

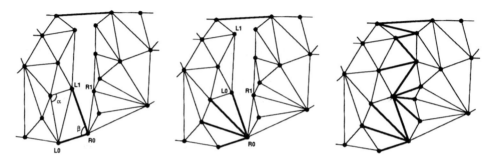

Figure 15.8   Edges between T_L and T_R are constructed and invalid edges are swapped.

15.4.2 **Merging the blocks**

After each block is triangulated, they are merged one by one until a continuous network covers the area. The basis for the merge is formed by two adjacent triangular meshes that meet the criterion for Delaunay triangulation. When the process is completed, the resulting network also has to be in accordance with the circle criterion. In this section T_L and T_R represent two adjacent Delaunay triangulations, and T is T_L and T_R merged together.

When two separate triangular networks are merged, it is likely that some changes will occur in the boundary area between the two meshes. These changes depend on the point distribution along the boundary, and how the areas are triangulated.

In the *build-and-swap algorithm* (Midtbø, 1993b) (Fig. 15.8) new edges are constructed to connect T_L and T_R. Subsequently some of the old edges from T_L and T_R are detected by the maximum angle-sum criterion and swapped until T meets the circle criterion for Delaunay triangulation.

The *merge-and-swap algorithm* (Midtbø, 1993b) only reorganises existing edges (Fig. 15.9). This is possible because the edges on the boundary of T_L and T_R are kept identical. These edges should cover the whole boundary between T_L and T_R. The two identical edges are melted together and the boundary edges are tested and swapped by the maximum angle-sum criterion until T becomes a valid Delaunay triangulation. It is important to keep the triangular meshes 100 per cent in accordance with the circle criterion to make this kind of merge operation

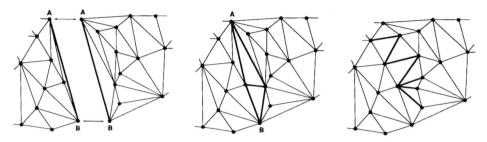

Figure 15.9 T_L and T_R are merged together by edge swapping only.

possible. Consequently, every point along the boundary is a vertex of a triangle, even if it lies on the boundary edge. In this case, the three vertices of the triangle lie on a line and represent a triangle with area = *0*.

Running time for the *build-and-swap algorithm* is about $O(n\log n)$, where *n* is the number of points in touch with the merging. The worst case running time for the *merge-and-swap algorithm* is $O(n^2)$. On the contrary, the simple *merge-and-swap algorithm* is very stable and makes the handling of additional points inside the triangulation more consistent (Midtbø, 1993b).

15.4.3 Storage of large TINs on disk

For many purposes it is advantageous to store the TIN structure itself on disk. This is specially true when fast access to a huge model is essential. A large model is hard to handle, and it will be necessary to divide the TIN into smaller units before it is stored on the disk. Only parts of the TIN need to be loaded into the computer memory for future processing and calculations. The size of the *TIN storage blocks* may depend on the capacity of the computer, the shape of the surface and the purpose of the model.

It is obvious that nationwide terrain models of high resolution consume a lot of disk space. This research proposes to store a generalised TIN model where additional points are listed for each triangle. When we need certain areas of higher resolution, it is possible to refine the network for the required blocks. This method makes it possible to store TIN structures for large areas, avoiding too much data being stored on the disk. Further, it is faster to make a TIN of high resolution than if the TIN has to be constructed from the beginning.

Example. In coastal areas ships need information about the bottom of the sea. Together with an accurate navigation system, a digital terrain model of the sea floor is an important part of a modern sea map system. Figure 15.10 shows a ship under way while TIN storage blocks, representing the bottom of the sea, are loaded into the computer's memory on the ship. The TIN storage blocks are, for example, rough networks which are refined when there are shallow waters in front of the ship. When the ship has passed a TIN storage block, the computer memory allocated to this block is released, and a new block can be included ahead of the ship.

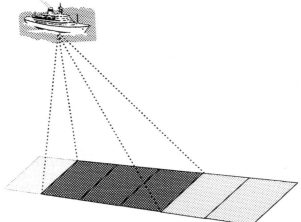

Figure 15.10   Use of TIN blocks.

15.5 Incremental Delaunay tetrahedrisation

In digital terrain modelling we talk about 2.5 dimensional models. This means terrain models where coordinates for three dimensions (x, y and z) are used for the construction of a TIN. However a TIN is always projected on the x, y plane. As overhangs are seldom in terrain surfaces, the 2.5-dimensional model is efficient and well suited for terrain representation. In comparison, a real three-dimensional model for handling terrain surfaces, which often consists of large data sets, will probably be too heavy and inefficient.

There are still a lot of other fields where a 3D representation is necessary. Underground deposits, for example, are difficult to describe by a 2.5D model. An irregular triangular network in 3D is actually a network that consists of tetrahedra, and Edelsbrunner *et al.* (1986) argue that 3D tetrahedrisation is central to a number of applications in numerical computing and in solid modelling. The solution of partial differential equations by the finite element method and the structural analysis of complex physical solids require the decomposition of a given spatial domain into elementary cells. In their simplest form these cells are tetrahedra. Field (1986), Field and Smith (1991) and Schroeder and Shephard (1988, 1990) use Delaunay tetrahedrisation for the decomposition of solids into finite element meshes for analysis by the Finite Element Method. Here it is important to divide the solids into equal-sized tetrahedra. Vertices of the elements are consequently chosen during the tetrahedrisation. Delaunay tetrahedrisation is used because it is able to create well-proportioned elements. Baker (1989) partitions solids into tetrahedra in an attempt to study turbulent air flows around aircraft. He also refers to the use of tetrahedral models in electromagnetic problems.

Although there are numerous algorithms for triangulation, 3D tetrahedrisation algorithms have been given less attention. However, authors such as Edelsbrunner and Shah (1992), Field (1986), Joe (1989), Kanaganathan and Goldstein (1991), Midtbø (1993a), and Watson (1981) have explored the subject.

15.5.1 Geometrical conditions for 3D Delaunay networks

The geometry is described by a tetrahedral irregular network T_n constructed from a given set S_n of n points in 3D space. When Delaunay triangulation is extended to one more dimension, this results in Delaunay tetrahedrisation. As three points are necessary to describe a plane in space, at least four points describe a volume element in space. The most elementary volume element is a tetrahedron, which is given by four points. In two dimensions there are no vertices from the network situated inside the circumscribing circle of any triangle. The corresponding situation in three dimensions is called the *sphere criterion*:

> *Definition 2.* A Delaunay network in three dimensions consists of non-overlapping tetrahedra where no points in the network are enclosed by the circumscribing spheres of any tetrahedra.

15.5.2 Data structures for a tetrahedral irregular network

It is necessary to use a dynamic data structure, where local reorganisation of the network is efficient in terms of time, to handle an incremental algorithm. For this purpose an extension of Heller (1990) dual-edge structure into three dimensions is appropriate. This *double twin-edge* structure consists of edges that have pointers to the endpoint, the next edge in the triangle, a twin edge in the same tetrahedron *and* a pointer to the twin edge in the adjacent tetrahedron.

15.5.3 Incremental generation of a tetrahedral network

A 3D version of the incremental algorithm can be used for the construction of the network. During the construction it is advantageous to use a dynamic data structure where the additional points are stored in lists as attributes to their enclosing tetrahedra. In this structure no search for the enclosing tetrahedron is necessary, and it is possible to calculate efficiently if additional points make any contribution to the data model. In other words, we have got an adaptive data model where it is possible to choose the level of precision.

Before the incremental tetrahedrisation starts, an initial network has to be constructed. This 'network' may, for example, be a large tetrahedron enclosing every point in the 3D space. This tetrahedron becomes a base for further division into smaller tetrahedra. The algorithm can be briefly formulated as the following.

1　Construct a circumscribing tetrahedron for the point set.
2　Insert a point into the mesh by:
　　Splitting the points circumscribing tetrahedron into four new tetrahedra.
　　Reorganising edges close to the inserted point, until every tetrahedra meets the sphere criterion. The circumscribing sphere of the tetrahedra having the inserted point as a vertex, is examined. No opposite point of an adjacent tetrahedron can lie inside this sphere.
3　After invalid edges are deleted and new ones are inserted, the mesh becomes a Delaunay network again, and the remaining points can be included in the network by step 2.

15.5.4 Qualified selection of points

Section 15.3 describes how qualified point selection reduces the amount of data in a logical way. Using an incremental tetrahedrisation and a convenient point handling structure, this feature can be extended into three dimensions. The result is an adaptive data modelling by tetrahedral irregular networks.

Example. A lot of temperature measurements have been sampled for a sea area. If the sampling of data is easy, a great part of the measurements are probably redundant. For each sample x, y, depth and temperature are measured. A tetrahedral data structure may be generated to store the information. If the temperature of some samples inside a tetrahedron can be calculated from the four vertices of the tetrahedron, it is not necessary to store these samples in our data structure.

It is of course possible to split the tetrahedron in the point which makes the most significant contribution to the model. Like in the triangulation (Eqs 15.1 and 15.2) we can indicate a threshold value for the inclusion of the point. The contribution of the point is determined by Eq. (15.4). When $dist_{4D} = 0$ the point makes no contribution to the model.

$$dist_{4D} = \frac{ax_i + by_i + cz_i + dv_i + e}{\sqrt{a^2 + b^2 + c^2 + d^2}} \tag{15.4}$$

The size and type of λ_{3D} will depend on the problem to solve and the precision of the given data set. In a large data set, with a considerable part of redundant or 'almost redundant' data, this feature may be very useful. It is also possible in 3D to construct generalised models of tetrahedral networks. These networks can be refined by using a smaller λ_{3D}, or in other words, by zooming the '3.5 dimensional' model.

15.6 Conclusion

Efficient basis algorithms for inserting and removing points pave the way for models of changeable resolution. In this model rough analysis and simulations can be executed for large areas, and points can be included to increase the precision in selected areas for further analysis. When appropriate merging routines exist it is possible to maintain nationwide surface models by using a block-based management of the TIN. Certain blocks can be selected and used for necessary calculations and analysis. Methods for a qualified selection of points can be extended to a three-dimensional Delaunay network of tetrahedra. However, this feature depends on an efficient data structure where the appropriate points can be selected and inserted into the model in an incremental manner.

References

AHO, A. V., HOPCROFT, J. E. and ULLMAN, J. D. (1987). *Data Structures and Algorithms*, Reading M.A.: Addison-Wesley.

BAKER, T. J. (1989). Automatic mesh generation for complex three-dimensional regions using a constrained Delaunay triangulation, *Engineering with Computers*, **5**, 161–75.

BJØRKE, J. T. (1988). Quadtrees and triangulation in digital elevation models, in *International Archives of Photogrammetry and Remote Sensing*, Vol. **27**, part B4, pp. 38–44, Kyoto: The Committee of the 16th International Congress of International Society for Photogrammetry and Remote Sensing.

EDELSBRUNNER, H. and SHAH, N. R. (1992). Incremental topological flipping works for regular triangulations, in *Proceedings of the Eighth Annual Symposium on Computional Geometry*, pp. 43–52, New York: The Association for Computing Machinery.

EDELSBRUNNER, H., PREPARATA, F. P. and WEST, D. B. (1986). *Tetrahedrizing Point Sets in Three Dimensions*, Technical Report UIUCDCS-R-86-1310, University of Illinois: Department of Computer Science.

FIELD, D. A. (1986). Implementing Watson's algorithm in three dimensions, in *Proceedings of ACM Symposium Computational Geometry*, pp. 246–59, New York: The Association for Computing Machinery.

FIELD, D. A. and SMITH, W. D (1991). Graded tetrahedral finite element meshes, *International Journal for Numerical Methods in Engineering*, **31**(3), 413–25.

GUIBAS, L. and STOLFI, J. (1985). Primitives for the manipulation of general subdivision and the computation of Voronoi diagrams, *ACM Transactions on Graphics*, **4**(2), 74–123.

HELLER, M. (1990). Triangulation algorithms for adaptive terrain modelling, in *Proceedings of the 4th International Symposium on Spatial Data Handling*, pp. 163–74, Zürich: University of Zürich and Commission on Geographic Information Systems.

JOE, B. (1989). Three-dimensional triangulations from local transformations, *Siam Journal on Scientific and Statistical Computing*, **10**(4), 718–41.

KANAGANATHAN, S. and GOLDSTEIN, N. B. (1991). Comparision of four-point adding algorithms for Delaunay-type three-dimensional mesh generators, *IEEE Transactions on Magnetics*, **27**(3), 3444–51.

LAWSON, C. L. (1972). Transforming triangulations, *Discrete Math.*, **3**, 365–72.

LAWSON, C. L. (1977). Software for c1 surface interpolation, in Rice, J. (Ed.) *Mathematical Software III*, Academic Press, New York.

LEE, D. T. and SCHACHTER, B. J. (1980). Two algorithms for constructing a Delaunay triangulation, *International Journal of Computer and Information Sciences*, **9**(3), 219–42.

McCULLAGH, M. J. and ROSS, C. G. (1980). Delaunay triangulation of a random data set for isarithmic mapping, *The Cartographic Journal*, **17**(2), 93–9.

MIDTBØ, T. (1993a). Incremental Delaunay tetrahedrization for adaptive data modelling, in *Proceedings of EGIS '93*, pp. 227–36, Utrecht: EGIS Foundation.

MIDTBØ, T (1993b). 'Spatial modelling by Delaunay networks of two and three dimensions', Ph.D. thesis, Trondheim: The Norwegian Institute of Technology.

MIDTBØ, T. (1994). Removing points from a Delaunay triangulation, in Waugh, T. C. and Healey, R. G. *Proceedings of the 6th International Symposium on Spatial Data Handling*, pp. 739–50, London: Taylor & Francis.

SAMET, H. (1984). The quadtree and related hierarchical data structures, *Computing Surveys*, **16**(2), 187–260.

SCHROEDER, W. J. and SHEPHARD, S. (1988). Geometry-based fully automatic mesh generation and the Delaunay triangulation, *International Journal for Numerical Methods and Engineering*, **26**, 2503–15.

SCHROEDER, W. J. and SHEPHARD, M. S. (1990). A combined octree/Delaunay method for fully automatic 3-D mesh generation, *International Journal for Numerical Methods in Engineering*, **29**, 37–55.

WATSON, D. F. (1981). Computing the *n*-dimensional Delaunay tesselation with applications to Voronoi polytopes, *The Computer Journal*, **24**(2), 167–72.

CHAPTER SIXTEEN

Shape-based Reconstruction of Natural Surfaces: An Application to the Antarctic Sea Floor

MICHELA SPAGNUOLO

16.1 Introduction

Surface modelling is an important aspect of Geographical Information Systems (GIS) which has to deal with the definition of suitable computational structures able to represent surfaces in digital form. Generally, the surfaces handled within GIS may vary from terrain to temperature or wind fields and can be informally defined as bivariate functions whose main characteristic is that they depend on two coordinates representing a location in the real world. As far as this chapter is concerned, surface modelling is discussed in the specific case of topographic surfaces, that is, function whose value is the elevation of the coordinate points.

Surface data are usually integrated within the general database of a GIS as particular records. However, their nature requires a more complex approach. For example, to study the evolution of a natural phenomenon over a given area, one may wish to evaluate some continuous functions (for example, the physical laws representing the phenomenon) over the terrain model, rather than retrieving and sorting single data items. This aspect of surface modelling is also discussed by Včkovski in Chapter 25 in this volume. Včkovski suggests the idea of virtual data sets to encapsulate in an integrated structure both the measured surface data and the methods to compute approximate values where the data are not available. Moreover, in many GIS the intrinsic three-dimensional nature of spatial data has often been disregarded and data is treated as if projected onto a flat surface. Three-dimensional modelling techniques could be used instead, to provide more effective models of geographical surfaces and to broaden the spectrum of spatial relationships to a fully three-dimensional reality (Cambray, 1994).

The field of surface representation in GIS has to deal with several problems. Unlike the context of computer graphics, where surfaces can be generated using an equation which formalises their behaviour, topographic surfaces are generally defined only by sets of points belonging to the surface with no further information on the behaviour of the surface. Given the data points, a suitable model has to be constructed which has a twofold purpose: on the one side it should store the collected data efficiently and on the other side it should provide a sensible approximation scheme, supporting surface estimation where data are not available. This has given rise to interesting research issues on the definition of the best interpolation models (McCullagh, 1988) and has caused the development of several tools for generating grid models from raw data. The reason for the initial success of grid models lies primarily in their simplicity and suitability for interpolation purposes. Moreover, the wide availability of data in grid form, such as remotely sensed data, has led to an extensive use of grid models which have served as containers for the enormous amount of captured data. However, not enough attention has been given to the actual information content of the data and the quality of the produced models. As a consequence, in the recent past, it has been generally stated that more flexible models should be used which can represent surfaces at different levels of detail, supporting irregular sampling density according to the complexity of the surface morphology. The best-known representative in the class of adaptive models is triangulation, the advantages of which over regular grid models have been highlighted by several authors (Brändli and Schneider, 1995; De Floriani *et al.*, 1985).

Even if adaptive, the triangulation model is still lacking in providing an effective reconstruction of the surface: edges between data points are drawn mainly according to aesthetic criteria (for example, equiangularity) and may distort the actual shape of the surface. Other approaches seem more promising for discrete surfaces, such as, for example, the *constrained triangulation*, where the constraints are often defined by surface features which can be deduced by the raw data distribution. An interesting example of this approach can be found in Tang (1992), where a method has been proposed to produce high-quality three-dimensional surface models from contour data.

The methodology presented in this chapter focuses on the importance of a reasoning process on raw data with the aim to extract, whenever possible, knowledge about the shape of the real surface which should be used as a skeleton for the effective reconstruction of the surface. The main points which are addressed concern the organisation of raw data into classes of morphological importance and the delineation of structural surface features which can be deduced by exploiting the geometry of the sampling process. The proposed approach has been defined within the National Project of Research in Antarctica (PNRA) for the specific sub-project on 'Three-dimensional reconstruction of the Antarctic sea floor' led by the CNR Institute for Applied Mathematics. For this reason, a detailed description of the proposed methodology is given in the specific context of Antarctic sea floor reconstruction.

The chapter is organised as follows. First, surface modelling is addressed, comparing the traditional techniques with shape-based approaches. Then the requirements imposed by the specific context of Antarctica and the steps to reach a shape-based model are described including the generalisation of data in profile form, the extraction of profile specific points, the delineation of structural features among profiles and the resulting model for the sea floor. Finally, some conclusions are briefly drawn.

16.2 Shape-based surface modelling

The need to represent curves and surfaces arises in two cases: in modelling synthetic objects, where no direct reference to any physical object has to be preserved and in modelling existing objects, for example a terrain or a statue. The modelling methodologies used in the two cases are generally different (Foley *et al.*, 1990). In the first case, the object is created in the modelling process, that is, the object matches its representation exactly. To create the object, the user may mathematically describe the shape properties of the surfaces, which results in a different choice of the generating equations. In the latter case, the mathematical description of the surface may be unavailable and the object model has to be defined on the basis of a sampling set. Keeping these data as fixed, the model is intended to define an approximation of the real object, which usually is constructed by fitting pieces of planes or other shapes to the set of points.

The case of natural surface modelling obviously falls into the second category and data about the real world are acquired using some technique such as ground surveys, photogrammetry, automatic digitalisation of maps or, as in the case of the sea floor, by direct bathometric measures along predefined routes. Having collected the necessary data, a suitable structure has to be chosen, which usually sets up a topology and geometry among the points. In this context, topology generally means adjacency relationships, that is, the topology defines which points have to be considered neighbouring, while the geometry establishes the shape of the approximating surface. For example, in a regular grid, a point is usually connected to its four direct neighbours and the geometry may fit a bilinear approximation in each square grid. Regular grids are mainly used for the ease of storing and retrieving data. However, the assumption of a regular distribution of the data points does not fit well with the aim of having an accurate and efficient surface reconstruction. A variable sampling density is, indeed, more appropriate to follow the changes in the terrain morphology, allowing a lower sampling density in regions of gentle relief and a higher density in rough areas (Falcidieno *et al.*, 1992; Weibel and Heller, 1990).

Thus, the problem should be addressed to find an appropriate computational structure to represent irregular distributions of data points. One solution, which has been used quite often, is to reconstruct a grid even if the data are irregularly distributed. This process makes use of interpolation to compute surface values at grid nodes, which generally do not coincide with data points. Several methods have been developed which can be classified as global and local methods: global methods fit some continuous function to the entire data set, with a computational complexity which is usually $O(n^2)$ where n is the number of data points. Local methods are based on the hypothesis that only points at a certain distance can influence the shape of the surface at a given point, that is, the function has to fit only a local neighbourhood of a given point (Lam, 1983; McCullagh, 1988).

This approach, however, does not respond to the requirement of efficiency, since the advantage of having variable sampling density vanishes as soon as grid points are computed. Moreover, regular grids do not support editing facilities in the sense that, if new points are measured, their insertion into the model would require a complete rearrangement of the whole structure. Geometric models with an irregular topology, such as the triangulation, are much more suitable for representing irregular distributions of data points and offer plenty of solutions to merge locally new point sets, at the cost of a more complex handling of the structure.

Given a set of points, their triangulation is not unique and the problem of finding a *good* triangulation has been the topic of research for many years. In surface approximation problems, an acceptable triangulation is one in which most of the triangles are as equiangular as possible in order to limit the occurrence of long thin triangles. In this condition, a limit can be given to the maximum approximation error. The equiangularity criterion is satisfied by the *Delaunay triangulation* (Preparata and Shamos, 1985), which has been widely considered a standard for surface representation.

When using the Delaunay triangulation for surface reconstruction, the approach generally used consists in triangulating the data projections onto the XY plane and then in reassigning the elevation to the data, bringing the triangulation back to the three-dimensional space. Two problems arise when using this method: first, the neighbourhood relationships concern the data projection and not the points in space; thus points which are reasonably near on the plane may be quite distant in the space. Second, equiangularity is a good property for approximation purposes but it can distort the shape reconstruction of the surface. In Fig. 16.1, a simplified example is sketched to show a typical situation where these problems could occur. The surface presents a ridge which has been sampled with relatively few points with respect to the neighbouring region; when looking at the point projections it happens that points on the left and the right of the ridge are nearer, thus, they will be joined by an edge of the triangulation which distorts the reconstruction of the surface shape, since a concavity is introduced where in reality there was a ridge.

In order to prevent this kind of problem, new approaches to the definition of triangulation methods have been proposed which, for example, minimise mathematical quantities such as the curvature (Brown, 1991; Dyn *et al.*, 1990). This procedure, often called *data-dependent* triangulation, is not directly applicable to the context of natural surface modelling, since it makes no sense to impose a behaviour such as smoothness to a real surface, where continuity is the only mathematical property which can be reasonably required. It is extremely useful, instead, to introduce the concept of *constrained triangulation*, that is, a triangulation in which a set of edges are constrained to be drawn between particular vertices (Pienovi and Spagnuolo, 1994). The concept of constrained triangulation has been used by several authors and definitions have been given which merge the Delaunay within the constrained triangulation (De Floriani *et al.*, 1985; Preparata and Shamos, 1985). Using the constraints, it is possible to force particular edges to belong to the final surface model, for example, those which give rise to features in the surface shape. In this manner, a better surface reconstruction can be obtained.

This approach poses the problem of developing methodologies to recognise shape features given a set of surface samples. Interesting work has been done in this context by Tang (1992), on the definition of high-quality surface models from contour data. The idea was that many surface features could be deduced by looking at the shape of the contours using the same method that cartographers use: peaks and pits correspond to contours which do not contain any other contour, while ridges and ravines correspond to lines with nested concavities. To detect the presence of these features, the medial axis transformation has been proposed

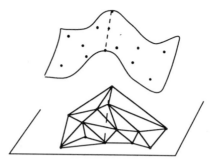

Figure 16.1 The Delaunay triangulation does not preserve the ridge line.

which reduces the shape of the contours to a simplified shape whose characteristics can suggest surface features. Once having extracted the characteristic points and lines, a constrained Delaunay triangulation is used to produce the final model, where elevations for the extracted features are deduced with consideration about the behaviour of neighbouring regions.

The work presented by Tang can be extended to a general approach to shape reconstruction which can be used whenever the surface is sampled along *privileged* directions, such as, for example, parallel sections of the surface. The steps involved in this process can be summarised as follows: first, raw data should be analysed and implicit information should be exploited, whenever possible, in order to construct a morphological framework for the reconstruction of the surface; then the surface model should be defined, constrained to the features, if any, which have been extracted. This procedure can be used quite often, because it is likely to happen that data are not acquired in a random fashion, but rather following some privileged directions. In the next section, a detailed example will be given of the advantages of a careful preprocessing of the raw data in the context of the reconstruction of the Antarctic sea floor.

16.3 The approach to the reconstruction of the Antarctic sea floor

Since 1995, our Institute has been actively participating in the National Project of Research in Antarctica (PNRA) which involves several scientific activities related to the Antarctic environment, such as, for example, glaciology, climate, oceanography, biology and, with special attention, cartography of the Antarctic sea floor. Within the PNRA, the Institute for Applied Mathematics is part of the sub-project entitled 'Hydrography' whose main aim is the production of bathymetric maps for coastal zones in Antarctica at different scales. Our contribution mainly consists in the definition of three-dimensional models for an efficient and effective representation of the sea floor and the definition of methods for handling these models (construction, visualisation, updating, simulation, interrogation and so forth).

The examples shown in this chapter refer to the data collected during the hydrographic activity in the marine zone adjacent to the Italian station at Terra Nova Bay. Bathymetric data have been collected during surveys carried out along parallel courses by a ship equipped with an echo-sounding instrument which measures the depth of the sea floor at almost regular intervals. Depending on the characteristics of the surveyed zone, it is possible to have either plenty or lack of data: in Antarctica, navigation is possible where the water is free of ice, while it can be impossible in other zones due to the continuous presence of floating ice. Moreover, data are collected according to rules corresponding more to the safety of surface navigation than to the needs of the three-dimensional sea floor reconstruction. Thus, the spatial distribution of data is irregular and unbalanced, with redundant points along the navigation routes and the absence of measurements between adjacent routes. It can then be assumed that the sea floor is measured along profiles which correspond to almost vertical sections of the surface.

Computations and analysis performed on these data have produced bathymetric maps useful to position technical instruments and to navigate on the sea surface. However, these maps do not contain enough information to be the reference model for more complex applications, such as submarine robotics and oceanography. As a consequence, the problem has been addressed of building a complete three-dimensional model of the sea floor, which should serve as a support and reference for the research activity in Antarctica. With regard to the modelling aspects, the requirements which have been considered are the following:

- *closely fitting original data*, in the sense that the model must keep their content of information, mainly in the critical zones;

- *easy to be updated locally*, when additional data are collected;

- *predictive*, that is, able to simulate the surface also in parts lacking of information, taking advantage of different sources (behaviour of the coastal terrain, features of neighbouring sea beds and so forth).

Shape-based modelling has several advantages which help to satisfy the previously listed requirements. The idea is to use the hypothesis of spatial coherence of natural surfaces, in particular of the sea floor, in the sense that reasonably near locations have similar geographical attributes. Based on this assumption, the flow of operations used to construct an effective and efficient sea floor model is depicted in Fig. 16.2.

The first step consists of a profile analysis which is aimed at filtering measurement errors and identifying the profile characteristic points, which will be used as pointers to surface features for the subsequent steps. The next step is the recognition of surface features, more precisely of ridges and ravines, whose presence can be deduced by similar configurations of characteristic points among adjacent profiles. Based on ridges and ravines, the final model is constructed as a triangulation constrained to the extracted shape features. These steps will be described in detail in the following sections.

16.3.1 Profile characterisation and simplification

The purpose of profile filtering and analysis is twofold: discard measurements' errors and balance the spatial distribution of data points, since redundant points along profiles may distort the reconstruction of the sea floor and, most importantly, make the model too bulky. Large quantities of data do not necessarily mean a good information level: the idea is to then reduce the number of points needed to

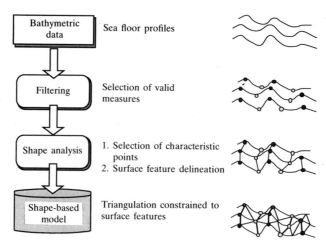

Figure 16.2 The shape-based approach to the reconstruction of the Antarctic sea floor.

represent each profile, preserving the points with a high information content (for example, points of minimum and maximum) and discarding those points which do not effectively contribute to defining the profile shape.

This simplification procedure is known in cartography as line generalisation and several methods have been proposed in the literature to provide fast and accurate algorithms. Different approaches to the evaluation of line-simplification algorithms have been adopted over the years and several measures have been used by McMaster (1986) who shows that the method of Douglas and Peucker (1973) is 'the most cartographically sound' in the sense that it produces the least areal and vector displacement from the original line. The basic idea of this method is the use of a tolerance rectangle whose height is related to the scale at which details can be considered irrelevant: all points falling within this rectangle can be discarded without loss of information. This method has been applied by Orgolesu (1994) for the simplification of the profiles of the Antarctic sea floor, with a generalisation to the three-dimensional space, where a tolerance cylinder substitutes the tolerance band. As in the two-dimensional case, the method starts by joining with a straight line the first and last points of the profile. Then, a cylinder is considered having its axes along the segment and having the radius equal to a predefined tolerance value (Fig. 16.3). If all the points of the profile fall into the cylinder, then the straight line segment is considered adequate for representing the line in simplified form. Otherwise, the point with the maximum offset from the cylinder axes is selected and the profile is subdivided at that point.

The described selection procedure is recursively repeated for the two parts of the line, until the tolerance criteria is satisfied. The selected points are finally chained to produce a simplified line.

This algorithm gives good approximation results but it has the great disadvantage of being strongly dependent on the choice of the first and last points, that is on the orientation of the tolerance cylinder. Moreover, the method does not respect the requirements imposed by a simplification process which aims at selecting a subset of points with the best information content from the point of

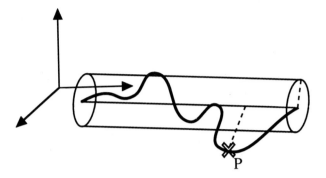

Figure 16.3 The generalisation of the Douglas and Peucker (1973) algorithm to the three-dimensional space. The tolerance cyclinder replaces the tolerance band.

view of sea floor reconstruction. Thus, the problem of profile generalisation has been split into the following subproblems:

- The characterisation of the profile shape by identifying the specific points such as maxima and minima;
- The classification of the recognised points as points having *global* and *local* importance, according to the scale parameter at which the simplification should be carried out;
- Perform the line simplification, using the modified Douglas and Peucker (1973) algorithm, constrained to the selected points with global importance.

The first two steps have been solved by analogy with multiresolution analysis in signal processing: signals are compared with profiles while points with global or local importance can be seen as the different frequencies composing the signal. In particular, wavelet analysis is a very powerful mathematical tool for hierarchically decomposing functions into a sequence of overall shapes plus details which range from broad to narrow (Daubechies, 1992; Stollnitz *et al.*, 1995). The distinction between points having a global or local importance is made looking at the wavelet transform of the line. At each level of the filtering process, the insignificant details at this particular level are discarded. In Fig. 16.4, an example is shown of a line processed with the wavelets technique: in Fig. 16.4b it can be seen how spike details are discarded after the first transformation step; in Figure 16.4c, small fluctuations are linearised and, finally, in Fig. 16.4d the overall shape of the original line is delineated quite clearly. Further transformations produce further shape simplifications, the last level being simply a straight line. If the shape generalisation shown in Fig. 16.4d is considered satisfactory, then the characteristic points of the line at that level identify the *global* characteristic points of the original line. Conversely, the level at which a characteristic point represents a *local* detail is the level at which it is lost during the wavelets processing.

For the simplification of the seabed profiles, the Daubechies basis of wavelets has been used (Daubechies, 1992) and the results of the process on a set of profiles are shown in Fig. 16.5.

Once having selected the profile characteristic points, the simplification can be done in the intervals defined by the points in the original line. In our application,

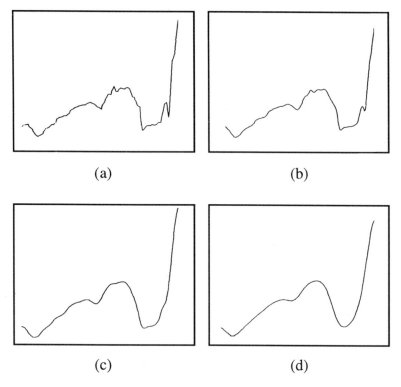

(a) (b)

(c) (d)

Figure 16.4 Example of simplification using the wavelet transform. In (a) the original line is depicted while in (b), (c) and (d) the results are shown of three iterations of the process.

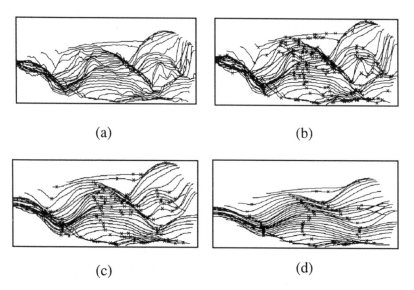

(a) (b)

(c) (d)

Figure 16.5 The results of the wavelet technique for the simplification of sea floor profiles. In (a) the original profiles are shown; some intermediate results are depicted in (b) and (c); in (d) the final simplification level is shown.

the Douglas and Peucker (1973) method has been chosen, modified to the three-dimensional domain. This procedure can be seen as a semi-local method as the result of the generalisation is not any longer dependent on the whole line but only on the characteristic points.

With regard to the computational complexity of the proposed approach, the part concerning the wavelet transform has a $O(n\log n)$ complexity, while the Douglas and Peucker algorithm is $O(n^2)$, where n is the number of points in the line. The method seems to be very flexible and has the great advantage of yielding directly a hierarchical organisation of the simplified line, according to different scales.

16.3.2 Delineation of structural features

The second step of sea floor modelling consists of finding the morphological features of the seabed, ridges and canyons, using a geometric reasoning technique on the set of simplified profiles. More precisely, it is possible to identify ridges and canyons of the sea floor whose direction is almost orthogonal to the profile direction. Indeed, a ridge (respectively, a canyon) of the sea floor will give rise to a maximum (respectively, a minimum) on a profile which has been measured along a course almost orthogonal to the ridge (or canyon).

The basic idea is to look for patterns of similar shapes occurring across adjacent profiles: for example, a ridge line can be identified as a pattern of maxima points having a similar shape. Let us assume we have defined a measure of morphological similarity between critical points. With reference to Fig. 16.6, the algorithm for feature recognition starts by selecting a seed critical point on a profile as the current point for the reconstruction process (the point P in Fig. 16.6a). Then, the adjacent profile is scanned to find the critical points having a shape similar to the current point (points Q_1 to Q_5 in Fig. 16.6b). Among these similar points, Q_4 is selected as being more similar and reasonably near to the current point. If such a point is found, then a line is drawn between the current and the new point (the line from P to Q_4 in Fig. 16.6c), and the new point will become the current point when the procedure is repeated on the remaining profiles. To evaluate the distance among the current point on a profile and the similar points on the adjacent profile, the idea of *influence cone* is used: a cone is considered with the vertex in P and the axis in the direction of the last feature segment. Only the points falling into the cone are selected for further analysis, which are Q_3 and Q_4 in the example shown. In other words, the amplitude of the influence cone determines the distance at which two critical points can be considered as belonging to the same feature.

With regard to the measure of morphological similarity, the curvature of the profile at the critical point has been used in the prototype implementation of the described approach. Other measures are currently being considered, which may include some more specific parameters. In this context, an interesting work is presented by Plazanet (Chapter 17 in this volume) for the cartographic generalisation of line elements in maps: several measures for the shape of line segments have been defined, which are used to classify and segment a line into a collection of simpler shape elements. This approach could probably be used to develop a set of rules to define the similarity of critical points also in the context of sea bed profiles.

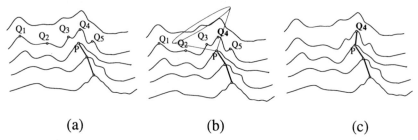

(a)   (b)   (c)

Figure 16.6 The method for reconstructing surface features from the simplified profiles: in (a) the current point P is shown with the points $Q1$ and $Q5$ on the adjacent profile, which have shape characteristics similar to P; in (b) the influence cone is depicted with the vertex in P; in (c) the most similar point, $Q4$, is selected and joined to the current point P.

The proposed approach for surface features recognition has some limitations inherent in the technique used to acquire the data: a line feature which occurs in a direction almost parallel to the profile direction cannot be recognised using our technique. Moreover, the proposed approach does not take into account branching features. This problem, however, is mainly related to the early stage of the prototype implementation and there are no theoretical impediments for considering more complex situations.

16.3.3 Construction of the sea floor model

The structural features which have been extracted define a kind of skeleton used as constraint to build the triangulation of the data. An example is given of the recognised skeleton (Fig. 16.7a) and of the final sea floor model, defined as a constrained Delaunay triangulation (Fig. 16.7b). The choice of the Delaunay triangulation has to be considered as a preliminary step, since other optimisation methods are currently being studied, which will probably offer better properties in the general context of shape-based modelling.

The triangulation constrained to shape features is a suitable model with respect to the requirements listed in Section 16.3. Since the important data points are kept as vertices of the triangulation, the sea floor model closely fits the original data. The insertion of new points can be done in a quite standard way, for example, locating the points within the existing triangles and performing a subdivision of that triangle to include the new points. Moreover, methods are currently being studied to define a complete set of procedures to handle merging and splitting triangulations so that entire sets of data can be considered as single model units.

Finally, with regard to the property of being predictive, the constrained triangulation offers a good support for this task. Let us imagine that two sets of profiles are available for a given area while a gap of data is between them (Fig. 16.8). If a ridge has been located in both the two sets of profiles, and if it can be reasonably assumed that the morphology of the sea floor cannot change too rapidly, then it could be assumed that the ridge also continues in the gap. Having drawn this probable line, the triangulation could be constructed to somehow fill the gap.

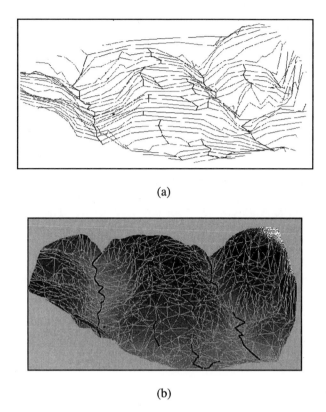

(a)

(b)

Figure 16.7 The final sea floor model. In (a) the profiles are shown with the extracted characteristic lines; in (b) the triangulation is constructed constrained to the recognised surface features.

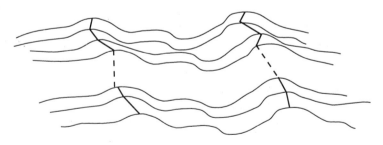

Figure 16.8 Filling lack of information based on morphological continuity.

Obviously, the described procedure makes sense only when the model is completed with the help of specialists and it should be clear that the accuracy and fidelity of the approximation in critical areas is extremely low. However, this approach may at least help the generation of good graphic rendering.

Another advantage of the proposed approach is that it limits the effect of the sampling anisotropy during the reconstruction of the sea floor. Indeed, if the original profiles are used without any simplification preprocessing, then the density of samples along routes forces the definition of edges along profiles which may distort the reconstruction. Moreover, the insertion of constraints in directions which are transversal with respect to the main sampling direction also has the effect of balancing the spatial distribution of the data.

16.4 Conclusions

This chapter has given an overview of the ongoing research activity at the CNR Institute for Applied Mathematics in the field of surface modelling. The concept of shape-based modelling has been introduced and its application to the reconstruction of the Antarctic sea floor has been shown. The proposed modelling strategy has the advantage of producing a surface model which is effective and efficient: effective because it explicitly keeps the main shape features of the surface, and efficient because it is constructed discarding irrelevant details.

This approach gives rise to an approximate model, whose error is defined as the distance of the discarded points from the approximated surface. This permits its modification by introducing those data points that give the highest error. In this way, the proposed model defines a hierarchical organisation of the data, according to their information content.

Acknowledgements

I would like to thank all the people working in the computer graphics group of IMA-CNR, especially Antonella Sanguineti, Corrado Pizzi and Stefano Orgolesu for their excellent work within the project.

I would also like to thank Professor Robert Laurini, who is patiently supervising my Doctorate Thesis at the Institut National des Sciences Appliquées, Lyon. Thanks are also given to Dr Bianca Falcidieno for the encouragement to write the chapter and all the people involved in the Antarctica Project.

References

BRÄNDLI, M. and SCHNEIDER, B. (1995). Shape modeling and analysis of terrain, *International Journal of Shape Modeling*, **1**(2), 167–89.

BROWN, J. L. (1991). Vertex based data dependent triangulations, *Computer-Aided Geometric Design*, **8**(3), 239–51.

CAMBRAY, B. (1994). 'Etude de la modélisation, de la manipulation et de la représentation de l'information spatiale 3D dans les bases de données géographiques', unpublished PhD thesis, Paris: University of Paris.

DAUBECHIES, I. (1992). *Ten Lectures on Wavelets*, CBMS-NSF Regional Conference Series in Applied Mathematics, Philadelphia, Pennsylvania: SIAM.

DE FLORIANI, L., FALCIDIENO, B. and PIENOVI, C. (1985). Delaunay-based representation of surfaces defined over arbitrarily-shaped domains, *Computer Vision, Graphics and Image Processing*, **32**(1), 127–40.

DOUGLAS, D. H. and PEUCKER, T. K. (1973). Algorithm for the reduction of the number of points required to represent a digitised line or its caricature, *The Canadian Cartographer*, **10**(2), 112–22.

DYN, N., LEVIN, D. and RIPPA, S. (1990). Data dependent triangulations for piecewise linear interpolation, *IMA Journal of Numerical Analysis*, **10**, 137–54.

FALCIDIENO, B., PIENOVI, C. and SPAGNUOLO, M. (1992). Discrete surface models: constrained-based generation and understanding, in Falcidieno, B., Herman, I. and Pienovi, C. (Eds), *Computer Graphics and Mathematics*, pp. 245–61, Berlin: Springer-Verlag.

FOLEY, J. D., VANDAM, A., FEINER, S. K. and HUGHES, J. F. (1990). *Computer Graphics: Principles and Practice*, Reading, MA: Addison-Wesley.

LAM, N. S. (1983). Spatial interpolation methods: A review, *The American Cartographer*, **10**(2), 129–49.

MCCULLAGH, M. J. (1988). Terrain and surface modelling systems: Theory and practice, *Photogrammetric Record*, **XII**(72), 747–79.

MCMASTER, R. (1986). A statistical analysis of mathematical measures for linear simplification, *The American Cartographer*, **13**(2), 103–17.

ORGOLESU, S. (1994). 'Analisi di forma mediante wavelet per la generalizzazione di superfici naturali', unpublished Ms Thesis, Genoa: University of Genoa.

PIENOVI, C. and SPAGNUOLO, M. (1994). Handling discrete surfaces by analysis and simulation, *Computer&Graphics*, **18**(6), 785–93.

PREPARATA, F. P. and SHAMOS, M. I. (1985). *Computational Geometry*, New York: Springer-Verlag.

STOLLNITZ, E. J., DE ROSE, T. D. and SALESIN, D. H. (1995). Wavelets for Computer Graphics: A primer, Part 1, *IEEE Computer Graphics and Applications*, May, 76–84.

TANG, L. (1992). Automatic extraction of specific geomorphological elements from contours, in Bresnahan, P., Corwin, E. and Cowen, D. (Eds), *Proceedings of the 5th International Symposium on Spatial Data Handling*, Vol. 2, pp. 554–66, Columbus: International Geographical University.

WEIBEL, R. and HELLER, M. (1990). A framework for digital terrain modeling, in Brassel, K. and Kishimoto, K. (Eds), *Proceedings of the Fourth International Symposium on Spatial Data Handling*, pp. 219–29, Zürich: Department of Geography, University of Zürich.

CHAPTER SEVENTEEN

Modelling Geometry for Linear Feature Generalisation

CORINNE PLAZANET

17.1 Introduction

Current GIS offer only limited capabilities for generalising data. Typically the user can utilise a single-line generalisation algorithm which may not suit his or her needs. So far more than a dozen line simplication and smoothing algorithms have been proposed. None of them provides a generic and systematic solution for simplication and smoothing line features, while some of them, or sequences of them, may be appropriate in some instances, but not for a whole data set. This chapter highlights the importance of modelling the geometry of linear features in order to generalise them. It proposes a constructive approach by segmenting lines and qualifying sections detected according to different criteria at several levels ('hierarchical' process). The aim of such an approach is twofold:

- to guide the choices of adequate sequences of generalisation operations and algorithms;
- and to provide basic tools for assessing the quality of the results of generalisation alternatives in terms of shape maintenance.

Cartographic generalisation schematically consists of selecting the features to be maintained at the targeted scale, simplifying non-relevant characteristics, enhancing significant shapes, displacing without defacing global and local shapes and finally harmonising the final aspect (see Chapters 12 and 13 in this volume). This chapter addresses issues related to linear feature generalisation (especially simplification and enhancement of road features) considered independently from their cartographic context.

A primary requirement for cartographic generalisation, as well as for data generalisation, is to preserve in the best possible way geometrical properties, spatial and semantical relations, when coming to a coarser resolution (Lagrange, Chapter 12, this volume). For cartographic generalisation, this change in resolution is due to scale reduction, graphical limitations and symbolisation while it is motivated by data reduction in the case of data generalisation. So far, according to many studies such as: Beard, 1991; Buttenfield, 1991; McMaster, 1989; and

McMaster and Buttenfield, Chapter 13 in this volume. Existing automated tools for linear feature generalisation are not efficient enough because of numerous constraints to take into account variations in the quality of their output according to the characteristics of the line and the subjective aspect of generalisation (for example, how to decide whether or not to omit or maintain, simplify or enhance a local shape within a linear feature).

These studies suggest that generalisation requires first that the object under scrutiny be segmented, so that each of its local characters can be processed by adequate algorithms. This point of view is also reinforced by approaches in other fields of generalisation. For example, Weibel (1989), proposed in respect to relief generalisation, a segmentation and characterisation stage based on geomorphological principles before applying specific algorithms. In our studies, we need first to segment the linear feature according to geometrical criteria. Therefore, we propose to construct a hierarchical model, based on recursive segmentation and geometrical qualification stages. Such a descriptive process should provide us with basic and complementary information valuable for guiding appropriate global and local decisions of generalisation, facilitating the choice of the most suitable generalisation algorithms and tolerance values on segmented line sections.

Three kinds of issues need to be tackled in order to achieve good generalisation results for linear features: eliciting geometric knowledge on linear features; formalising rules for generalisation decisions of operations and algorithms; and assessing the quality of generalisation operations in terms of shape maintenance. This chapter focuses largely on the first point, the second and third points are addressed in the concluding sections.

17.1.1 Cartographic constraints

Cartographers, when generalising manually, have a global and continuous perception of each line, clearly applying cartographic knowledge that has not been formalised as rules anywhere (Plazanet, 1995). Road-network generalisation, especially in mountainous areas, raises spatial problems solvable by simplification or enhancement. To eliminate a hairpin bend from a winding road (French), cartographers generally proceed by maintaining the first and last bends, choosing *the one* bend to eliminate (often the smallest) and emphasising the others (Weger, 1993).

What kind of knowledge guides the cartographer to take decisions? So far, some cartographic rules which guide intrinsic linear feature generalisation have been identified by looking at the cartographer's own procedures. These include to:

- preserve intrinsic topology;
- take into account symbol width;
- take into account semantic nature of the features;
- focus on the message of each particular feature; and
- respect geometric shapes.

The cartographic environment of linear shapes and the notion of scale are essential, providing subjective decisions by the cartographer. One of the challenges in automated generalisation consists in devising approaches which account for shape generalisation with respect to scale and shape environment.

17.1.2 Operations and algorithms

Many algorithms for linear feature generalisation, especially line simplification, have been developed such as the famous Douglas algorithm. However, several studies have shown how difficult it is to apply them to a variety of linear features (Beard, 1991; Buttenfield, 1991; Jenks, 1989; McMaster, 1989). According to Herbert *et al.* (1992).

> The use of line simplification algorithms on a data set within an existing GIS can be a very uncertain process. Effectively, a user is forced into a position of experimenting with different tolerance levels to try and get the results he or she wants, repeating a 'generalise . . . display' cycle until he or she achieves his or her objective (p. 555).

Furthermore, the quality of the result seems to vary according to the line characteristics (Buttenfield, 1991). Also, a complex sequence of generalisation operations may be needed in order to generalise a road (McMaster, 1989).

Linear generalisation encompasses more than simplification operations. Shape emphasising or bends elimination within a series of consecutive bends of same shape (operation of schematisation) may also be required (see the proved examples at the end of the chapter).

The problem lies clearly in the choice of the sequences of algorithms and tolerance values to apply according to their geometric characteristics. Furthermore, the geometric characteristics in many cases vary along the line. We think it is necessary to segment lines into sections sharing similar characteristics in order to choose the most suitable operation or sequence of operations of generalisation (simplification, caricature and schematisation) and corresponding sequences of algorithms for each line section.

17.2 A descriptive model of linear feature geometry

Our approach is in fact a hierarchical segmentation and qualification of linear features process analogous to Ballard's strip-trees (1986) and to the method proposed by Buttenfield (1991) where series of measures are computed for each line. In 1987 Buttenfield proposed a very similar process for the classification of cartographic lines also dealing with segmentation measurements, based on the assumption that 'a line is composed of a trend line and details that bifurcate from it' (p. 8). Our approach is different in that we do not consider local changes of direction, but variations according to shape criteria (at a 'more global' level).

17.2.1 The sinuosity notion

According to McMaster (1995): 'Individuals seem to judge the shape of the line on two criteria: the directionality of the line and the basic sinuosity of the line' (p. 166). Our study confirms that sinuosity is a qualifier of greater importance in describing lines, particularly roads. A line that contains many changes of direction is generally qualified as sinuous in a basic sense. Sinuosity may also be qualified according to the semantic type of objects. For instance a river trace is quite often irregular, rough, while a road may follow 'zigzags', hairpins, and closed bends (Plazanet, 1995). Road features are human constructions, designed from regular mathematical curves. The directionality may be seen as a mean line that

Figure 17.1   Trend line (smoothing) of an IGN-BDCarto® line.

Buttenfield (1987) calls the *trend line*: 'Conceptually, the trend line is the smoothest possible approximation of the cartography, a generalisation to the *n*th degree' (p. 9).

At the global level, the sinuosity may be qualified by the sinuosity of the trend line and the sinuosity of the original line considered here very roughly (Fig. 17.1).

At the local level, we may qualify the sinuosity of lines, looking locally within homogeneous sections at bends making spirals, hairpins and so on. Such portions of curves can be described according to shape (kind of curvature: sine curves, spirals, rectangles), amplitude base symmetry direction and size (linked with the notion of resolution).

17.2.2 Descriptive model and process

Our assumption is that, in fact, different levels of perception imply different shapes of a line, from the most global level down to the most local. At the global level, we can look at the entire line and appreciate the shape of the trend line, while individual bend shapes appear at a more local level of perception. In this chapter, we use the following terminology for global to local analysis levels (Fig. 17.2): the global level, corresponding to the entire line; the intermediate levels, corresponding to a line section; and the local level, corresponding to a bend.

Then our perception of a line is generally based upon the different ways we may make explicit different criteria of sinuosity at different levels of perception (shape of the trend line, local shapes of bends, differences in shapes, that is heterogeneity, density of bends and complexity).

In our approach, homogeneity is required as the first criterion adopted to describe lines. A line may be called non-homogeneous when it contains sections that seem different according to sinuosity. At each level of perception, the line (or line section) is considered from the point of view of homogeneity, and then possibly segmented according to this intuitive criterion. Basically, the process consists, therefore, in examining line sections recursively (starting from the entire line), and at each step, by either segmenting or further analysing the line section. The test on homogeneity dictates further computation if the line is non-homogeneous, it is segmented at a lower level, or else a detailed classification is determined through closer analysis.

As a result, the descriptive model built is a description tree corresponding to a hierarchical analysis. The nodes of the tree correspond to line sections and carry descriptive attributes, and the transition(s) between a node and its child(ren) may be the result of segmentation or analysis step(s) undertaken. The root of the tree

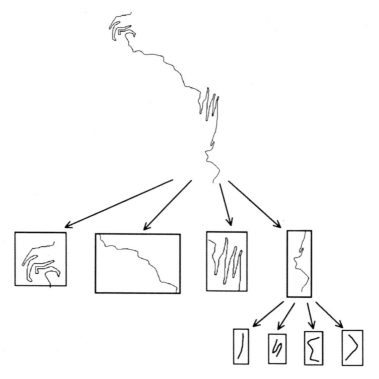

Figure 17.2 An example of natural line segmentation (IGN BDCarto® road).

corresponds to the original line and gets as attribute a rough sinuosity qualification, while the leaves may correspond to highly homogeneous sections, or individual details (for example, a single bend defined as a constant sign curvature section – for a given level of analysis – delimited by two inflection points) and carry a shape class attribute.

The lower the node in the tree is, the more detailed the analysis which corresponds to this node. At the root level, homogeneity is considered only very roughly in order to determine if the line must be segmented into significantly different sections. Inversely, a leaf corresponding to an individual detail such as a bend can be qualified according to the above-described criteria.

Once the description tree is built, a bottom–up analysis is required in order to check the coherence of processed information between the different nodes (for example, to avoid further segmentation when each segmented section belongs to the same class at a given level). An example of how the description tree may be used for generalisation is presented in Fig. 17.10.

17.3 Computational methods for the description

The segmentation and qualification stages involve different processes according to the varying levels of analysis. Techniques based on the location of characteristic points have been investigated, which aim at providing a set of measurements as well as qualitative descriptions for each analysis level.

17.3.1 The use of characteristic points

The line segmentation and geometry description stages are based on characteristic point detection. In the literature, be it psychology or computer vision or more recently cartography, shape detection is based on the detection of characteristic points (Attneave, 1954; Hoffman and Richards, 1982; Plazanet *et al.*, 1995; Thapa, 1989). For instance Spagnuolo (Chapter 16, this volume) characterises relief from points of maxima of curvature on profiles in order to reconstruct the Antarctic sea floor.

In Euclidean geometry, characteristic points are curvature extrema: vertices and inflection points. However, as argued by Freeman (1978), this definition is too restrictive: 'We shall expand the concept of critical points to include also discontinuities in curvature, end-points, intersections (junctions) and points of tangency' (p. 161). According to our needs (no discontinuities in curvature for the considered features), we define the following points as characteristic points: start and end-points, minima of curvature (inflection points), maxima of curvature (vertices) and critical points (which are a subset of inflection points). Among the characteristic points, the critical points are used to divide a line into sections. The other points are used for the precise qualification of bends.

The inflection point detection. Because of the acquisition process, lines frequently contain spurious micro-inflections which are not characteristic. Moreover, according to the level of analysis, only the main inflection points are of interest. A process has been implemented and tested to detect significant inflection points from a smoothed line (Plazanet *et al.*, 1996). According to Babaud *et al.* (1986) the Gaussian filter g_σ:

$$g_\sigma(k) = \frac{1}{\sigma\sqrt{2\Pi}}\, e^{-k^2/2\sigma^2}$$

is 'the only kernel for which local maxima always increase and local minima always decrease as the bandwidth of the filter is increased' (p. 27).

Therefore, the convolution of the curve with the filter is computed in a discrete space, where the line has first been re-sampled. Every point (x,y) of the re-sampled line has an homologous point (x_l,y_l) on the smoothed line so that:

$$x_l = \sum_{k-4\sigma}^{4\sigma} x(i-k) * g_\sigma(k), \text{ and } y_l = \sum_{k-4\sigma}^{4\sigma} y(i-k) * g_\sigma(k)$$

Studying the vectorial product variation along the smoothed line allows for the detection of characteristic inflection points: around these points, there is a significant change in the value of the vectorial product.

The value of σ (number of neighbouring points taken into account to compute an average position) characterises the smoothing scale. The higher the σ value, the stronger the smoothing. Thus, the choice of the σ value depends on the level of analysis.

Progression in the analysis is ensured by the tuning of inflection point detection: The lower the level in the tree, the lower the value of σ. This way a more and more detailed series of inflection points is used as the basis for segmentation and sinuosity classification. We are currently using the following empirical rule $\sigma = k*\text{length(line)}*ln(k*\text{length(line)})$ where *ln* is the neperian logarithm and *k* is a constant (1/1000 for our data set presented below) for a first level of analysis. More generally, such a tolerance value should be considered as being proportional to the scale, and probably also to the line sinuosity.

17.3.2 Segmentation: a process for a first level

Hoffman and Richards (1982) suggest that curves should be cut at the minima of curvature, which are particular characteristic points. In our process, critical points, that is points at which a line is segmented, are considered as being a subset of inflection points. The shape criteria are obviously difficult to model, as they remain rather fuzzy and may be considered from different points of view. As a first attempt for a rough level segmentation, we consider the following homogeneity definition based on the variation of the distances between consecutive inflection points:

Let $IP(\lambda)$ be the set of inflection points detected for a level of analysis λ.

Let $M(IP(\lambda))$ be the mean of distances $d(IP_i,IP_i+1)$ between consecutive inflection points.

Let $\Delta(IP(\lambda))$ be the sequences of deviations $D_i = d(IP_i,IP_{i+1}) - M(IP(\lambda))$. In fact, only the signs S_i of these D_i are retained here:

Then the inflection points IP_i, such as $S_{i-2} = S_{i-1} \neq S_i$ define potential locations for the segmentation of lines. These points are retained as being critical points (Fig. 17.3). A line where the inter-distances between inflection points vary significantly is qualified as being non-homogeneous. Such a definition is applicable more particularly when lines are quite non-homogeneous. Thus we are considering whether to add a test for the homogeneity of lines (by tuning an indicator of homogeneity) in order to process the segmentation stage only when necessary.

Besides, it is clear that homogeneity encompasses more criteria than just inter-distances between inflection points. Nevertheless, the above definition is applicable for the first segmentation stage, particularly thanks to the Gaussian filtering that is used to retain the inflection points within highly sinuous sections (that are highly persistent to smoothing) and to quickly eliminate details in other, lower sinuous sections (Fig. 17.4).

To ensure further segmentation at a lower level, we are considering whether to add complementary criteria used in the same way as the inter-distance of inflection points.

17.3.3 Qualification

Attneave (1954) has proved that 'information is further concentrated at points where a contour changes direction most rapidly' (p. 185), that is vertices. The qualification stages in our process are largely based on quantitative measurements and qualitative representations starting on characteristic point location.

17.3.3.1 Classification of first-level segmented sections

Since original lines are segmented in homogeneous sections (between two critical points) at a more or less rough level of analysis λ, a set of measurements can be

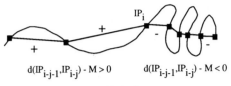

$$d(IP_{i-j-1}, IP_{i-j}) - M > 0 \qquad d(IP_{i-j-1}, IP_{i-j}) - M < 0$$

IP_i is retained as a critical point.

Figure 17.3 A rough segmentation stage.

Figure 17.4 An example of significant difference in the detection of inflection points according to the sinuosity type.

calculated based on the location of characteristic points on these sections, which can then be classified into shape classes. Actually, such a classification may consider not only a sinuosity criterion, but other criteria as well. But the measurements are chosen so as to represent first facets of sinuosity. For every bend defined as a fraction of curve between two consecutive inflection points (Fig. 17.5), it is possible to calculate:

- height h
- Euclidian distance between the inflection points, *base*
- curve length between inflection points I1 and I2, l
- angle between $\phi1$ and $\phi2$

These first measurements allow us to define line classes at a rough level. We are considering the addition of more complex measurements in order to provide further details in the description:

- area between the curve and the segment (I1,I2)
- area of the triangle (S,I1,I2)
- projection of the vertex S on the segment (I1, I2).

For a section, we may calculate the mean (or median) value, variance, minimum and maximum values of each of these measurements.

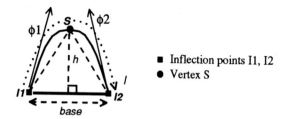

Figure 17.5 Measurements on a bend.

Additional measurements based on the line connecting the inflection points may include:

- total number of bends
- total absolute angles of the inflection points line
- ratio of the curve length of the line connecting the inflection points and the curve length of the original line.

A lot of measurements are possible from our hierarchical representation or from those previously proposed (Buttenfield, 1991; McMaster, 1986; Plazanet, 1995) that can be reused or refined for our representation such as, for instance, the area of the bounding rectangle of each segmented section. The difficulty lies in choosing subsets appropriate for the different analysis tasks and analysis levels, so that every subset contains normalised and non-correlated (or loosely correlated) measurements.

These first measurements bring out a definition of sinuosity but they do not allow us to precisely qualify bend shapes. For instance, the ratio between the area of the curve and the segment (I1,I2) and the area of the triangle (S,I1,I2) gives an indication of the complexity of the bend shape, but does not qualify it. A proposition of a qualitative description method is proposed below.

17.3.3.2 *Future research: a qualitative description of bend shape*

Very sinuous sections, especially mountain roads, may correspond to a series of complex bend shapes that need further qualitative description, added to the basic rougher classification. Bends in the example in Fig. 17.6 are complex, containing more inflection points than the delimited ones, detected at a high level of analysis. The inflection points delimiting a bend are the ones which are highly persistent to the Gaussian smoothing. A hierarchy in the importance of the inflection points of a bend is required so that they can be described in a more detailed manner.

Thus a complex bend could be described as a combination of simple bends. Existing representations of sections of curves such as 'codon' descriptors in computer vision: Hoffman and Richards (1982); Mocktarian and Macworth (1992); Rosin (1993); Ueda and Suzuki (1990) seem to propose far too restricted classes of curves. Due to the complexity of geographic feature shapes, we need to extend the 'codon' notion. Having predefined bend shape classes of codon descriptors (as in the first attempt proposed in Fig. 17.7), matching the codon descriptor of a particular bend may provide a qualitative structured description of bends.

The main difficulty for an implementation is likely to remain the determination

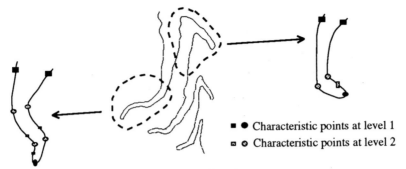

■ ● Characteristic points at level 1
□ ⊙ Characteristic points at level 2

Figure 17.6 An example of complex shape bend.

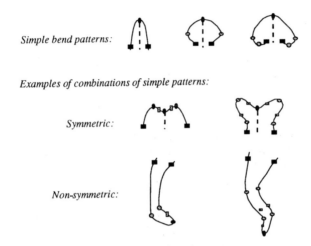

Simple bend patterns:

Examples of combinations of simple patterns:

Symmetric:

Non-symmetric:

Figure 17.7 Examples of different kinds of bends.

of the corresponding characteristic points at different levels of smoothing (required in order to build the hierarchy in the importance of the inflection points). Inflection points are not regularly located when applying the different smoothings necessary to identify them.

17.4 Experimental results

A first series of part of experiments have been conducted on a set of lines taken from the BDCarto® IGN database, based on the measurements described above, and using a classical cluster analysis software. The first objective consists in splitting lines into straight/sinuous/strongly sinuous sections.

17.4.1 Characteristic points detection and segmentation

Figure 17.8 presents some examples of inflection point and critical point detection on a series of 5 m resolution BDCarto® roads (scale 1:50 000). Vertices between two inflection points have been simply estimated for this experiment at the point which is the farthest from the anchor line defined by the two inflection points (Fig. 17.5).

The tolerance value σ of the smoothing is determined automatically for each line. These experiments show that even with a rough homogeneity criterion, variations in distances between inflection points are inspected to determine a first-level set of homogeneous sections.

17.4.2 Clustering homogeneous sections

Following this first stage, an experiment is conducted to qualify and classify the sinuosity of each segmented section. To start with, we chose a small set of measurements based on the simple measurements: h, l and *base* (Fig. 17.5). For each segmented section, we calculated the median values:

- ratio l/*base* between the curve length l from I1 to I2 and the Euclidean distance *base*;
- ratio h/*base* between the height h and the Euclidean distance *base*.

A set of 40 lines have been segmented and classified using the S-PLUS® cluster analysis package. The results show some deficiencies but, still, we come to a relevant first level of classification into barely sinuous/sinuous/highly sinuous sections (Fig. 17.8).

The main limitation is that different shapes (from a perceptual point of view) produce similar values (see in Fig. 17.8 the top left – section S4–vs. the bottom right – section S2 – of line 4376). A closer analysis of each section is required in order to differentiate their geometrical characteristics.

Yet another problem is related to the local positions of the inflection points, and thus of the critical points, most notably with complex bend shapes. Local refinement of these positions is required so that they can be used reliably and predictably within the given theories of both segmentation and measurement.

17.5 Potential outcomes

The resulting geometrical description is a tree, which needs first to be validated by means of a bottom–up analysis. From a theoretical point of view, we can consider that this tree provides us with the knowledge base from which we will be able to choose adequate generalisation operations. In order to guide such choices, a rule base is now needed. Among the factors that we have to consider, there are initial and targeted scales, targeted map symbolisation and geometry (quantitative measurements and qualitative description). For instance, looking at two series of bends that perceptually seem to share similar geometrical characteristics, it has to be remembered that if the symbol widths or their semantic natures are different, generalisation operations can be different. Moreover, if two series of bends belong to the same geometric class, and if their sizes are very different, different generalisations might be required: little details might be eliminated, while bigger ones are emphasised. How can we take such problems of size into account?

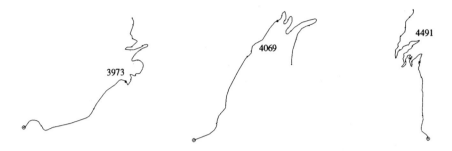

Examples of results of automatic segmentation of BDCarto® roads

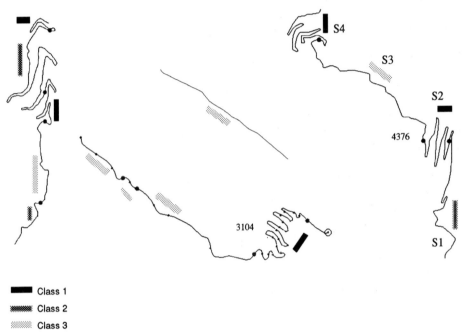

■ Class 1
▓ Class 2
░ Class 3

Figure 17.8 Classification on segmented roads. Small dots are the detected inflection points, big dots are critical points.

Our segmentation and analysis process actually is not linked with generalisation algorithms, but rather with human visual perception. Will it be possible, through formalised rules, to directly link the geometric classes to the generalisation operations, or even to the algorithms? In order to fully access these questions, further studies are required.

Assessing the effects of generalisation algorithms and validating the quality of generalisation alternatives. We need computational methods for assessing the effects of generalisation algorithms on the geometry of linear feature (shape degradation, topological degradation, quantitative measurements and comparison to manual generalisation). Besides, a series of tests is required for each geometrical type in order to determine correctly which algorithms or sequences of algorithms are suited to each operation.

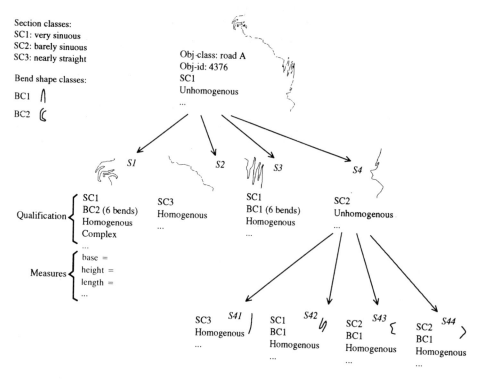

Figure 17.9 An example of a descriptive tree of a BDCarto® road.

In terms of shape maintenance, a potential outcome of the description tree might be to compare two versions of a line in terms of shapes: it should be easier to compare the description trees, instead of the raw geometries of the line and its generalised version.

The results distribute correctly at the first-level line sections into 'rather straight', 'smoothly sinuous' and 'significantly sinuous' classes, so that we know where to apply the Douglas algorithm or smoothing using a Gaussian filtering with tolerance values in relation to the sinuosity type. It is clear that a problem is raised for handling curves at critical points, when using different algorithms or sequences of algorithms on a segmented line.

Formalising rules for generalisation decisions (operations and/or algorithms). In addition, there is a need to formalise cartographic rules from geometrical knowledge. Thus, in order to extract necessary information from the line description at several levels (Plazanet, 1995), further analysis of the tree is required such as: shape levels (relative levels of shapes within shapes), shape environment (relative positions of shapes within shapes, how shapes fit into the more global shapes of upper levels), shape or bend repetitions, intrinsic conflict areas of the line.

Some other cartographic rules clearly intervene in generalisation decisions related to map specifications: symbolisation, scale-reduction factor, theme of the map. We need to formalise this knowledge by means of rules. A promising way to acquire generalisation procedural knowledge by means of techniques of computational intelligence is presented in Weibel *et al.* (1995).

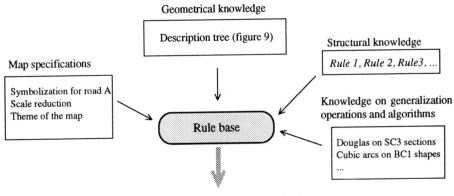

Geometrical knowledge

Description tree (figure 9)

Structural knowledge

Rule 1, Rule 2, Rule3, ...

Map specifications

Symbolization for road A
Scale reduction
Theme of the map

Knowledge on generalization
operations and algorithms

Rule base

Douglas on SC3 sections
Cubic arcs on BC1 shapes
...

Choices of adequate local generalization

One possible solution :

SECTION	OPERATIONS	ALGORITHMS
S1 ⇨	Schematization	⇨ ???
S2 ⇨	Simplification + Smoothing	⇨ Douglas (80 m.) + Gaussian smoothing
S3 ⇨	Schematization	⇨ ???
S41 ⇨	Smoothing	⇨ Gaussian smoothing ?
S42 ⇨	Enhancement	⇨ ???
S43 ⇨	Simplification + Smoothing	⇨ Douglas + Gaussian smoothing
S44 ⇨	Smoothing	⇨ Gaussian smoothing ?

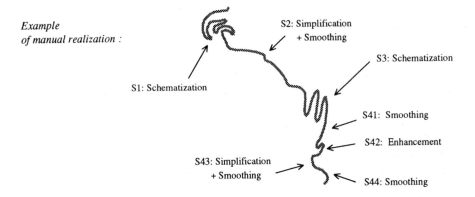

*Example
of manual realization :*

S2: Simplification
+ Smoothing

S3: Schematization

S1: Schematization

S41: Smoothing

S42: Enhancement

S43: Simplification
+ Smoothing

S44: Smoothing

Figure 17.10 Example of use of the description tree.

We may also formalise simple rules for a particular bend or series of bends delimited by two inflection points (example of section S42 in Fig. 17.9). For example, Rule 1: If the distance between these points is smaller than symbolisation width, then an internal conflict is detected. Rule 2: If these points determine an isolated hairpin bend inside a globally straight section, then this bend is a salient bend which must be maintained. Rule 3: A salient bend which gives rise to an internal conflict must be amplified, if it is a road.

An example is proposed in Fig. 17.9 showing the use of the description tree to choose sequences of generalisation operations and adequate algorithms. Each node of the tree carries a descriptor composed of qualitative criteria values and measurements, a section class and a bend-shape class code that may both be zero.

Figure 17.10 illustrates the use of this geometrical knowledge in cartographic generalisation decisions where map specifications and knowledge on generalisation operations and sequences of algorithms are integrated.

17.6 Conclusion

The approach addressed in this chapter for automating linear feature generalisation consists in the elicitation of complementary knowledge on the geometry of linear features (more particularly roads) in order to guide the choices of adequate generalisation sequences of operations and algorithms. Such an approach is based on the assumption that guiding the local choices of generalisation tools from the complementary information on the geometry, and the semantic, spatial and cartographical constraints will hopefully provide adequate (or at least acceptable) solutions. In our sense, such a way to proceed will not do without the quality assessment stage for which the geometrical description will also be an important and helpful basic information.

In the current stage of the research, we are experimenting with dedicated sets of measurements for different steps of segmentation and geometry analysis. The next stage will address the generalisation process by implementing a rule-base prototype, taking into account the hierarchical description proposed. Further work will focus on the definition of assessing tools.

References

ATTNEAVE, F. (1954). Some informational aspects of visual perception, *Psychological Review*, **61**(3), 183–93.

BABAUD, J., WITKIN, A., BAUDIN, M. and DUDA, R. (1986). Uniqueness of the (G)aussian (K)ernel for (S)cale-(S)pace (F)iltering, *IEEE Transaction on Pattern Analysis and Machine Intelligence*, **8**(1), 26–33.

BALLARD, D. H. (1986). Strip trees: A hierarchical representation for curves, *Communications of the Association for Computing Machinery*, **B**, 2–14.

BEARD, K. (1991). Theory of the cartographic line revisited, *Cartographica*, **28**(4), 32–58.

BUTTENFIELD, B. P. (1987). Automating the identification of cartographic lines, *The American Cartographer*, **14**(1), 7–20.

BUTTENFIELD, B. P. (1991). A rule for describing line feature geometry, in Buttenfield, B. P. and McMaster, R. B. (Eds), *Map Generalization*, Chapter 3, pp. 150–71, London: Longman Scientific & Technical.

FREEMAN, H. (1978). Shape description via the use of critical points, *Journal on Pattern Recognition*, **10**, 2–14.

HERBERT, G., JOAO, E. M. and RHIND, D. (1992). Use of an artificial intelligence approach to increase user control of automatic line generalisation, in Harts, J., Ottens, H. and Scholten, H. (Eds), *Proceedings of EGIS '92*, **vol. 1**, pp. 554–63, Utrecht: EGIS Foundation.

HOFFMAN, D. D. and RICHARDS, W. A. (1982). Representing smooth plane curves for visual recognition, *Proceedings of American Association for Artifical Intelligence*, pp. 5–8, Cambridge, MA: MIT Press.

JENKS, G. F. (1989). Geographic logic in line generalisation, *Cartographica*, **26**(1), 27–42.

MCMASTER, R. B. (1986). A statistical analysis of mathematical measures for linear simplification, *The American Cartographer*, **13**(2) 103–17.

MCMASTER, R. B. (1989). The integration of simplification and smoothing algorithms in line generalization, *Cartographica*, **26**(1), 101–21.

MCMASTER, R. B. (1995). Knowledge acquisition for cartographic generalization: Experimental methods, in Müller, J. C., Lagrange, J. P. and Weibel, R. (Eds), *GIS and Generalization: Methodology and Practice*, pp. 161–80, London: Taylor & Francis.

MOCKTARIAN, F. and MACKWORTH, A. K. (1992). A theory of multi-scale, curvature based shape representation for planar curves, *IEEE Transactions on Pattern Analysis and Machine Intelligence*, **14**, 789–805.

PLAZANET, C. (1995). Measurements, characterisation and classification for automated linear features generalisation, *Proceedings of AutoCarto 12*, **vol. 4**, pp. 59–68, Charlotte CA: ASPRS and ACSM.

PLAZANET, C., AFFHOLDER, J. G. and FRITSCH, E. (1995). The importance of geometric modelling in linear feature generalisation, in Weibel, R. (Ed.), *Cartography and Geographical Information Systems Journal Special issue on Map Generalization*, **22**(4), pp. 291–305.

ROSIN, P. L. (1993). Multiscale representation and matching of curves using codons, *Computer Vision: Graphics and Image Processing*, **55**(4), 286–310.

THAPA, K. (1989). Data compression and critical points detection using normalized symmetric scattered matrix, *Proceedings of AutoCarto 9*, pp. 78–89, Baltimore, MA: ASPRS and ACSM.

UEDA, N. and SUZUKI, S. (1990). Automatic shape model acquisition using multiscale segment matching, *Proceedings of the 10th International Conference on Pattern Recognition*, pp. 897–902.

WEGER, G. (1993). *Cours de cartographie*, Paris: IGN Internal report.

WEIBEL, R. (1989). Konzepte und Experimente zur Automatisierung der Reliefgeneralisierung, PhD Thesis, Zürich: Geographisches Institut. Universitat, Zürich.

WEIBEL, R., KELLER, S. and REICHENBACHER, T. (1995). Overcoming the knowledge acquisition bottleneck in map generalization: The role of interactive systems and computational intelligence, in Frank, A. and Kuhn, W. (Eds), *Proceedings of COSIT '95 Conference*, pp. 139–56, Berlin: Springer.

CHAPTER EIGHTEEN

Representations of Data Quality

KATE BEARD

18.1 Introduction

Although some users may wish to remain oblivious to data quality, it is nonetheless important that users of GIS have the opportunity to inform themselves about data quality should they so choose. Within the scientific community, data documentation, which includes information on data quality, is becoming a mandatory part of the research process. Several programs such as the US Global Change Program require data documentation and sharing as conditions of a grant award. Data sharing through such efforts as the Spatial Data Clearinghouse (Tosta, 1994) and research under way on digital spatial libraries (Smith *et al.*, 1994) creates additional demands for metadata and data quality information. Thus, there is a growing need to address the issue of data quality representation within the overall context of system design such that quality information will be easy for producers to incorporate, easy for users to access and comprehensive enough to serve a range of analytical needs.

Adding data quality information to the database raises a number of issues. It has the potential to at least double the volume of the database with potentially greater increases as more rigorous analyses are required. Errors, particularly in resource and environmental modelling, are not simple nor are the set of possible relationships between geographic phenomena and their quality descriptions straightforward. Conventional GIS have become reasonably successful at representing spatial data as exact objects (points, line and areas) or discretised fields but they have several problems with certain forms of complexity, scale differences, generalisation and accuracy (Burrough and Frank, 1995).

Neither the Spatial Data Transfer Standard (SDTS) (Fegas *et al.*, 1992; US Department of Commerce, 1992) nor the Content Standard for Geospatial Metadata (Federal Geographic Data Committee, 1994) specify formats for storing and managing data quality. The first implementations of quality reports were generated in the form of written documents which accompanied the data files. This format evolved to an electronic version in which a separate ASCII file formed the quality report accompanying the data (often a README file). More recent work has focused on storing quality data in a relational DBMS (Blott and Včkovski,

1995; Hiland *et al.*, 1993). In this form the quality information is maintained separately from the data, but is accessible for search and query through standard database functionality. Some current implementations have combined data and metadata including quality descriptions within a relational database. For example, an attribute may be followed by an estimate of its standard error. In this form the quality documentation can be more easily revised as data are updated. Some of the most recent work has proposed an object-oriented approach for representation of metadata including some components of data quality (Anderson and Stonebraker, 1994).

Other options for representing and communicating data quality have included implicit incorporation of quality within the data structure. Dutton's (1989) quaternary triangular mesh (QTM) is an example in which locational data can be stored within the QTM hierarchy based on their level of precision. While this method is innovative, the structure is deterministic and manages only a single component of quality. Lastly, visual representations have been proposed as methods for storing and communicating quality information to users (Beard and Mackaness, 1994; Buttenfield and Beard, 1994; Goodchild *et al.*, 1993a; MacEachren, 1993; Ramlal, 1991). Visual methods may make quality information more accessible to users as well as provide a compact and efficient form for storing spatially structured quality information. The drawbacks of visual methods are that they are most often implicit, non-quantitative and subject to individual user interpretation.

Most of the above solutions for representing data quality have been approached as add-ons to an existing model. The problem with many existing data models is that they were not designed with data quality representation in mind. One of the still dominant models, the map or object model, has well-documented deficiencies as a model for representing data quality (Burrough, 1986; Goodchild and Gopal, 1989). The meaning of map model as used in this chapter assumes some phenomena have been mapped and that the data model is then developed from the map representation rather than the original characteristics and behaviours of the phenomena. Figure 18.1 contrasts a conceptual data model generated from a mapped version of reality (a) with option (b) which models reality directly.

One problem in modelling the map is that the map conventions and abstraction are incorporated in the model. This is not a problem in all circumstances, but is with respect to modelling data quality. Since the map includes generalised objects for which the generalisation process is not reconstructable there is no way to reconstruct let alone quantify the imprecision or uncertainty in the result. Also,

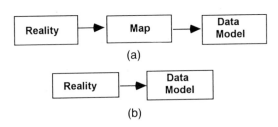

Figure 18.1   In (a) the data model models the mapped version of reality; in (b) reality is modelled directly.

the abstractions of the map may not be appropriate representations for real-world objects. A standard area class or choropleth map depicting homogeneous units with well-defined boundaries is often not a realistic representation of geographic phenomena, nor does it provide a basis to assess the reliability of the representation. Adding quality information to the database may ultimately require replacing current data models with models that can more directly support data-quality information. The goal should be incorporation of sufficient information to allow assessment of fitness for use.

18.2 Requirements for data-quality representation

Given that existing models exhibit limitations for representing data quality we should first ask what are some basic requirements for representations of data quality. We can list a number of these including the following which, while listed as individual bullets, are not necessarily independent.

- *Integration* – the quality descriptions should not be separate from the data but integrally linked with them.
- *Multidimensionality* – quality is a complex description that should include information to assess positional, thematic and temporal adequacy of the data.
- *Dynamic* – as data change through various processing and updates the quality information should be updated accordingly.
- *Accessible* – quality information or processes to extract it from the data should be readily available to users.
- *Flexible* – the model should be flexible enough to handle quality reporting in various formats including map, graphic, or tabular formats.

18.2.1 Integration of data and quality information

Some people have argued that the separation between data and metadata is an arbitrary one. In data quality, as a component of metadata, there is no physical reason for its separation as there may be in storing metadata for a catalogue or data directory (Blott and Včkovski, 1995). There is potentially great benefit from an integral association of data with descriptions or measures of its quality. Approaches which separate quality descriptions from the data risk reducing ease of access. With the increased interest in data sharing and the ensuing increase in data transfers, an assurance that quality information will travel with the data becomes important. Integration of data with its quality description will also be important in maintaining consistency as updates are made to the data.

18.2.2 Multidimensional aspects of data quality

Data by themselves do not contain all the necessary information to specify data quality. Beard *et al.* (1991) describe data quality as a point in a three-dimensional space of data goodness, application and purpose (Fig.18.2). To provide a

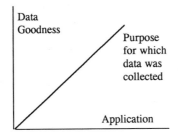

Figure 18.2 Multidimensional data-quality space.

	Location	Theme	Time
Accuracy			
Resolution			
Consistency			
Completeness			
Lineage			

Figure 18.3 Matrix of data-quality components. The intersection of a row and column would provide a measure for a specific quality component (e.g. thematic resolution which could translate as the detail or refinement of a classification system).

comprehensive representation of quality the users' knowledge of the application and its purpose must be combined with the characteristics of the data. Paradis and Beard (1994) offer an example of a data-quality filter which allows users to express their data-quality requirements and view these in relation to the available data.

The axis of data goodness itself is multidimensional. Not only must the spatial positions and relationships among the data be adequately documented but reliability of the thematic and temporal content should be conveyed as well. A matrix for data quality, identified in Buttenfield and Beard (1994) provides a structure for organising the multidimensional components of quality. This matrix crosses the SDTS quality components (accuracy, consistency, completeness and lineage with the addition of resolution) with the three dimensions of spatial observations (location, theme and time) as suggested by Sinton (1978), Fig. 18.3. This matrix provides a checklist of components which may be of concern to users although they may not have equal interest in all components at any one time. For some applications locational accuracy may be of little concern while thematic completeness may be critical. This matrix was used as a framework for access to data-quality information in Paradis and Beard (1994).

18.2.3 Dynamic data quality

The quality of GIS data is not static. As several processes are applied to data in the course of an application the original accuracy and consistency of the data may be altered. Operations which change the geometry, relationships, resolution or classification of the data change the quality. These alterations to the data should be tracked to keep users informed of possibly significant changes to the data which may affect their applications. Routine updates to the data may change the quality, but if database integrity checks apply the effects on quality from such changes should be minimal. It is more likely that updates or additions to the database will improve rather than degrade quality. Updates should make the information more current and new additions to the database, such as new samples, can improve the accuracy of GIS products such as interpolated surfaces generated from a set of samples. Positive improvements in data and products should be documented along with degradations, but for dynamic data-quality monitoring, changes resulting from GIS processes will be more problematic than updates.

18.2.4 Improved access to data quality

Data quality as it is currently handled is not readily accessible to the conventional GIS user. Separate quality reports can be time-consuming to read through and assess and thus can be easily ignored by users. Quality information should ideally be stored in a database to provide access through standard query processing. Incorporation of quality information in a database also supports integrity checks to maintain consistency.

A user interface design supporting access to quality is equally important. The interface should provide a framework that makes access easy and the information understandable so users are encouraged to examine the quality of the data in the context of their own work. Quality information should be made accessible in the course or context of an application so that it informs the work flow and does not interrupt it.

18.2.5 Flexibility in data-quality representation

Users may want to examine data quality in different forms. Some users may want very detailed descriptions of a data set with well-defined confidence intervals for research purposes. Others may be satisfied with brief summary indices or assessments. Report formats for data quality should be flexible to accommodate a range of users' needs. Some users may find graphs and visual representations of quality more informative than numerical or statistical summaries. The ability to report quality in different formats is thus desirable.

18.3 A conceptual model for data quality

This chapter proposes an object-oriented model for representing data quality. While each of the five characteristics discussed above is desirable, this chapter focuses specifically on two; the integration of data with quality information and support for dynamic data-quality management. The chief advantage of an object-oriented approach is that data quality can become a more integral part of the data description rather than an *ad hoc* addition. The chapter presents three versions of a conceptual model that supports increasingly comprehensive representations of quality. The first case describes a model which is static and deterministic. In this version data quality is assumed to be constant. The second case describes a model for data quality which is dynamic. The third case presents a model for data quality which is stochastic.

Data quality is fundamentally dependent on the data collection or observation techniques as well as any subsequent data-processing methods (Hunter, 1991; Hunter and Beard, 1992). For example, the type of positioning instrument used to collect a location reference (GPS or compass and tape), the number and density of samples, the calibration of a laboratory device, or the software for rectifying an image can all impact on the quality of the data. In the proposed model, the primary objects are observations or raw measurements. This approach overcomes the limitations of the scenario described in Fig. 18.1a, in which valuable information is discarded in the process of compiling a map and hence does not appear in the conceptual model. With the proposed approach, the behaviour and characteristics of the observation devices and methods can be incorporated in the model. It also provides a foundation to track changes in quality introduced by data-processing techniques and changes in the database over time.

For each of the three versions of the model, the example of satellite imagery is used to illustrate the concept. The basic objects for the satellite image example are as shown in Fig. 18.3. This schema diagram uses object classes from the Spatial Archive and Interchange Format (SAIF) (Surveys and Resource Mapping, 1994). SAIF is a standard format developed in British Columbia, Canada, expressly as a means for sharing spatial and spatiotemporal data. It was designed to handle virtually any kind of data. The Sequoia 2000 project used SAIF to create a metadata schema for satellite images (Anderson and Stonebraker, 1994) but without focusing on the data quality components specifically. This model adapts SAIF to explicitly incorporate data quality.

Fig. 18.4 gives the class hierarchy for standard SAIF objects which are later modified to develop the satellite imagery schema. Within SAIF, class hierarchies begin with an AbstractObject or an EnumerationObject. A GeographicObject is a subclass of AbstractObject which represents parts of the real world. A Coverage is a subclass of GeographicObject which describes some geographic phenomena distributed over an arbitrary region of space. It maps a position in space to a value according to some function, in this case a simple assignment of image values to grid positions. A Raster is a type of coverage in which the geometry is restricted to a grid and the values pertaining to the grid are provided by content values. A GeneralRaster is a subclass of Raster which consists of Channels, an IndexScheme, an ImageGrid which defines the geometry and ContentValues which describe the actual values stored for each grid location. ImageGrid is a subclass of grid which defines a row ordered grid with the grid origin and reference point for each cell assumed to be the upper left-hand corner. The Channels object consists of a list of the instances of Channels. The Channels object has attributes consisting of channel number, channel name, channel description and units. Channels correspond to image layers in a database. For example, instances of channels may be the seven bands of a TM scene or the result of statistical or algebraic algorithms applied to an image layer. The GeneralRaster will be the primary object used to develop the data-quality schemas. The abstraction relationships and dependencies are as shown by the key.

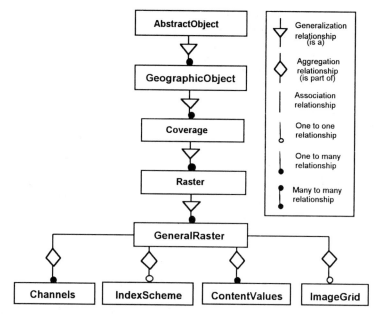

Figure 18.4   Schema of SAIF objects which forms the basis for the satellite imagery schema.

18.4 The static deterministic model

The static deterministic model is the simplest to implement but supplies the least comprehensive information on data quality. This model assumes measures of positional accuracy, attribute accuracy, or other quality components are documented for a data set. Examples of deterministic measures include the circular map accuracy standard, epsilon bands or the per cent correctly classified statistic for classified remotely sensed imagery. These values would be arrived at through some testing procedure (for example, deductive or in comparison with ground truth observations). We assume quality information is generated by the producer and attached to the data but no subsequent changes are made to this quality information. This simplifies the model since there is no need to know the specifics of how individual processes affect data quality.

To describe this model we assume a spatiotemporal observation which consists of the tuple $\{X, Y, Z, T, A_i\}$, where X, Y and Z identify the horizontal and vertical spatial reference of an observation, T is the temporal reference or time at which the observation was made and A_i $i = 1, \ldots, m$ are the m measured characteristics of the location. Each of these elements of the tuple may have one or more quality descriptors associated with them. For example X, Y and Z may each have RMSE estimates. T may have an accuracy estimate and similarly each A_i may have an accuracy estimate. This notation assumes each attribute A_i was measured at the same time. We can document a number of examples for which this would be the case. In soil surveys, samples are often taken along transects where at each sample location, soil colour, texture, depth to water table, depth to bedrock or other variables are measured. If each of these elements of the tuple occurs in the

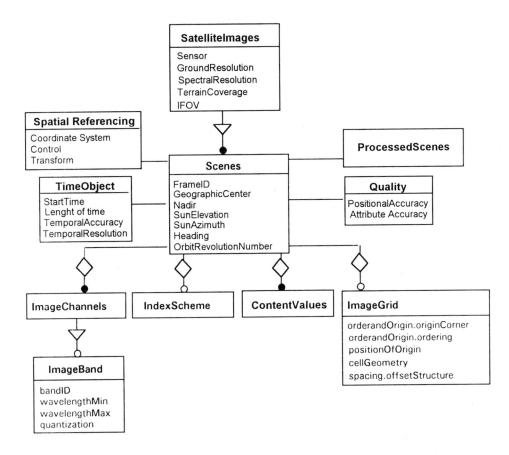

Figure 18.5 Schema for the static deterministic model of data quality for satellite imagery.

database, each with a single quality descriptor, then the quality information will increase the size of the database by a factor of $m + 4$.

Figure 18.5 shows an example of the static deterministic model for satellite imagery. In this schema SatelliteImages is a subclass of the SAIF Raster class and Scene is a subclass of the SAIF class GeneralRaster. The SatelliteImages subclass has attributes which describe a specific sensor type, ground and spectral resolution and other characteristics specific to a sensor. The Scenes subclass also contains attributes specific to a satellite and sensor system along with inherited attributes from the GeneralRaster subclass including ImageChannels (a subclass of Channels) an IndexScheme, ContentValues and an ImageGrid. ImageBand is a subclass of ImageChannels which provides specific information on each satellite band.

Scenes have an association relationship with a TimeObject, a SpatialReferencing object, a Quality object and a ProcessedScenes object. The TimeObject provides a start time, duration and temporal accuracy and resolution information for Scenes. The SpatialReferencing object identifies a coordinate system, control information and transformation information for Scenes. Instances of the Scenes subclass could include raw images, or images with some preprocessing for geometric corrections. Quality attributes are generated by the producer to describe a scene. In this static case the quality attributes are inherited intact by instances of the ProcessedScenes subclass. The benefits of developing a data-quality model to this level would be that information about the data-collection system would be incorporated in the model and a producer's assessment of quality would be directly attached. The limitations are that it would not accurately reflect the quality of the data with respect to processes applied nor would it provide confidence intervals for any particular measure.

18.5 Dynamic data-quality model

The dynamic model assumes that quality is not static and changes with time. There are a number of possible variations on a dynamic model, all of which may change the database. The possibilities include new attributes which may be collected over time at the same locations, new samples (new x, y, z and attribute data) which may be collected and added to the database and processes which may be applied to the original data set. In this chapter we focus on the latter case and assume that quality only changes as processes are applied to the data set. As described earlier, changes produced by processes are the most problematic in describing changes to data quality. To model this we need two types of time. We can use the common terminology of database time and event time. Database time is the time when transactions take place within an information system and event time is when events occur in the application domain. Using the tuple notation again we now have a version which includes two representations of time and indexes processes applied to the data. The tuple includes: $\{X, Y, Z, A_i, ET, P_j, DTj\}$ where as before X, Y and Z are the spatial references, A_i $i = 1, \ldots, m$ is a vector of attributes, ET is event time or the time at which the data were recorded, P_j, $j = 1, \ldots k$ is a set of processes applied to the data and DTj is the set of database times associated with each process.

Figure 18.6 shows the modified schema for the dynamic deterministic case. ProcessedScenes appear again in an association relationship with Scenes. ProcessedScenes also have an association relationship with a SpatialReferencing object and a ProcessTimeObject. In this schema, positional and attribute accuracy which were previously attributes of the Quality object and in an association relationship with Scenes become attributes of ProcessChannels. Instances of ProcessChannels can be resampled images, rectified images, classified images or filtered images. In this case each instance inherits a quality attribute and updates it as a function of the process which created the new instance. Revised quality estimates may be arrived at by deduction from knowledge of the process and particular parameter values or through testing (Chrisman, 1983).

The dynamic version of the model does not significantly increase the storage requirements over the static model. The major difference results from quality components now being stored for each process. Under this model storage at a minimum increases by the number of quality components (q) times (k) the number of processes.

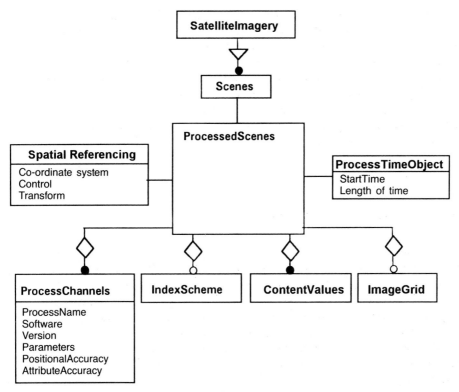

Figure 18.6 Schema for the dynamic deterministic version of the data-quality model for satellite imagery.

18.6 Stochastic data-quality model

A stochastic model for data quality is one in which phenomena to be represented are subject to uncertainty and governed by the laws of probability. A stochastic process is defined by specifying the joint distribution function for any finite collection of random variables (Ripley, 1981). Goodchild *et al.* (1992) define and develop an error model as a stochastic process capable of generating a range of distorted versions of the same reality. Under this approach the values of spatial attributes or coordinates are represented by a range of possible values rather than a single deterministic value. The advantage over deterministic measures of accuracy is that the stochastic model is capable of generating parameters for a distribution and hence confidence intervals.

The stochastic model is based on the generation of several sample representations (images) which are realisations of a stochastic process. In the general case the stochastic model for a spatial attribute is:

$$\{A(x): x \in D\};$$

and a realisation of this is $\{a(x): x \in D\}$ where $x \in R^d$ is a generic location in d-dimensional Euclidean space and A is a random quantity at spatial location x

(Cressie, 1991). For the geometric component we can assume a stochastic model:

$$\{X{:}X \in D\}$$

where X is a vector of random variables x and y designating location. Heuvelink (1993) provides an example for quantitative spatial attributes. He addresses several scenarios based on error perturbed inputs. If the n outcomes of running process P_j are u_i $i = 1, \ldots, n$ the model results form a random sample from the distribution U. The parameters of $f(u)$ such as the mean and variance can be estimated from a random sample from U. The distribution of U is approximated to any level of precision by repeating a Monte Carlo simulation a sufficiently large number of times. Goodchild *et al.* (1992) provide a similar example for categorical random variables. Their model requires (1) a vector of given probabilities for each cell of a raster which determines the proportion of realisations in which the cell is assigned a given class and (2) a correlation parameter which determines the level of dependence between outcomes in neighbouring cells. Confidence limits are produced by again simulating several realisations of the model. The outcomes of both of these models provide information on the reliability of both positions and attributes. A purely geometric model based on a raster representation would require a probability vector of locations for a cell corner and, depending on the process and data being modelled, a correlation coefficient.

Using the tuple format as given for the two previous cases, for the stochastic case we include $\{X_{js}, Y_{js}, Z_{jr}, A_{ijt}, ET, P_j\}$. Theoretically there is some deterministic value for each element of the tuple, however, uncertainty about them allows us to treat them as the outcome of some random mechanism. Thus each element of the tuple is now a random variable. As before X, Y, Z are the spatial references, A_i, $i = 1, \ldots, m$ is the set of attributes, ET is the time recorded for each observation, P_j, $j = 1 \ldots k$ are the processes applied to the data and the subscripts s, r, t index a set of n realisations associated with each of the P_j processes. Distribution parameters for X and Y are assumed equal and uncorrelated so both may be indexed by s.

For the satellite imagery example the central object is a subclass of General-Raster called a RealisationSet. A RealisationSet, as a subclass of General Raster, inherits the subclass Channels now specialised as RealisationChannels. A RealisationSet has an association relationship with the subclass ProcessedScenes. In other words, a set of realisations exist for each process applied to an image. Under the stochastic model, ProcessChannels includes at least two special instances: one called a probability channel which defines the probability distribution of outcomes for a specific process and the other called a parameter channel. Using the example of the Goodchild *et al.* (1992) model the probability channel would contain the probability distribution of class assignments for a specific cell and the parameter channel would contain the spatial auto-correlation parameter r. By storing this parameter as a channel it is allowed to vary from place to place.

Instances of RealisationChannels are the n realisations of a process. Each takes the form of a grid with content values which are realisations for a given process, probability distribution and correlation coefficient if applicable. For example, these may be the n outcomes of a classification process, the n outcomes of resampling, or the n outcomes of rectification. It should be noted that for a

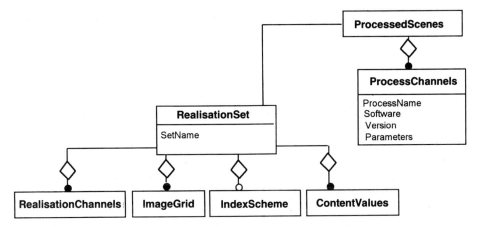

Figure 18.7 Schema for the stochastic version of the data-quality model for satellite imagery.

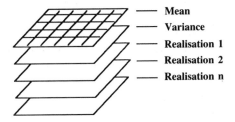

Figure 18.8 Instances of the RealisationChannels subclass.

rectification process the ImageGrid would have n realisations of its geometry. Two additional instances of RealisationChannels are the mean and variance for the realisation set as shown in Fig. 18.8. These are respectively a grid of mean values and variances for each cell over the realised set. Once the parameters for a distribution have been estimated, there is no requirement to store the n realisations permanently unless they are desired to help visualise the spatial distribution of the uncertainty (Goodchild *et al.*, 1993b).

Under the stochastic model the storage and processing requirements increase substantially since a number of simulations are required for each process. Various estimates for the size of n have been presented (Bierkins and Burrough, 1993; Openshaw *et al.* 1991; Peck *et al.* 1988; Smith and Hebbert, 1979). Peck *et al.* (1988) indicated that for ground water modelling, $n = 100$ was sufficient to obtain the mean, $n = 1000$ was required for the variance and $n > 10\,000$ was required for the 1 per cent quantile. Other additions include the probability channel which is required to store the probability values for outcomes of the various processes and in general at least one additional parameter channel. If all realisations are stored then the storage increases by $(n + 2) \times (j + 2)$. This points to a formidable increase in processing and overwhelming storage demands even with the current pace of technology. If realisations are only computed and not permanently stored the storage overhead is less burdensome $(2 * (j + 2))$ but the required processing is still intense.

18.7 Conclusions

The quality of GIS output is a function of the quality of the inputs plus error propagation through a set of GIS operations. While the demand for data-quality information is growing the ability to provide it to users in the form of confidence intervals can be a complex process. This chapter presents requirements for the representation of data quality and proposes three options for representing data quality. Each of the options requires incrementally more comprehensive information on data and the processes applied to them with associated increases in required computing and processing. With the increased overhead comes more robust information on quality. The static model makes the assumption that quality information remains constant. Deterministic quality descriptions (which may originate with the data producer) are stored with the original data. The overhead is minor but no information is available on the reliability of intermediate or final GIS products. The dynamic case improves this situation by updating deterministic measures based on knowledge about each process. This additional information generates some increase in storage overhead. The stochastic version can provide the most comprehensive information on quality in the form of confidence intervals for any intermediate or final GIS product but at the cost of a sizeable increase in overhead. Several challenges still lie ahead in making quality information generally available to users in a comprehensive and robust form.

References

ANDERSON, J. T. and STONEBRAKER, M. (1994). Sequoia 2000 metadata schema for satellite images metadata in video images, *ACM SIGMOD Record*, **23**(4), 42–8.

BEARD, M. K. and MACKANESS, W. (1993). Visual access to data quality in Geographic Information Systems, *Cartographica*, **30**(2 and 3), 37–45.

BEARD, M. K., BUTTENFIELD, B. and CLAPHAM, S. (1991). 'Visualisation of spatial data quality', Scientific Report for the Specialist Meeting, Technical Paper 91-26, Santa Barbara, CA: NCGIA.

BIERKINS, M. F. P. and BURROUGH, P. (1993). The indicator approach to categorical soil data, *Journal of Soil Science*, **44**, 361–8.

BLOTT, S. and VČKOVSKI, A. (1995). Accessing geographical metafiles through a database storage system, in Egenhofer, M. and Herring, J. (Eds), *Advances in Spatial Databases*, pp. 117–31.

BURROUGH, P. (1986). *Principles of Geographical Information Systems for Land Resources Assessment*, Oxford: Clarendon Press.

BURROUGH, P. and FRANK, A. U. (1995). Concepts and paradigms in spatial information: Are current Geographical Information Systems truly generic? *International Journal of GIS*, **9**(2), 101–16.

BUTTENFIELD, B. P. and BEARD, M. K. (1994). Graphic and geographic components of data quality, in Hearnshaw, H. and Unwin, D. (Eds), *Visualisation in Geographic Information Systems*, pp. 150–7, London: Belhaven Press.

CHRISMAN, N. (1983). The role of quality information in the long-term functioning of a GIS, in Weller, B. S. (Ed.), *Proceedings of Auto-Carto 6*, pp. 303–12, Falls Church, VA: American Society of Photogrammetry and American Congress on Surveying and Mapping.

CRESSIE, N. (1991). *Statistics for Spatial Data*, New York: John Wiley.

DUTTON, G. (1989). Modelling locational uncertainty via hierarchical tessellation, in

Goodchild, M. and Gopal, S. (Eds), *Accuracy of Spatial Databases*, pp. 125–40, London: Taylor & Francis.

FEDERAL GEOGRAPHIC DATA COMMITTEE (1994). *Content Standard for Digital Geospatial Metadata* (8 June), Washington DC.: Federal Geographic Data Committee.

FEGAS, G. R., CASCIO, L. and LAZER, A. (1992). An overview of FIPS 173, The spatial data transfer standard, *Cartography and Geographic Information Systems*, **19**(5), 278–91.

GOODCHILD, M. and GOPAL, S. (1989). *The Accuracy of Spatial Databases*, London: Taylor & Francis.

GOODCHILD, M., BUTTENFIELD, B. and WOOD, J. (1993a). Visualising data validity, in Hearnshaw, H. and Unwin, D. (Eds), *Visualisation in Geographic Information Systems*, pp. 141–9, London: Belhaven Press.

GOODCHILD, M., CHI-CHANG, L. and LEUNG, Y. (1993b). Visualising fuzzy maps, in Hearnshaw, H. and Unwin, D. (Eds), *Visualisation in Geographic Information Systems*, pp. 150–7, London: Belhaven Press.

GOODCHILD, M., GUOQUING, S. and SHIREN, Y. (1992). Development and test of an error model for categorical data, *International Journal of Geographic Information Systems*, **6**, 87–104.

HEUVELINK, G. B. (1993). Error propagation in quantitative spatial modelling, Published Ph.D. dissertation, University of Utrecht.

HILAND, M. W., WAYNE, L. D., CARPENTER, M., MCBRIDE, R. and WILLIAMS, S. J. (1993). Louisiana coastal GIS network: A cataloguing framework for spatial data, pp. 312–22, *Proceeding GIS/LIS*. Minneopolis: ACSM, AM/FM, URISA, ASPRS.

HUNTER, G. (1991). Processing error in spatial databases: the unknown quantity, *Proceedings of the Symposium of Spatial Database Accuracy*, pp. 203–14, Melbourne, Australia: Department of Surveying and Land Information, The University of Melbourne.

HUNTER, G. and BEARD, M. K. (1992). Understanding error in spatial databases, *The Australian Surveyor*, **37**(2), 108–19.

MACEACHREN, A. (1993). Approaches to truth in geographic visualisation, in Pequet, D. J. (Ed.), *Proceedings Auto-Carto 12*, **4**, 110–18, Falls Church: American Society of Photogrammetry and American Congress on Surveying and Mapping.

OPENSHAW, S., CHARLETON, M. and CARVER, S. (1991). Error propagation: a Monte Carlo Simulation, in Masser, I. and Blakemore, M. *Handling Geographic Information*, pp. 78–101, Essex: Longman Scientific and Technical.

PARADIS, J. and BEARD, M. K. (1994). Visualisation of spatial data quality for the decision maker: A data quality filter, *URISA Journal*, **6**(2), 25–34.

PECK, A. J., GORRELICK, S. M., DE MARSILY, G., FOSTER, S. and KOVALEVSKY, V. S. (1988). *Consequences of Spatial Variability in Aquifer Properties and Data Limitations for Groundwater Modelling Practice*, JAHS Publication No. 175, Wallingford: International Association of Hydrological Sciences Press.

RAMLAL, B. (1991). Communicating information quality in a GIS environment, unpublished thesis, Enschede, The Netherlands: ITC.

RIPLEY, B. D. (1981). *Spatial Statistics*, Chichester: John Wiley.

SINTON, D. (1978). The inherent structure of information as a constraint to analysis; mapped thematic data as a case study, in Dutton, G. (Ed.), *Harvard Papers on Geographic Information Systems*, vol. 7, Reading MA: Addison-Wesley.

SMITH, R. E. and HEBBERT, R. H. B. (1979). A Monte Carlo analysis of the hydrological effects of spatial variability on infiltration, *Water Resources Research*, **15**, 419–29.

SMITH, T., GOODCHILD, M., MITRA, S. K., IBARRA, O. H. and DOZIER, J. (1994). Proposal to the National Science Foundation: The Alexandria Project: towards a distributed digital library with comprehensive services for images and spatially referenced information, Santa Barbara: University of California Santa Barbara.

Surveys and Resource Mapping Branch (1994). *Spatial Archive and Interchange Format (SAIF)*. Province of British Columbia; Ministry of Environment, Lands and Parks, April 1994.

Tosta, N. (1994). Data data: partners, *Geo Info Systems*, **4**(10), 24–5, 53.

US Department of Commerce. (1992). *Spatial Data Transfer Standard (SDTS)*, Federal Information Processing Standard 173, Washington DC: Department of Commerce, National Institute of Standards and Technology.

PART FOUR

Concepts and Paradigms

CHAPTER NINETEEN

Concepts and Paradigms of Spatial Data

PETER FISHER

19.1 Introduction

Learning the paradigms related to an area of scientific endeavour is the essential rite of passage from student to practitioner. The role of researcher is to probe the usefulness of those paradigms and to contribute to the progression of the subject area from one paradigm to another. This chapter sets out some of the historical paradigms which currently permeate the area of spatial data processing, especially the Boolean map, the layer-based view of the world, the raster and vector controversy, objects and fields and metrics of space. The shortcomings of all these are established in discussion, and recent developments which may be prefacing changes to a new paradigm are reviewed. Kuhn (1970) states that a paradigm is 'an accepted model or pattern' (p. 23). He goes on, however, to 'appropriate' the word (p. 23) and to imbue it with the sense it brings to modern science. Specifically, the paradigm has come to mean the belief system of a particular science at a particular time. Kuhn (1970) sees science as advancing through a punctuated evolution (to employ a Darwinian metaphor) from one set of agreed principles (paradigm) to another, where the focus of the science on one set of principles pervades the discipline for a considerable (but not necessarily a long) time, and is followed by a very rapid change to another set of principles which then make up the belief system. At any time the paradigm or paradigms of a discipline effectively define it. Of course, Popper (1965) has proposed a more democratic model of the advancement of science, but this can most easily be seen in effect below the paradigms of that science, contributing to those paradigms and to their eventual overthrow.

It is so easy to identify paradigms at many different levels of scientific investigation that it seems impossible to refute Kuhn's model. To quote just a few paradigms which the GIS community touches on, but not those at the core of the subject, we can see the monolithic Darwinian paradigm, which is the quintessential scientific paradigm. Darwin's *On the Origin of Species by Means of Natural Selection* came as the culmination of a literature based on sound scientific investigation which was ignored by the establishment, both scientific and otherwise. That one book, the culmination, overthrew the pre-existing paradigms which, in the Christian world, were based on more or less literal interpretations of the Bible, and within a few short years of being introduced it had become itself the accepted paradigm (Desmond and Moore, 1991), to the extent that it now pervades many subject areas; everything from animals to society to economy and even scientific thought is said to *evolve*, frequently by a process similar to natural selection. It is amusing to notice that the popular understanding of Darwinianism is misidentified with evolution, since Darwin actually suggested the mechanism for evolution, not evolution itself. Closer to our own time, Chaos theory is another good example of a paradigm whose significance spread beyond the immediate scientific discipline. The work of Lorenz, Mandlebrot and others started the change from the deterministic world of Newtonian physics. They were a fringe movement, but during a short period in the 1980s Chaos was on everyone's lips and is now perceived to be one of the primary paradigms of mathematics (Stewart, 1989).

Paradigms 'gain their status because they are more successful than their competitors in solving a few problems that the group of practitioners has come to recognise as acute' (Kuhn, 1970, p. 23). In effect 'The study of paradigms . . . is what mainly prepares the student for membership in the particular scientific community with which he will later practise' (p. 11). It is the creed of a discipline. Effectively without a paradigm a discipline cannot exist in any real sense, although the large disciplines typifying the domains of many university departments may actually possess multiple paradigms intricately interweaving. The advantage of paradigms is that they 'focus attention on a small range of relatively esoteric problems' forcing 'the scientists to investigate some part of nature in detail and depth that would otherwise be unimaginable' (p. 24). On the other hand, they also have the undesirable consequence of channelling most academic endeavour onto those topics at the expense of other possible areas, and non-conformist opinion has trouble being heard or published.

The realm of spatial information science (Goodchild, 1992) is the realm of many different disciplines as Worboys (1994) so clearly states. It is therefore in the position of borrowing paradigms from those different disciplines, and one cause for the misunderstandings between practitioners is that they are trained in different paradigms, bringing along different belief systems and different vocabularies. Indeed, much of the critical tirade of some of the authors contributing to *Ground Truth* (Pickles, 1995) and equally critical reviews (Unwin, in press) is the epitome of the misunderstanding between paradigms within geography; a discipline with more paradigms than most. The technical aspects of the paradigms we deal in may not become so widespread as Chaos theory and Darwinianism, but it is nonetheless important to recognise them if at all possible and to question whether they are appropriate.

In this chapter I recognise a number of paradigms which have had, and continue to have, profound influences on the way we work with spatial information systems. These range from the pre-GIS paradigm of the Boolean map, through the concept of layers, to the raster-vector debate, and its implications via objects and fields to metrics of space. Others may see other paradigms but these are the ones I find most prevalent. I believe that all are flawed, as is so often the case, and that in a very short time we see an increasingly widespread breakdown of these paradigms to be replaced by others. Indeed, I draw attention to a few recent and not so recent articles which are pointing out the insufficiency of these existing paradigms. Other paradigms will emerge if this momentum for change increases, and these new paradigms, it is hoped, will create a greater uniformity of understanding among those researching and practising GIS, whatever their background, than is currently the case. In the nature of paradigms it is not possible to predict what they will be, only to recognise them with some hindsight.

It is worth noting therefore that the term has come in for some abuse in the existing GIS literature. Lanter and Veregin (1992) put forward what they term a paradigm, something which is not possible under the Kuhnian model, although they may reflect the original dictionary definition more closely. The recent paper by Burrough and Frank (1995) uses the term more conventionally. This chapter echoes many of the opinions of their former paper, but it is hoped that it presents a different view of the same issues.

19.2 The paradigm of the Boolean map: an inherited impedance

One of the profoundest influences upon the design and use of Geographical Information Systems is the paradigm of the Boolean categorical map. Although not actually endorsing the paradigm it is endemic in introductory texts (Burrough, 1986; Laurini and Thompson, 1992). This conceptualisation of space has a number of antecedents. Primary among these is cartographic production. The view of map production is that information should be presented in as simple a way as possible. In the case of the topographic map, this is relatively straightforward. Roads seem to have clear-cut edges, and houses are easily definable in most peoples' experience. Naturally Boolean phenomena fit well with the Boolean map concept. Thus property ownership in the Western world is inherently Boolean, although ownership in common is identified, but usually still with clear boundaries if multiple owners. The nationalistic and political maps are also largely Boolean. In the representative democracies of the Western world, a single elector has the right to vote for a defined set of candidates. In the purest form of constituency-based electoral system there is also only one representative who will end up in the elected house of government (Parliament), although in systems of proportional representation there is usually some variety in the representatives to whom an elector might petition.

By contrast, the world we inhabit, the world we live in, as opposed to that in which we are counted, taxed and governed, is one without definite boundaries. It is entirely vague, and may vary from moment to moment. Whether I am prepared to walk somewhere or choose to drive is a decision about the world I inhabit; a decision which will vary from time to time. Thus on one occasion when I have plenty of time, or the weather is fine, I may choose to walk, but on another occasion when I am expecting to bring back more than I could carry, or it is pouring with rain, I may choose to drive. On the other hand, there are many locations I would never even think of walking to, unless it was for some higher motive than day-to-day existence such as a hiking holiday, a protest march or for charity fundraising. In other countries and cultures, the thresholds of convenience and time may be very different. Similarly, the concept of ownership of land is different in other areas, and is commonly actually reversed in many aboriginal cultures, so the group is owned by the land, and in return for the land providing the essentials for living, the group looks after the land (Turnbull, 1989; Young, 1992). Indeed, within these societies boundaries may not actually exist, only a sense of some piece of land owning the group, and that ownership may be defended and fought over, while other foraging land may not be a matter for belligerent confrontation.

The mapping paradigm passed to us from the geologists fits well with the concept of a Boolean map. A geological map is essentially a record of the litho-stratigraphy, and each rock is seen as fitting into a single litho-stratigraphic unit, and once in that unit it can only fit into it and no other unit. That unit itself maps into one and only one higher-level stratigraphic unit in a hierarchical sequence up to the whole of geological time. Geological mapping seems to have a harder time, however, when single attributes are mapped, or attempts are made to detect likely mineral prospects (Bonham-Carter, 1994). The vegetation map is less satisfactory (Brown, 1995). The concept of vegetation associations such as grassland and heathland, or even natural woodland, is not clear-cut, but are rather points on a continua (in n-dimensional space).

In the context of computerised information systems we can identify the equally profound influence of database management systems which are almost exclusively designed around the concept of the Boolean spatial object in the Boolean Class, that is, any object has sharply defined transitions to other objects in all directions, and, furthermore, that the class that object belongs to may be decided unequivocally. Although alternative conceptualisations of classes do exist in database management theory, they have received a trivial amount of attention either in that literature or from GIS researchers (Kollias and Voliotis, 1991). Therefore, almost any real-world phenomena, attributes of which are to be stored and accessed in a database management system, has to be forced into classes of measurable Boolean information. This approach is relatively easy for library information systems, for example, where the book is the unit of reference, itself physically a Boolean object, but look at the complexity of a system designed to search by subject (a more complex, less obviously Boolean concept); it either produces far too many references, or it omits the few you do know are relevant and available. Similarly, personnel management is based around the individual person, but when it comes to including information on contentious issues such as socio-economic class, the alternatives increase dramatically, and one reason for the large number of competing classification schemes is that this is inherently not a Boolean phenomenon.

In short, the idea that the world is composed of objects with clear-cut boundaries may be acceptable for the purpose of producing clear maps, but it is a poor representation of reality. Those phenomena which do match this representation may actually be the exception rather than the rule. At a first look the assumptions do match some of the major classes of use of GIS, such as cadastral information systems and facilities management and mapping, but even here, as in library systems, the approach frequently breaks down when more diverse questions are asked.

Much research in both applications and theory have accepted the importance of this paradigm. Thus habitat modelling is commonly conducted in a Boolean world (Agee et al., 1989); animals live here not there, and here is suitable while there is not. More recent developments admit to some uncertainty in the model, recognising some uncertainty in as much as it is placed in a probabilistic environment (Aspinall, 1992). Similarly, almost all work on topology of spatial relations has assumed the Boolean model of reality (Egenhofer et al., 1994;

Egenhofer and Franzosa, 1991, 1992), although that by Clementini and Di Felice (1996) and Cohn and Gotts (1996) does address the generalisation into three valued situations as opposed to binary ones.

The fuzzy data model based on Zadeh (1965) provides the most obvious alternative to the Boolean model. It has been espoused by various authors, including Altman (1994), Burrough (1986), Fisher (1996a), Leung (1988) and Robinson (1988). Every element in a set is assigned a value between 0 and 1 which records the degree to which it belongs to that set. Fisher (1994) has applied this in the context of visibility analysis, and the clarity of object within the visible area, while Burrough (1989) has particularly applied it to the area of land evaluation and the meaning of words such as stony in the context of soils information. Lagacherie *et al.* (1996) examines the nature of soil boundaries, and Brown (1995) examines the use in mapping vegetation associations. Much remains to be done to further the use of fuzzy sets in spatial information processing; an approach which, at least at a theoretical level, promises to provide a more realistic alternative to the Boolean paradigm, although its ascension does not seem imminent.

19.3 Layer-based raster or vector models: paradigms in transition

Within Geographical Information Systems the idea that the world can be broken up into its constituent themes which can be treated independently of each other is endemic, and again characterises the discussion in much of the introductory literature (Burrough, 1986; Laurini and Thompson, 1992). The exact origin of this idea is not entirely clear (Burrough, 1986), but it is widely used in geographical analysis and landscape architecture (McHarg, 1969), and was certainly a major factor of the Dane County Land Records Project, itself a profound influence on discussion of Multipurpose Cadastre from which an often-reproduced illustration of the concept originated (Chrisman and Niemann,1985; WLRC, 1987). This layer-based approach is endemic in many of the software packages in use today, under some name. It is seen as having the advantage of simplifying a complex world.

There may well be some natural breaks in the continuity and interaction of themes, but there have been two directions of movement away from the layer-based model in recent data modelling. Both strands show that the layer-based paradigm may actually act as a hindrance to modelling data effectively. It should be noted that in some areas of geographic data storage and processing the layer-based approach is still attractive (Hadzilacos, 1996).

The layer view suffers from some major shortcomings. If two different classes of polygonal information share a boundary in common, then registering the boundary as the same physical object in both layers is very difficult, and likely to lead to generation of erroneous sliver polygons. Updating such boundaries may require the updating of both data layers, when there is no information to inform the editor working on one layer as to which other layers might be affected by the change. This verges on a nightmare scenario for maintenance of database integrity (a nightmare being dreamt by many in operational GIS). It can be partly overcome by object-oriented databases where the boundary may be one set of objects and the polygons with their different attributes as others. Updating the boundary line will necessarily update it in all related effective layers. Layer-based systems may regain importance when we start to deny the existence of boundaries (Hadzilacos, 1996).

Another feature of all introductory textbooks on GIS is the supposed dichotomy between raster and vector data (Burrough, 1986; Laurini and Thompson, 1992). These two views of the world dominate the form of the spatial data, whether the attribute data structure is flat file, hierarchical, relational, or even object-oriented. Software vendors make much of the vector and raster processing capabilities of their systems. Much has been talked of the integration of the two, but as yet few truly integrated systems seem to exist, where the nature of the spatial information is only metadata used by the system in deciding how to address the function required. Most software systems known to the author require the user to convert the data from one format to another for supposedly integral processing, or have very limited capabilities in one area or another. Academics have suggested many schemes for integration. Burrough (1986), for example, suggested a practical data structure for integrated raster-vector data without developing a new conceptualisation.

Raster and vector data structures have both advantages and disadvantages which are widely reviewed in the literature. For example, Fisher (1996b) examines some of these as they relate to the raster data, quoting such issues as pixelised/rasterised data, not dwelling on the cartographic steps quoted by many as a major disadvantage, but rather working on the implicit assumption that the world is actually in a pixel structure. Some have read the debate about raster and vector as one between technological factions convinced that one or the other was best. The fact was, and remains, that the argument was over which structure gave the best approximation of reality, with the factions supporting the two abstractions of reality. Neither claimed that their approximation was right, or was even correct. Indeed, since any database structure is an abstraction of reality, it can only ever approximate that reality, and so can only have advantages over another. Those advantages may be so compelling that they override the disadvantages of another, but with raster and vector data this is obviously not the case, or both would not have had such longevity in the subject.

Goodchild (1989, 1992) has proposed the alternative view of objects and fields, which originally shared much in common with vector and raster, but in the most recent definition presents a more useful alternative, whereby a field is a database which includes an exhaustive description of the plane. This implies that you can never have more than one phenomenon at a location within a data type or layer. This is a serious inhibition for many, but does include both raster and vector databases. An object database is one where objects are defined, but they exist in their own right in an empty space. Therefore, many objects within any one theme may exist at any location and, furthermore, objects can be coded by membership functions or as Boolean phenomena. In short, both objects and fields include classic raster and vector data structures.

Raper and Livingstone (1995) advise that modelling geographical processes (particularly environmental models) is limited by use of any generic spatial data model. In attempting to develop a data model for coastal land forms they found a need for three different types of object: material, location, and process. The lack of any of these effectively resulted in a limited view of the problem. With an object-oriented system it is quite possible to execute the data model they propose; a layer-based system simply cannot do it without development of extensive specialised programme modules for modelling which develops an effective version of their data model. It seems likely that the problems noted by Raper and

Livingstone (1995) are true in many other situations including socio-economic applications. The problem with their approach is that the database is no longer generic. A change in the application from coastal processes to habitat modelling is hard to conceive with the same database, but it may only be a matter of formalising a change which we all implicitly execute in making such a switch in application.

In summary, recent research has shown various shortcomings in the layer-based data model characteristic of so much GIS work. The more recent object-oriented model of data may provide some advantages. The historically acknowledged problems of both raster and vector may also be addressed by adopting an object-oriented view. These seem to be paradigms in transition.

19.4 The alternative metrics of space

Embedded in almost everything we do within GIS is the Euclidean metric of space. Distance is measured in linear units against a standard such as metres, yards, kilometres or miles. This scientific measurement of space is widely adopted, has been used for centuries and is the accepted standard. Thus a road has a measurable length, a stream a measurable discharge at a particular time, and a skyscraper a measurable height. Not only are these properties measurable in terms of dimensions, length and time, but their measurement is subject to quantifiable errors depending on repeatability and precision. This is, if you like a positivist's view of the world, depending on repeatability and stable prediction from observation.

The world we, as human beings, occupy is not so easily quantified, however. Limited research in GIS, linking to behavioural geography and cognitive sciences, has, for example, examined the meaning of vague spatial concepts such as NEAR, and NORTH-OF, and even ABOVE (Albert, Chapter 21, this volume; Altman, 1994; Fisher and Orf, 1992; Frank, 1996; Gapp, 1994; Subramanian, 1995; Yuan, Chapter 22, this volume). NORTH-OF can be defined as having a precise geometric meaning, but that is not what we really mean. One object being north of another does not mean that at least one point within the first must be due north of at least one point within the other. In one situation an object which is Northeast of another may still be considered north of it. Meaning is imbued by context, and application. This is equally true of all such terms; they are highly subjective and may carry different meanings between different individuals, in different languages, and at different times.

These and many other concepts of space and space-related phenomena are not directly quantifiable. For example, the quality of a view is hard to measure. Indeed, as with the other concepts, it is so poorly defined it may not be clear what should be measured, and yet it is a phenomenon which someone using a GIS may wish, and feel they should be able, to assess. The region to be assessed may be poorly defined, and so unmeasureable (Fisher, 1996a). Thus, the neighbourhood I live in is never well defined, and frequently varies according to where I am going, whom I have met, and the nature of my journey.

What is common among all these concepts (and many others) is that they are parameters of the lived, perceptual world we belong in, occupy and live in. All are bases for query and use of spatial information. I do not want to know the bars which are within 1 km of my home. At its simplest such an inquiry may yield one which is the other side of the railway line when the nearest crossing point (bridge or level crossing) is 5 km. Rather, I want to know only those bars which are walking distance (whatever that may mean). If today I am prepared not to drink alcohol and so to drive, then walking is no longer a criteria but driving time is. In crowded roads this too is not related to Euclidean distance, and not even to absolute time. Thus, knowledge of the length of the road to the bar, and the details of the route (wait times at junctions, average speeds on roads and so forth) is never really enough to clearly define whether any individual considers the bar to be within driving distance of his or her present location.

The idea of a non-positivist view of spatial metrics relates back to the problem of the Boolean map identified above. Similar to the assertion that a road is a certain length is the concept that the road as an object is definable, maps to a particular class, and possesses precise spatial (and temporal) limits. This assertion is frequently true of the physical expression of the built environment. More often than otherwise, however, it is not true of the natural environment and of our perception of the environment either natural or built.

19.5 The cartographic and image paradigms

All the above existing paradigms can perhaps be viewed as contributing to higher-level paradigms showing the history and background of GIS from either cartography or image processing. The prevalence of the cartographic paradigm is perhaps best exemplified by two phenomena which it seems impossible to shake people out of. First is the habit of referring to the scale of a digital database. Scale in this context is, of course, irrelevant and completely unimportant. It does conveniently convey metainformation about the database, but scale is a profoundly unhelpful term in the digital context. Second, the use of the term 'map' to describe a part of a database (commonly a layer) remains endemic among GIS users. Perhaps it is a term in transition, but it seems to reflect the conceptualisation of the data as hard copy objects. The belief that the hard copy map is the ultimate product of a GIS (however produced) is also prevalent, in spite of the developments in scientific visualisation (Fisher *et al.*, 1994).

The image paradigm emerges from remote sensing and is based on determination of information from satellite data. System designers and users widely refer to data layers as images, and it is undoubtedly true that the raster data structure provides a clear link with this parent.

19.6 Conclusion

Paradigms by their very nature come into existence, find proponents and detractors within the field and as a result they become the focus of research and eventually go into decline, because they are found to be wanting (failing to represent exactly the phenomena they are supposed to explain). The paradigm usually will, for a period, enter the belief system of the discipline concerned. That period may be anything from tens of years to centuries. Within GIS we have a number of widely recognised paradigms. As reviewed above these include the conception of the layer of data being necessarily either vector or raster, data belonging to Boolean categories, and having measurable parameters. These can perhaps all be recognised as being part of embracing paradigms based on traditional cartography or image processing. The increased use and availability of GIS means that all of these have been examined in detail by wide groups of users as well as spatial information scientists. All would appear to be in some state of transition, moving into new realms of integration, object-orientation, multi-valued logic and visualisation.

One remaining paradigm within the area of GIS is that identified by Huxhold and Levisohn (1995). Namely the belief that implementing GIS within an organisation will have numerous possible benefits, whatever the organisation. It is rather different from the others reviewed here, but like those is undergoing scrutiny and uncritical belief in it will lead to problems (see Part Two in this volume).

References

AGEE, J. K., STITT, S. C. F., NYQUIST, M. and ROOT, R. (1989). A geographic analysis of historical grizzly bear sightings in the North Cascades, *Photogrammetric Engineering and Remote Sensing*, **55**(11), 1637–42.

ALTMAN, D. (1994). Fuzzy set theoretic approaches for handling imprecision in spatial analysis *International Journal of Geographical Information Systems*, **8**(3), 271–90.

ASPINALL, R. (1992). An inductive modelling procedure based on Bayes' theorem for analysis of pattern in spatial data, *International Journal of Geographical Information Systems*, **6**(2), 105–21.

BONHAM-CARTER, G. F. (1994). *Geographic Information Systems for Geoscientists: Modelling with GIS*, Oxford: Elsevier.

BROWN, D. (1995). 'Characterizing Continuity in Presettlement Forest Types for Mapping and Analysis', Paper presented at the ESF-GISDATA and NSF-NCGIA 1995 Summer Institute in Geographic Information, Wolfe's Neck, Maine, 26 July–1 August 1995.

BURROUGH, P. A. (1986). *Principles of Geographical Information Systems for Land Resources Assessment*, Oxford: Oxford University Press.

BURROUGH, P. A. (1989). Fuzzy mathematical methods for soil survey and land evaluation, *Journal of Soil Science*, **40**(3), 477–92.

BURROUGH, P. A. and FRANK, A. U. (1995). Concepts and paradigms in spatial information: are current geographical information systems truly generic? *International Journal of Geographical Information Systems*, **9**(2), 101–16.

CHRISMAN, N. R. and NIEMANN, B. J. (1985). Alternative routes to a multipurpose cadastre: merging institutional and technical reasoning, in *Proceedings of Auto-Carto 7*, pp. 84–94, Bethesda, MD: American Congress on Surveying and Mapping.

CLEMENTINI, E. and DI FELICE, P. (1996). An algebraic model for spatial objects with undetermined boundaries, in Burrough, P. A. and Frank, A. (Eds), *Natural Objects with Indeterminate Boundaries*, London: Taylor & Francis, pp. 155–169.

COHN, A. G. and GOTTS, N. M. (1996). The 'egg-yolk' representation of regions with undetermined boundaries, in Burrough, P. A. and Frank, A. (Eds), *Natural Objects with Indeterminate Boundaries*, London: Taylor & Francis, pp. 171–187.

DARWIN. C. (1859). *On the Origin of Species by Means of Natural Selection*, London: John Murray.

DESMOND, A. and MOORE, J. (1991). *Darwin*, London: Michael Joseph.

EGENHOFER, M. and FRANZOSA, R. (1991). Point-set topological spatial relations, *International Journal of Geographical Information Systems*, **5**(2), 161–74.

EGENHOFER, M. and FRANZOSA, R. (1992). On the equivalence of topological relations, *International Journal of Geographical Information Systems*, **9**(2), 133–52.

EGENHOFER, M., CLEMENTINI, E. and DI FELICE, P. (1994). Topological relations between regions with holes, *International Journal of Geographical Information Systems*, **8**(2), 161–74.

FISHER, P. F. (1994). Probable and fuzzy models of the viewshed operation, in Worboys, M. (Ed.), *Innovations in GIS 1*, pp. 161–75, London: Taylor & Francis.

FISHER, P. F. (1996a). Boolean and fuzzy regions, in Burrough, P. A. and Frank, A. (Eds), *Natural Objects with Indeterminate Boundaries*, London: Taylor & Francis, pp. 87–94.

FISHER, P. F. (1996b). The pixel: a snare and a delusion, in Pan, P. (Ed.), *Integration of Remote Sensing and GIS*, Remote Sensing Society Monograph, (in press).

FISHER, P. F. and ORF, T. (1991). An investigation of the meaning of near and close on a University campus, *Computers, Environment and Urban Systems*, **15**(1), 23–35.

FISHER, P. F., DYKES, J. D. and WOOD, J. (1994). Visualization and map design, *The Cartographic Journal*, **30**, 136–47.

FRANK, A. U. (1996). Qualitative spatial reasoning: Cardinal directions as an example, *International Journal of Geographical Information Systems*, **10**(3), 269–290.

GAPP, K.-P. (1994). A computational model of the basic meaning of graded composite spatial relations in 3D space, in Molenaar, M. and De Hoop, S. (Eds), *Advanced Geographic Data Modelling*, Publications on Geodesy, New Series Number 40, pp. 66–79, Delft: The Netherlands Geodetic Commission.

GOODCHILD, M. F. (1989). Modeling error in objects and fields, in Goodchild, M. and Gopal, S. (Eds), *Accuracy of Spatial Databases*, pp. 107–13, London: Taylor & Francis.

GOODCHILD, M. F. (1992). Geographical information science, *International Journal of Geographical Information Systems*, **6**(1), 31–45.

HADZILACOS, T. (1996). Are layers a good choice for fuzzy objects? in Burrough, P. A. and Frank, A. (Eds), *Natural Objects with Indeterminate Boundaries*, London: Taylor & Francis, pp. 237–255.

HUXHOLD, W. E. and LEVISOHN, A. G. (1995). *Managing Geographic Information System Projects*, Oxford: Oxford University Press.

KOLLIAS, V. J. and VOLIOTIS, A. (1991). Fuzzy reasoning in the development of geographical information systems: FRSIS: a prototype soil information system with fuzzy retrieval capabilities, *International Journal of Geographical Information Systems*, **5**(2), 209–23.

KUHN, T. S. (1970). *The Structure of Scientific Revolutions*, 2nd Edn, Chicago: Chicago University Press.

LAGACHERIE, P., ANDRIEUX, P. and BOUZIGUES, R. (1996). The soil boundaries: from reality to coding in GIS, in Burrough, P. A. and Frank, A. (Eds), *Natural Objects with Indeterminate Boundaries*, London: Taylor & Francis, pp. 275–287.

LANTER, D. P. and VEREGIN, H. (1992). A research paradigm for propagating error in layer-based GIS, *Photogrammetric Engineering and Remote Sensing*, **58**(6), 825–33.

LAURINI, R. and THOMPSON, D. (1992). *Fundamentals of Spatial Information Systems*, London: Academic Press.

LEUNG, Y. C. (1988). *Spatial Analysis and Planning under Imprecision*, New York: Elsevier.

MCHARG, I. L. (1969). *Design with Nature*. New York: Doubleday/Natural History Press.

PICKLES, J. (Ed.) (1995). *Ground Truth: The Social Implications of Geographic Information Systems*, New York: Guilford.

POPPER, K. (1965). *The Logic of Scientific Discovery*, New York: Harper.

RAPER, J. F. and LIVINGSTONE, D. E. (1995). Development of a geomorphological spatial model using object-oriented design, *International Journal of Geographical Information Systems*, **9**(4), 359–83.

ROBINSON, V. B. (1988). Some implications of fuzzy set theory applied to geographic databases, *Computers, Environment and Urban Systems*, **12**(2), 89–97.

STEWART, I. (1989). *Does God Play Dice?* Oxford: Blackwell.

SUBRAMANIAN, R. (1995). *Processing Semantically/ill-defined Spatial Operators in Geographic Databases: A Query Processing Model and Implementation*, Paper presented at the ESF-GISDATA and NSF-NCGIA 1995 Summer Institute in Geographic Information, Wolfe's Neck, Maine, 26 July–1 August 1995.

TURNBULL, D. (1989). *Maps are Territories*, Geelong: Deakin University Press.

UNWIN, D. J. (1996). Review of Ground Truth, *International Journal of Geographical Information Systems*, **10**(2), 237–238.

WLRC (1987). *Modernizing Wisconsin's Land Records*, Wisconsin Land Records Committee, Madison.

WORBOYS, M. F. (1994). Object-oriented approaches to geo-referenced information, *International Journal of Geographical Information Systems*, **8**(4), 385–400.

YOUNG, E. (1992). Hunter-gatherer concepts of land and its ownership in remote Australia and North America, in Anderson, K. and Gale, F. (Eds), *Inventing Places: Studies in Cultural Geography*, pp. 255–72, Melbourne: Longman Cheshire and New York: Halstead Press.

ZADEH, L. A. (1965). Fuzzy sets, *Information and Control*, **8**, 338–53.

CHAPTER TWENTY

Cognitive Perspectives on Spatial and Spatio-temporal Reasoning

DAVID MARK

20.1 Introduction: the importance of cognition

Knowledge about human cognition is important to the social and behavioural sciences because human behaviour is based on the beliefs and knowledge that people have about their worlds. An understanding of human spatial cognition and reasoning will undoubtedly be useful in explaining and predicting human behaviour in geographic space. It should also contribute to the development of fundamental theory in geographic information science and cognitive science. Finally, formalising these theoretical advances should aid in the design of improved information systems to support human decision-making in geographic space.

This chapter reviews some of the aspects of spatial cognition and related topics that are most relevant to the advancement of geographic information science. It attempts to paint a broad picture of the field, admittedly from this author's perspective. Details cannot be included, but there are pointers to some key references for further reading. The chapter may raise more questions than it answers, but questions are fundamental to progress in basic research.

20.2 Cognition and cognitive science

Artificial Intelligence (AI) is a key area of computer science, and has been applied to geographic information in a number of ways (Couclelis, 1986; Smith, 1984). AI is a diverse field, but two major subareas can be distinguished by their very different goals. The field of Expert Systems (ES) seeks to build computational systems that perform tasks formerly thought to require human intelligence. In ES, it is not necessary or even desired that the machine perform the task in the same way that humans do; although studies of human performance may lead to implementable models, other methods of designing Expert Systems are also important. In essence, the goal of ES is engineering, that is, building things that work well.

The other side of AI has more scientific objectives, attempting to explain or account for aspects of human intelligence, or of 'intelligence' in general. Normally, this side of AI uses computer implementations as a method for approaching understanding, and as a forum for formal modelling. Cognitive science includes this kind of AI, along with related disciplines from the behavioural and social sciences.

> Cognitive science is a new field that brings together what is known about the mind from many academic disciplines: psychology, linguistics, anthropology, philosophy, and computer science. It seeks answers to such questions as: What is reason? How do we make sense of our experiences? What is a conceptual system and how is it organized? Do all people use the same conceptual system? If so, what is that system? If not, exactly what is there that is common to the way all human beings think? The questions aren't new, but some recent answers are.   (Lakoff, 1987, p. xi)

Spatial cognition forms an important part of cognitive science. Lakoff and Johnson (1980) suggested that spatial situations form the basis or source domain for many metaphors used to structure more abstract conceptual domains. Cognitive science provides geographic information science with research methods and philosophical stances that help ground theories of spatial information and spatial cognition. Geographic information science can in turn provide cognitive science with new and perhaps more sophisticated formalisms for representing spatial entities and relations.

The role of formalism in geographic information science and cognitive science deserves some special attention. Computer programs involve formal models of the phenomena that the program is modelling or representing. Explicit formal models increase the chance that the program will behave consistently and correctly. Formal models also help in the development of theories, and in designing hypotheses to test. Formalisation is certainly essential to software engineering. It is also a very important part of a cognitive science approach to spatial information and cognition. Scientific explanations go beyond the formalisation of the problem and its elements, and normally involve empirical evidence that supports, or at least fails to refute, theoretical statements.

20.3 Experiential realism

Positivist approaches to explanation and understanding are based on implicit or explicit claims that the world has an objective nature, independent of human observation. Science is involved with discovering, uncovering, or revealing the inherent structure of this world. Scientists do not always agree, and thus multiple and contradictory explanations may exist for a while. However, proponents of each explanation would probably agree that only one of them can be correct. People trained in the sciences often assume that this search for refutable theoretical descriptions of the real world is the only basis for explanation and understanding. However, there are other philosophical stances that propose quite different methods of inquiry and understanding. For example, solipsism asserts that the human mind does not have direct access to the world at all – people only have 'beliefs', based on what their senses reported. In this extreme view, the world may not even exist outside our own minds.

Clearly, there is a huge gulf between objectivist and solipsist accounts of explanation. However, each view appears to have some merit. Experiential realism is a philosophical stance that appears to bridge this gap by asserting that mental models of the world come from experience with that world. Even though the mind cannot experience that world 'directly', the indirect experiences of that world are shaped in consistent ways by the physical nature of the world, by the physical nature of our individual senses, and by the physical nature of our own bodies and how they interact physically with that world. George Lakoff introduced experiential realism in great detail in his 1987 book *Women, Fire, and Dangerous Things*. The model is, however, well described in the much earlier book by Lakoff and Johnson (1980), and further elaborated by Johnson (1987). It appears to have strong potential as a basis for computational models of the human mind. In geography and spatial cognition, cognitive models based on experiential realism have been discussed in several papers (Couclelis, 1988; Frank and Mark, 1991; Mark and Frank, 1989, 1996).

Categories form a keystone of experiential realism. Much of the organisation we see in the world is structured by categories. Once a novel experience has been classified, we can infer a lot about it from the characteristics of the category we have placed it in. Science typically models categories using classical set theory. However, Rosch (1973, 1978) and other cognitive psychologists have shown convincingly that categories often have 'fuzzy' boundaries, cores, prototypes, best examples, internal radial structure and so forth. For example, although from a technical point of view, all 9000 species belonging to class *Aves* are equally birds, when people are asked to give an example of a kind of bird, they almost always give 'typical' ones such as sparrow or robin, not familiar yet atypical species such as duck or penguin. Experiential realism includes this sort of model of categories, and can readily be applied to spatial relations and cognition as well as to geographic entity types. For reviews of these ideas in a geographic context see Mark (1993a, and b), and Fisher (Chapter 19 in this volume).

20.4 Spatial cognition

As noted above, categorisation is a basic human cognitive activity. Classification is also widely used in science. Classification is almost always a useful step toward understanding, even if the classes are extreme simplifications of the complexity actually present in the world or in our mental models of it. Spatial cognition is a complex field of study that has been examined by psychologists, geographers, urban planners and many others. Aspects of spatial cognition can be classified along many dimensions or axes, and a variety of such models have been used. In this section, several different categorisations of spatial cognition are presented. The dimensions and basics of the classifications are given in more detail by Mark and Freundschuh (1995). One basic dimension for differentiation of spatial cognition is the kind or type of spatial knowledge. Another is *scale*[1] or size of the environment or objects involved. A third basis for classification is the *source* of spatial information, particularly which human senses are involved.

20.4.1 Classification of kinds of spatial knowledge

The *kind* or type of spatial knowledge is a fundamental basis for classification. Division of knowledge into procedural and declarative has a long history in artificial intelligence (Barr and Feigenbaum, 1982). The distinction seems to hold particularly well for geographic cognition and knowledge, although with some modification. Much of spatial reasoning is transformations among these forms.

Procedural knowledge concerns how to get around in geographic space, the information that forms the basis for navigation and wayfinding. In its most pure form, this knowledge is inaccessible to conscious introspection. However, it often can be reasoned about consciously, and we may be able to tell others how to get from place to place by accessing our own procedural knowledge of the route. But we may still be unable to estimate distances between points along familiar routes, especially by straight line distances if the route is sinuous.

Declarative knowledge is simply facts about geographic space and the entities and phenomena in it. Athens is the largest city in Greece, Maine is adjacent to New Hampshire, Strasbourg is on the Rhine, the Nile is the longest river in Africa, and Mount Everest is the highest mountain – these are all geographic facts. They may be known in relative isolation. One could know those listed above without having any idea of the distance between Athens and Mount Everest, or even which continent the latter is on.

Spatial cognition research has often presented findings about a third category of spatial knowledge, namely *configurational knowledge*. Configurational knowledge of a geographic space is essentially 'map-like', and includes knowledge of relative positions, distances and angles. This can be treated as a special type of declarative knowledge, and certainly falls on the declarative side of the procedural/declarative dichotomy. However, it is peculiarly spatial, and thus differs from declarative knowledge of the more general type.

20.4.2 Scale-based classification of spatial knowledge

As noted above, *scale* is another basis for classifying spatial knowledge. Geometry and physics are essentially constant over a large range of spatial and temporal scales; only around the size of the individual atom and smaller, or as we approach the speed of light, do Newtonian physics and Euclidean geometry have to be abandoned in favour of newer twentieth-century models. Many views of spatial cognition have naïvely accepted this, and assumed that the 'same' models of geometry apply on the table, in the city, or over continents and hemispheres. However, it seems that human spatial concepts may vary with scale, in particular with size relative to the human body. This should be no surprise if experiential realism is more or less correct, since the human body, senses, and power to manipulate form the cognitive anchor points for the external world.

Researchers have identified several scales, or scale-based typologies of kinds of spatial knowledge. Two particular scales come out in most such analyses. One kind of space is defined by *manipulable spaces* and entities, roughly human-sized and smaller. These can be perceived holistically, through several senses at once, and are fundamentally three-dimensional. The other kind of space is much larger than the human body, and is normally experienced part by part during immersion within them. These are best termed *geographic spaces*, but have also been termed transperceptual spaces (Downs and Stea, 1977). Geographic spaces are 'assembled' in memory through spatial reasoning.

Manipulable and geographic spaces differ in a number of ways. Manipulable spaces are populated by three-dimensional entities, most of which are moveable. When they are moved, their attributes such as size, shape, and colour are not expected to change. Euclidean geometry provides very good descriptions of manipulable spaces. Such spaces are often described using observer-centred references frames (left–right, and so forth). Geographic spaces, on the other hand, contain many entities with indistinct boundaries. Geographic entities are normally parts of the Earth's surface, and thus not moveable. If they do 'move' (the spread of a wild fire, for example), then their attributes are also expected to change. Euclidean geometry does not seem to be natural for describing geographic spaces; traditional maps, which themselves are manipulable entities, play a key role in assembling cognitive information about geographic spaces into coherent Euclidean mental models (Freundschuh, 1991; Lloyd, 1989). Cardinal directions are often used for describing relative positions in geographic space.

There are exceptions to some of the above tendencies. For example, many people seem to have trouble using cardinal directions even in familiar outdoor spaces, yet some cultures use cardinal directions even for table-top spaces (Pederson, 1993). Montello (1993) has proposed a more elaborate subdivision of kinds of spaces based on scale, and Couclelis (1993) has suggested that the noted raster-vector 'debate' for geographic information systems may relate to the differences between manipulable spaces (entities, vector) and geographic spaces (fields, raster).

20.4.3 Classification of spatial knowledge based on sources

A third basis for classification is the *source* of spatial information, particularly where human senses are involved. Mark (1992) suggested that information based on different senses also represents a hierarchy for metaphorical structuring. *Haptic* spatial knowledge, acquired primarily from touch and body movement, is the most basic. This would appear to apply only in manipulable space, and is probably well developed before a child reaches the age of one year. Mark claimed that *pictorial* spaces, based on the remote sensing of the immediate environment, mainly through sight and sound, are conceptually structured in terms of concepts from haptic space. However, the range of pictorial space is unlimited, and is bounded only by barriers to vision and hearing that are provided by the environment. Many of the 'spatialisation metaphors' discussed by Lakoff and Johnson (1980) fall into this category, or structure more abstract conceptual domains in these terms. Third, the sense-based classification identified *transperceptual* (geographic) spaces, which are structured partly by pictures (maps, mental maps). Mark noted that while there are metaphorical links across the hierarchy defined by haptic, pictorial, and transperceptual spaces, each can also form the basis for human-computer interaction, in general as well as for GIS. Haptic concepts lead to direct manipulation interfaces, pictorial to pan and zoom metaphors, and transperceptual to wayfinding metaphors.

20.4.4 Primitives of spatial cognition

Having discussed some general models of spatial cognition from a 'top–down' perspective, it is appropriate to turn briefly to look at some primitives or building blocks for spatial cognition. In his influential and controversial book, *The Child's Conception of Space* (Piaget and Inhelder, 1956), Jean Piaget proposed that 'the perception of space involves a gradual construction and certainly does not exist ready-made at the outset of development' (p. 6). Piaget claimed that by the age of about 4–5 months, the child can recognise and reason about five spatial relations. In order from the most elementary, these are: proximity, separation, order (or spatial succession), enclosure (or surrounding) and continuity (pp. 6–8). Piaget went on to relate these five spatial relations to elements of Gestalt theory, and to the bases of topology, concluding that these are evidence that topology is very basic to perception and cognition (p. 9). Golledge (1995) proposed a somewhat higher-level set of primitive concepts on which models of spatial knowledge can be built. Golledge proposes that identity, location, magnitude, and time are first-order primitives, with distance, angle and direction, sequence and order, and connection and linkage as fundamental derived concepts.

Once self has been differentiated from surrounding, and the world has been divided into 'things', two topological primitives seem essential and distinct: containment and contact. These can form the basis of formal models of topological spatial relations, as shown by Pantazis (Chapter 23 in this volume). The 9-Intersection model of spatial relations (Egenhofer and Herring, 1994) is also based on these two, but includes just one spatial relation *per se*, namely intersection (superimposition), and models contact as an intersection of boundaries. Another spatial primitive, part-whole decomposition, is used to divide each spatial entity into an interior and a boundary. Mark and Egenhofer (1994a and b) have shown that this model can account for much of the variation in how subjects apply higher-level terms such as 'crosses' or 'enters' to spatial relations between roads and parks in geographic space. The use of such models for other aspects of spatial cognition and reasoning awaits further studies.

20.5 Evidence: ways of studying spatial cognition

The cognitive sciences, being interdisciplinary, include a variety of methods for reaching 'explanation'. In a scientific approach, objective 'evidence' is normally required to support theories or other explanations, but different fields favour different kinds of evidence, or different standards. Evidence can come from mathematical proofs, from successful computational implementations, human subjects experiments, observation of natural human behaviour or its effects, from the nature and structure of natural language, through introspection, and in other ways.

One important methodological tension is between observing natural human behaviour and its results, or performing experiments. Rigorous experimental control is seldom possible in the real world. Thus it is difficult to rule out the possibility that extraneous factors are influencing the results. However, well-controlled experiments in human cognition normally involve highly simplified situations that have been abstracted from reality, and thus results may not provide convincing explanations of natural behaviour. The practical question is either, how can researchers achieve control in experiments conducted in the real world, or how can they obtain measures of natural behaviour in laboratory experiments? If neither of these goals can be achieved, then what are the appropriate trade-offs between control and natural behaviour? These are important questions for anyone planning to design experiments with human subjects.

Human natural language can be a particularly interesting site for studying spatial cognition. Many aspects of natural language can be used to get at principles of spatial reasoning and cognition. These include the grammar and syntax of languages, the lexicons of languages and their etymologies, as well as their semantics, pragmatics and use. All of these can provide valuable information and insights about human spatial cognition. Cross-linguistic studies are especially valuable.

The relation of language to cognition is, however, controversial. The Sapir–Whorf hypothesis asserts that people think 'in' (using?) their natural language. If this is true, then it seems that the nature and structure of that language can influence thought. In an extreme form of 'Whorfianism', language *determines* and *limits* thought. More frequently, current proponents of this approach support a linguistic relativism, where certain choices of reasoning or expression are more likely for speakers of one language than for another. Many linguists, recently and notably Stephen Pinker (1994), reject any form of linguistic relativism, asserting that language is used for communicating ideas but not for formulating them. Scientific conservatism suggests that linguistic relativism should not be assumed without evidence, but the design of systems based on an 'all-people-think-like-we-do' principle seems like cultural imperialism. Thus work to resolve the presence or degree of linguistic relativism in spatial relations is of high priority.

Research approaches that combine formalisms and human-subjects testing appear to be especially valuable, since each approach can enrich the other. A sound formal model promotes critical experimental design, drawing attention to which aspects of experimental stimuli should be manipulated and which aspects should be held constant. On the other hand, experimental results can indicate where the formal model needs to be elaborated and where it should be abstracted. Also, multiple modes of testing or experimental protocols, such as verbal description tasks, graphic prototypes, groupings, or agreement tasks, should all be employed, as they focus on different aspects of conceptual definitions (Mark *et al.*, 1995).

20.6 Time in geographic space

Time is a somewhat abstract concept that is often structured both cognitively and linguistically through its relation to space. For example, most terms for temporal relations have origins as spatial prepositions or equivalents. Time and space are linked through motion, and time also manifests itself through change. Some researchers have suggested that time in geographic space is different from time in manipulable space. However, evidence for this distinction is not as clear as it is for scale-based kinds of spatial entities and spaces. Two time topics will be discussed in this section: time geography and fictive motion.

20.6.1 Time geography

In the 1960s, Swedish geographer Torsten Hägerstrand developed a number of theoretical models of time and process in geography. Perhaps the most famous of these models are his work on spatial diffusion processes (Hägerstrand, 1970). However, another important part of Hägerstrand's work is now known as time geography. Time geography provides a conceptual model of how time and velocity limit spatial interaction and planning. A central idea is that of the time-space 'prism'. In time-geography diagrams such as this, geographic space is commonly collapsed to one dimension on the horizontal axis, and time is plotted vertically. If a person (or other object) is not moving, then a vertical line appears on the diagram. An oblique line represents motion (space and time are both changing), and the more nearly horizontal the line is, the faster the travel. Thus, if a person has an opportunity to move in a fixed-time period, but with a maximum possible speed, their movements are restricted to a diamond-shaped polygon referred to as a prism. (If space were plotted two-dimensionally with time as a third dimension, and if travel were equally fast in all directions, then such a prism would appear as a double cone.) The parts of the space-time prism that are farther from the start and end point represent places where there is less time available for non-travelling actions. The time-geography model is even more interesting in the way it exposes constraints when two or more people, based at different locations, try to meet for some joint activity. The meeting must occur at locations and times that fall within the intersections of their space-time activity prisms.

The time-geography model has had important impacts on geographic thought (Parkes and Thrift, 1980; Pred, 1981), but usually as a conceptual model for formulating hypotheses and constraints, rather than as an empirical or computational model. Perhaps this is because methods in logic programming and inference were not widely known to geographers when the time-geography model was most popular. Time geography represents an important potential way for time to be integrated into spatial analysis for GIS, as well as a link to artificial intelligence programmes for spatio-temporal reasoning. Miller (1991) reports an implementation of the time-geography model in a well-known commercial GIS.

20.6.2 Time and motion in geographic space

The last point in this section regards the issue of whether space and motion might be more appropriate primitives or basic concepts than space and time. We learn in science, especially physics, that time and space (distance) are dimensional primitives ($V = L/T$, not $T = L/V$), and that velocity, motion and change are derived properties. But does this reflect the way that people naturally think about spatial processes, and reason about space, time and motion? Leonard Talmy has made some interesting generalisations about a cognitive and linguistic phenomenon that he calls 'fictive motion' (Talmy, 1996). In fictive motion, objects that do not move in the literal (physical) sense are talked about as if they move. A good example is 'The wall runs from the ridge to the valley', which is 'fictive' (not true), because the wall does not move at all with respect to the landscape, nor was it necessarily constructed in an end-to-end sequence. Rather, our hypothetical 'gaze', or the attention of our cognition, moves along it in order to describe or understand its configuration.

Talmy has suggested that fictive motion is a language universal: every known language uses fictive motion. The pervasive use of fictive motion in human language suggests that in some ways motion may be more fundamental or basic even than space itself, never mind time. If motion is really more basic to cognition than time or space in isolation, what are the implications for spatio-temporal reasoning systems or for temporal data in GIS? It might still be best to represent space and time as separate primitives, and then infer motion, process, and change from them. Is such a choice just an implementation detail, as long as a process-based 'view' of databases and analytical models is available to users? How would we evaluate the relative advantages and disadvantages of different spatio-temporal systems? These again are fundamental research questions, being addressed in artificial intelligence, cognitive science, and geographic information science.

20.7 Summary and priorities

This chapter proposes that human spatial cognition is an important topic in its own right, and provides basic insights that are relevant to geographic models, geographic software design, and geographic information science in general. Methods from psychology, artificial intelligence, linguistics, and other parts of cognitive science can fruitfully be brought to bear on problems of cognition in and about geographic space. A philosophical position termed experiential realism seems to be a useful background for spatial cognition research, although there are alternatives. Categories are central to this model, and are a useful focus for research on cognition. Categorisations can apply to kinds of things, to attributes, to processes, or to relationships.

Spatial cognition and spatial knowledge can be categorized according to kinds of knowledge, or according to scale (size) of objects, or according to the sources of spatial information. These classifications are by no means independent, but highlight different aspects and may be relevant for different purposes. Also, primitives of spatial cognition have been identified and reviewed.

A particular point of emphasis in this chapter is kinds of evidence. Both a strength and a problem for the cognitive science is the multiplicity of kinds of evidence and ways of knowing. Some people think that nothing has been established unless it has been implemented on a computer, while others feel that computer implementation is merely application. Some do not believe a conclusion unless a well-controlled experiment has been run with at least 30 subjects, whereas other rely on introspection, or on a small group of informants cultivated over months or years of participant anthropology. If alternative methods are brought to bear on the same problem, more complete answers may be revealed that convince a wider range of scientists and applications specialists.

Time is considered to be like space in some cultures and cognitive frameworks. For certain questions or models, time may be treated separately, but this can cause problems for the modelling of process and motion. Time and space combine though constraints on velocity, to limit the ranges of human activities that are possible. Time geography provides formal methods for showing how available time constrains spatial choice. Also, language often connects space and time by conceptualising static objects as if they were moving, or moving objects as if they were static.

I believe that some key priorities for research in this area are to perform tests with human subjects to verify every assumption of the models being applied. Also, formal descriptions of all of the assumptions underlying the models or experimental designs must be given. This will allow differences in empirical results to be accounted for, and should eventually lead to unified models of the universals of human spatial and spatio-temporal cognition, and methods for highlighting individual differences and differences between human societies.

Acknowledgements

This chapter is a result of research at the US National Center for Geographic Information and Analysis, supported by a grant from the National Science Foundation (SBR-88-10917); support by NSF is gratefully acknowledged. Comments from the anonymous reviewers were helpful in the revision of this chapter.

Note

1   I adopt the conventions from the physical and social sciences, that scale is used to refer to the characteristic sizes or lengths of things. Large-scale entities, processes, or events effect large pieces of the earth's surface, whereas smaller-scale things take up less space. This may seem obvious, but it brings the term into apparent conflict with usage in cartography, where small-scale maps are used to portray large-scale things, and vice versa. Some cartographers suggest using 'scope' to refer to the extent of entities in the real world.

References

BARR, A. and FEIGENBAUM, E. A. (Eds), (1982). *The Handbook of Artificial Intelligence*, vol. 2, Los Altos, CA: William Kaufmann, Inc.

COUCLELIS, H. (1986). Artificial intelligence in geography: Conjectures on the shape of things to come, *The Professional Geographer*, **38**, 1–11.

(1988). The truth seekers: Geographers in search of the human world, in Golledge, R., Couclelis, H. and Gould, P. (Eds), *A Ground for Common Search*, pp. 148–55, Santa Barbara, CA: The Santa Barbara Geographical Press.

(1993). People manipulate objects (but cultivate fields): Beyond the raster-vector debate, in GIS, in Frank, A. U., Campari, I. and Formentini, U. (Eds), *Theories and Methods of Spatio-Temporal Reasoning in Geographic Space*, Lecture Notes in Computer Science No. 639, pp. 65–77, Berlin: Springer-Verlag.

DOWNS, R. M. and STEA, D. (1977). *Maps in Minds: Reflections on Cognitive Mapping*, New York: Harper & Row.

EGENHOFER, M. and HERRING, J. (1994). Categorizing topological spatial relations between point, line and area objects, in Egenhofer, M. J., Mark, D. M. and Herring, J. R. (Eds). *The 9–Intersection: Formalism and its Use For Natural-Language Spatial Predicates*, Santa Barbara, CA: National Center for Geographic Information and Analysis, Report 94–1.

FRANK, A. U. and MARK, D. M. (1991). Language issues for GIS, in Maguire, D., Goodchild, M. and Rhind, D. (Eds), *Geographical Information Systems: Principles and Applications*, vol. 1, pp. 147–63, London: Longman.

FREUNDSCHUH, S. M. (1991). 'Spatial knowledge acquisition of urban environments from maps and navigation experience', unpublished PhD dissertation, Buffalo, New York: Department of Geography, State University of New York at Buffalo.

GOLLEDGE, R. (1995). Primitives of spatial knowledge, in Nyerges, T. L., Mark, D. M., Laurini, R. and Egenhofer, M. (Eds), *Cognitive Aspects of Human-Computer Interaction for Geographic Information Systems*, pp. 29–44, Dordrecht: Kluwer Academic Publishers.

HÄGERSTRAND, T. (1970. What about people in regional science? *Papers of the Regional Science Association*, **14**, 7.

JOHNSON, M. (1987). *The Body in the Mind: The Bodily Basis of Meaning, Imagination and Reason*, Chicago: University of Chicago Press.

LAKOFF, G. (1987). *Women, Fire, and Dangerous Things: What Categories Reveal About the Mind*, Chicago: University of Chicago Press.

LAKOFF, G. and JOHNSON, M. (1980). *Metaphors We Live By*. Chicago: University of Chicago Press.

LLOYD, R. (1989). Cognitive maps: Encoding and decoding information, *Annals of the Association of American Geographers*, **79**(1), 101–24.

MARK, D. M. (1992). Spatial metaphors for human-computer interaction, *Proceedings, Fifth International Symposium on Spatial Data Handling*, vol. 1, pp. 104–12, South Carolina: International Geographical Union, Commission on GIS.

MARK, D. M. (1993a). A theoretical framework for extending the set of geographic entity types in the US spatial data transfer standard (SDTS), Proceedings of GIS/LIS'93, Minneapolis, November 1993, vol. 2, pp. 475–83, Maryland: American Society for Photogrammetry and Remote Sensing and American Congress on Surveying and Mapping.

MARK, D. M. (1993b). Toward a theoretical framework for geographic entity types, in Frank, A. U. and Campari, I. (Eds), *Spatial Information Theory: A Theoretical Basis for GIS*, Lecture Notes in Computer Sciences No. 716, pp. 270–83, Berlin: Springer-Verlag.

MARK, D. M. and EGENHOFER, M. J. (1994a). Calibrating the meanings of spatial predicates from natural language: Line-region relations, in Waugh, T. C. and Healey, R. G. (Eds). *Proceedings, Spatial Data Handling 1994*, vol. 1, pp. 538–53, London: Taylor & Francis.

MARK, D. M. and EGENHOFER, M. J. (1994b). Modeling spatial relations between lines and regions: Combining formal mathematical models and human subjects testing, *Cartography and Geographic Information Systems*, **21**(4), 195–212.

MARK, D. M. and FRANK, A. U. (1989). Concepts of space and spatial language, *Proceedings, Ninth International Symposium on Computer-Assisted Cartography (Auto-Carto 9)*, pp. 538–56, Baltimore, Maryland: American Society for Photogrammetry and Remote Sensing and American Congress on Surveying and Mapping.

MARK, D. M. and FRANK, A. U. (1996). Experiential and formal models of geographic space, *Environment and Planning B.*, **23**(1), 3–24.

MARK, D. M. and FREUNDSCHUH, S. M. (1995). Spatial concepts and cognitive models for geographic Information use, in Nyerges, T. L., Mark, D. M., Laurini, R. and Egenhofer, M. (Eds), *Cognitive Aspects of Human-Computer Interaction for Geographic Information Systems*, pp. 21–8, Dordrecht: Kluwer Academic Publishers.

MARK, D. M., COMAS, D., EGENHOFER, M. J., FREUNDSCHUH, S. M., GOULD, M. D. and NUNES, J. (1995). Evaluating and refining computational models of spatial relations through cross-linguistic human-subjects testing, in Frank, A. U. and Kuhn, W. (Eds), *Spatial Information Theory: A Theoretical Basis for GIS*, Lecture Notes in Computer Sciences No. 988, pp. 553–68, Berlin: Springer-Verlag.

MILLER, H. (1991). Modelling accessibility using space-time prism concepts within geographical information systems, *International Journal of Geographical Information Systems*, **5**, 287–301.

MONTELLO, D. (1993). Scale and multiple psychologies of space, in Frank, A. U. and Campari, I. (Eds), *Spatial Information Theory: A Theoretical Basis for GIS*, pp. 312–21, Berlin: Springer-Verlag.

PARKES, D. and THRIFT, N. (1980). *Times, Spaces, and Places*. New York: John Wiley.

PEDERSON, E. (1993). Geographic and manipulable space in two Tamil linguistic systems, in Frank, A. U. and Campari, I. (Eds), *Spatial Information Theory: A Theoretical Basis for GIS*, Lecture Notes in Computer Sciences No. 716, pp. 294–311, Berlin: Springer-Verlag.

PIAGET, J. and INHELDER, B. (1956). *The Child's Conception of Space*, London: Routledge & Kegan Paul.

PINKER, S. (1994). *The Language Instinct*, New York: William Morrow and Company.

PRED, A. (1981). *Space and Time in Geography: Essays Dedicated to Torsten Hagerstrand*, Lund Studies in Geography, B, 48, Lund, Sweden: GWK Gleerup.

ROSCH, E. (1973). On the internal structure of perceptual and semantic categories. in Moore, T. E. (Ed.), *Cognitive Development and the Acquisition of Language*, New York: Academic Press.

ROSCH, E. (1978). Principles of categorization, in Rosch, E. and Lloyd, B. B. (Eds), *Cognition and Categorization*, Hillsdale, NJ: Erlbaum.

SMITH, T. R. (1984). Artificial intelligence and its applicability to geographic problem solving. *Professional Geographer*, **36**(2), 147–58.

TALMY, L. (1996). Fictive motion in language and 'ception', in Bloom, P., Peterson, M., Nadel, L. and Garrett, M. (Eds), *Language and Space*, pp. 211–276, Cambridge, MA: MIT Press (in press).

CHAPTER TWENTY-ONE

The Role of Spatial Abilities in the Acquisition and Representation of Geographic Space

WILLIAM ALBERT

21.1 Introduction

One of the many goals of The National Center for Geographic Information and Analysis (NCGIA, 1989) is to develop a unified theory of spatial relations. One aspect of this goal is to understand better how individuals cognitively represent geographic space and how this ultimately manifests itself in the form of spatial decision-making and behaviour. In order to accomplish this goal, many issues must be considered. These issues might include spatial knowledge acquisition and representation, the role of language in the development of spatial representations and the development of artificial intelligence models of spatial relations, to name but a few. The research presented in this chapter examines spatial knowledge acquisition and representation: specifically, the role played by individual cognitive abilities in this process. This study is carried out in a broader context of examining the external (environment) and internal (cognitive) factors which influence spatial knowledge acquisition and representation.

In the last two decades, researchers from a variety of disciplines have investigated people's ability to acquire and represent spatial or geographic information. A large proportion of this research has focused on navigation and was carried out explicitly or implicitly within the theoretical frameworks of environmental cognition put forward by developmental and environmental psychologists. Researchers have generally agreed that spatial knowledge is represented in three general forms: landmark or declarative; route or procedural; and configurational or survey. Other terms have been used to describe these general representations of spatial knowledge, typically in the context of developmental or sequential stages of learning. Tolman (1948) refers to spatial knowledge occurring in the form of strip maps and comprehensive map knowledge. Piaget and Inhelder (1956) categorise spatial learning in terms of topological, projective and Euclidean knowledge. Kuipers (1978) suggests spatial learning can take the form of sensorimotor, topological and metrical knowledge.

21.1.1 Landmark knowledge

Lynch (1960) defined landmarks as features in a landscape which make a strong and immediate impression on the senses and have a high probability of being incorporated into a cognitive map. Thorndyke (1981) asserts that landmarks occur in the form of perceptual icons stored in visual memory. Perceptual icons are distinct visual images of a scene consisting of a single object or collection of objects organised in a salient manner according to the navigating individual. These objects may include artificial structures such as unique buildings or signs and natural features such as rivers and hills. Many declarative attributes may be associated with a particular landmark. For example, location, orientation, functionality, physical dimensions and event-specific activity help define a landmark. All of these attributes play a role in the development of a landmark's salience; however, the location of a landmark ultimately helps to build an overall cognitive map.

21.1.2 Route knowledge

Procedural or route knowledge allows an individual to follow routes and narrate sequences of observed landmarks (Seigel and White, 1975). Using route knowledge the navigating individual has access to spatial information at a local level. As experience along a route increases, segments of a route may be chunked together to form a more compact knowledge structure (Allen and Kirasic, 1985). Kuipers (1978) proposed that route knowledge consists of a sequence of view-action pairs. The navigating individual is aware that a certain action (either a turn or move) will result in a certain view of the environment. A sequence of such view-action pairs enables an individual to navigate a route or compute what can be seen if a certain action is executed at a particular place. Lloyd (1982) proposed that route knowledge may be represented as a set of verbal codes which includes information regarding the sequence of actions that are necessary for travelling a specific route.

21.1.3 Survey knowledge

Survey knowledge is an overall representation of an environment. The transition from route to survey knowledge is marked by two distinguishing features (Chown et al., 1995). First, an objective frame of reference must be developed to form survey knowledge. Second, the individual is able to determine spatial relationships between objects not close in space. Survey knowledge consists of the individual having a global access to all spatial relations simultaneously. Couclelis et al. (1987) used the anchor-point theory to describe the transition from route to survey representations. The anchor-point hypothesis proposes that route knowledge develops into survey knowledge through an integration of primary and secondary anchors into an overall configuration. Individuals placed in a novel environment quickly select landmarks or nodes that 'anchor' the ensuing route knowledge. These landmarks 'anchor' an individual's initial knowledge in the sense that this knowledge is directly linked to the internal representation of the landmark. Later, less important landmarks, paths and areas become associated with the nearest primary anchor. The overall result is the formation of a cognitive map of the environment containing a linked hierarchical spatial structure.

21.2 Factors influencing spatial knowledge acquisition and representation

Many factors may influence spatial knowledge acquisition and representation. Three general types of factors may include internal 'cognitive' factors, external 'environmental' factors and the medium in which the environment is presented or experienced. This chapter examines one aspect of cognitive factors, mainly spatial abilities, however, a general review is provided to place this research in a broader theoretical framework.

21.2.1 Internal factors

Internal or human cognitive factors include individual cognitive abilities, central information-processing capacity, perceptual or sensory limitations and attitudes. Very little research has focused on the relationship between cognitive abilities and environmental or spatial knowledge. Pearson and Ialongo (1986) found that four spatial ability tests accounted for a small, yet significant amount of the variance of errors in a landmark location task and route knowledge task. This finding suggests that individual spatial abilities play at least a modest role in spatial knowledge acquisition and representation.

The role of attention or central information processing in the acquisition and representation of spatial knowledge is significant (Lindberg and Gärling, 1983). The information-processing approach assumes that an individual must have allocated a certain amount of central information capacity and this capacity is limited at any given moment (Kahneman, 1973). Lindberg and Gärling (1983) investigated this issue in a full and divided attention condition task. Subjects in the divided attention condition were required to count backwards while learning the spatial relations between route segments and reference points. Subjects in the full attention condition did not have to perform the counting task while performing the same spatial learning tasks. Reducing the amount of processing capacity resulted in poorer performance on the spatial tasks. In addition, subjects were unable to access spatial relations, yet information on the locomotion path was still accurate. This is consistent with findings in the field of face recognition which suggests that attention is required to encode spatial relations, yet is not required to encode individual parts of a configuration (Reinitz et al., 1994).

Internal factors such as self-reports on sociability and self-acceptance were found to correlate with pointing errors, when other possible mediating variables were controlled (Bryant, 1982). This finding suggests that an individual's attitude towards the environment predicts (to a modest degree) the spatial knowledge that is acquired and represented.

21.2.2 External factors

External factors which may influence spatial knowledge acquisition and representation focus on environmental characteristics. Characteristics such as the existence or lack of landmarks, the complexity of the layout, the scale or scope of the environment (Montello, 1993) and the richness or density of features all play a significant role. Freundschuh (1992) found that regular-gridded environmental layouts promote survey knowledge more than irregular environments. Golbeck (1985) distinguishes between two broad categories of environmental characteristics: structural and organisational. Structural characteristics include the presence of landmarks, proximity of features and presence of barriers. Organisational characteristics refer to the relation between the structural features such as the clustering of features, orientation of features and saliency of objects in an environment.

21.2.3 Medium of presentation

An individual may acquire spatial knowledge from a wide variety of sources. These sources may include graphical media (maps, scale models, floor plans and aerial photographs), textual media (description of routes and overall environments), direct perception (panoramic or limited views), and actual navigation in the environment. Each of these sources influences how spatial knowledge is represented, the accuracy and content of the representation and which tasks may be successfully performed. When an individual navigates a route or is told how to get from one place to another, the individual has only access to information along a particular route. In other words, the spatial relations in a navigation task are implicitly defined. Conversely, when an individual has access to the overall layout (survey representation) of an environment (via maps, aerial photographs, or panoramic views) all spatial relations between places are explicitly defined. The seminal work by Thorndyke and Hayes-Roth (1982) highlights this distinction. They found that subjects who studied maps and those that have had extensive navigational experience in the same environment performed differently on similar spatial tasks. Map subjects were superior on judgements of relative location and straight-line distances, while navigation subjects were superior for estimating route distances and orienting oneself to unseen objects. As navigational experiences increase, so does access to global spatial relations.

21.3 Individual cognitive abilities

This section highlights how specific cognitive abilities associate with specific spatial tasks used to measure the acquisition and representation of spatial knowledge. As an individual's knowledge structure of a particular environment develops, so does his/her repertoire of tasks that he/she is able to perform accurately. Table 21.1 provides a theoretical framework for identifying which cognitive abilities may be required to complete specific spatial tasks and their associative knowledge structures.

21.3.1 Visual recognition memory

The first stage of spatial knowledge acquisition is the recognition of landmarks from either a different perspective or over multiple exposures from the same perspective. Visual recognition memory may be a significant predictor in this task since it involves encoding and recalling a complex visual pattern. Once an

Table 21.1 The relationship between knowledge structures, spatial tasks and cognitive abilities

Knowledge structure	Spatial task	Cognitive abilities
Landmark	Recognition of landmark	Visual (recognition) memory
Unordered view-action pairs	Correct actions at choice points	Associative memory
Ordered view-action pairs	Sequencing view-action pairs	Memory span and associative memory
Linked landmarks + metric	Route distance estimation	Visual (spatial) memory and memory span
Survey configuration	Direction estimation	Visual (spatial) memory and spatial orientation
Survey configuration	Euclidean distance estimation	Visual (spatial) memory and spatial orientation
Survey configuration	Triangulation between points	Visual (spatial) memory and spatial orientation
Survey configuration	Location of landmarks	Visual (spatial) memory and spatial orientation

individual is able to recognise a landmark, semantic information may help provide a richer knowledge base.

21.3.2 Associative memory

When an individual navigates, a set of procedures are constructed in the form of view-action pairs. These procedures are initially represented as unordered productions (Thorndyke, 1981). Unordered productions are a set of associations between actions and environmental scenes independent of one another. Associative memory should predict success at this stage of navigation since it involves the ability to recall one part of a previously learned but otherwise unrelated pair of items when the other is presented (Educational Testing Service, 1976). Associative memory also includes the ability of recognition. This ability may be enhanced by strategies such as mnemonic techniques.

21.3.3 Memory span

The next stage in spatial acquisition is linking together and ordering a set of view-action pairs. Once an individual holds in long-term memory the correct sequence of view-action pairs, relative distances between landmarks can be computed. Thorndyke (1981) refers to this stage as ordered productions. Memory span (along with associative memory) should best predict this stage of navigation since it is the ability to recall a number of distinct elements in a sequence. In the case of navigation, distinct elements represent the view-action pairs. For example, ordered view-action pairs may be knowledge that the church is before the gas station and after the hotel and the appropriate action at each choice point.

21.3.4 Visual spatial memory

Prior to this point, metric properties have been absent from any spatial representation. The encoding of metric properties along a route allows for distance estimation between landmarks. Two primary cognitive abilities may play a significant role in the formation of a metric component of route knowledge: memory span and visual memory. Memory span ensures that the correct sequence of landmarks are remembered. Visual (spatial) memory involves the ability to remember the location of objects. In order to estimate distances between pairs of landmarks along a route, absolute or relative location of landmarks must be encoded into memory. Once this is done, some simple geometric techniques may be used to compute the distance between locations of landmarks. This chapter examines route-distance estimations as one aspect of overall route knowledge.

21.3.5 Spatial orientation

Once an individual is able to integrate route knowledge from different paths into an overall configuration, survey knowledge is said to occur. This transition process is largely based on establishing an objective frame of reference. The cognitive abilities of visual (spatial) memory and spatial orientation should predict success at this stage of spatial knowledge acquisition and representation. Visual (spatial) memory is helpful in that it provides an understanding of the location of landmarks in relation to one another. Spatial orientation is critical since it involves the ability of the individual to remain unconfused by the changing orientation of the route. This will help establish the overall configuration of the environment (based on a fixed reference system). Spatial orientation is critical in tasks such as estimating the directions between landmarks and straight-line distance estimations between landmarks. Generally, the more serpentine the route, the more difficult not to get confused by the changing orientation along the route.

Survey representations by definition must have a metric quality. However, this knowledge may be in varying degrees of completeness. In any case, this quality is essential to complete successfully specific spatial tasks such as direction estimations between landmarks, Euclidean or straight-line distance between landmarks, locating landmarks, and computing the shortest paths between points (for example, triangulation between three points). Success at these spatial tasks is largely based on visual (spatial) memory and spatial orientation. The remainder of the chapter examines two of these spatial tasks.

21.4 Methods

To examine the relationship between a limited set of cognitive abilities and spatial knowledge acquisition and representation an experiment was designed to include three spatial judgement tasks (route–distance estimation, Euclidean distance estimation and direction estimations) and two types of individual difference tests: spatial abilities and visual memory. According to Table 21.1, visual memory should predict accuracy on both route knowledge (route–distance estimations) and survey knowledge (direction and straight-line distance estimations). Conversely, spatial orientation should predict better success on survey knowledge than route knowledge.

21.4.1 Subjects

A total of 40 Undergraduate ($n = 29$) and Graduate ($n = 11$) students at Boston University volunteered to participate as subjects. Seventeen subjects were female and 23 were male. Data from one subject was discarded because this person did not understand the instructions.

21.4.2 Individual difference tests

The Educational Testing Service (1976) kit of factor-reference tests provided measures of two general types of cognitive abilities: spatial abilities and visual memory. Spatial abilities may be subdivided into two distinct categories: spatial orientation and spatial visualisation (McGee, 1979). Spatial orientation is the ability to remain unconfused by changing orientations of a whole object. Spatial visualisation is the ability to manipulate mentally the internal parts of a stimuli. Golledge *et al.* (1995) suggest that spatial visualisation is less useful for examining spatial knowledge acquisition. This may be due to either increased task complexity and/or the lack of relevance. Since navigation does not involve rotating internal parts of the environment, but rather only oneself with respect to an objective frame of reference, spatial visualisation should not figure prominently. Visual memory tests the subject's memory of the location and orientations of landmarks in a spatial environment. The following four tests were chosen to provide measures of both spatial factors and the visual memory factor.

1 Card rotation tests the ability to see a difference in figures. Subjects must compare cards and decide if one card has been rotated or flipped in some manner. This test loads on the factor of spatial orientation.

2 Cube comparison involves deciding if two cubes are the same (but rotated) or different given a pattern of letters or symbols on each cube. This test loads on the factor of spatial orientation.

3 Form board tests the subjects' ability to tell what pieces can be put together to complete a figure. Subjects are given five geometric pieces and must indicate whether each piece may be used to complete a figure. Subjects are allowed to rotate, but not flip the pieces. This test involves both mental rotation and serial operations that are required to decide if each piece may be used to complete the figure. The form board test loads on the factor of spatial visualisation.

4 Building memory tests the subjects' ability to remember the configuration, location and orientation of visual landmarks on a map. Subjects study a map showing street and landmarks. They are tested on the absolute location of each landmark on a blank street map. The subject is not required to store the relations between landmarks, only their absolute location in the environment.

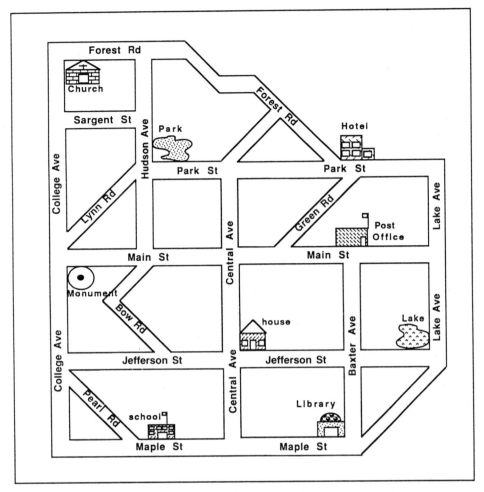

Figure 21.1 Street network and landmarks for Map A.

21.4.3 Stimuli for spatial learning task

Two maps were used in this study. Each depicted a fictitious town consisting of streets and landmarks (Fig. 21.1). Landmarks were all standardised in size and line type and were simplified in their design, yet easily identifiable. Streets were standardised in their lengths. All north–south running streets were named as 'Avenues' and east–west running streets were 'Streets'. All diagonal streets were named as 'Roads'. Map A contained five diagonal street segments and Map B contained four diagonal street segments.

21.4.4 Procedure

Subjects participated in two sessions. During the first session subjects completed a battery of four individual difference tests. During the second session they learned

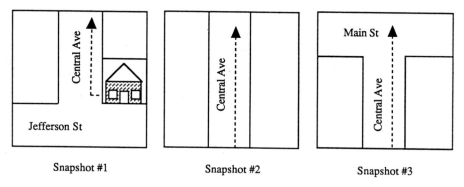

Figure 21.2 Example of three map 'snapshots' from Map A.

about and were subsequently tested on, a spatial environment. Each session lasted approximately one hour. The spatial learning session consisted of a 'mental walk' through a fictitious town using a series of small map 'snapshots' (Fig. 21.2). Small segments of the map were presented to subjects in 5-second intervals. The subjects only saw one segment of the map at a time. Each map segment spatially overlapped the next by one-third to allow the subjects to make easy connections between the slides. The presentation of Map A contained 51 slides and Map B contained 55 slides. Each walk was presented three times consecutively, followed by a map test. This method of presentation allowed for control of the information that was presented in terms of landmarks and inadvertent access to survey level information. After each slide sequence subjects received three spatial judgement tests. Subjects were presented with random pairs of landmarks, and then made one of the judgements described below.

1   Route–distance estimations were made on a 0–100 point scale, where 100 represented the maximum distance of the route (begin point to end point) and 0 indicated the same location. Subjects judged the total distance that would be travelled on the street route from one of the landmarks to the other. Accuracy was measured as the absolute difference between the estimated and actual distance (measured in a 100-point scale).

2   Euclidean or straight-line distance estimations were also made on a 0–100 point scale. Subjects were instructed to form a 'mental map' of the town based on their walk and then determine the two furthest landmarks. They were told that these landmarks were 100 units apart in Euclidean distance; they were asked to base their distance judgements relative to this distance. Subjects then estimated Euclidean (rather than route) distances between landmark pairs.

3   Direction estimations involved estimating the direction of travel from one landmark to another. Subjects were given a compass rose in 5° intervals and instructed to estimate the direction between randomly selected pairs of landmarks. Subjects were told to estimate what direction one landmark was relative to the other. Subjects responded to the closest 5°. For the purpose of comparison between the other two measures, the direction estimation error was converted to a similar 100-point scale.

21.5 Results and discussion

There was no difference in performance between map A and map B. This is not surprising since the various difficulty dimensions are approximately equal between the maps. The distribution of scores on three of the four individual difference tests are approximately normal. The distribution of the building memory test scores is slightly skewed, indicating a moderate ceiling effect. On the four individual difference tests there is evidence to support a gender difference. Results from a difference of means test showed that males significantly out-performed females on all four individual difference tests (at the alpha $= 0.05$ level) (Table 21.2). However, there was no gender difference on the three spatial judgement tests (Table 21.3). There was not a significant correlation between age and any of the spatial or individual difference scores.

21.5.1 Map performance

Performance on route–distance estimations was superior to Euclidean distance estimations and direction estimations. The differences between route distance and Euclidean distance estimation were statistically reliable, $t(76) = -8.32$, $p = 0.0001$, as were the differences between route distance and direction estimations, $t(76) = -4.17$, $p = 0.0001$. However, there was no significant difference between Euclidean and direction estimations, $t(76) = 0.94$, $p = 0.35$. This is logically consistent since both Euclidean distance and direction estimations require the use of survey knowledge, while route–distance estimations require the use of route knowledge. This is consistent with findings in other studies which suggest route knowledge is superior to survey knowledge with limited navigation experience (Thorndyke and Hayes-Roth, 1982).

Table 21.2 Gender differences on the individual difference tests as measured by performance scores

	Card rotations	Cube comparisons	Form board	Building memory
Male	58.8	26.4	56.1	10.3
Female	46.0	17.4	44.6	8.3
p-value	0.018	0.005	0.053	0.030

Table 21.3 Gender differences on the spatial judgement tests as measured by per cent error

	Route distance error	Euclidean distance error	Direction error
Male	14.8	25.9	22.6
Female	16.3	27.3	27.0
p-value	0.39	0.51	0.27

Table 21.4　Correlation matrix between all study variables

	Card rotation	Cube comparison	Building memory	Form board	Route	Euclidean
Cube comparison	$0.80^{\dagger}$					
Building memory	0.32^{*}	$0.44^{\dagger}$				
Form board	$0.50^{\dagger}$	$0.48^{\dagger}$	0.16			
Route	-0.32^{*}	-0.37^{*}	$-0.50^{\dagger}$	-0.08		
Euclidean	$-0.40^{\dagger}$	$-0.44^{\dagger}$	-0.35^{*}	-0.13	0.42^{*}	
Direction	$-0.49^{\dagger}$	$-0.55^{\dagger}$	$-0.47^{\dagger}$	-0.19	0.57^{*}	0.55

$^{*}P < 0.05$　　$^{\dagger}P < 0.01$

21.5.2　Correlations

The Pearson Product-Moment correlation matrix offers a summary of the relationship between the three spatial judgement tasks and the four individual difference tests (Table 21.4). Visual memory is hypothesised to be correlated with both route and survey knowledge (all three spatial tasks), while spatial orientation (cube comparisons and card rotations) is hypothesised to correlate with survey knowledge only (Euclidean distance and direction scores).

21.5.2.1　*Correlations between intelligence scores and spatial judgements*

Three of the four individual difference measures (cube comparisons, building memory and card rotations) are significantly correlated with performance on all three of the spatial judgement tasks. Consistent with the hypothesis, visual memory is correlated with both route and survey knowledge. Inconsistent with the hypothesis, spatial orientation significantly correlates with both route and survey knowledge. As hypothesised, the form board test (spatial visualisation) is not significantly correlated with any of the spatial judgement tasks.

Overall, these results show strong similarities between route and survey knowledge. Specifically, factors that predict route–distance estimation accuracy also predict Euclidean distance and direction judgement accuracy; furthermore, they predict them equally well. While overall performance in route and survey knowledge is different, the same individual difference tests are equally correlated with all three spatial judgement variables.

21.5.2.2　*Correlations between spatial estimation tasks*

Scores on the three spatial estimation tasks are significantly intercorrelated. This indicates that each test is not testing a unique portion of spatial knowledge. This may imply that route and survey judgements involve the use of similar underlying representations. In future studies, other tests will be employed which tap into unique aspects of route and survey knowledge.

21.5.3 Multiple and partial regressions

While the correlation matrix offers a summary of the relationship between the study variables, it does not take into account the covariation that exists between the variables. A multiple and partial regression analysis were computed in order to measure the unique variance in spatial judgement performance accounted for by each individual difference test.

The multiple regression analysis showed that the individual difference measures predicted a significant percentage of the total variance for all three spatial judgements. Specifically, they accounted for 21.1 per cent of the variation of average route error, 15.6 per cent of the variance of average Euclidean error and 31.1 per cent of the variance of average direction error. This strongly suggests that spatial abilities and visual memory predict performance on spatial judgement tasks.

21.5.3.1 Route distance error

The building memory (visual memory) test significantly explained the variance of average route error, while all other individual difference variables were held constant. Building memory uniquely accounted for 16.9 per cent of the route–error variance. This is consistent with the hypothesis that visual memory is critical in developing route knowledge since it requires subjects to remember the locations of individual items. Also, neither spatial orientation score predicted average route–error variance, when all other variables were controlled. This is consistent with the hypothesis since the orientation of the route should not matter in the estimation of distances along a route.

21.5.3.2 Euclidean distance error

No individual difference variable significantly explained the variance of average Euclidean distance error, while all other individual difference variables were held constant. This is inconsistent with the hypothesis which predicted both spatial orientation and visual memory as significant predictors of Euclidean distance error. The increased complexity of this spatial task may be one explanation for this result.

21.5.3.3 Direction error

Building memory test significantly explained the variance of average direction error, while all other individual difference variables were held constant. Building memory uniquely accounted for 9.8 per cent of the direction error variance. This is consistent with the hypothesis that visual memory is critical in developing survey knowledge since it requires subjects to remember the locations of individual items. However, neither spatial orientation variables explained direction error, when all other variables were held constant. This is inconsistent with the hypothesis.

21.6 Conclusions

Two primary questions have been examined in this study. First, do individual cognitive abilities influence spatial knowledge acquisition and representation? Second, if cognitive abilities do play a role in this process, can certain cognitive abilities predict success for specific spatial tasks?

Results of the study showed that the individual difference test battery accounted for a significant amount of the total variance in performance on the three spatial judgement tasks. This helps reconfirm the significant role cognitive factors play in the acquisition and representation of geographic space.

While the answer to the first question may seem obvious, examining how specific cognitive abilities predict success at specific spatial learning tasks is less answerable. This study found that visual memory was a significant predictor of route–distance error and one aspect of survey knowledge (direction error), while controlling for possible covariation. This points to the great significance that visual memory plays in the acquisition and representation of geographic space. Encoding the location of landmarks is one of the principal steps in the cognitive mapping process. However, inconsistent with the hypothesis, spatial orientation did not significantly predict performance on any of the three spatial tasks. Several reasons may be possible. Perhaps the route did not have enough turns and most subjects were not confused by the changing orientation. The strong intercorrelation between the individual difference tests also suggests that each test may not be measuring a unique aspect of intelligence. In addition, the strong intercorrelation among the spatial learning tests may suggest that each test is not measuring a unique aspect of spatial knowledge, but rather varying degrees of the same type of spatial knowledge. The problems found in this study are also problematic of many studies involving individual difference measures related to spatial abilities. Future research in this area must address these issues in a comprehensive manner. One possible way to alleviate this problem is to offer a greater range of individual difference tests and spatial learning tests in a more realistic environment.

References

ALLEN, C. L. and KIRASIC, K. C. (1985). Effects of the cognitive organisation of route knowledge on judgements of macrospatial differences, *Memory and Cognition*, **13**(3), 218–227.

BRYANT, K. J. (1982). Personality correlates of sense of direction and geographical orientation, *Journal of Personality and Social Change*, **43**, 1318–24.

CHOWN, E., KAPLAN, S. and KORTENKAMP, D. (1995). Prototypes, location and associative networks (PLAN): Towards a unified theory of cognitive mapping, *Cognitive Science*, **19**, 1–51.

COUCLELIS, H., GOLLEDGE, R. G., GALE, N. and TOBLER, W. (1987). Exploring the anchor-point hypothesis of spatial cognition, *Journal of Environmental Psychology*, **7**, 99–122.

EDUCATIONAL TESTING SERVICE (1976). *Manual for Kit of Factor-Referenced Cognitive Tests*, Princeton, NJ: Educational Testing Service.

FREUNDSCHUH, S. M. (1992). Is there a relationship between spatial cognition and environmental patterns? in Frank, A. U., Campari, I. and Formentini, U. (Eds), *Theories and Methods of Spatio-Temporal Reasoning in Geographic Space*, pp. 288–304, Berlin: Springer-Verlag.

GOLBECK, S. L. (1985). Spatial cognition as a function of environmental characteristics, in Cohen, R. (Ed.), *The Development of Spatial Cognition*, pp. 225–255, Hillsdale, NJ: Erlbaum.

GOLLEDGE, R. G., DOUGHERTY, V. and BELL, S. (1995). Acquiring spatial knowledge: Survey versus route-based knowledge in unfamiliar environments, *Annals of the Association of American Geographers*, **85**. 134–58.

KAHNEMAN, D. (1973). *Attention and Effort*, Englewood Cliffs, NJ: Prentice Hall.

KUIPERS, B. (1978). Modelling spatial knowledge, *Cognitive Science*, **2**, 129–53.

LINDBERG, E. and GÄRLING, T. (1983). Acquisition of different types of locational information in cognitive maps: Automatic and effortful processing? *Psychological Research*, **45**, 19–38.

LLOYD, R. (1982). A look at images, *Annals of the Association of American Geographers*, **72**, 532–48.

LYNCH, K. (1960). *Image of the City*, Cambridge, MA: MIT Press.

McGEE, M. G. (1979). Human spatial abilities: Psychometric studies and environmental, genetic, hormonal and neurological influences, *Psychological Bulletin*, **86**, 889–918.

MONTELLO, D. R. (1993). Scale and multiple psychologies of space, in Frank, A. U. and Campari, I. (Eds), *Spatial Information Theory: A Theoretical Basis for GIS. Lecture Notes in Computer Science 716*, pp. 312–21, New York: Springer-Verlag.

NATIONAL CENTER FOR GEOGRAPHIC INFORMATION AND ANALYSIS (1989). The research plan for the National Center for Geographic Information and Analysis, *International Journal of Geographic Information Systems*, **3**, 117–36.

PEARSON, J. L. and IALONGO, N. S. (1986). The relationship between spatial ability and environmental knowledge, *Journal of Environmental Psychology*, **6**, 299–304.

PIAGET, J. and INHELDER, B. (1956). *The Child's Conception of Space*, London: Routledge & Kegan Paul.

REINITZ, M. T., MORRISSEY, J. and DEMB, J. (1994). Role of attention in face encoding, *Journal of Experimental Psychology: Learning, Memory and Cognition*, **20**, 161–8.

SEIGEL, A. W. and WHITE, S. H. (1975). The development of spatial representations of large-scale environments, in Reese, H. (Ed.), *Advances in Child Development and Behaviour*, pp. 9–55, New York: Academic Press.

THORNDYKE, P. W. (1981). Spatial cognition and reasoning, in Harvey, J. H. (Ed.), *Cognition, Social Behaviour and the Environment*, pp. 137–49, Hillsdale, NJ: Erlbaum.

THORNDYKE, P. W. and HAYES-ROTH, B. (1982). Differences in spatial knowledge acquired from maps and navigation, *Cognitive Psychology*, **14**, 560–89.

TOLMAN, E. (1948). Cognitive maps in rats and men, *Psychological Review*, **55**, 189–208.

CHAPTER TWENTY-TWO

Modelling Semantical[1], Spatial and Temporal Information in a GIS

MAY YUAN

22.1 Introduction

Recent interests in the incorporation of temporal information into a GIS have emphasised the properties and structures of time in terms of representation and indexing in accordance with space. Proposed schemes include models of snapshots (Armstrong, 1988), space-time composites (Langran and Chrisman, 1988), spatio-temporal objects (Worboys, 1992), temporal objects in three domains of semantics, time and space (Yuan, 1994) and event-based temporal intervals in a triad representational framework of object, location and time (Peuquet, 1994; Peuquet and Duan, 1995). The modelling of time in a GIS seems to launch into an object-oriented paradigm. The transition from concepts of time layers to time objects implicates the needs for the separation of temporal information from spatial information in data modelling regardless of practical issues associated with these proposed models. Time has its own identity and represents an individual, absolute, or relative concept in both the three-domains model and the triad representational framework, instead of being an attribute of location, as in the snapshot model, or an integral part of spatial entities, as in the space-time composites and spatio-temporal objects.

Many advantages result from modelling semantics (thematic attributes), time and space (location), separately in a GIS. In addition to an advance from a world history database to a process database in GIS as elaborated in Peuquet (1994), such a separation allows semantical, temporal and spatial components of geographic information to be handled dynamically and intelligently in terms of representation, analysis and reasoning. With respect to representation, the three types of geographic information consist of different but comparable properties. These properties or their relationships may change independently, such that changes in spatial properties of an entity may or may not imply any changes in its semantics; for example, Kesseler's farm may be still owned by Edward Kesseler even if the farm has been expanded to the east field. The separation will avoid possible duplication of semantical information, which is very likely to occur in the models of snapshots, space-time composites and spatio-temporal objects. This is because semantical changes, such as land-use change, usually involve changes in spatial configuration and spatial object identities according to a reconstruction of spatial primitives in these models unless data are stored in a raster framework.

In analysis, space (location), time and semantics (attributes) are three components of geographic information (Berry, 1964). The three components are handled differently in geographic analysis, in which we control one of the components, fix another and then measure the other (Sinton, 1978). This separation facilitates a direct support of correspondent elements of space, time and semantics to suit the needs for geographic analysis. In terms of reasoning, the three types of semantical, temporal and spatial information fit different logic systems since their inherited characters differ from one another. The traditional predicate logic may suit well for reasoning about the context of semantics or attributes. It is also used in spatial problems, although some spatial problems may not fit well with the monotonic reasoning scheme. Fuzzy logic may be better for spatial reasoning since geographic phenomena or concepts are often continuous and transitional without well-defined boundaries. For example, a place located 20 miles from a shopping mall may be considered equally close to the mall as a location 21 miles away from the mall. On the other hand, temporal problems had better apply temporal or modal logic in querying and problem-solving. Modal logic encounters parallel and branching structures of time and embeds connectives to reason possibly and necessarily true conditions. The separation of the three types of information is helpful in incorporating different logical systems with semantical, temporal and spatial domains.

This chapter addresses two issues. The first issue concerns how to model semantics, time and space as objects in three separate but interlinked domains in a GIS, including the interrelations and interactions among these three types of objects within and between domains. The second issue is how to utilise these interrelations and interactions in order to facilitate geographic analysis and problem-solving. The basic idea behind this chapter is to model semantics, time and space separately and then to link semantical, temporal and spatial objects accordingly. As such, this framework enables GIS data to share objects in the three domains and eliminates data redundancy. For example, a semantical object may be linked to multiple temporal objects if the semantical object occurs more than once at a location. This chapter defines the conceptual data model and shows how to link objects from the three domains to represent spatio-temporal information in GIS; however, the incorporation of logic systems to the model is outside the scope of the chapter. The chapter starts with definitions of semantical, temporal and spatial domains together with operations and rules to ensure integrity of the data model. The discussion continues with interactions and inter-relations of semantical, temporal and spatial objects so as to determine links among objects from the three domains. Representational capabilities of the three domains model are later tested through six basic types of spatio-temporal information.

22.2 Definitions

A data model defines types of data objects and a framework to organise and manage them. It consists of three components:

1 a set of object types which defines a set of basic building blocks;

2   a set of operations which provides a means for manipulating object types in a database; and

3   a set of integrity rules which constrains the valid states of the databases and ensures that they conform to the data model (Date, 1983).

Modelling geographic information involves defining the three components and designing a data framework to structure data objects for the support of geographic queries by semantics (meanings, attributes, or entities), time (instant, duration, or periodicity) and space (shape, size, site, or location). This research models geographic information by independent but inter-related objects in semantical, temporal and spatial domains. Types of data objects, operations and integrity rules in each of the domains define aspatial and atemporal attributes, temporal characteristics and spatial properties of geographic entities, respectively. This section aims to define the three domains and their associated object types, basic operations and integrity rules. These definitions formulate the three-domain model and provide a foundation for designing correspondent operational languages in follow-up research.

22.2.1 Semantical domain

A semantical domain is composed of semantical objects for concrete entities or abstract concepts (Definition 1). Every semantical object has a unique identifier. Semantical objects are stored according to classes and are organised in hierarchies. Examples of these hierarchies are taxonomies for plants or soils, and administrative units for fire management.

Sowa (1984) pointed out four main relationships in conceptual modelling: classification, association, generalisation and aggregation. The four relationships have been applied to object-oriented programming and data modelling and will be helpful in structuring semantical objects in a semantical domain. A semantical domain may include both a knowledge base of isa hierarchies or semantic networks and a database of attribute tables or data frames. Data models, relational or object-oriented, developed in information sciences can be applied to form and organise semantical objects in a semantical domain.

> *Definition 1   A semantical domain* (U) represents the topical interests in a Universe of Discourse (UoD). Every semantical domain includes a set of real or abstract objects about which knowledge is being expressed. Semantical objects, O_1 and O_2, are in a semantical domain if and only if they are related by at least one of the following relationships: classification, generalisation, aggregation or association.

22.2.2 Temporal domain

A temporal domain is conceptualised as a universal time arrow (Definition 2). The time arrow starts somewhere, perhaps infinity, in the past and approaches to the future. There are two types of temporal objects: temporal points representing instances and temporal lines representing intervals. A temporal point is defined by its location on the universal time arrow in a temporal domain. A temporal line is defined by two temporal points, which mark the starting and the ending instances, plus a duration shown by the distance between these two temporal points. Every temporal object is defined by a set of time points and/or time lines along the universal time arrow. Temporal objects are either simple or complex. Simple temporal objects are points or non-overlapping temporal lines (Definition 3). A complex temporal object consists of a set of contiguous or discrete simple and/or complex temporal objects. Complex temporal objects can be used to represent temporality of recurring events or phenomena and they are useful to calculate frequency and period of events. Time scale can be defined by a time line representing the scope of time involved in a study. Time granularity can be determined by the length of time lines associated with an event.

Definition 2   A temporal domain, (T), is an arrow starting at a time point or infinity in the past and extending to the future, plus a temporal tree of simple and/or complex objects. The arrow is composed of a finite set of simple or complex temporal objects, t_s or t_c. Simple temporal objects are mutually exclusive. A complex temporal object is composed of a set of simple and/or complex temporal objects: $T \equiv \{0\ \infty : t_c \cup t_s / \text{whereas } t_{s1} \cap t_{s2} = \phi\}$

Definition 3   Temporal overlap. Let t_1 and t_2 be two complex temporal objects and q be a simple or complex temporal object in a temporal domain. If $t_1 \cap t_2 = q$ and $q \neq f$, then t_1 and t_2 are overlapped at q in the temporal domain. If $t_1 \cap t_2 = \phi$, then t_1 and t_2 do not have overlaps in the temporal domain.

Additional temporal objects may be introduced into a temporal domain during data input or processing and they may overlap with pre-existing simple temporal objects. The temporal domain thus needs to be reconstructed under the constraint of linear enforcement to ensure temporal topology by forcing all temporal objects situated along a universal temporal arrow. As a result, no overlaps will exist among simple temporal objects and temporal relationship, such as before or after, can be calculated on the time arrow. Temporal topology among complex temporal objects can be inferred based on their composed simple temporal objects. Temporal reconstruction is to decompose overlapping temporal objects to non-overlapping simple temporal objects and rebuild complex temporal objects (Definition 4). A temporal tree is constructed to hold parenthood relationships among old simple temporal objects and the new ones generated by temporal reconstruction (Fig. 22.1). Temporal trees make it possible to keep unique identifiers for all temporal objects in a temporal domain and maintain object persistency to facilitate temporal queries and modelling.

Definition 4   Temporal reconstruction. Let a temporal arrow consist of a set of simple temporal objects, $T = \{t_1,\ t_2,\ t_3, \ldots, t_n\}$. If a newly introduced temporal object x overlaps with temporal object t_1, then t_1 is decomposed into t_{n+1} and t_{n+2}, and x is decomposed into t_{n+1} and t_{n+3}, whereas t_{n+1} is the overlapping temporal object. The temporal domain is reconstructed and redefined by a new set of simple temporal objects, $T' = T - t_1 + t_{n+1} + t_{n+2} + t_{n+3}$. A temporal tree of $t_1 \rightarrow \{t_{n+1}, t_{n+2}*\}$ holds the parenthood relationship among temporal objects in the past and the current configuration of temporal object sets, whereas * indicates the overlapping temporal object between t_1 and x.

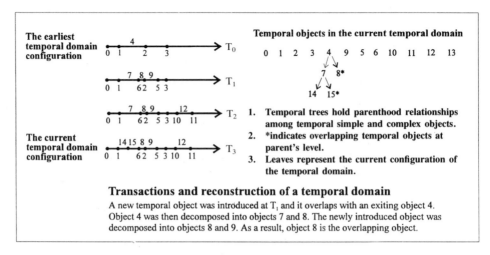

Figure 22.1 An example of a temporal domain and temporal trees.

22.2.3 **Spatial domain**

A spatial domain is a three-dimensional space, which consists of a finite set of simple and complex spatial objects (Definition 5). Basic spatial object types include points, lines, areas, cells and volumes. However, they can be extended according to spatial object types defined in the Spatial Data Transfer Standard (SDTS). Spatial points are defined by coordinates of locations. A spatial line is defined by two spatial points, a spatial polygon is defined by a closure of spatial lines and a spatial volume is defined by a closure of spatial polygons. Simple spatial objects, 0D, 1D, 2D, or 3D, are non-overlapping objects with shared geometry. For simplicity, this chapter only discusses 0D, 1D and 2D objects. Simple spatial objects of lines and polygons are constrained by planar enforcement so that a point will be generated if two lines cross each other. Complex spatial objects are composed of simple objects and they may overlap with each other (Definition 6). Spatial objects are organised in hierarchies of simple and complex spatial objects with shared geometry. Simple spatial objects are non-overlapping objects. Geometry is shared by simple spatial objects. These simple objects constitute the framework of the current space. Complex spatial objects are composed of simple objects and they may overlap with each other. Spatial properties of complex spatial objects can be computed from their compound simple objects by spatial aggregation. Spatial properties of complex spatial objects can be computed by aggregating properties from their simple spatial objects.

Each spatial object has a unique and persistent identifier. A spatial domain needs to be reconstructed whenever overlay occurs among simple spatial objects (Definition 7). If a new spatial object is introduced and has overlay with an existing spatial object, both objects will be decomposed into non-overlapping simple spatial objects. The pre-existing spatial object keeps its identifier but it changes from a simple spatial object to a complex spatial object. The newly introduced spatial object will also be a complex object of two or more simple spatial objects. If a spatial object is removed from a spatial domain, its identifier is kept in a queue

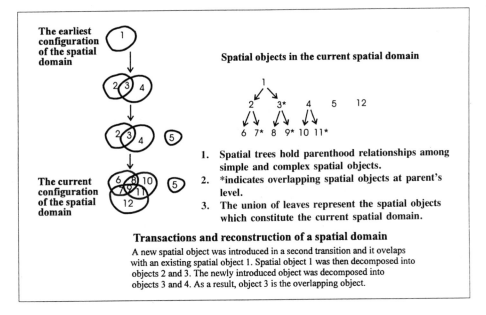

The earliest configuration of the spatial domain

Spatial objects in the current spatial domain

The current configuration of the spatial domain

1. Spatial trees hold parenthood relationships among simple and complex spatial objects.
2. *indicates overlapping spatial objects at parent's level.
3. The union of leaves represent the spatial objects which constitute the current spatial domain.

Transactions and reconstruction of a spatial domain

A new spatial object was introduced in a second transition and it ovelaps with an existing spatial object 1. Spatial object 1 was then decomposed into objects 2 and 3. The newly introduced object was decomposed into objects 3 and 4. As a result, object 3 is the overlapping object.

Figure 22.2 An example of a spatial domain and spatial trees.

and is ready to be assigned to a new spatial object. Once an identifier is assigned to a spatial object, it is bound to the object until the object is removed. If a pre-existing spatial object is decomposed into two or more simple objects, the spatial object becomes a complex spatial object but still has its original identifier. A spatial tree is built to hold the parenthood relationship among old and new simple spatial objects (Figure 22.2). No spatial reconstruction is needed for cell objects, however.

Definition 5 A spatial domain. S, is a three-dimensional space, which consists of a finite set of simple and complex spatial objects. Types of spatial objects include point, p, lines, l, polygons, g, cells, c and volume, v: $S = \{p, l, g, c, v\}$. Simple spatial objects of points, lines and polygons are constrained by planar enforcement. Simple volumes are constrained by volumetric enforcement. Cells are arranged in a matrix and their locations are addresses in the matrix. A matrix may contain cells of different sizes.

Definition 6 Spatial overlap. Let s_1 and s_2 be two complex spatial objects and q be a simple or complex spatial object in a spatial domain. If $s_1 \cap s_2 = q$ and $q \neq \phi$, then s_1 overlaps s_2 at q in the spatial domain. If $s_1 \cap s_2 = \phi$, then s_1 and s_2 do not have overlaps in the spatial domain.

Definition 7 Spatial reconstruction. Let a set of simple spatial objects represent a spatial domain, $S = \{s_1, s_2, s_3, \ldots, s_n\}$. If a newly introduced spatial object x overlaps spatial object s_1 then s_1 is decomposed into s_{n+1} and s_{n+2} and x is decomposed into s_{n+2} and s_{n+3}, whereas s_{n+2} is the overlapping spatial object of x_1 and x. The spatial domain is reconstructed and redefined by a new set of spatial objects, $S' = S - s_1 + s_{n+1} + s_{n+2} + s_{n+3}$. A spatial tree $(t_1 \rightarrow s_{n+1}, s_{n+2}{}^*)$ holds the parenthood relationship among temporal objects in the past and current temporal object sets, whereas * indicates the overlapping spatial object between s_1 and x.

22.3 Rules of integrity: topological constraints in spatial and temporal domains

Topological constraints have been identified as the uniform representation schema for both spatial and temporal concepts (Dutta, 1991). They are applied to ensure validity of the spatial and temporal domains and to hold topological relationships among simple objects.

22.3.1 GIS data modelling at a primitive object level versus at a feature level

There is a need to distinguish GIS data modelling at a primitive data object level from a feature level. All objects considered at a primitive data object level are simple objects. These simple objects constitute the basic blocks for building a GIS database, whereas objects modelled at a feature level present a one-to-one mapping from geographic entities to GIS objects. Therefore, a feature level may include both simple or complex objects. Rules of planar enforcement and space exhaustion are applied to ensure topological integrity in modelling a 2D space at a primitive object level in a GIS database. Planar enforcement constrains that a node is produced whenever two lines cross each other. As a result, there are no over-passing lines allowed at a primitive data object level. Non-intersecting interstate highways and local roads, for example, are represented by complex line objects at a feature level. Space exhaustion excludes the concept of empty space and, therefore, determines polygon configurations and topology at a primitive data object level. Both planar enforcement and space exhaustion make no ambiguity about linear and polygonal topology at a primitive data object level, whereas the two rules do not apply to GIS feature levels (Langran, 1992; van Roessel and Pullar, 1993).

22.3.2 Temporal objects versus spatial objects

Time is usually conceptualised as one dimension in nature. Similar to constructs used in modelling space, points and lines are used to represent temporal concepts at a primitive object level. However, temporal concepts can be linear, branching, parallel or cyclic at a feature level. As such, the rule of planar enforcement is applicable to simple objects in both temporal and spatial domains. Even if multiple temporal dimensions are discussed, for example, transaction time, database time and institutional time, a temporal domain contains only temporal objects of points or lines along or bifurcating from a line of synchronicity (Barrera *et al.*, 1990). A temporal polygon will be formed when analysing two local events by using event time rather than absolute time coordinates. For example, coexistence of thunderstorms and tornadoes can form temporal polygons if we use the duration of thunderstorms as the x axis and of tornadoes as the y axis. In modelling such multiple events, the rule of space exhaustion can be applied to the temporal domain to ensure the integrity of polygon topology at a primitive object level. Therefore, periods of thunderstorm-but-no-tornado, thunderstorms-and-tornadoes and so on, can be represented in a temporal domain. Contrarily to polygons in a spatial domain, temporal polygons have no explicit meanings in terms of area measure but show coexistence and possible temporal correlation of two events.

In summary, both spatial and temporal domains maintain the integrity of topology imposed by the rules of linear or planar enforcement and space/time exhaustion at a primitive data object level. Both rules are applied during spatial or temporal reconstruction to decompose overlapping simple objects and to build spatial or temporal trees for keeping track of object identifiers before and after a reconstruction. The configuration of a spatial or temporal domain represents topological relationships among simple objects at a primitive object level. Spatial or temporal trees are in fact representing hierarchical or containment relationships at a feature level in space or time, where objects structured in the trees can be complex or simple and overlaps are allowed (Figs 22.1 and 22.2).

22.4 Linking spatial and temporal objects to represent spatio-temporal information

When semantics, time and space are modelled as separate domains in a GIS database, links or pointers are required to associate these objects to describe an entity's thematic properties and its spatial and temporal attributes. Associations of semantics, time and space can be outlined by the three following statements, which are generally true but not exhaustive in a physical world:

1 An entity can only appear at a location at a given point in time; that is we can determine an entity by a given location and a given time.

2 A location can be only occupied by an entity at a given point in time; that is we can determine a location by a given entity at a given time.

3 A point in time may have various entities at various locations, or an entity may appear at the same location at different points in time; that is we cannot determine a point in time by a given entity at a given location.

Therefore, there are one-to-one mappings from semantical and temporal objects to spatial objects and from spatial and temporal objects to semantical objects (Fig.

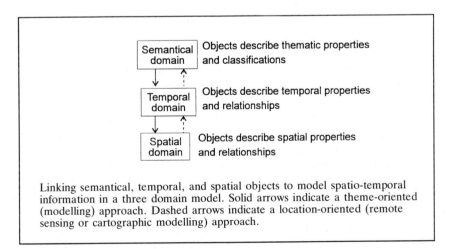

Linking semantical, temporal, and spatial objects to model spatio-temporal information in a three domain model. Solid arrows indicate a theme-oriented (modelling) approach. Dashed arrows indicate a location-oriented (remote sensing or cartographic modelling) approach.

Figure 22.3 Links among semantical, temporal and spatial domains.

22.3). However, mappings from semantical and spatial objects to temporal objects may be many-to-many. Therefore, links among the three domains should be either from semantical through temporal to spatial objects, or from spatial through temporal to semantical objects, to eliminate data redundancy and ambiguity. Nevertheless, linkages from temporal objects to semantical or spatial objects can be inferred by these two directional links. These links should be dynamic in order to facilitate different views and also be flexible in order to support introduction of new objects in the three domains.

Spatial objects and temporal objects represent spatial and temporal properties of geographic phenomena. There are no topological constraints used to manage relations among objects from spatial and temporal domains, since any given geographic phenomenon can experience changes in its spatial properties independently from its temporal ones and vice versa. Spatial and temporal objects are topologically independent, though they interact with each other to represent concepts of spatio-temporal variations and/or behaviours of geographic phenomena.

Changes occur to attributes of a phenomenon, environmental settings, behaviours of an event, or mechanisms of a processes. However, change detection is determined by sampling intervals and data resolutions in relation to changes in reality. This chapter classifies six major types of spatial and/or temporal changes in geographic information (Table 22.1).

1   For a given site, the occurrences and duration of events or attributes may change during a certain period of time; analysis is done by fixing location, controlling attribute and measuring time.

2   For a given time, a certain phenomenon may change its characteristics from site to site; analysis is done by fixing time, controlling attribute and measure location.

3   For a given time, attributes may change from site to site; analysis is done by fixing time, controlling locations and measuring attributes.

4   For a given event, its characteristics or processes may change from site to site during a certain period of time; analysis is done by fixing attributes, controlling locations and measuring time.

5   For a given site, attributes may change from time to time; analysis is done by fixing location, controlling time and measuring attributes.

6   For a given event, its location may change from time to time; analysis is done by fixing attribute, controlling time and measuring location.

Type 1 changes only involve variations in attributes over time; there is no variation in spatial properties. Therefore, this kind of data modelling and analysis can be totally handled in a semantical domain as historical transactions in a relational or object-oriented database-management system. Type 2 changes describe static spatial distribution of a geographic phenomenon, such as topography or air pressure. Techniques, such as contouring, choropleth mapping and dasymetric mapping are often used to present such information, while current GIS store this kind of information in forms of vector or raster layers. Changes of types 3 to 6 alter geometry or topology of spatial and/or temporal properties. This study defines changes of types 3 and 4 as spatial changes, whereas 5 and 6 as temporal changes.

Table 22.1   Six types of spatio-temporal changes (information) and analysis

	Change type	Fix	Control	Measure	Representation and analysis
Semantics (analyse differences in temporal or spatial properties among different things or entities)	1	Location	Attribute	Time	Link a spatial object (Buffalo, NY) to a set of temporal objects (days in 1994) and to a set of semantical objects (snowing or sunny days). Analyse occurrences and duration of the semantical objects at that location.
	2	Time	Attribute	Location	Link a set of semantical objects (contours of 100 ft and 120 ft to a temporal object (1995) and to a spatial object (the location of the contour in Wichita mountain). Analyse spatial objects (locations) of the semantical objects (these contours).
Space (compare differences in semantical or temporal properties from site to site)	3 (static)	Time	Location	Attribute	Link a set of spatial objects (Cleveland County and Oklahoma County) to a set of temporal objects (1994 and 1995) and then to a semantic object (population). Compare value difference in the semantical object (difference in population) at difference locations.
	4 (transitional)	Attribute	Location	Time	Link a set of semantical objects (lows in highs of temperature in 1995) to a set of temporal objects (days of lows and highs), and then to a set of spatial objects (New York City and Tokyo). Analyse differences in temporal objects (occurrences and duration of lows and highs in temperature).
Time (compare differences in terms of semantical or spatial properties from time to time)	5 (mutation)	Location	Time	Attribute	Link a spatial object (a burn site), to a set of temporal objects (1980, 1985, and 1990) and to a set of semantical objects (prescribed fire, natural brush fire). Compare difference in the set of semantical objects (fire intensity or vegetation survivorship).
	6 (movement)	Attribute	Time	Location	Link a semantical object (a wildfire) to a set of temporal objects (time in a day) and to spatial objects (locations of burns). Compare differences in spatial objects (fire spread from one location to another).

Constructions of links among the three domains to represent type 3 to type 6 changes will be elaborated in the following subsections.

22.4.1 Spatial changes

Spatial changes refer to variations across space at a given time or in a period, in which comparisons are made between two or more sites according to data of the same vintage. This study distinguishes two types of spatial changes: static (type 3) or transitional (type 4). Static spatial changes concern variations of a geographic phenomenon at a snapshot, whereas transitional spatial changes compare states of an event or a process at different sites (Fig. 22.4). For example, we can compare

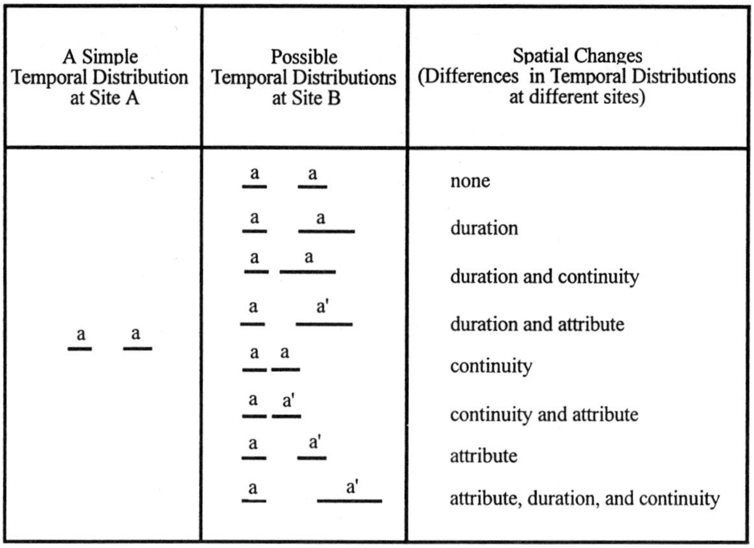

A Simple Temporal Distribution at Site A	Possible Temporal Distributions at Site B	Spatial Changes (Differences in Temporal Distributions at different sites)
a ___ a ___	a ___ a ___	none
	a ___ a ___	duration
	a ___ a ___	duration and continuity
	a ___ a' ___	duration and attribute
	a ___ a ___	continuity
	a ___ a' ___	continuity and attribute
	a ___ a' ___	attribute
	a ___ a' ___	attribute, duration, and continuity

Figure 22.4 Types of spatial changes.

an attribute (land-use types) at different sites (site A and site B) to acquire static spatial variations of the attribute at a given time: site A is in an agricultural zone and site B is a wetland in 1992. Another example, central and western India experience heavy rainfalls in El Niño years, while northern India usually has a deficit in precipitation during these periods. Type 3 static changes describe the spatial distribution of a geographic phenomenon at a given point in time and such changes can be shown by mappings from a temporal object to a sequence of spatial objects.

Type 4 transitional changes describe variations of spatial properties for a given event or process in a time series. Such changes can be shown by linkages from a set of temporal objects to a set of spatial objects. Three basic parameters in measuring spatial changes are attribute, duration and continuity of a phenomenon, event or process. Eight subtypes of spatial changes arise from variations in attributes, duration and continuity (Fig. 22.4). For example, we can compare transitional changes of land-use types at site A with what has occurred at site B from T_1 to T_2. Or, we can compare impact of the monsoon trough on rainfall processes in El Niño years at these two sites.

Both type 3 and type 4 changes can be represented by linking a set of spatial objects to a set of temporal and semantical objects. If only two sites are considered, the spatial set will include two simple or complex spatial objects. Each of the spatial object sets will link to a set of temporal objects, respectively. Analysis is performed within each set of temporal objects to detect changes in duration or continuity. If semantical objects are included, these temporal objects can further link to correspondent semantical objects to examine changes in attributes.

22.4.2 Temporal changes

Temporal changes are changes occurring at different points or periods in time and they are recognised by changes in spatial properties and/or locations from time to

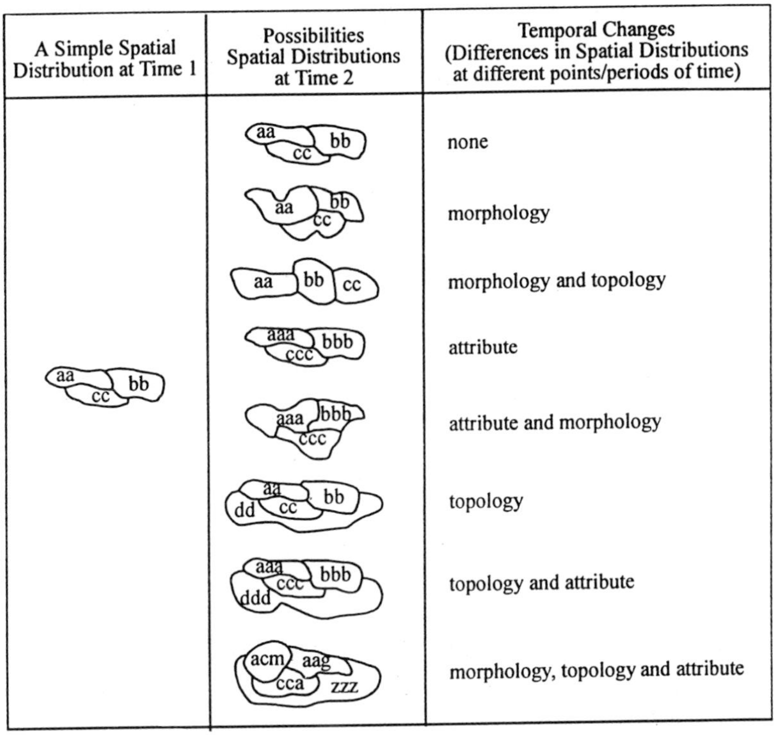

A Simple Spatial Distribution at Time 1	Possibilities Spatial Distributions at Time 2	Temporal Changes (Differences in Spatial Distributions at different points/periods of time)
	aa bb cc	none
	aa bb cc	morphology
	aa bb cc	morphology and topology
	aaa bbb ccc	attribute
	aaa bbb ccc	attribute and morphology
aa cc bb	aa dd cc bb	topology
	aaa ccc bbb ddd	topology and attribute
	acm aag cca zzz	morphology, topology and attribute

Figure 22.5 Types of temporal changes (Armstrong, 1988).

time (Fig. 22.5). Two types of temporal changes are identified in this research: mutation (type 5) or movement (type 6). Mutation refers to changes occurring to the internal mechanisms of an event or a process, or to the interactions between events/processes and their environments. Movement concerns the travel of an event or an entity from one place to another and the event or entity may or may not involve changes in spatial properties other than location. For example, the construction of canals in the Everglades National Preservation in southern Florida may have mutated mechanisms of the local hydrological processes. Also, a practice of artificial rainfall in a region may mutate the patterns of precipitation and water resources in its surrounding area. Therefore, a comparison of frequency, period and severity of an event in an area at different periods of time may suggest type 5 temporal changes in the area. Type 6 temporal changes are movement of a feature, event, or spread of a phenomenon. Examples for movement include spread of wildfire, insect infestation and animal movement.

Both type 5 and type 6 changes can be shown by linking a series of semantical and temporal objects to correspondent spatial objects. Type 5 changes describe the mutation of a type of processes or events (semantical objects) by two sets of temporal objects; each of them is linked to a set of spatial objects. Comparisons are made between the two sets of spatial and temporal objects to show how a process mutates its attributes, temporal properties and spatial characteristics in the two sets of time series. On the other hand, type 6 changes denote a single event or process and can be represented by linking a semantical object to a series of temporal objects and then to a set of spatial objects to show the movement of this event during the period constituted of these temporal objects. Eight subtypes of temporal changes result from combinations of changes in attributes, morphology and topology (Armstrong, 1988).

22.5 Conclusion and future research

This chapter has presented a new approach for modelling spatial and temporal information by objects in semantical, temporal and spatial domains. Data modelling defines the data objects, frameworks to structure these data objects and rules of integrity to ensure validity of data in the data model. This chapter outlined a conceptual data model, a three-domain model, for structuring and representing spatio-temporal information in a GIS context with an emphasis on data objects and integrity rules in semantical, temporal and spatial domains. The merits of modelling semantical, spatial and temporal concepts as independent but inter-related sets of objects in three domains are shown by: (1) the simple rules of planar enforcement and space exhaustion common to both spatial and temporal domains; (2) persistent and unique identifiers for all simple and complex objects in semantical, temporal and spatial domains, and (3) the abilities to link objects dynamically from the three domains to represent six basic types of spatio-temporal changes in geographic information. However, identification of semantical objects and numerous links among the three domains may appear challenging in building a very large spatial database.

Spatial and temporal objects are comparable in terms of their geometric properties and topological relations. Linear and planar enforcement will ensure 1D/2D topological relations at primitive object levels in both spatial and temporal domains. Space exhaustion applies to 2D space and 1D time so as to guarantee no empty space and no null time at primitive object levels. However, neither rule is applied to objects at a feature level, where overlapping and/or complex objects are allowed. Spatial and temporal trees organise and maintain parenthood relationships among complex and simple objects in spatial and temporal domains. All simple or complex objects, therefore, hold persistent identifiers in spatial or temporal domains.

Modelling semantical objects plays an important role in representing spatio-temporal information. A semantical object can be linked to a set of temporal objects and these temporal objects can further be linked to a set of spatial objects. In such a way, the system can represent an event moving or spreading across space during periods of time with varying rasters. Moreover, a semantical object can be linked to one temporal object but this temporal object in turn links to a set of spatial objects. In this case, the system represents an event occurring at multiple locations at the same time. For example, a wildfire can burn multiple places at any given instance. Therefore, central to the research are the fundamental questions of 'what is a geographic entity?' and 'how to identify semantical object?' because geographic meanings may vary in perspectives. In addition, issues of data accuracy can be discussed, separately or together, in terms of attributes in a semantical domain, time in a temporal domain, or locations in a spatial domain. Problems of different spatial and temporal scales may define or link objects in the three domains differently and may sometimes want to share views. Further research should work on defining geographic entities to construct semantical objects in a

GIS context and developing a framework to describe data accuracy with respect to semantics, time and space.

Note

1   The word 'semantics' or 'semantical' is, in the context of data modelling, the meanings of objects or concepts in the physical or abstract world. Here, it refers to aspatial and atemporal properties of geographic features. The word is chosen instead of thematical or topical because people interpret reality somewhat subjectively in ways meaningful to their cognition and perspectives.

References

ARMSTRONG, M. P. (1988). Temporality in spatial databases, *Proceedings: GIS/LIS'88*, Fall Church, Virginia: American Congress on Surveying and Mapping, **2**, 880–9.

BARRERA, R., FRANK, A. and AL-TAHA, K. (Eds), (1990). *Temporal Relations in Geographic Information Systems*, Technical Paper 91-4, Santa Barbara: NCGIA.

BERRY, B. J. L. (1964). Approaches to regional analysis: a synthesis, *Annals of the Association of American Geographers*, **54**(1), 2–11.

DATE, C. J. (1983). *An Introduction to Database Systems*, 4th Edn, vol. 2, Reading MA: Addison-Wesley.

DUTTA, S. (1991). Topological constraints: a representational framework for approximate spatial and temporal reasoning, *Proceedings: SSD'91*, pp. 161–80, Berlin: Springer-Verlag.

LANGRAN, G. (1992). *Time in Geographic Information Systems*, London: Taylor & Francis.

LANGRAN, G. and CHRISMAN, N. R. (1988). A framework for temporal geographic information, *Cartographica*, **25**(3), 1–14.

PEUQUET, D. J. (1994). It's about time: a conceptual framework for the representation of temporal dynamics in geographic information systems, *Annals of the Association of American Geographers*, **84**(3), 441–62.

PEUQUET, D. J. and DUAN, N. (1995). An event-based spatio-temporal data model (ESTDM) for temporal analysis of geographical data, *International Journal of GIS*, **9**(1), 7–24.

SINTON, D. (1978). The inherent structure of information as a constraint to analysis: mapped thematic data as a case study, in Dutton, G. (Ed.), *Harvard Papers on GIS*, vol. 7, Reading MA: Addison-Wesley.

SOWA, J. F. (1984). *Conceptual Structures: Information Processing in Mind and Machine*, Reading, MA: Addison-Wesley.

VAN ROESSEL, J. and PULLAR, D. (1993). Geographic region: a new composite GIS feature type, *Proceedings: Auto-Carto 11*, pp. 145–56, Minneapolis: ASPRS/ACSM.

WORBOYS, M. F. (1992). A model for spatio-temporal information, *Proceedings: The 5th International Symposium on Spatial Data Handling '92*, pp. 602–11, Berlin: Springer-Verlag.

YUAN, M. (1994). Wildfire conceptual modeling for building GIS space-time models, *Proceedings: GIS/LIS'94*, pp. 860–9, Phoenix, Arizona: ASPRS.

CHAPTER TWENTY-THREE

CON.G.O.O.: A Conceptual Formalism for Geographic Database Design

DIMOS PANTAZIS

23.1 Introduction

This chapter presents the basic concepts of a new formalism called CON.G.O.O. (from the French: CONception Géographique Orientée Objet) for geographic database design. This formalism has been developed in the Laboratory SUR-FACES of the University of Liège (Belgium) by the author in the context of his PhD thesis (Pantazis, 1994a and b, 1995). New research projects aiming at its further development and planned for the year 1995–6 will also be set out.

CON.G.O.O. proposes some solutions to the specific conceptual modelling problems of geographic data and is set up on the object-oriented analysis and design as presented by Coad and Yourdon, 1991a and b). In addition, various modelling techniques have been included for the final development of CON.G.O.O's first version. A specific CASE (Computer Assisted Systems/Software Engineering) tool supporting CON.G.O.O. is also in development. CON.G.O.O formalism operability and usefulness have already been tested in four real-scale projects: the Luxembourg Land Planning Ministry geographic database development, the cartographic generalisation research project of the MET (Ministry of the Equipment and Transport of Wallonie-Belgium), the geographic database design of Namibia General Survey and Mapping Administration, and the geographic database design of the Hangzhou's GIS (China) for tourism and land planning in the context of a bilateral scientific cooperation project between China and Belgium. CON.G.O.O. formalism is a part of the first Geographic Information Systems Design method named ME.CO.S.I.G. (from the French MEthode de CONception de Systèmes d'Information Géographique) (Pantazis, 1994a and b).

This chapter is organised as follows. The next section discusses the principal modelling formalisms used for 'conventional' database design, their applications in geographic database design and their deficiencies related to the conceptual modelling of geographic data. It is followed by two sections dedicated to the presentation of CON.G.O.O.'s basic concepts and their graphic representation respectively, while the final section provides some conclusions and areas for further work.

23.2 Conceptual formalisms for non-spatial and spatial databases design

The most important elements of an Information System (IS) are its databases. The development of a database usually requires three different development phases: the conceptual (design) phase, the logical phase and the physical phase. Different types of models are used in each phase. To create these models, we need different kinds of formalisms. At the design phase we use 'conceptual models'. The conceptual models are independent of the technological aspects, since they define 'what' without considering at all 'how'. Those models are constructed with the help of conceptual formalisms as Chen's entity-relationship (Chen, 1976), Remora's formalism (Rolland *et al.*, 1986), Coad and Yourdon (1991a and b) object-oriented analysis and design formalism '. . . which are formal languages made up of a restricted number of concepts with graphic notation and rules of use. . . ' (Caron and Bédard, 1993, p. 337). So we use a formalism to produce a model. Sometimes in the literature relative to databases, the term 'formalism' (as defined here) is replaced by the term 'model' and the term 'model' by the term 'schema'. With those terms we use a 'model' to produce a 'schema'. Hereafter we will only use the terms 'formalism' and 'model'; we use a formalism to produce a model.

At the logical level we transform the conceptual model into a 'logical model' (its type depends on the kind of database we want to create: hierarchical, network, relational, object-oriented); at the physical phase we implement the logical level on a specific software and hardware platform (for example, MS-DOS and DBASE IV on PCs and UNIX and ORACLE on workstations).

Over the past 15 years, the Entity–Relationship Approach (E/R) (generally modified and improved in multiple ways in order to increase its efficiency) has evolved as an almost *de facto* standard for conceptual and logical design of databases and information systems (Pernul and Tjoa, 1992). Moreover, different object-oriented (OO) conceptual modelling approaches have also been developed and are operational today (Booch, 1991; Coad and Yourdon, 1991a and b; Shlaer and Mellor, 1990).

Several formalisms combine the E/R approach with the OO approach, for example, Morejon and Oudrhiri (1994), Spaccapietra and Parent (1989). Chen (1992) notes that '. . . ER and OO can learn from each other. What needs to be done is to specify the details on how to make the ER model "active"' (p.2).

Some authors using the 'Event' concept in different ways, for example, Caron (1991), Caron and Bédard (1992), Morejon and Oudrhiri (1994), combine the processes models (as for example, the Data Flow Diagrams or Petri Networks) with different types of data models, producing integrated (data + processes) modelling solutions.

A few years ago the design of a geographic objects database did not resort to any conceptual formalism. (By geographic objects we mean the 'objects' [events, concepts, phenomena] that have a graphic representation spatially/geo-referenced.) Digitisation of paper maps (task realised at the physical level) without any significant work during the two others phases (conceptual and logical modelling) often resulted in 'problematic' and uncompleted databases.

The specific character and properties of the geographic objects, the complexity of the topological and other relationships between them, and their usually large initial number are only a few of the particular problems that have led to the use of conceptual data models.

Conceptual formalisms useful in the context of 'classic' databases (for example, E/R formalism) are used for the development of geographic databases (for example, Pantazis, 1993). However, they are inefficient. Some university-based research projects have developed specific conceptual formalisms for the modelling of geographic data (Laval University, Federal Polytechnic School of Lausanne and Claude-Bernard University of Lyon). The major part of these efforts has focused on the Entity Relationship conceptual formalism (Chen, 1976) that has been adapted and improved several times in order to take into account the specific aspects of the geographic data. Caron and others (Université Laval, Québec) have proposed some promising improvements to the E/R formalism that are particularly important, for example, the adding of specific spatial and temporal reference pictograms to deal with geographic entities and to manage their life cycle. The E/R formalism completed with these improvements and modifications has taken the name of MODUL-R formalism and represents today one of the best formalisms for geographic data modelling at the conceptual level (Caron, 1991; Caron and Bédard, 1993; Caron et al., 1993; Gagnon, 1993).

However, some deficiencies of MODUL-R and other formalisms related to the conceptual modelling of geographic data have been underlined. The main ones follow (Pantazis, 1994a and b).

1　Insufficient distinction between the different kinds of geographic objects. MODUL-R is limited to the distinction provided by the 'spatial pictograms' without specific theoretical concepts concerning the different kinds of geographic objects (Caron and Bédard, 1993; Caron et al., 1993). The other formalisms (E/R, Coad and Yourdon OO and so forth) do not provide a better theoretical framework for the geographic objects modelling.

2　Insufficient distinction between the entities (objects) and the classes of these entities, for example, the E/R formalism (Chen, 1976). However, this distinction is partially offered by different kinds of object-oriented formalisms (Coad and Yourdon, 1991a and b) but is not adapted to the specific nature of the geographic objects. Such a distinction will allow the identification and description of topological and other relationships, which are different for groups of geographic objects than from those for geographic objects.

3　Insufficient distinction among the different kinds of classes of objects. The only distinction generally offered by the object-oriented formalisms is between general and specialised classes, for example, Coad and Yourdon (1991a and b).

4　Insufficient graphic notation of the topological relationships. The only possibility proposed until now is that offered by the E/R model: representation of all the relationships by ellipses and lines. This solution is insufficient for two reasons. First, the big graphic overloading, preventing an easy and quick understanding; second, the limited capacity for representing non-obligatory topological relationships (relationships that may exist between the geographic objects).

5   Insufficient distinction between the topological and other types of relationships. After a detailed bibliographic research we have not found formalisms that distinguish topological from other relationships linking geographic objects.

6   Partial representation of topological constraints between the entities (objects). With the use of the classic formalisms, the topological constraints (topological relationships that have to exist or that must not exist) cannot be represented in their totality without graphically overloading the model.

7   Dichotomy of processes (treatments) models and data models, for example, in the E/R formalism, ERC+ formalism (Entity Relationship Complex plus generalisation) (Spaccapietra and Parent, 1989). The OO formalism generally offers integrated modelling solutions but it does not distinguish the processes (treatments) concerning the objects from those concerning the classes (= sets, groups) of objects.

8   Absence of one integrated approach allowing the use of relational or object-oriented DBMS (Database Management System). The conventional conceptual formalisms do not provide models that fit well with the object-oriented DBMS (Rolland and Brunet, 1992).

9   Absence of one formalism taking into account the two worlds physically but also conceptually different (raster/surface, vector/object).

10   Absence of specific techniques and concepts for the conceptual modelling of objects with coarse boundaries.

Figure 23.1 represents a conceptual model concerning geographic and non-geographic objects made with the E/R model with spatial pictograms, to illustrate those deficiencies.

As we can see, the only distinction between the entities (equals objects in this case) that we can do with the help of the spatial pictograms is between geographic entities (for example, forest) and non-geographic entities (for example, person). But it is obvious that the urban area is a geographic entity (composed entity) different from the house (simple entity). The E/R formalism does not provide specific concepts for the modelisation of the geographic objects.

In addition, the different kinds of relationships between the entities have no specific graphical representation. However, the relationships are not all similar, the 'whole-part' relationship (urban area [= whole]/road [= part]/parcel [= part]) is not the same as the topological relationship 'adjacent' or 'superimposition'. The E/R formalism does not provide specific concepts and graphical notations concerning the topological relationships between geographic objects. Even if it is possible to use specific verbs or adjectives (for example, adjacent or superimposed to), we miss a general agreement on the meaning of those verbs or adjectives when applied between geographic objects. For the representation of the topological relationships we use ellipses as for the other kinds of relationships. This means insufficient distinction between the topological and other types of relationships. On the other hand, the representation of the topological relationships among all the geographic objects is not only insufficient but very difficult to read. The topological constraints such as a forest or a lake could not be superimposed to an

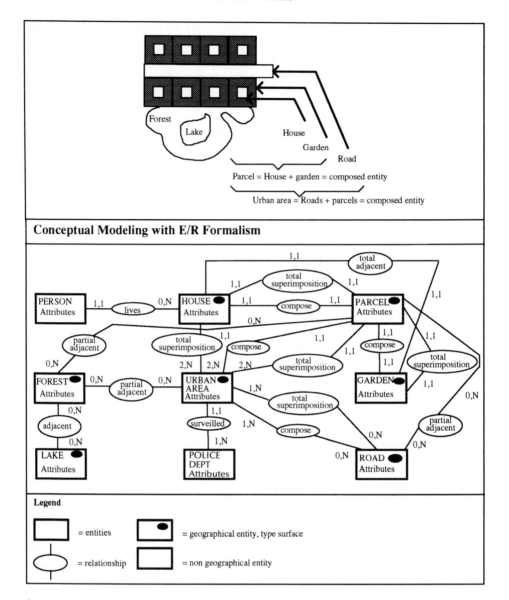

Figure 23.1   An E/R model for geographic and non-geographic entities.

urban area, and can by no means be represented. Each box represents the instances, and not the totality of each entity. The concept of class (= group, set of objects) does not exist. The treatments that could be applied to the entities do not appear. It is well known that the E/R formalism provides data models, but not treatment models. This means dichotomy between these two kinds of models. This model oriented to relational database does not warrant an easy passage to object-oriented DBMS.

Most of these deficiencies concern not only the E/R formalism but all of the existing formalisms for database design. In Section 23.3, we will introduce the new formalism CON.G.O.O. and present the solutions it provides to these deficiencies.

Because of lack of space, the answers derived from CON.G.O.O. concerning the integration of the two worlds physically but also conceptually different (raster/surface, vector/object), and the specific techniques and concepts for the conceptual modelling of objects with coarse boundaries, will not be presented.

23.3 CON.G.O.O. basic concepts and terminology

CON.G.O.O. defines and distinguishes classes (= groups of one kind of objects) and objects. Once again, the term class does not have the same meaning as in OO. languages. It must be understood as a set, a group of objects, or the totality of one similar kind of object (for example, forests). The description of an object is provided by properties usually called attributes; behavioural characteristics are usually called treatments and/or methods (= complex treatments), and structural characteristics concerning the topological and other types of relationships between the objects, classes, layers and sublayers.

We distinguish the following object categories:

- *The geographic objects (GO)*. The geographic objects are objects having a graphic representation spatially georeferenced.
- *The non-geographic objects (NGO)*. These concern all the other kinds of objects (alphanumerical, aerial photos or satellite images without geographic reference, video data and so forth).

Three categories are sufficient to cover the whole range of geographic objects at the conceptual level. They are:

1 *The simple geographic objects (SGO)*. We may consider three different sub-categories:
   (a) type point (SGO, type P)
   (b) type line (SGO, type L)
   (c) type surface (SGO, type S)
   Figures 23.2a, b and c show some examples of these kinds of objects.

2 *The composed geographic objects (CGO)*. This kind of object stems from the union or the division of simple geographic objects of the same type, belonging to the same or different classes. According to whether the SGO belong to the same class or not, we obtain HCGO (Homogeneous CGO) or ECGO (hEterogeneous CGO) respectively. Figure 23.2d, e, f, g and h show some examples of such kinds of objects.

3 *The complex geographic objects (CXGO)*. This category is made of geographic objects that stem from the union of different types (points + lines, lines + surfaces and so on) of geographic objects (simple, composed or complex or a combination of them). The same distinction as above can be made between Homogeneous CompleX Geographic Objects (HCXGO) and hEterogeneous CompleX Geographic Objects (ECXGO). Figure 23.2i and j give some examples of these kinds of objects.

This classification of the geographic objects combines in a very powerful and original manner the geometric and semantic characteristics of the geographic data and allows a clear distinction between the different kinds of geographic objects.

The semantic power of those concepts allows for the first time a precise and specific characterisation of each geographic object.

Table 23.1 summarises the different categories of geographic objects proposed by the CON.G.O.O. formalism.

Attributes. An attribute is an element of the CON.G.O.O. formalism that is associated with a particular object or a particular class of objects, for example, the type of the spatial distribution of the geographic objects that belong to the class. It describes a characteristic of the object or the class of object and allows a value to be assigned to that characteristic. Every object may have two kinds of attributes: the 'graphic' attributes (which concern its type and its graphic representation) and the 'logical' attributes (which concern its identity description). The non-geographic object and the classes have only logical attributes.

The classes are sets of objects that have common characteristics expressed by their logical attributes and/or their common name (semantic identification). The graphic characteristics and the implementation mode (point, line, surface, combination) can be different for objects belonging to the same class. Here the term class means the totality of the objects that share some common semantic characteristics independently of their geometric or graphic representation. The 'class' concept in the frame of CON.G.O.O. formalism allows the modelling of the topological and other relationships not only between single objects (instances) but also between groups of objects, and the distinction of the treatments concerning either single objects or groups of objects.

We distinguish the following categories of classes.

1   *Simple class (SC)*. Class that does not contain geographic objects (GO) having multiple types of implementation modes or graphic representations and are not specialised in other classes (Fig. 23.3a).

2   *General class (GC)/Specialised class (SC)*. The general class is a class that is specialised in other classes. The specialised class is a sub-group of objects of a general class (Fig. 23.3b). This classification stems from the generalisation/specialisation relationship application (see further for this term).

3   *Class with objects having multiple implementation modes* (objects of different types: points, lines and so on). The class that contains objects with the same semantic aspect (rivers, forests, houses, . . .) but with different implementation modes, for example, rivers as lines, rivers as surfaces (Fig. 23.3c).

4   *Class with objects having multiple graphic representation*. This class contains objects with the same semantic aspect but with different graphic aspects such as red houses and blue houses (Fig. 23.3d).

This characterisation of classes is the CON.G.O.O. answer to the insufficient distinction between the different kinds of classes of geographic objects. Separating implementation mode, graphic representation and semantic aspects of geographic objects allows CON.G.O.O. to establish a clear distinction between the different kinds of every class.

Layers are sets of classes of objects that have in common different kinds of relationships. During the construction of the conceptual model we put together the geographic objects that have common topological and/or other relationships. For example, if we know that the forests, lakes and urban areas cannot be superimposed (overlaid), we put them together in a layer called 'land cover'. The

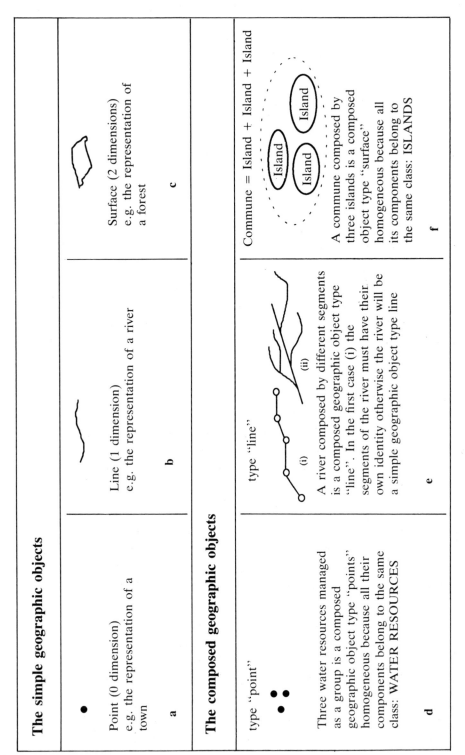

The simple geographic objects

Point (0 dimension)
e.g. the representation of a town

a

Line (1 dimension)
e.g. the representation of a river

b

Surface (2 dimensions)
e.g. the representation of a forest

c

The composed geographic objects

type "point"

Three water resources managed as a group is a composed geographic object type "points" homogeneous because all their components belong to the same class: WATER RESOURCES

d

type "line"

(i)

(ii)

A river composed by different segments is a composed geographic object type "line". In the first case (i) the segments of the river must have their own identity otherwise the river will be a simple geographic object type line

e

Commune = Island + Island + Island

A commune composed by three islands is a composed object type "surface" homogeneous because all its components belong to the same class: ISLANDS

f

Figure 23.2 Examples of CON.G.O.O.'s class categories.

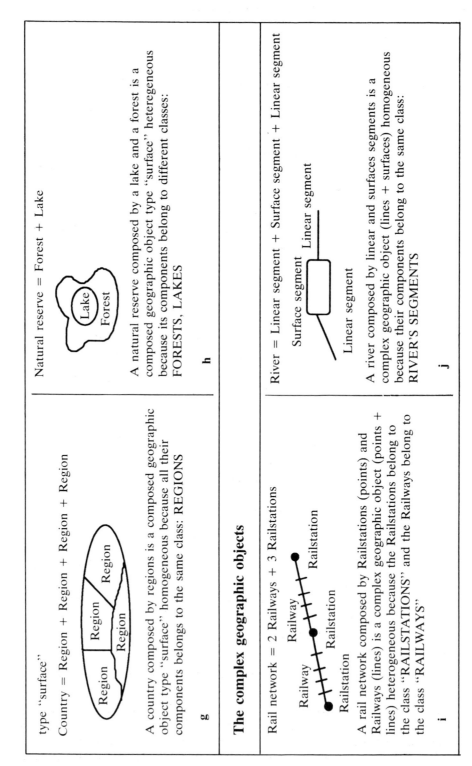

type "surface"

Country = Region + Region + Region + Region

A country composed by regions is a composed geographic object type "surface" homogeneous because all their components belongs to the same class: REGIONS

g

Natural reserve = Forest + Lake

A natural reserve composed by a lake and a forest is a composed geographic object type "surface" heteregeneous because its components belong to different classes: FORESTS, LAKES

h

The complex geographic objects

Rail network = 2 Railways + 3 Railstations

A rail network composed by Railstations (points) and Railways (lines) is a complex geographic object (points + lines) heterogeneous because the Railstations belong to the class "RAILSTATIONS" and the Railways belong to the class "RAILWAYS"

i

River = Linear segment + Surface segment + Linear segment

A river composed by linear and surfaces segments is a complex geographic object (lines + surfaces) homogeneous because their components belong to the same class: RIVER'S SEGMENTS

j

Figure 23.2 (continued)

Table 23.1   CON.G.O.O.'s geographic objects categories

Geographic objects	Examples
Simple geographic objects (SGO)	Type point Type line Type surface
Composed geographic objects (CGO)	Type point (1object) Type line (1object) Type surface (1object)
Composed geographic objects (CGO)	Homogeneous (all their component objects belong to the same class) Heterogeneous (their component objects belong to different classes)
Complex geographic objects (CXGO)	Type points + lines (1object) Type points + surfaces (1object) Type lines + surfaces (1object) Type points + lines + surfaces (1object)
Complex geographic objects (CXGO)	Homogeneous (all their component objects belong to the same class) Heterogeneous (their component objects belong to different classes)

common topological relationship (topological constraint in this case) of those objects is the interdiction of the superimposition between them (Fig. 23.4a).

Sublayers are sets of classes of objects of the same layer that have in common different kinds of relationships. If we know for example that the lakes may be partially or totally adjacent to the forests, we will create a sublayer with those two classes: lakes and forests (Fig. 23.4b).

Relationships. We distinguish three kinds: structural, topological and logical relationships.

1 *Structural relationships*. This term comes under: the *Generalisation-Specialisation* (*Gen-Spec*) and the *Whole-Part* types.

- *The gen-spec relationship*. Less formally this relationship is known as 'is a kind of' (for example, Transport-[Motor Vehicle, Aircraft]). It is not applied between objects, but between classes. The application of this relationship results in creating two categories for classes: the generalised classes and the specialised classes.

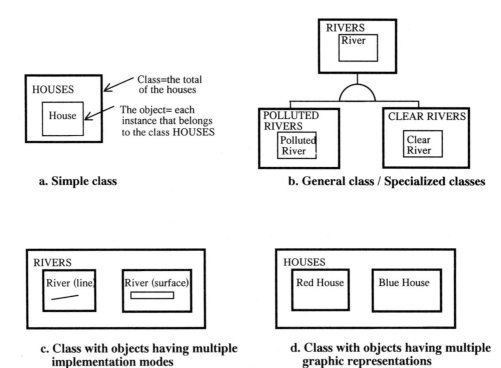

Figure 23.3 CON.G.O.O.'s classes categories.

- *The whole-part relationship.* This relationship concerns the representation of the structure of the composed or complex objects and/or
  class/sublayers/layers and applies to objects, objects and classes, objects and
  layers, objects and sublayers. As far as we know, for the first time in the
  frame of a conceptual formalism, 'whole-part' relationships are proposed
  not only between single objects, but also between objects and groups of
  objects or between groups of objects; for example, the object 'country' is
  composed of the layer 'land cover' that contains the classes of objects:
  forests, urban areas, lakes and so on.

2 *Topological relationships.* Every geographic object and every class of
  geographic object is described in space compared with other objects and/or
  other classes of objects, sublayers, layers with two topological relationships: the
  'adjacency' (A) and the 'superimposition' (S) relationships. In spite of some
  other approaches (Egenhofer, 1991; Egenhofer and Franzosa, 1995; Molenaar
  et al., 1994; Sharma; *et al.*, 1994), we believe that these two concepts cover the
  whole range of topological expressions such as: disjoints, meets, equals,
  overlaps, contains and covers. Moreover, we distinguish the 'obligatory' (must
  be and/or forbidden) and the allowed (may be; permitted) topological
  relationships and we distinguish three application levels: total (*t*), partial (*p*),
  non-existent (*ne*). Table 23.2 shows the topological relationships between
  Simple Geographic Objects (see this table in comparison to Fig. 23.5). We
  mention that in some cases we have to add the direction of the application of
  those relationships (relationship applied from which object to which object

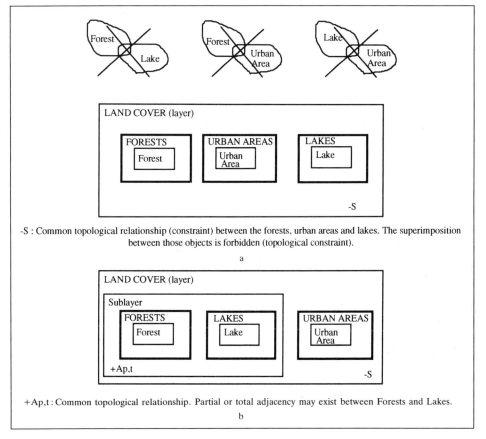

Figure 23.4   CON.G.O.O.'s layer and sublayer concepts.

because the application level may be different according to the direction, for example, surface-to-line adjacency relationship). CON.G.O.O formalism provides the possibility for a total representation of an 'integrated topology', meaning that all intersections (= total or partial superimpositions), containment (total superimposition, total adjacency) and adjacencies are represented. Representation at the conceptual level means easy identification at the logical/physical level, and also gives the possibility for real-time or *a posteriori* control of the database.

A topological relationship matrix (adjacency/superposition matrix) gives all the obliged and allowed topological relationships among all geographic objects of a database. Figure 23.6 presents a simple example of such a kind of matrix.

3   *Logical relationships.* The logical relationships concern all the relationships that are not topological or structural relationships, for example, the fact that a factory (geographic object represented by a point) discharges its waste water to a river (geographic object represented by a line) is information that will relate the two geographic objects (river and factory) but is not a topological relationship (the factory point and the river line have no common topological relationship) nor a structural relationship (as defined in this chapter).

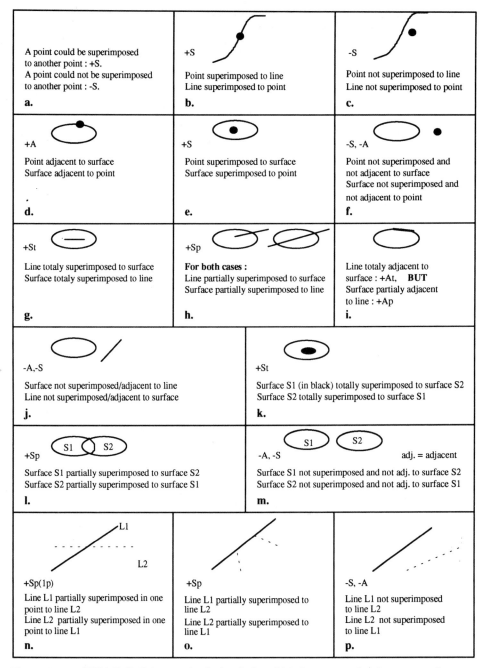

Figure 23.5   CON.G.O.O.'s topological relationships between simple geographic objects.

Each category of relationships can be applied among all kinds of objects/ classes/ sublayers/ layers but the Gen-Spec relationship can be applied only between classes.

Table 23.2 Topological relationships between SGO (symmetrical matrix)

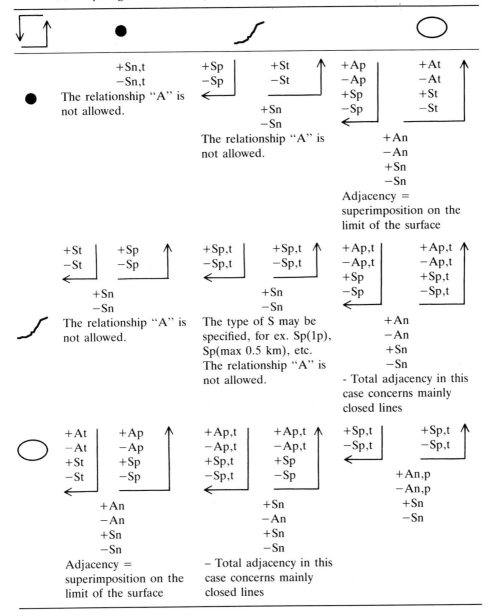

	●	𝒮	○

●	+Sn,t
	−Sn,t
	The relationship "A" is not allowed.
		+Sp +St
		−Sp −St
		←
		  +Sn
		  −Sn
		The relationship "A" is not allowed.
			+Ap +At
			−Ap −At
			+Sp +St
			−Sp −St
			←
			  +An
			  −An
			  +Sn
			  −Sn
			Adjacency = superimposition on the limit of the surface

𝒮	+St +Sp
	−St −Sp
	←
	  +Sn
	  −Sn
	The relationship "A" is not allowed.
		+Sp,t +Sp,t
		−Sp,t −Sp,t
		←
		  +Sn
		  −Sn
		The type of S may be specified, for ex. Sp(1p), Sp(max 0.5 km), etc. The relationship "A" is not allowed.
			+Ap,t +Ap,t
			−Ap,t −Ap,t
			+Sp +Sp,t
			−Sp −Sp,t
			←
			  +An
			  −An
			  +Sn
			  −Sn
			- Total adjacency in this case concerns mainly closed lines

○	+At +Ap
	−At −Ap
	+St +Sp
	−St −Sp
	←
	  +An
	  −An
	  +Sn
	  −Sn
	Adjacency = superimposition on the limit of the surface
		+Ap,t +Ap,t
		−Ap,t −Ap,t
		+Sp,t +Sp
		−Sp,t −Sp
		←
		  +Sn
		  −An
		  +Sn
		  −Sn
		- Total adjacency in this case concerns mainly closed lines
			+Sp,t +Sp,t
			−Sp,t −Sp,t
			←
			  +An,p
			  −An,p
			  +Sn
			  −Sn

The simple and complex treatments / processes (= 'methods' in 'object-oriented language') are 'integrated' in every class and every object. Consequently the 'classic' dichotomy of data/processes models is avoided (Fig. 23.7). This encapsulation packs the structure and treatments concerning the objects and classes together, and at the same time distinguishes the treatments concerning the group of objects (for example, delete all the group of the objects) and the class itself (for example, create class, delete class) from those concerning the instances of objects (for example, create object, edit object, delete object).

23.4 Graphic representation of the basic CON.G.O.O. concepts

Figure 23.8 shows the graphic representation of the basic concepts of CON.G.O.O. formalism. The symbol for *an object* is a box. The geographic dimension/reference of the object is expressed by one or more points, lines, surfaces, or a combination of these elements (composed objects, complex objects). *The object geographic reference* (its type) is indicated with the help of special pictograms (a point, a line, a surface or a combination) or a textual description, for example, 'Point', or 'SGO tp = Simple Geographic Object type point'. The pictogram or the textual description is put in the upper left corner of the box.

A *class* is represented by a box that surrounds the object. Should only one object exist in the class, the four corners of the two boxes are linked together by four lines.

The graphic representations of *Gen-Spec, whole-part and logical relationships* follow the rules proposed by Coad and Yourdon (1991a and b).

The obligatory topological relationships are shown by a line cutting a box in its middle. Inside the box we notify the letter that indicates the obligatory relationship. For example 'S' for 'superimposition' or 'A' for 'adjacency'. The 'allowed' and 'obligatory' topological relationships are transformed in 'integrity constraints' for every class/sublayer/layer. These constraints constitute spatial properties of the objects and describe their spatial behaviour with regard to the other objects and classes of objects. The allowed topological constraints are designed with a (+) in the lower left corner of the box of the class/sublayer/layer and the obligatory topological constraints are notified with a (−) or a (+) in the lower right corner of the class/sublayer/layer. Letters are used to distinguish the three application levels of these constraints: total (*t*), partial (*p*), not existent (*ne*). This original manner to represent the topological relationships makes it possible to create models that reflect the topology of any geographic database without reduction of the readability of the model or overloading.

> *In the case of semantic conflicts between the different types of topological relationships, the applied topological relationship will be the one shown by a line cutting a box.*

Figure 23.9 presents a CON.G.O.O. application example. The model concerns the same geographic and non-geographic objects as in Fig. 23.1. This CON.G.O.O. model is much easier to read and understand. As we can observe, all the topological relationships and constraints between all the geographic objects are represented with only two concepts: adjacency and superimposition, in a very easy manner. Each geographic object has its own identity: simple, composite and so forth. This identification will help to avoid double work during the digitisation phase, for example, the parcel (composed geographic object) will not be digitised, because it is composed of gardens and houses (simple geographic objects). The classes of objects (together with their names and their identifiers = single alphanumerical element used to identify each class and each object) are noted with the topological relationships of the geographic objects that belong to them. The distinction between objects and classes will also help to avoid problems with respect to confusion between names and identifiers of objects and classes. The different kinds of relationships (structural, logical, topological) are each drawn in a different manner so it is easy to distinguish between them. This is very important for the realisation of the geographic database and its *a posteriori* or real-time

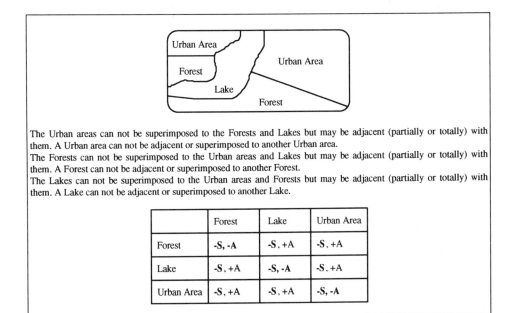

The Urban areas can not be superimposed to the Forests and Lakes but may be adjacent (partially or totally) with them. A Urban area can not be adjacent or superimposed to another Urban area.

The Forests can not be superimposed to the Urban areas and Lakes but may be adjacent (partially or totally) with them. A Forest can not be adjacent or superimposed to another Forest.

The Lakes can not be superimposed to the Urban areas and Forests but may be adjacent (partially or totally) with them. A Lake can not be adjacent or superimposed to another Lake.

	Forest	Lake	Urban Area
Forest	-S, -A	-S, +A	-S, +A
Lake	-S, +A	-S, -A	-S, +A
Urban Area	-S, +A	-S, +A	-S, -A

Figure 23.6 Topological relationships matrix. The obligatories topological relationships are notified by bold characters and the allowed by normal.

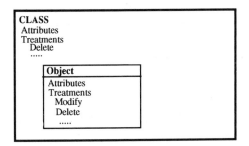

Figure 23.7 CON.G.O.O.'s integrated data/treatments modelling.

control. The treatments are described in each object and each class. The classic dichotomy between treatments and data models is avoided, and a clear distinction is made between the treatments and attributes concerning groups of objects and those concerning aspects and instances of objects.

23.5 Conclusions – perspectives

The basic concepts of a new object-oriented formalism has been presented for the conceptual design of geographic databases.

The new CON.G.O.O. formalism combines the advantages of the conventional semantic models with some of the basic concepts of the object-oriented approach

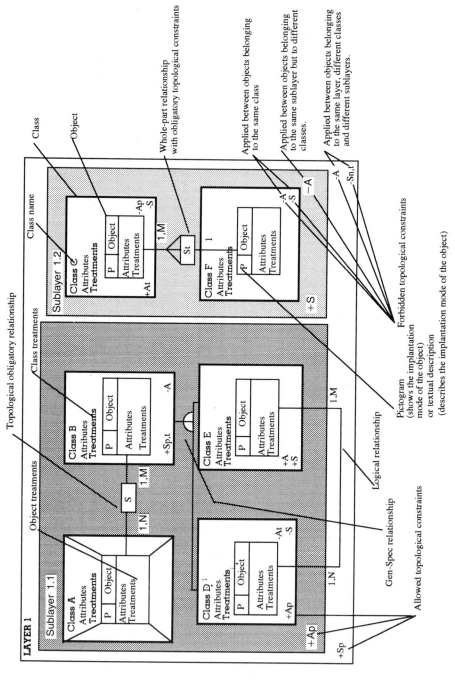

Figure 23.8   Graphic representation of CON.G.O.O.'s basic concepts.

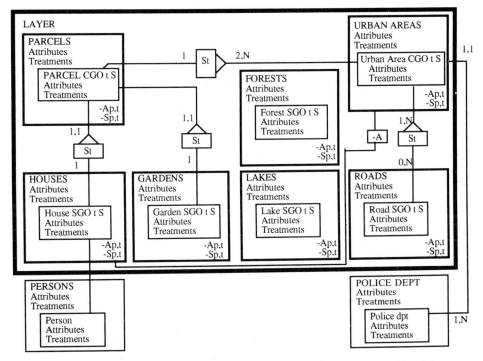

Figure 23.9 A CON.G.O.O.'s model for geographic and non-geographic objects.

but most of its central ideas are totally new. These ideas give solutions to the deficiencies of existing conceptual formalisms. The most important are:

- the geographic objects classification in different types
- the classes classification in different types
- the distinction of the different types of relationships between the objects
- the identification and representation of all the topological relationships between all the geographic objects of a database with only two concepts: adjacency, superimposition
- the establishment of relationships not only between the objects but also between objects and classes, objects and layers; and
- the classification of attributes and treatments in object attributes and object treatments, and classes attributes and classes treatments, and encapsulation, respectively, in the objects and classes.

The CON.G.O.O. formalism provides models that facilitate the communication among users, programmers and analysts of geographic databases because of their capacity to describe easily and clearly very complex environments (as is the case in geographic databases). The communicability and readability of such models (tested in the frame of real GIS projects), are undoubtedly improved by comparison to the conventional conceptual models. The integration of data, treatments/processes and other basic object-oriented concepts (for example, inheritance of the treatments/attributes through the structure of generalisation/specialisation relationships) that could be implemented to relational or object-oriented DBMS avoids the classic dichotomy of data and process/treatments models and fills the gap between object-oriented conceptual modelling and object-oriented databases development. The automatic conversion to the other databases development phases will also benefit from this work. This conversion is offered today for conventional databases by CASE-tools.

Some important characteristics of the CON.G.O.O formalism, which have not been presented here, will nevertheless be taken into consideration. They are principally the layer-to-layer relationships, the complex relationships (for example, topological and structural relationships applied at the same time on the same objects/classes/layers), the complex geographic objects topological relationships, the inheritance of graphical, logical attributes and treatments, the integrity controls of the CON.G.O.O.'s models, CON.G.O.O.'s constraints and rules.

Further research programmes are already planned to extend CON.G.O.O.'s development. This research concerns the following issues: additional conceptual development concerning the modelling of objects with indeterminate boundaries, specific functions in CON.G.O.O. supporting CASE tool for the automatic modelling of such objects, automatic transition from the conceptual level to the logical/physical level, time and moving objects modelling.

CON.G.O.O. concepts will also be used for the development of a GIS-user interface for spatial queries. Given the fact that database retrieval (a function among the most important ones provided by GIS-software) is requested by a query formulated in a language generally quite difficult for the end-user to learn, we do believe that the CON.G.O.O. concepts are a good base for the creation of a graphic interface spatial query language. The basic idea is that a CON.G.O.O. conceptual model (= schema diagram) will graphically represent the physical structure of the geographic database. The user could formulate database queries and updates graphically, by manipulating the model. Therefore, the proposed work will extend the graphical interfaces based on database models for queries and updates of conventional databases such as those created by Auddino *et al.* (1992) and develop them using the concepts discussed in this chapter.

References

AUDDINO, A., AMIEL, E., DENNEBOUY, Y., DUPONT, Y., FONTANA, E., SPACCAPIETRA, S. and TARI, Z. (1992). *Database Visual Environments based on Advanced Data Models*, Technical Report No. 92/13, Lausanne, Suisse: Laboratoire de bases de données, Ecole Polytechnique Fédérale.

BOOCH, G. (1991). *Object-Oriented Design with Applications*, Redwood City, CA: Benjamin/Cummings.

CARON, C. (1991). 'Nouveau formalisme de modélisation conceptuelle adapté aux SIRS', Unpublished MSc dissertation, Québec: Centre de Recherche en Géomatique Faculté de Foresterie et de Géomatique, Université Laval.

CARON, C. and BÉDARD, Y. (1992). MODUL-R : Un nouveau formalisme permettant de mieux décrire le contenu d'une base de données à référence spatiale, *Conférence Canadienne sur les SIG-92*, pp. 329–41.

CARON, C. and BÉDARD, Y. (1993). Extending the individual formalism for a more complete modeling of urban-spatially referenced data, *Computer, Environment and Urban Systems*, **17**(4), 337–46.

CARON, C., BÉDARD, Y. and GAGNON, P. (1993). MODUL-R : un formalisme individuel adapté pour les SIRS, *Revue de géomatique*, **3**(3), 283–306.

CHEN, P. (1976). The entity-relationship model: toward a unified view of data, *ACM Transactions on Database Systems*, **1**, NB 1, 9–36.

CHEN, P. (1992). ER vs. OO, Entity-relationship approach E/R '92, *Proceedings of 11th International Conference on the Entity-Relationship Approach*, pp. 1–2, Berlin: Springer-Verlag.

COAD, P. and YOURDON, E. (1991a). *Object-Oriented Design*, Englewood Cliffs: Prentice Hall.

COAD, P. and YOURDON, E. (1991b). *Object-Oriented Analysis*, 2nd Edn, Englewood Cliffs: Prentice Hall.

EGENHOFER, M. (1991). Reasoning about binary topological relations, *Proceedings of Advances in Spatial Databases, 2nd Symposium, SSD'91*, pp. 143–60, Berlin: Springer-Verlag.

EGENHOFER, M. and FRANZOSA, R. (1995). On the equivalence of topological relations, *International Journal of Geographical Information Systems*, **9**(2), 133–52.

GAGNON, D. P. (1993). *MODUL-R Version 2.0, Internal Report*, Québec: Centre de Recherche en Géomatique, Faculté de Foresterie et Géomatique, Université Laval.

MOLENAAR, M., KUFONIYI, O. and BOULOUCOS, T. (1994). Modeling topologic relationships in vector maps, in Waugh, T. C. and Healey, R. G. (Eds), *Advances in GIS Research*, **1**, pp. 112–25, London: Taylor & Francis.

MOREJON, J. and OUDRHIRI, R. (1991). Le modèle EA2: Entité-Association, in Morejon J. (Ed.), *MERISE vers une modélisation orientée objet*, pp. 234–44, Paris: les éditions d'organisation.

PANTAZIS, D. (1993). Conception d'un systéme d'information spatiale pour le contrôle de la qualité des eaux fluviales. Application à un laboratoire cantonal, *Bulletin de la Société Belge de Photogrammétrie de Télédétection et de Cartographie*, mars–juin N° 189–90, 41–55.

PANTAZIS, D. (1994a). La méthode de conception de S.I.G. ME.CO.S.I.G. et le formalisme CON.G.O.O. (CONception Géographique Orientée Objet), in Harts, J., Ottens, H. and Scholten, H. (Eds), *Proceedings in EGIS/MARI '94*, pp. 1305–14, Utrecht: EGIS Fondation.

PANTAZIS, D. (1994b). 'Analyse méthodologique de phases de conception et de développement d'un système d'information géographique', Thèse de doctorat non publiée, Liège: Université de Liège.

PANTAZIS, D. (1995). CON.G.O.O. formalism basic concepts and three conversion examples, *Proceedings of Joint European Conference and Exhibition on Geographical Information (JEC)*, pp. 223–4, Basel: AKM.

PERNUL, G. and TJOA, A. M. (Eds) (1992). Foreword, Entity-relationship approach E/R '92, *Proceedings of 11th International conference on the Entity-Relationship Approach*, pp. v–vi, Berlin: Springer-Verlag.

ROLLAND, C. and BRUNET, J. (1992). Object database design, paper presented at the European Joint Conference on Engineering Systems design and analysis (ESDA), Istanbul.

ROLLAND, C., FOUCAULT, O. and BENCI, G. (1986). *Conception des systèmes d'information: la méthode REMORA*, Paris: Eyrolles.

SHARMA, J., FLEWELLING, D. and EGENHOFER, M. (1994). A qualitative spatial reasoner, in Waugh, T. C. and Healey, R. G. (Eds), *Proceedings of the Sixth International Symposium on Spatial Data Handling*, vol. 2, pp. 665–81, London: Taylor & Francis.

SHLAER, S. and MELLOR, S.-J. (1990). Object lifecycles: Modeling the world in states, in Morejon, J. (Ed.) (1994) *MERISE vers une modéleisation orientée objet*, pp. 202–17, Paris: les éditions d'organisation.

SPACCAPIETRA, S. and PARENT, C. (1989). About entities, complex objects and object oriented datamodels, internal report, Databases Laboratory, Polytechnic School of Lausanne, Switzerland.

CHAPTER TWENTY-FOUR

Agent-based Pattern Identification in Imagery: Exploring Spatial Data with Artificial Life

MARK V. FINCO

24.1 Introduction

Patterns found in maps and images often give geographers insight about processes which are acting to create the physical and social landscape. It has been suggested that the primary purpose of geographic observation is to record the traits and location of phenomena with accuracy which is sufficient for pattern visualisation and analysis (MacEachren, 1992). What is a pattern seems almost intuitive to human beings; however, the literature very rarely describes what pattern is, or how it is used in spatial analysis. Indeed, Tobler (1979) suggested that spatial analysis is best done by humans and that spatial data should be presented as completely as possible for human interpretation. Perhaps the superior ability of humans to identify and interpret patterns is a reason why there is no comprehensive pattern definition or theory.

Getis and Boots (1978) attempted to develop a spatial pattern theory using spatial process and geometric primitives (points, lines and areas) as an organisational framework. This work and numerous research papers (Aspinall, 1992; Culling, 1985; de Boer et al., 1993; Openshaw, 1994; Schalkoff, 1992) suggest that the definition of 'pattern' is relative to (1) the objectives of the investigation, (2) the phenomena being studied, (3) the scale of the study and (4) the data structure used to represent the phenomena. For example, shopping malls can be represented as areas to evaluate store access, or as points to compare their locations with regional demographics. Similarly, rivers can be represented by lines to evaluate stream order, but are better represented as areas or surfaces if stream-bed morphology is of interest. It is this multiplicity of definitions which makes development of a single definition and method for analysing pattern difficult. Regardless of the application, however, pattern always has spatial heterogeneities with a degree of regularity (Levin and Sengel, 1985) and features are arranged in space where similar features are proximal. This definition, however vague, provides a common starting point into the investigation of how to find pattern and why pattern is meaningful. While giving the problem structure, it gives the investigator flexibility to appropriately define both 'space' and 'similar'.

Patterns are important because the non-random arrangement of features suggests that processes are acting somehow to organise space. Patterns suggest more than mere spatial coincidence and are evidence of spatial process(es) (Bunge, 1966). Scientific inquiry using spatial pattern and morphology has been referred to as spatial logic by Dobson (1992). Spatial logic considers spatial pattern as evidence of interaction among many processes, and the landscape is the integration of these processes over both time and space. For example, the distribution of volcanoes in the Hawaiian islands is spatial evidence that a spatial process (a linear thinning or fissure in the earth's crust) is acting to produce islands distributed along a line. Likewise, the first hypothesis of continental drift came from observations that the continental edges of South America and Africa fit well together (Dobson, 1992). Both of these examples use patterns to develop a hypothesis, even though the mechanisms responsible for patterns may not be immediately evident.

Ever since the time of Plato and Aristotle two major scientific research approaches have existed: deductive and inductive. Every deductive investigation has been biased by previous experience. Data obtained for inductive study frequently contains as much information about the people and technology used to gather the data, as about the phenomena being studied.

Deductive geographic investigation has been the primary approach in the past and has focused on confirmatory spatial data analysis (Openshaw, 1991) modelling expected relationships between understood phenomena. The predominance of confirmatory analysis makes sense in a world with limited geographic data and untested theories.

This chapter, however, is interested in geographic investigation where the use of pattern and spatial logic is decidedly inductive. Digital geographic data are becoming increasingly available.[1] This availability together with electronic data highways is creating a data-rich environment which will require new methods of looking for pattern. The new style of analysis will need to be data-driven and exploratory, looking to the data for explanation while minimising the influence of pre-existing theories or biases (Openshaw, 1994).

24.2 A model of geographic investigation

Geographic investigation is proposed to be a three-step process (Fig. 24.1). The serial progression from raw data to explanation is pure induction, whereas feedback in the process represents the introduction of deductive elements in the investigation. The steps in geographic investigations are (a) spatial re-mapping, (b) pattern recognition and (c) theory and model building.

24.2.1 Spatial re-mapping

Goodchild (1992) notes that re-working spatial data may be required before patterns diagnostic of the underlying laws become clear. Re-working geographic data is often a simple re-mapping based on a different distance metric. The re-mapping process is very familiar to geographers. For generations geographers have mapped and re-mapped the spherical earth on to flat projections. It is well

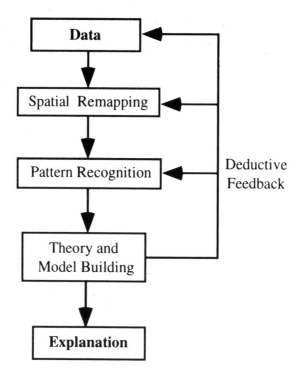

Figure 24.1 Diagram of the geographic investigation process.

known that each sphere-to-paper projection has properties relating to distance, shape and area and that these differences can be used to highlight or distort different properties of spatial data sets. In terms of the earlier definition of pattern, re-mapping defines how 'space' is measured. The following examples show that this definition of space is crucial to revealing patterns. Re-mapping can either draw out pattern where there apparently is none, or can eliminate a pattern which is an artefact of the data-collection method. In either case, it is a critical step in the geographic investigation process.

Consider the case of points distributed along a ray, each point representing 'home' and various 'cities'. If Euclidean distance is used the cities appear evenly distributed and without pattern (Fig. 24.2a). However, if the cities are connected by an unseen road network and travel time is used as the distance metric, the points might be rearranged. Cities which are connected by highways could become closer in travel time than cities which in a Euclidean sense are closer, but are connected by dirt roads. With this re-mapping, a pattern in the cities' locations is evident (Fig. 24.2b).

It is also possible for a pattern to appear when there is actually none. Data collection via remote sensing (for example, aerial photography) routinely represents the irregular terrain of the earth's surface as a flat image. This re-mapping can produce patterns. Consider an evenly distributed forest in rough terrain (Fig. 24.3). If viewed from above, segments A and C appear to be low-density forest and segment B appears to be high-density forest. If surveyed from the ground, however, the surface distance between the trees is equal for all three segments.

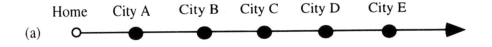

Figure 24.2 Re-mapping of cities along a ray.

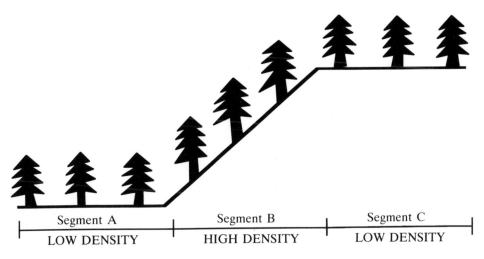

Figure 24.3 Re-mapping terrain in remote sensing.

Because of terrain differences, the view from above showed a variable density forest in an evenly distributed stand of trees.

24.2.2 Pattern recognition

Once re-mapped, pattern recognition is the next step in a geographic investigation. The previous definition of a pattern stated that they are arrangements of features in space where similar features are grouped. Patterns are the spatial equivalent of taxonomic classification. Where taxonomies are typically based on attributes alone, patterns have defining attributes and spatial contextual characteristics. The value of classification has long been recognised by the scientific community as central to scientific understanding. John Wesley Powell, a pre-eminent explorer and scientist of the nineteenth century, once stated: 'every stage in the progress of knowledge is marked by a stage in the progress of classification' (Rabbitt, 1969, p. 7). Given the close relationship between classification and pattern, it is not surprising that patterns are central to geographic investigation. Unfortunately, pattern recognition is not always as easy as the 'cities' example shown in Fig. 24.1. As Tobler (1979) eluded, humans identify patterns in spatial data very well. The human mind works well with multiple attributes, spatial context and variable scale. Value variations in black-and-white images are easily identified by humans and assignment of one of the three primary colours (red, green and blue) to attributes generally makes pattern identification in three or fewer attribute dimensions possible by humans. Data sets with more than three attributes, however, pose problems for humans.

Pattern recognition in N-dimensional spatial data generally requires computer assistance. While re-mapping defines how space is measured, definition and use of the spatial concepts 'near', 'far' and 'similar' needs to be resolved by the pattern recogniser. Current approaches to the pattern-recognition problem are incomplete. Patterns have both attribute and spatial contextual characteristics which are infrequently used together. The literature shows that remote sensing techniques often neglect the spatial context of the data, while pattern-recognition algorithms rarely use more than one attribute dimension.

24.2.3 Theory and model building

Once pattern has been identified, theory and model building is the final step in a geographic investigation. Explanation requires knowledge of the subject in conjunction with the pattern to suggest spatial process models. Process understanding is often derived by finding correlates with the pattern (that is, other variables which coincide with the heterogeneities in the space) that may explain their existence (Goodchild, 1992). Results from the process models are, in turn, validated by comparison to empirical reality (for example, patterns found on the landscape).

24.3 Related research

24.3.1 Pattern analysis techniques

Many methods have been developed to recognise or analyse pattern in spatial data sets. Developments in this field have been contributed from many speciality areas in geography, most notably spatial analysis and remote sensing.

Spatial autocorrelation is a mathematical construct which describes the degree to which close objects are similar or dissimilar. Indeed, geographers would have a difficult time imagining a world without spatial autocorrelation, for without spatial autocorrelation every place would be independent and words like 'region' and 'area' would have no meaning.

Spatial autocorrelation is an extension of time series data analysis techniques (Goodchild, 1986). Spatial autocorrelation works with two distinctly different types of data: attribute data and locational data and the general form for spatial autocorrelation measures is

$$\sum_i \sum_j c_{ij} \, w_{ij} \tag{24.1}$$

where c_{ij} is a measure of similarity between object i and object j and w_{ij} is a closeness measure. Spatial autocorrelation measures are used in geographic analysis to compare spatial data sets and to refine models. Spatial autocorrelation is not used to delineate patterns, but only to suggest that pattern is present. In addition, these measures typically use only one attribute dimension.

Remote sensing's contribution to pattern recognition comes in the form of classification techniques. Unsupervised clustering is one of the most common image-classification techniques and in intent comes close to achieving the unbiased pattern recognition required for inductive geographic investigation. Unsupervised clustering uses natural groupings of raster cells in spectral space to determine the location and membership of each cluster. It is an *aspatial*, exploratory data-analysis technique. Many algorithms have been developed to build the clusters (Jensen, 1986; Richards, 1986). Most are iterative and require the analyst to parameterise the analysis based on the number of clusters desired, the minimum distance between clusters and a convergence threshold. The classes which result from unsupervised clustering are spectral classes. Because the clustering techniques are aspatial, any pattern in the classified image is the result of the spatial earth-system processes which acted to create the pattern. If the original image was spatially randomised (that is, each raster cell retains its spectral properties, but not its location) the results from unsupervised clustering would be exactly the same.

Contextual and image segmentation classifications use both spectral and spatial attributes to find pattern in spatial data sets. The techniques used to incorporate spatial information have included the parameterised growth of regions within an image (Woodcock and Harward, 1992), knowledge-based systems which are based on a set of heuristics (Johnsson, 1994) and the use of non-spectral data to segment and enhance the classification accuracy (Lozano-Garcia and Hoffer, 1993). The pattern-recognition technique described in the following sections uses the spatial nature of remotely sensed data together with a better understanding of spatial processes to classify imagery.

24.3.2 The nature of spatial processes

When considering the processes which create patterns, an important taxonomic difference between patterns is whether they result from a centrally controlled process (that is, governed by a single set of rules or equations), or emerge from many, interacting decentralised processes.

Centralised process models are very common in scientific literature and, more importantly, are commonly used by scientists as a problem-solving approach. Centralised models either explicitly or implicitly assume some type of central controlling mechanism or physical law. For example, consider the advance of a landslide and its characteristic tapered-tear drop shape. Changes in the shape of a landslide can be modelled by a system of differential equations which use mass, energy and momentum balances to estimate the progression of the leading edge of the landslide. The phenomenon is then modelled as a single entity with the boundary defined as the edge of the landslide at any particular point in time. Mass and energy do not move across the boundary because changes in these parameters define the landslide boundary. In fact, many of the centralised models have their origins in cause-and-effect models developed in physics (Resnick, 1994). As have most scientists, geographers have tended to model spatial phenomena, whether they be physical or social, in a similar manner.

A new 'decentralised' modelling paradigm has recently been recognised by the scientific community. Ideas for decentralised models come from theoretical biology and not from classical physics (Bonabeau and Theraulaz, 1994) and many of the decentralised concepts are germane to geographic studies. Decentralised models create patterns through local interactions between decentralised and autonomous components. In the landslide example each boulder would be modelled as a component of the complex system 'landslide'. Each boulder would have mass, potential energy (governed by its height on the surface), kinetic energy (based on its speed and direction) and inertial factor(s). The landslide would be triggered by converting potential energy to kinetic energy and propagated by a similar mechanism. Likewise, the proliferation of a plant species in an area might be modelled using a seed-by-seed distributed model with factors like seed-distribution pattern (based on prevailing winds and/or animal habits), germination success rate, nutrient and water availability and so on. Decentralised models are often less mathematically sophisticated and more computationally intensive than their centralised counterparts. In essence, the mathematical complexity and constraining assumptions which were required for an analytic solution can be traded for computational power, simpler models and fewer constraints. Heavy computational requirements have restricted decentralised model development and encouraged the predominant use of centralised models. Recent advances in computing, specifically the increasing availability of massively parallel processors, have eliminated the computational bottleneck and encouraged the development of algorithms, like decentralised models, which are parallel by their nature.

Like spatial processes, pattern-analysis techniques can also be considered as either centralised or decentralised. Centralised techniques look at patterns from the top–down, typically using group or population statistics in the pattern recognition (for example, unsupervised clustering) and are most prevalent in the literature. Recognising that many spatial processes are decentralised, however, has encouraged development of a decentralised pattern-recognition method based on the following modelling techniques.

24.3.3 Decentralised process models

24.3.3.1 *Cellular Automata Models (CA)*

Cellular automata are dynamic models which operate in raster data spaces. States within the raster space can be defined as binary (for example, alive/dead), or as discrete, multilevel (for example, 10 levels of population density). State transitions are governed by local, uniform transition rules. 'Local' refers to the fact that the state transition of cells is based on the current state of each cell and its immediate neighbours. 'Uniform' indicates that the transition rules are consistent for the entire raster space. State transitions can also be a function of environmental factors in the neighbourhood. Cellular automata are dynamic systems whose behaviour is completely specified by local relationships (Toffoli and Margolus, 1987).

The best-known example of a cellular automation model is Conway's game of 'Life' (Couclelis, 1985; Gardner, 1970). 'Life', as defined in Conway's game, takes place in a binary raster space seeded with ones ('alive' cells) in a field of zeros ('dead' cells). Evolution in the model is defined by three simple transition rules: (1) alive cells die if they have less than two or greater than three alive neighbours; (2) dead cells become alive if they have exactly three neighbours, and (3) cells stay in the same state if neither (1) or (2) are true. Much of 'Life's' notoriety is due to the range of complex patterns which are generated using these very simple transition rules (Couclelis, 1985). The patterns also have complex evolutionary behaviours which are functions of the initial conditions (Eigen and Winkler, 1981). For example, patterns may occur in oscillating sequences, or may die out completely depending on the initial configuration of 1s and 0s.

Most cellular automata research has been performed in computer science and physical science. Many physical science applications are presented in Toffoli and Margolus (1987). Among the examples are: diffusion and equilibrium models (for example, self-diffusion and mean free path) and fluid dynamics models (for example, sound waves and flow around an obstacle). Only a limited number of geographic applications are in the literature. Tobler (1979) first discussed cellular models in geography as a modelling approach which could implicitly account for spatial interaction. Couclelis (1985, 1988) revisited cellular automata as a method to model human systems. White and Engelen (1994) used an urban growth and land use CA model to explore the general principles of 'urban spatial organisation'. A regional CA model for a small Caribbean island was then built as a pragmatic extension of the urban growth model.

24.3.3.2 *Artificial Life (AL) Models*

Like CA models, AL models seek to understand complex systems. The AL approach, however, does so by building systems of autonomous adaptive life forms (hereafter called creatures). Artificial life research originated in CA research (Langton, 1986), but has since evolved significantly. The initial motivation for AL research was to see if the emergent behaviour observed in CA models could explain lifelike behaviours. Artificial life models differ from CA models in that AL models have separate rule sets for each entity, whereas CA models have uniform rule sets. They also differ in that the artificial creatures can adapt to their environment as time passes (non-stationary). For example, the pseudo-code logic for a CA generally is:

```
For  every time step
        For every cell
                Make appropriate state transition
        End
  End,
```

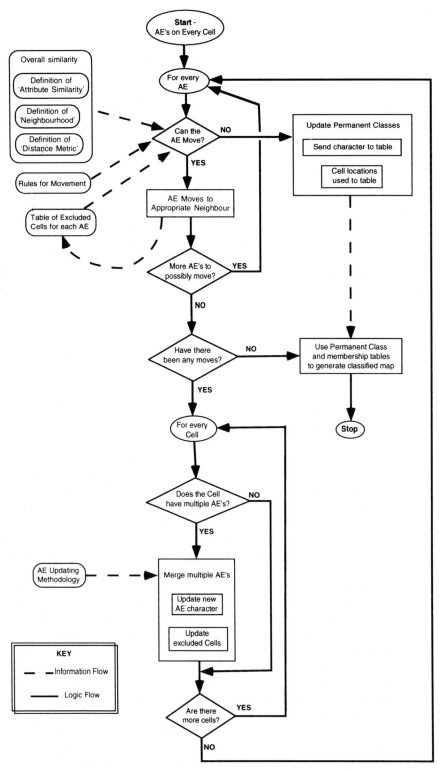

Figure 24.4   Logic–flow diagram for the artificial explorer (AE).

whereas, the pseudo-code for an AL model is,

> For every time step
>> For every *creature*
>>> Take appropriate action (for example die, procreate, eat, learn, adapt, mutate, move and so on)
>> End
> End.

At first it may seem that AL research has nothing to do with geographic analysis; however, Openshaw (1994) demonstrated the use of AL creatures as 'intelligent life-form(s) that can explore the GIS database looking at evidence of pattern in any or all of the three data spaces (attribute, space and time). It can be driven by a genetic optimisation procedure . . . feed on pattern and reproduce only if they are successful in the data environment in which they live' (p. 95). This article appears to be the only example of AL in either the geographic or pattern analysis literature.

24.4 Artificial explorer design

The design of a pattern-recognition technique using a decentralised analysis technique is an extension of Openshaw's (1994) work with artificial space-time-attribute creatures (STAC). Cellular automata (CA) and artificial life (AL) are primarily known as modelling methods, but on rare occasions (Openshaw, 1994) have been used for analysis.

Mapping by humans is used as the conceptual model to develop an artificial explorer (AE). If a person were asked to walk around a site and map important patterns, how would they do it (assume they move between the cells of a raster database and can navigate easily)? Analogous to the definition of pattern, the AE quantitatively understands how to measure geographic space (what is 'close') and compare attribute characteristics (what is 'similar') to find pattern in a data set. The AE(s) move around the data set based on the similarity of the neighbouring cells, gathering information about the locations which they have visited. Using this approach, every cell in the data set needs to be visited for the analysis to be complete. Initially there is an AE on each cell. As each AE moves, it updates the characteristics of the class of its initial cell. When two or more AEs move to the same cell they share their knowledge about the class which they are mapping using Bayesian updating, merge, and move on as one AE. The logic–flow diagram for the AE is given in Fig. 24.4.

The first loop of the logic (that is, 'For each AE') uses *a priori* definitions of 'attribute similarity', 'neighbourhood' and 'distance metric' to assess whether there are any similar neighbours. A table of previously visited cells accompanies each AE and is used to ensure those cells are not revisited. When an AE cannot move, the characteristics of that AE are stored in a table of permanent classes and the AE is removed from the landscape. If no AE moves then the classification is complete and a map is made from the table of permanent classes.

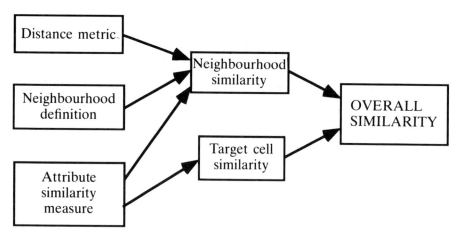

Figure 24.5 Diagram showing the relationships between distance metric, neighbourhood, similarity measure and overall similarity.

The definitions of 'distance metric' and 'attribute similarity' are central to what a pattern is and how the AE moves. The *distance metric* is defined by the re-mapping exercise and is necessary to compare adjacent and diagonal neighbours. *Attribute similarity* defines how closely matched two locations are based only on an attribute space (for example, spectral reflectance) comparison. *Overall similarity* uses the two previous definitions and a definition of *neighbourhood*, to compare the neighbouring locations in a spatial context (Figure 24.5). Overall similarity will allow a trade-off between attribute similarity and distance. For example, if two neighbouring cells are exactly the same (in attribute sense), but one is an rooks-case neighbour and one is a diagonal neighbour, then the rooks-case neighbour might be judged more similar in an overall sense because of its proximity.

In addition to these fundamental definitions, the AE must also know the *rules for movement*, and how to *update information* when it moves or merges. Movement to locations where the AE has already been is not allowed. This eliminates the possibility of an AE bouncing between two locations and will ensure that the AE gains as much information about the class it is mapping as possible. If a neighbour or neighbour(s) are within a certain tolerance of overall similarity, then the neighbour(s) are of like class. If the neighbours are outside the tolerance, they are of a different class. The most similar neighbour, which has not already been visited, is moved to and counted as a member of the initial cell's class.

There are two distinct pieces of information to be updated by the AE. The first is attribute updating. This is information about the attributes of all the cells visited to create that AE. When two AEs are merged their information is weighted based on the number of cells used to create the AE. The second piece of information is the excluded cell list for each individual AE. Updating is performed by a simple union of the two lists of excluded cells.

The AE approach is similar to the artificial STAC introduced by Openshaw (1994), but with two very important exceptions. Openshaw's STAC was allowed to adapt its rule sets based on the environment in which it lived. STACs evolved based on the data environment, each STAC being its own entity, living by its own set of rules. A human or artificial explorer, however, needs neither to adapt to its environment nor to have different decision rules at different locations. Indeed, it would not be desirable for two AEs to have different decision rules at different places. Uniformity of movement criteria makes the classified map methodologically consistent. Because the rules are the same across the study area, the artificial explorer can use the stationary and uniform rule sets of cellular automata inside an artificial life-research framework.

24.5 Progress, extensions and implementation

The research described above is progressing in several areas. Conceptual and methodologic design is complete and work continues on the AE. Imagery for the final testing of the AE is currently being prepared for analysis. Computer code and data structures for the AE are being developed at this time.

The AE computer code will be modular so that more elaborate spatial data-analysis methods can be investigated in the future. For example, genetic algorithms could be incorporated into the AEs, or the rules of movement could be changed such that other-than-raster data structures could be analysed. The modular AE computer code will also be an advanced starting point for the development of artificial life-based spatial models.

This research is expected to produce both tangible products (for example, computer code) and, more importantly, a better understanding of what pattern is and how it might be recognised at various scales using a decentralised technique. The image analysis will investigate the effects of various attribute-similarity measures, neighbourhoods and overall similarity algorithms on the patterns which are identified. If the reasons for the different patterns can be explained, then some insight might be gained into how humans identify pattern at multiple scales from a single-scale image. The ultimate goal of the research is to develop a better conceptual understanding and a practical methodology which will advance the use of spatial data for exploratory geographic investigations.

Note

1   NASA's Earth Observing System (EOS) alone will be collecting terabytes of information daily about earth-environment conditions. EOS Program Scientists have personally expressed concern over how much of the data would ever be analysed without better tools to 'explore' the data (Asrar, 1994).

References

ASPINALL, R. (1992). An inductive modelling procedure based on Bayes' theorem for analysis of pattern in spatial data, *International Journal of Geographical Information Systems*, **6**(2), 105–21.

ASRAR, G. (1994). Personal communication at NASA's Earth Science Summer School, Pasadena, CA.

BONABEAU, E. W. and THERAULAZ, G. (1994). Why do we need artificial life?, *Artificial Life*, **1**, 303–25.

BUNGE, W. (1966). *Theoretical Geography*, Lund, Sweden: C.W.K. Gleerup.

COUCLELIS, H. (1985). Cellular worlds: a framework for modeling micro-macro dynamics, *Environment and Planning A*, **17**, 585–96.

COUCLELIS, H. (1988). Of mice and men: what rodent populations can teach us about complex spatial dynamics, *Environment and Planning A*, **20**, 99–109.

CULLING, W. D. (1985). *Equifinality: Chaos, Dimension, and Pattern*, Geography Discussion Papers, New Series No. 19, London: London School of Economics and Political Science, Department of Geography.

DE BOER, R. J., VAN DER LAAN, J. D. and HOGEWEG, P. (1993). Randomness and pattern scale in the immune network: A cellular automata approach, in Stein, W. D. and Varela, F. J. (Eds), *Thinking about Biology*, pp. 231–52, Reading, MA: Addison-Wesley.

DOBSON, J. E. (1992). Spatial logic in paleogeography and the explanation of continental drift, *Annals of the Association of American Geographers*, **82**(2), 187–206.

EIGEN, M. and WINKLER, R. (1981). *Laws of the Game: How Principles of Nature Govern Chance*, New York, NY: Alfred Knopf.

GARDNER, M. (1970). The fantastic combinations of John Conway's new solitaire game of 'Life', *Scientific American*, **223**(4), 120–3.

GETIS, A. and BOOTS, B. (1978). *Models of Spatial Processes*, New York: Cambridge University Press.

GOODCHILD, M. F. (1986). *Spatial Autocorrelation*, Norwich, UK: Geo Books.

GOODCHILD, M. F. (1992). Analysis, in Abler, R., Marcus, M. and Olson, J. (Eds), *Geography's Inner Worlds*, pp. 140, New Brunswick, NJ: Rutgers University Press.

JENSEN, J. R. (1986). *Introductory Image Processing: A Remote Sensing Perspective*, Englewood Cliffs, NJ: Prentice-Hall.

JOHNSSON, K. (1994). Segment-based land-use classification from SPOT satellite data, *Photogrammetric Engineering and Remote Sensing*, **60**(1), 47–53.

LANGTON, C. G. (1986). Studying artificial life with cellular automata, *Physica D*, **22**, 120–49.

LEVIN, S. A. and SENGEL, L. A. (1985). Pattern generation in space and aspect, *Society for Industrial and Applied Mathematics Review*, **27**(1), 45–66.

LOZANO-GARCIA, D. F. and HOFFER, R. M. (1993). Synergistic effects of combined Landsat-TM and SIR-B data for forest resources assessment, *International Journal of Remote Sensing*, **14**(14), 2677–94.

MACEACHREN, A. M. (1992). Visualization, in Abler, R., Marcus, M. and Olson, J. (Eds), *Geography's Inner Worlds*, p. 104, New Brunswick, NJ: Rutgers University Press.

OPENSHAW, S. (1991). Developing appropriate spatial analysis methods for GIS, in Maguire, D. J., Goodchild, M. F. and Rhind, D. W. (Eds), *Geographical Information Systems*, vol. 2, pp. 389–402, New York, NY: John Wiley.

OPENSHAW, S. (1994). Two exploratory space-time-attribute pattern analysers relevant to GIS, in Fotheringham, S. and Rogerson, P. (Eds), *Spatial Analysis and GIS*, pp. 83–104, London: Taylor & Francis.

RABBITT, M. C. (1969). John Wesley Powell: Pioneer statesman of Federal Science, in *The Colorado River Region and John Wesley Power*, USGS Publication vol. 669, p. 7, Washington, DC: United States Government Printing Office.

RESNICK, M. (1994). Learning about life, *Artificial Life*, **1**, 229–41.

RICHARDS, J. A. (1986). *Remote Sensing Digital Image Analysis: An Introduction*, Heidelberg: Springer-Verlag.

SCHALKOFF, R. J. (1992). *Pattern Recognition: Statistical, Structural and Neural Approaches*, New York, NY: John Wiley.

TOBLER, W. R. (1979). Cellular geography, in Gale, S. and Olsson, G. (Eds), *Philosophy in Geography*, pp. 279–386, Dordrecht: D. Reidel Publishing Company.

TOFFOLI, T. and MARGOLUS, N. (1987). *Cellular Automata Machines: A New Environment for Modelling*, Cambridge, MA: The MIT Press.

WOODCOCK, C. and HARWARD, V. J. (1992). Nested-hierarchical scene models and image segmentation, *International Journal of Remote Sensing*, **13**(16), 3167–87.

WHITE, R. and ENGELEN, G. (1994). Cellular dynamics and GIS: Modelling spatial complexity, *Geographical Systems*, **1**, 237–53.

CHAPTER TWENTY-FIVE

Digital Representation of Continuous Random Fields

ANDREJ VČKOVSKI

25.1 Introduction

Data describing continuous fields is a major information type used in many natural sciences. A physical observable that is mathematically modelled as a function over a continuous space is a (continuous) field. Whenever data sets describing fields are used one has to deal with some consequences of the discretisation necessary when sampling the field (Kemp, 1993). We refer here to the discretisation of the definition domain or support of the field and not of the value domain. A field has to be described with a 'finite' set of samples. Each of these samples (the 'field values') is related to an element or subset of the support. This reference of field values will be subsequently called the 'index' for the corresponding field value.

Consider a data set consisting of annual mean surface air temperature over Switzerland sampled at irregularly distributed sites. The 'support' of the field is the surface of Switzerland, the 'indices' are the sample site and time, the field values are each site's mean temperature value averaged from quasi-continuous temperature measurements.

The typical way to digitally represent such fields is naturally related to the measurement of the field, that is, the digital representation is given by the collection of measured data values, the corresponding indices and some metadata. There are many standardised ways for such digital representations, for example, HDF and netCDF. These standards – designed to improve the organisation of large data repositories – often cover both low-level details for the binary encoding of the information and also higher-level definitions for the overall structure of the data.

For specific appliations, especially if the data are produced and used within the same discipline or institution, the usage of such standards is an effective way to organise data representing fields. However, if the data have to be shared over discipline boundaries, and if the sampling costs are very high and thus promote extensive reuse of the data, then these representations have several shortcomings. The more a specific data set is used, the more probable it is that the initial structure of the data set does not fit the needs of every application. For some applications, the set of indices (sites at which field values were sampled) does not match the set of locations at which data values are desired; for example, the field values need to be spatially and/or temporally interpolated. Other applications do not actually need the information provided by the data set, but they need the data set to derive some other information which cannot be assessed directly but modelled with a sufficient quality (for example, conversion into another system of measurement units as a very simple model). There is a need for a *generalised* representation of a field — a representation that not only contains a set of sampled data values but also derived information. That means, a representation that *models* the field.

The methods to derive such data (interpolation, modelling using other variables) depend strongly on the phenomenon described by the data set. An appropriate selection of these methods needs evaluation of explicit and implicit data characteristics and many data-unrelated information such as metadata and domain knowledge (Bucher and Včkovski, 1995). The selection of an appropriate method, for, example, spatial interpolation, and its application is therefore a task located in the 'problem domain' of the data set and not in the application's domain. This shows that from a viewpoint of object-oriented design (Booch, 1991) such methods 'belong' to the *data set* and not the application.

These findings and an approach for their implementation will be justified and discussed in the remainder of this chapter. First, the relation of a physical field to its samples is discussed. This discussion is necessary to understand what data values are used to describe a field, and how they relate to the field. Then, various possible extended field representations are discussed. The last section covers an overview of the virtual data set (VDS), which is an extended data repository suitable for a generalised modelling of continuous fields (and other information), and first-implementation approaches.

25.2 Continuous random fields and their sampling

25.2.1 Sampling of non-random fields

A field on a given support $\mathbf{I}$ is a function $z(\cdot)$ assigning every element s of the support a corresponding field value $z(s)$ from the value domain $\mathbf{B}$:

$$s \xrightarrow{z} z(s), \quad s \in \mathbf{I} \subset \mathbf{R}^N, \quad z(s) \in \mathbf{B} \tag{25.1}$$

For the sake of simplicity we will assume the value domain $\mathbf{B}$ to be 'real' and 'one dimensional', i.e. $\mathbf{B} \subset \mathbf{R}$ and $z(\cdot)$ is a scalar field. The generalisation to vector fields (dim $\mathbf{B} > 1$) is straightforward.

The notion of 'continuity' of the support $\mathbf{I}$ implies that there has to be a 'metric' $||\cdot||$ defined on $\mathbf{I}$, to express 'distance' $d(x, y) = ||x - y||$, $x, y \in \mathbf{I}$ between two elements of the support. Therefore, the support $\mathbf{I}$ is a (dense and compact) subset of a metric space. The support usually has a dimension in the range of 2 to 4, but there are also cases with higher dimension. Table 25.1 shows some examples of fields with one-dimensional value domain $\mathbf{B}$ (that is, scalar fields).

The sampling of a field describing a physical property (for example, surface air temperature) needs a measurement model for this property – a model to describe the property with a number z. The measurement model for fields needs in addition to that a model for the sampling of the support $\mathbf{I}$. The measurement of the field consists of a set of numbers z_i, $(i = 1, \ldots, n)$ which needs to be related to the

Table 25.1 Examples of continuous fields

Field	Dim (**I**)	Indep. variables (**I**)	Dep. variable (**B**)
$H(x, y)$	2	x, y: Spatial coordinates	H: Height of orography (elevation)
$P(x, y, z)$	3	x, y, z: Spatial coordinates	P is soil porosity
$\theta(\lambda, \phi, p, t)$	4	λ, phi coordinates (longitude, latitude), p is pressure level, t is time	θ: potential temperature at (λ, ϕ, p, t)
$I(x, y, z, t, \lambda)$	5	x, y, z: Coordinates of a point in atmosphere, t is time, λ is wavelength	I: Intensity of radiation at x, y, z at time t with wavelength λ

field $z(s)$. The relation of measurement values z_i and the field $z(s)$ can be modelled as a projection of the field onto the measured value z_i.

$$z_i = \int_{\mathbf{I}} \phi_i(s) z(s) \ ds \tag{25.2}$$

The selection function $\phi_i(s)$ is normalised:

$$|| \phi_i || = \int_{\mathbf{I}} \phi_i(s) \ ds = 1 \tag{25.3}$$

Usually, each selection function $\phi_i(s)$ is related to a subset of the support s_i, i.e. it is localised. In this case, we can identify the selection function through its localisation s_i (the 'index'). We then use a shorthand notation and say 'z_i is the value of $z(s)$ at s_i'. There are, however, important cases where $\phi_i(s)$ is *not* localised, for example, when the spectral characteristics of a field are measured. Every z_i then corresponds to a specific spectral mode. In this case it makes no sense anymore to use some $s_i \subset \mathbf{I}$ to identify z_i.

In many applications, however, the sampled field value z_i is a weighted average of the field $z(s)$ 'around' the index s_i where weights are given by the selection function $\phi_i(s)$. If s_i is a point and the measurement/selection retrieves the field value exactly at that point then $\phi_i(s)$ is a Dirac-distribution centred at s_i:

$$\phi_i(s) = \delta(s_i - s) \tag{25.4}$$

In many cases (such as remote sensing images) the index is a proper subset of **I**, for example, a square pixel. Then the selection function ϕ_i yields a corresponding aggregation of the field $z(s)$. A simple example is the mean value over s_i with a selection function given by:

$$\phi_i(s) = \begin{cases} \dfrac{1}{V(s_i)} & : \quad s \in s_i \\[2mm] 0 & : \quad s \notin s_i \end{cases} \tag{25.5}$$

where the volume $V(s_i)$ of s_i is

$$V(s_i) = \int_{s_i} ds \tag{25.6}$$

A typical data set describing sampling of a non-random field therefore consists of a set of indices s_i and corresponding measured field values z_i. It is important to note that the set of indices s_i is only a place-holder for the set of selection functions $\phi_i(s)$. In order to understand the measurements we have to know what the $\phi_i(s)$ are. Unfortunately, there are no conventions which would allow data users to derive the $\phi_i(s)$ given a specific set of indices s_i. This statement 'Temperature in Zürich yesterday was $20\,°C$' gives $s_i = \{yesterday, Zürich\}$ but leaves it undefined if it was a daily average, or an hourly average or whatever.

Assuming that it is known (for example, by additional metadata) what the $\phi(s)$ are (given some s_i) a representation of a continuous field can therefore be written as (M is metadata):

$$\mathbf{D} = \{M, \{s_i; z_i\}_1^n\}, \quad s_i \subset \mathbf{I}, \quad z_i \in \mathbf{R} \tag{25.7}$$

The set of indices $\{s_i\}$ is sometimes not given explicitly but implicitly if there is some regularity, for example, raster data.

25.2.2 Sampling of random fields

The previous section has shown the extension of the classical measurement model, how a field $z(s)$ 'at an index' s_i is characterised with one value z_i which is related to s_i through $\phi_i(s)$. For many real-world fields this extension is not sufficient because the random variations of the phenomena considered are too large to be neglected. These random variations are due to various reasons, including inherent natural variation of the phenomenon, all types of measurement uncertainties, and ignorance and complexity of the underlying processes. There are several ways to account for such uncertainties. For the sake of brevity we will present here just probabilistic approaches, that is, modelling each field value as a random variable. Other approaches are discussed in Včkovski (1995).

The relation in Eq. (25.1) is therefore extended to map every $s \in \mathbf{I}$ to a random variable $Z(s)$. ($Z(s)$ is sometimes called a random function (Isaaks and Srivastava, 1989).) The random variables for different s are not independent, so that we are actually considering a probability space over $\mathbf{B} \times \mathbf{I}$. For the subsequent discussion we will use the probability distribution function $F(z; s)$ to characterise the field as defined by (again for scalar field $z \in \mathbf{R}$):

$$F(z; s) = P(Z(s) < z) \tag{25.8}$$

It is clear that it is not sufficient to characterise $Z(s)$ or $F(z)$ by a single real number. However, it is often suitable to model $F(z)$ by a small set of numbers. For example, the expectation value $E[Z(s)]$ and variance $V[Z(s)]$, given by

$$E[Z(s)] = \alpha_1 = \int_{\mathbf{B}} z\,dF(z) \tag{25.9}$$

$$V[Z(s)] = \beta_2 = \int_{\mathbf{B}} (z - \alpha_1)^2\,dF(z) \tag{25.10}$$

are often used to describe $F(z)$. There are many well-established methods to estimate these moments α_1 and β_2 with descriptive measures such as mean value and standard deviation.

This example shows that for random fields we need a set of numbers to represent the field 'value' at an index s_i, that is, $\{z_{i,1}, \ldots, z_{i,m}\}$. The number of parameters used to describe the field value is usually small (for example, $m = 2$ and $z_{i,1} = \alpha_1$ and $z_{i,2} = \beta_2$ as in the example above). There are, however, cases where the sampling of the phenomenon provides enough information to describe $Z(s)$ in more detail. An example is the characterisation of $Z(s)$ using its empirical distribution function (histogram) with c classes:

$$p_j = P(a_j < Z(s) \leqslant a_{j+1}) = F(a_{j+1}) - F(a_j)$$

$$j = 1 \ldots c \tag{25.11}$$

$$a_1 = -\infty, \ a_{c+1} = \infty$$

A field value at s_i is therefore represented with $2c - 1$ values:

$$z_{i,1} \quad = p_1$$

$$\vdots$$

$$z_{i,c} \quad = p_c \tag{25.12}$$

$$z_{i,c+1} = a_2$$

$$\vdots$$

$$z_{i,2c-1} = a_c$$

Whereas in the case of non-random fields there is a single selection function ϕ_i which maps the field $z(s)$ to the sample z_i, random fields need a set of selection functions $\Phi_{i,j}$ with

$$Z(s) \overset{\Phi_{i,j}}{\rightarrow} z_{i,j} \tag{25.13}$$

It is common that all parameters $\{z_{i,1}, \ldots, z_{i,m}\}$ use the same spatio-temporal selection function ϕ_i, such that $\Phi_{i,j}$ can be separated into $\Phi_{i,j} = \phi_i \circ \psi_j$. The ϕ_is are discussed in the previous section (for example, (25.4)), and the ψ_j for example, the functionals that yield statistical moments as in (25.9) and (25.10).

A data set D_r as defined in Eq. (25.7) extends to:

$$D_r = \{M, \{s_i; z_{i,1}, \ldots z_{i,m}\}_1^n\}, \quad s_i \subset I, \ z_{i,j} \in \mathbf{R} \tag{25.14}$$

It should be noted that random field models are applied successfully in many areas of spatial analysis. Of particular interest are random field models for spatial and temporal interpolation problems, where a data set D_r is given and the general problem statement is to find the result of a selection function $\Phi_0 = \phi_0 \circ \psi_0$ for some ϕ_0 and ψ_0 not provided through D_r.

The next section will now discuss the types of applications of data sets such as D or D_r and some approaches for extended representations.

25.3 Extended representations

In Včkovski, 1995 it was shown that the requirements on a representation are motivated through the various types of queries that are raised against such a data set. That is, a representation should be defined in a way that user queries can be answered as well as possible. This requirement is particularly important for the management of continuous fields. While with many other kinds of data the information needed is a subset (that is, selection) of the information available, when using continuous fields, often data values are needed that are not directly available. This is a consequence of the discretisation of a continuous field as shown in the previous section that is inherently necessary when sampling such fields. Many applications are, at least theoretically, interested in the 'whole' field $z(s)$ or $Z(s)$, respectively, and not in the set of samples as given by D or D_r. Actually, the 'whole' field can never be assessed, rather applications are often interested in field values at unsampled locations and aggregations (using $\phi \notin \{\phi_i\}$), or using other characterisations of uncertainty (that is $\psi \notin \{\psi_j\}$). This is mainly a consequence of the incompatibilities arising from integrating various data sets for a specific application. The various phenomena ('variables') described by the data sets typically are not sampled at the same (spatio-temporal) sites. When these variables are needed for an integrated analysis, for example, simulating an (unsampled) physical property, then the relationships between the various variables are local, that is, connecting the variables at the same time and location. Relationships exposing explicit spatial or temporal characteristics usually are only within one variable, for example, as a differential equation in space and/or time.

As mentioned in the introduction, sampled field values therefore often have to be used to predict values at unsampled locations or with other uncertainty representation. One could now argue that this is application-specific and that the digital representation does not have to account for possible later transformations. I believe that this is true if a data set is used only a few times. There are, however, data sets that are used many times, for various applications with differing requirements. A typical example is a digital elevation model (DEM) which is used for many applications. The applications might need terrain height, slope, aspect or other derived data, and often not for the (grid) points used when sampling or digitising the DEM. The quality of the derived data is then usually affected by various factors, including the following.

- Data users might not have the expertise to use appropriate methods for tasks such as spatial interpolation of the DEM, numerical differentiation to determine slope/aspect and other methods to derive the data needed.

- It is quite common that such data sets are transformed several times in sequence. For example, if the original data is unavailable (or the data users are not aware of using already derived data) then it happens that new data is derived using previously transformed data, yielding results that are very questionable (for example, propagation of uncertainties).

Within well-defined projects (data repositories covering a certain region) it is a common procedure to establish a standard format for all data within the repository, requiring for example all continuous fields (soil type, temperature, terrain height and so forth) to be sampled at previously defined points, thus avoiding any compatibility problems. As soon as data sets are to be used for many different applications (for example, a DEM) such a standard format will be always too rigid, that is, requiring many future transformations performed by various data users.

An approach to minimise the need for such transformations and thus avoid unnecessary loss of data quality was presented in Stephan *et al.* (1993). It is called virtual data set (VDS) and its scope is far more general than the application for continuous fields, yet fields are a good example of its usefulness and basic concepts. The next section shows the concept of VDS, discusses various implementation alternatives, and finally sketches the approach taken for the first prototype system.

25.4 Virtual data set

25.4.1 Overview

The basic idea of VDS is the extension of a data set in a way that the data set can answer queries asking for derived data, that is, there is an agent that can answer such queries. For continuous fields this means that a VDS describing a field $Z(s)$ exposes the whole field $Z(s)$ and not a selection of values as given in D_r. Of course, not every field value is persistently stored within a VDS. Rather, a VDS is a representation that contains information which allows field values $z(s_0)$ at unsampled locations and other derived data to be calculated on request, that is, when these data are queried for. The calculations are hidden from the data user: the user just requests the data without actually knowing that the data was somehow calculated. The user gets 'virtual data' in the sense that it is data without persistence on secondary storage.

There are various ways to achieve such behaviour. Focusing on continuous fields, one could for example extend the representation given by D_r with some additional parameters that define how the data values in D_r should be interpolated, for example:

$$D_1 = \{D_r, \text{'interpolate it using splines with parameters } a,b,c'\} \qquad (25.15)$$

The application reading the data set D_1 then must understand what this parameter 'interpolate it using splines with parameters a,b,c' is and accordingly process the data values when such data is queried. This shows the major drawback of such an approach: there are lots of methods to transform, interpolate, and derive data so that it is impossible to have them all formalised in the same manner unless one is willing to accept limitations to its applicability.

Another approach is to approximate the field by a parameterised family of functions and represent the field by the set of best-approximating parameters. This can be established, for example, by Fourier- or Wavelet-Transformation, expansions into polynomials and other similar methods. A data set would then be represented by the collection of parameters, allowing the application to calculate field values, their derivatives and so on at any location of the field. This approach shares the same shortcoming as the example discussed before: it is not very flexible, especially because there are often simply derivable values which would preferably be offered within the VDS as well (not just the field value and its derivatives).

The VDS concept approaches this question in an object-oriented (Booch, 1991) way: the derivation of values is seen as the responsibility of the data set and not of the application using the data since it is a task belonging to the problem domain of the data. Consequently, the VDS concept does not define how a continuous field should be digitally represented, but it defines an interface that a VDS should expose. Details on the data model, the way derived data is calculated and so on are hidden from the application. An application needing data values therefore does not 'read some data file' but requests the data through the interface the VDS offers. Using object-oriented terminology, data (the 'state' of the VDS) is accessed through methods of the VDS. It is the responsibility of the VDS to return the requested data to the calling application, either calculating the data (if it is derived — 'virtual' — data) or retrieving it from secondary storage.

This structure is analogous to similar approaches in the domain of standard database systems. SQL (Structured Query Language), for example, is an approach to define a standardised query language which shields database users from technical details of the database system. Microsoft's ODBC (Open Database Connectivity) or Apple's DAL (Data Access Language) are examples of standardised interfaces for data access. ODBC is an interface to allow applications to access data in database-management systems using SQL (Open Database Connectivity, 1992). A small but important difference between VDS and approaches like ODBC is that VDS primarily defines its interface towards applications and does not expose an 'internal' interface. Moreover, the interface is 'fat' in the sense that it contains various methods not just to provide data somewhere from secondary storage but also to derive, that is, calculate data. Such software components are sometimes referred to as middle-ware.

Figure 25.1 shows the overall structure of VDS and their use. Starting from samplings (left side), data sets are extended ('virtualised') with a standardised interface. This provides applications with the requested information, using both the original data and methods within the virtualisation layer to derive information. Note that results of an application (numerical simulation) can be enhanced to a VDS as well (as the example of the modelling of evapotranspiration shows).

25.4.2 System architecture

The implementation issues discussed in the following show the design used for a first prototype of VDS. Design decisions during the evaluation of various alternatives for the implementation were influenced by various factors, including the following.

- *Simplicity*. The prototype system should be based on a simple structure, that is, no complicated libraries, protocols and so on.
- *Flexibility*. The VDS interface and its implementation should allow for a wide range of applications, reduce assumptions (and restrictions) on what and how VDS can or could be used.

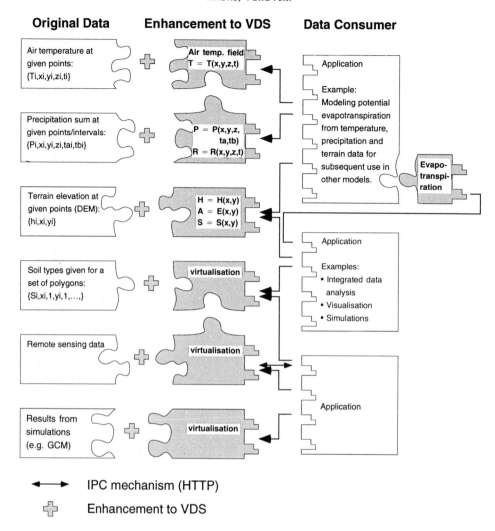

Figure 25.1   Overall structure and use of virtual data sets.

- *Standardisation*. The implementation path should follow well-established standards, both for the interface definition and for the coupling with samplings.
- *Interoperability*. Interoperability is one of the basic motivations for VDS (see also Bucher *et al.*, 1994).

The VDS structure indicates a client/server architecture for its implementation, where VDS are servers answering requests, and applications requesting data are clients. Out of many variants for an implementation it was first considered to use CORBA (Common Object Request Broker Architecture) from OMG (Object Management Group, (Common Object Request Broker Architecture, 1992)) as a framework. The high complexity of CORBA then lead to another design. One of the most important decisions in the design of client/server-systems is an appropriate choice of the communication protocol used between servers and clients. The factors influencing design decisions mentioned above lead to the choice of HTTP (Hypertext Transfer Protocol) as a base for the implementation of VDS. HTTP is the protocol used in the World-Wide Web (WWW) and has therefore gained a lot of attraction over the last years. The (literally) world-wide use of WWW established HTTP as a protocol for various applications. Its simplicity and availability on many platforms makes it a good starting point for a prototype implementation.

HTTP is message oriented: it is based on messages that clients and servers exchange. HTTP messages have a very simple structure and consist basically of a method and some arguments, depending on the method of the message. The HTTP-methods needed for the implementation of VDS are GET and POST. VDS are (at least in the first stage) seen as static entities, that is, there is no need for updates (VDS are not databases but data repositories). POST messages should therefore not be necessary. The POST method is nonetheless needed for the information exchange with the VDS broker. The VDS broker is a server that acts as a metadatabase for all available VDS. Clients needing data access the broker to search for a VDS. The broker then provides clients with the address of the required VDS. A VDS uses a POST message to register itself with the VDS broker by sending its meta-information, and the VDS broker stores the meta-information in its metadatabase (which is currently a simple file-system).

GET-messages are used for the queries that are sent to a VDS. A VDS therefore is a simple HTTP-server that only has to understand GET-messages. This makes it very simple for the extension of existing data sets to VDS since there is one single entry point – the handler for the GET-message – that has to be added (and of course the methods to calculate derived data). Figure 25.2 shows the HTTP request messages (reply messages are not shown) between applications and VDS. Note that a VDS may be a client, using another VDS as well. This is useful if a VDS needs other data to derive information, such as a temperature data set that needs elevation data (DEM) for spatial interpolation.

25.4.3 Architecture and design of VDS

Figure 25.2 shows that a VDS is seen from the outside as a HTTP-server. In this section a preliminary design of the interior structure of a VDS is discussed briefly. The components of a VDS are shown in Fig. 25.3 in more detail. The VDS consists of a set of procedural components and data components. The procedural components are as follows.

HTTP-server. This module handles all low-level details of the communication with clients. It acts as an entry point for all requests and queries. Queries for metadata and queries for cached data (data that was queried before) can be answered directly by this module.

Library of common methods. There is a need for a set of functions (or objects) that serve for all VDS. A (class) library of helper function is therefore referenced in every VDS.

Data set-specific methods. This is the core of a VDS and includes the handlers for all specific access methods for this VDS, for example, methods to derive values at unsampled locations. Typically, these methods will simply use

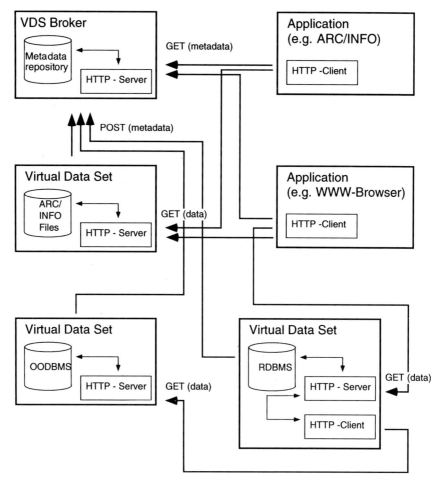

Figure 25.2 HTPP requests between VDS and applications.

functionality provided by the library discussed before, in an object-oriented environment they inherit the functionality provided by the library. In some special cases the methods contain specific algorithms which override the methods provided by the library.

These methods have to be provided by the data producer. For a successful operational use of such a concept it is therefore important that the creation of the methods is supported by corresponding tools, for example, visual programming environments and so on.

The 'data'-part of a VDS consists of the following.

Local cache. VDS have the particular ability to deliver virtual data, that is, data that is not persistently available. Since virtual data is computed on request this might lead to significant performance problems. It is therefore probably necessary to include some sort of local caching of derived values. This is particularly necessary if – and that is the objective of the VDS-concept – every access to the data goes through the VDS. Consider for example an application

Communication to the outer world

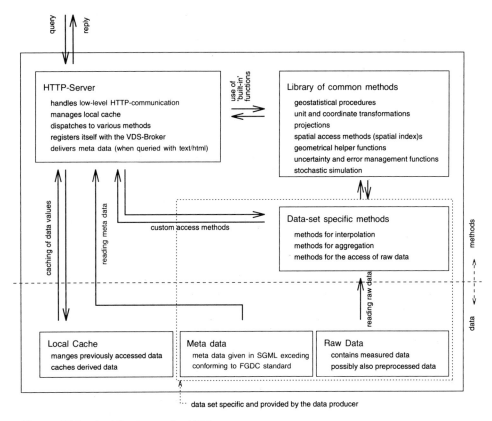

Figure 25.3 Architecture of a VDS.

with a graphical user interface that displays the data in a window. If the application does not cache itself the data, then it needs to read the data every time the particular window needs to be repainted.

Metadata.   Metadata is (or should be) an integral component of every data set. Because VDS uses WWW technology for message exchange it makes sense to encode the metadata in a way that corresponds to typical exchange formats used in WWW. One of these is HTML (Hypertext Mark-up Language) which provides a suitable way for the encoding of geospatial metadata (Blott and Včkovski, 1995). There are currently some proposals to standardise HTML-encoded metadata based on the recently accepted FGDC (Federal Geographic Data Committee) metadata content standard (Federal Geographic Data Committee, 1994).

Raw data.   This represents the actual measurement values as given by D or D_r. It is not necessary that the data conform to a specific representation and digital encoding, since the data are accessed through data-specific methods which can handle whatever is needed. It will be nonetheless useful to adopt some typical way of encoding the data (for example, as HDF-files) in order to make the data access simple through predefined access functions from the library of common methods.

The implementation of this concept offers many options. Currently, we are evaluating Java as a programming environment for VDS. Java is an object-oriented language that offers most of the features of modern object-oriented programming languages. But unlike for example C++, Java is not directly compiled into machine code but into an intermediate byte code (similar to perl or the UCSD-series of p-code languages). The intermediate code is then interpreted on the target platform. This code is designed in a way which tries to reduce the drawbacks of reduced execution speed due to interpretation, that is, the interpreter is computationally very efficient. The use of such an intermediate code offers many benefits, especially platform-independence.

The use of Java for the implementation on VDS offers some additional advantages:

- Java offers techniques for late binding – linking at runtime. This allows that some code, for example, for spatial interpolation, is transferred from a VDS to a client and executed within the client. This technique is currently used in some WWW-browsers such as HotJava from SunSoft, Inc. It allows small pieces of code, called applets, to be transferred to the client and enhance the functionality of the client, the WWW-browser.

- The previously mentioned WWW-browser HotJava is itself written entirely in Java. The source code is available and its use can therefore reduce the efforts necessary to implement VDS parts such as the HTTP-server module.

The future work will show if Java will meet the requirements of a development environment for the implementation of VDS.

25.5 Conclusion and outlook

The previous sections have shown that digital representations of continuous (random) fields have special requirements which are met by a client/server-architecture. A data set is not seen as an inactive collection of data values but as a service for various data requests. The usage of data representing continuous fields benefits from such an architecture since a VDS can simulate continuity. From the user's point of view, a VDS of a field $z(s)$ seems to describe the 'whole' field $z(s)$, for example, there is a value for every s. The methods to derive all these additional field values (and other information such as derivatives) are part of the data set and therefore can be expected to be more reliable than if data users would perform the transformations themselves. The shift of the responsibility for these transformations from data users to data producers should – following the paradigm of object-orientation – improve the quality of derived data. The transparent access to derived data also bears dangers along. Derived data are often predictions and affected by various sources of uncertainty. It is therefore necessary that VDS return quality information of derived data to the requesting application. This is emphasised by the fact that most fields are random fields and therefore their samplings and derived data have to include uncertainty measures. Furthermore, it is necessary that a VDS is informative about all transformations and derivations that were applied, that is, to act as a 'transparent black box'.

Future work will concentrate on a prototype implementation using the design presented in the previous section. The usage of a widely used protocol (HTTP) supports a rapid prototyping, particularly because there is a large collection of utilities and source code available. This complies with a general tendency within the software industry. Commercial as well as non-profit software technology moves towards systems built up from various interoperating software components ('Component-ware', for example, Microsoft OLE, SunSoft DOE, OpenDOC). This development also influences software areas that traditionally were populated by large, monolithic systems such as GIS, CAD and similar applications. There are now several research projects in the field of spatial data handling focusing on these topics, for example, the OGIS project (Buehler, 1994).

The objective of the prototype implementation is to analyse usability and improvement using data repositories built upon VDS compared with classical database management systems or even file-system-based repositories. The test bed will include in the first phase VDS of the digital elevation model of Switzerland as well as VDS covering a large collection of climate data (temperature, precipitation and so forth). Based on the prototype's experiences, the VDS structure could be adopted as an industry standard and extended to other application areas.

References

BLOTT, S. and VČKOVSKI, A. (1995). Accessing geographical metafiles through a database storage system, in Egenhofer, M. and Herring, J. R. (Eds), *Advances in Spatial Databases*, Lecture Notes in Computer Science, pp. 117–31, Berlin: Springer-Verlag.

BOOCH, G. (1991). *Object-Oriented Design with Applications*, Redwood City, CA: The Benjamin Cummings Publishing Company.

BUCHER, F. and VČKOVSKI, A. (1995). Improving the selection of appropriate spatial interpolation methods, in Frank, A. and Kuhn, W. (Eds) *Spatial Information Theory: A Theoretical Basis for GIS*, pp. 351–64, Berlin: Springer-Verlag.

BUCHER, F., STEPHAN, E.-M. and VČKOVSKI, A. (1994). Integrated analysis and standardization in GIS, in Harts, J., Otten, H. and Schotten, H. (Eds), *Proceedings of the ECIS'94 Conference*, **vol. 1**, pp. 67–75.

BUEHLER, K. A. (1994). *The Open Geodata Interoperability Specification: Draft Base Document*, Technical Report OGIS Project Document 94–025, OGIS, Ltd, Wayland: Open GIS Consortium, Inc.

CORBA. (1992). *The Common Object Request Broker: Architecture and Specification*, Object Management Group, OMG Document Number 91.12.1, Wellesley, MA: QED Publishing Group.

FGDC. (1994). *Content Standards for Digital Geospatial Metadata*, Reston: Federal Geographic Data Committee.

HDF. (Hierarchical Data Format): URL: http://www.ncsa.uiuc.edu/SDG/software/HDF/HDFIntro.html.

ISAAKS, E. H. and SRIVASTAVA, M. R. (1989). *An Introduction to Applied Geostatistics*. New York: Oxford University Press.

JAVA and *HotJava* DOCUMENTATION. URL:http //java. sun. com/documentation.html.

KEMP, K. (1993). *Environmental Modelling with GIS: A Strategy for Dealing With Spatial Continuity*, Technical Report. 93-3, Santa Barbara: NCGIA.

NETCDF. NetCDF Frequently Asked Questions. URL: gopher://groucho.unidata.ucar.edu:70/00/systems/netcdf/netcdf.FAQ.

OPEN DATABASE CONNECTIVITY. (1992). *Microsoft Open Database Connectivity Software Development Kit: Programmer's Reference*, Microsoft Corporation.

STEPHAN, E-M., VČKOVSKI, A. and BUCHER, F. (1993). Virtual data set: an approach for the integration of incompatible data, in *Proceedings of Auto-Carto 11 Conference*, pp. 93–102, Minneapolis: ASPRS/ACSM.

VČKOVSKI, A. (1995). Representation of continuous fields, in *Proceedings of Auto-Carto 12 Conference*, pp. 127–36, Charlotte, NC: ASPRS/ACSM.

PART FIVE

Spatial Analysis

CHAPTER TWENTY-SIX

GIS and Spatial Statistical Analysis

DAVID UNWIN

26.1 Institutions and precedents

I begin my chapter by examining four typical statements on the role of spatial analysis in geographical information systems (GIS).

> Spatial analysis will be applied to an ever increasing number of application areas. GIS data manipulation tools will become ever more sophisticated and easier to use. They are already today being included in office software packages such as spreadsheets. We are rapidly approaching the time when every desktop PC will be able to perform spatial analysis.
> Commission of European Communities, 1995, p. 6

This is a typically 'bullish' statement about the potential of spatial analysis in GIS. Similar statements have been made by numerous others. In his summary of a GISDATA Specialist Meeting on 'GIS and Spatial Analysis' held in Amsterdam in December 1993, Scholten (1994, p. 6) writes:

> There is wide agreement in both the GIS and the modelling communities that the future success of GIS technology will depend, amongst other factors, on the extent to which it can incorporate more powerful analytical and modelling capabilities. Spatial analysis, which in its widest sense is the description, explanation, and prediction of spatial and aspatial phenomena occurring in a spatial and/or space-time systems [sic], offers a wide range of methodologies and procedures which are highly relevant to GIS research. It is important to stress that spatial analysis is more the geo-statistics or spatial statistics.   [sic]

A year earlier in April 1992, the National Center for Geographic Information and Analysis (NCGIA) Specialist Meeting in San Diego on 'GIS and Spatial Analysis' led to the comment:

> Progress in this area is inevitable and . . . future developments will continue to place increasing emphasis upon the analytical capabilities of GIS.
> Fotheringham and Rogerson, 1993, p. 3

A year before that the UK Regional Research Laboratories held a workshop on 'GIS and Spatial Data Analysis' (SDA) which led to the idea that:

If GIS is to reach the potential implied by many of its definers and proponents as a general purpose tool for delivering a spatial perspective on data in a digital environment then GIS needs SDA. The view was also strongly expressed that SDA needs GIS if it is to become of wider use to the scientific community.

Haining and Wise, 1991, p. 1

There are three hidden assumptions in all this:

- that something called spatial analysis can be defined;
- that spatial analysis is useful in work using geographical information systems technology; and
- that it will be applied. Note the use of the future tense. Things *will* happen.

Discussions of the role of spatial analysis in GIS and the relationships between the two have frequently appeared in the geographical literature (see Anselin, 1989; Anselin *et al.*, 1993; Bailey and Gatrell, 1995; Fotheringham, 1992; Fotheringham and Rogerson, 1993, 1994; Goodchild, 1987; Goodchild *et al.*, 1992b; Openshaw, 1991). There is now also a substantial body of statistical theory about spatial data (see Cressie, 1991; Diggle, 1983; Haining, 1990; Ripley, 1981, 1988; Upton and Fingleton, 1985, 1989; Walden and Guttorp, 1992). With all this literature it is not easy to say anything original, but in this introduction I will concentrate on the three assumptions given above, examining in turn what we mean by spatial analysis, how it can help in GIS and vice versa and, in conclusion, whether or not it will be applied.

26.2 What do we mean by spatial analysis?

There are at least four variations on what is meant by the term *spatial analysis*, all of which seem to have at some time or other become conflated in the literature.

1 In the purely GIS literature, and especially in system manuals and brochures, the view seems to be that spatial analysis consists largely of what Bailey and Gatrell (1995), refer to as 'a general ability to manipulate spatial data' (p. 15), with a familiar set of largely deterministic functions to do this which includes basic spatial queries, buffering, overlay using simple map algebra and the calculation of derivatives on surfaces (slope, aspect). This might be termed *spatial data manipulation*, and, since it is precisely this ability to handle spatial data *spatially* that differentiates a GIS from any other database-management system, I regard it as essential to any information system claiming to be 'geographical'.

2 In statistics, there is a totally different view, best characterised by the notion that spatially distributed information can be regarded as the outcome of some stochastic process operating in the plane (Cressie, 1991; Unwin, 1981). Given a postulated process, we attempt to deduce its spatial outcomes and to examine whether or not an observed pattern is a plausible realisation of it. Alternatively, and much more difficult, given a spatial pattern the challenge is to identify the process and model it appropriately. This is *spatial statistical analysis*. Very little of it is to be found in existing GIS, except, perhaps, as additional code coupled

to a standard system by various software devices, as an integral part of a system developed for academic use such as IDRISI (Eastman, 1990), or as user contributions to a public domain system such as GRASS. Software for more specialised spatial statistical analysis is now readily available more or less in the public domain (Anselin, 1990; Diggle and Rowlingson, 1993; Rowlingson and Diggle, 1991). Recently INFOMAP (Bailey, 1990) has become available at very low cost (Bailey and Gatrell, 1995) and includes analysis methods such as density estimation, kriging and K-function computation, all of which would be very high on the 'wish list' of any spatial statistician. Much of the debate to date consists of calls for further incorporation of these techniques into the standard GIS tool kit and how this might be achieved. With the exceptions of techniques which replace or significantly extend existing GIS functions (see below), I am by no means certain that this is what either the marketplace or research users of GIS actually need. First, these methods have been developed in quite specific contexts and thus may lack the generality that is implied by the idea of them becoming part of any tool kit of functions. Second, the evidence to date of attempts by system vendors to incorporate them into their systems does not convince me that the vendor developers have sufficient knowledge or experience to do the job properly. Third, I am similarly unconvinced about the abilities of GIS users in both industry and the academy to use these methods properly. Given these difficulties, the issues of how to couple them to a GIS are a total irrelevance.

3   A third view of what spatial analysis might be is that held in academic geography and my separation of this view from that held by statisticians working in what at first sight might seem to be the same areas is quite deliberate. The geographical view is essentially data- and cartography-driven, concerned with the recognition and description of spatial patterns and their representation on maps. This is a data-driven view that might usefully be termed *spatial data analysis* and its origins lie in the quantitative geography of a couple of decades ago. The methods employed vary from allegedly simple statistics dating from the 1950s (notably the spatial autocorrelation tests of Cliff and Ord, 1973), through the direct use of visualisation (Haslett *et al.*, 1990; Hearnshaw and Unwin, 1994) to automatic machines (Openshaw, 1994) and artificial life forms (Openshaw, 1995) for pattern detection. There is some overlap here with (2), but nowhere near as much as might at first be imagined. There is also a tendency, particularly among statisticians working in spatial statistical analysis, to assume that spatial data analysis is in some sense trivial, a watered-down form of the real thing. There is also a tendency within academic geography to regard spatial data analysis as outmoded and incapable of providing useful geographical insight (Pickles, 1995). Both tendencies should be resisted; the problems and potentials of spatial data analysis remain as clear and as useful as they ever have. The difference now is that computing technology and data availability make it possible to tackle these same problems in new ways.

4   A fourth view is that of spatial analysis as *spatial modelling* either as part of a decision-support system (Clark, 1990; Densham, 1994) or within a dynamical modelling environment (Goodchild *et al.*, 1993; Van Deursen, 1995). These topics are covered elsewhere in this volume. Suffice it to note that the major

distinguishing feature of this approach to spatial analysis is that it is almost invariably deterministic. Some of the desired functions exist in proprietary GIS, such as in the ARC network module, *Tactician*'s retail gravity models, and in the recent integration of a generic dynamic modelling language within a raster GIS (van Deursen, 1995).

These are very different views of the nature of spatial analysis involving different traditions and approaches to knowledge generation. In all probability they will require different data models for their efficient implementation in software and equally different relationships with what we now know as GIS. I doubt that it will ever be sensible to integrate all four views within the same system.

For spatial data manipulation existing data models seem to be almost adequate, at least for commercial use, and existing systems have most, if not all, of the functions required. On the other hand, spatial statistical analysis may well be best accomplished outside of a GIS, but with relatively seamless data transfer from the statistical to the geographical software. Spatial data analysis requires a highly flexible, interactive computing environment which seems to imply relatively tight coupling of the data analysis function with the GIS, and, particularly, with advanced visualisation facilities. SPIDER/REGARD (Haslett *et al.*, 1990) and more recent developments of exploratory spatial data analysis software such as XLisp-Stat (Brunsdon and Charlton, 1995) and the Cartographic Data Visualiser (Dykes, 1995) represent important steps along this road. Notice that, unlike SPIDER/REGARD, the more recent systems have been developed making use of modern CASE tools that make the required integration relatively easy to achieve. Finally, for spatial modelling the needs seem to be rather different. Here it is necessary to structure the geography of interest and express problems in manners which arise naturally in the applications areas concerned rather than in any general GIS framework. It follows that standard GIS functions are likely to be incorporated within the modelling software rather than vice versa, as, for example, already happens in atmospheric general circulation models, hydrological models, and spatial decision-support systems. Alternatively, they will be provided by vendors as customised systems built from highly disaggregated GIS tool kits.

26.3 Putting spatial statistics into GIS

Despite the comments made in (2) above, there is good reason to add some better spatial statistical analysis to existing GIS. Primarily, this is because many GIS functions currently thought of as deterministic are intrinsically statistical in nature and hence to be treated probabilistically. An obvious example is map overlay which lies at the heart of many GIS operations (Unwin, 1996). Currently, most systems treat this as a deterministic operation involving the superimposition of layers made up of definite spatial objects with clear-cut boundaries measured without error according to some equally clear cut, but generally arbitrary, Boolean logic. Almost never in any real application are things as clear cut as this, and a variety of probabilistic alternatives have been evaluated. To some, many types of overlay are essentially a problem in linear modelling (Wang and Unwin, 1992). To others, the same problem can be formulated as one in Bayesian probability analysis (Agterberg, 1989; Bonham-Carter, 1991) and this has been implemented, with a very clear guide to the methods and concepts involved, in IDRISI (Eastman *et al.*, 1993).

Similarly, the issue of error and uncertainty in GIS results, although it often arises as a consequence of the ways in which the geography of interest is represented digitally (Unwin, 1995), is best resolved by adopting a suitable statistical model of the spatial variation involved and then estimating the parameters of this model. For spatially continuous field data this implies that use be made of the theory of regionalised variables and random fields. By far the best treatment casting this literature into a GIS framework is that in Heuvelink (1993). Any error model used must account for both the attribute error and those introduced by the mapping procedure used and its parameters can only be estimated if the spatial variability of both are properly defined: for example, using geostatistical interpolation (kriging) in one of its various forms (Atkinson, 1995; Oliver and Webster, 1990). It follows that good geostatistical tools must be high on any wish-list for statistical functions in GIS.

Although categorical map error can be addressed using the indicator approach from geostatistics, when dealing with error in objects life is usually rather more difficult and the statistical theory less mature, so that resort is usually made to simulation often conditioned by the observed data (Goodchild, 1989; Goodchild *et al.*, 1992a). It would be useful if general tools for examining how errors propagate in GIS were to be available (Heuvelink and Burrough, 1993; Heuvelink *et al.*, 1989), and recently progress has been made in the close coupling of an error- propagation tool called ADAM and a raster GIS (van Deursen, 1995). It can be argued that this type of facility should become part of any standard generic tool kit.

In addition to Bayesian estimation, geostatistical and error-propagation tools there is a long list of possible statistical functions that might be added to improve the 'functionality' for spatial statistical analysis of existing GIS. We all have our special favourites but, on the basis of the methods I am generally able to persuade my MSc and PhD students to adopt, my own list would add the following.

- All the methods for the analysis of point events developed by statisticians in the 40 years since that dreadful 'nearest neighbour' statistic was first proposed (Clark and Evans, 1954). Diggle (1983) and Ripley (1988) would be good places to start.

- Density estimation in the manner of Silverman (1986) as demonstrated by Gatrell (1994).

- Generalised linear-modelling tools, in the manner of GLIM (Aitkin *et al.*, 1989).

- A series of local statistical indicators of spatial association and inhomogeneity of the type proposed by Anselin (1993), Anselin and Getis (1992), Boots (1994) and Getis (1994).

- Decent exploratory tools based largely on visualisation and highly interactive engagement with the data as a legitimate and proper way to analyse spatial data (Unwin, 1994). With the possible exception of surface visualisation tools available in some raster systems, notably the '3D' extensions to GRASS, most GIS are woefully inadequate in this respect.

There may well be others, but the general nature of my list should be clear. It attempts to update what most geographers think of as spatial statistical analysis and thus correct the evident inability of many to take advantage of almost all of the developments in statistics, spatial statistical analysis and computing since the 1960s.

26.4 The concept of GISable statistics

A concept that has been developed by Openshaw and his colleagues (Openshaw and Clarke, 1996) is that of *GISable* statistics. By this they mean analytical and other approaches that are suited to a world in which computer power, very large data sets, and the availability of GIS should be taken for granted. The concept is a useful one, since, as they point out, it helps define a research agenda for developing new methods and draws attention to the fundamentally unsatisfactory, even unsound, nature of the traditional methods of statistical analysis when applied to spatial data. For example, almost all of the spatial statistical methods I included in *Introductory Spatial Analysis* (Unwin, 1981) should have very little place in our present research environment. Almost without exception they can be replaced by more recent techniques or, as is Openshaw's predilection, by a variety of computer-intensive procedures. The 10 'basic rules for identifying future GISable spatial analysis technology' put forward by Openshaw and Clarke, 1996, together with my own comments, are as follows.

Rule 1.   A GISable spatial analysis method should be able to handle large and very large N values. This seems self-evident in the days of widespread data availability. Openshaw and Clarke (1996) discuss it largely in terms of the implementation of functions for very large matrices. Somewhat inconsistently, although they argue for an ability to handle large N, they conclude that any statistical method that involves the use and manipulation of several N by N matrices is probably not sensible to include in a GIS.

Equally important, the traditional notions of sampling and confidence become irrelevant at very large N and are to a large extent a distraction that may serve to confuse rather than clarify. First, at very large N any global patterns that are detected (for example by a correlation between two variables) will almost certainly prove to be 'significant' and the issue of what population such large data sets 'sample' becomes very real. Second, as the area studied enlarges, or the resolution at which we view gets finer, so it becomes probable that any global statistic will mix together large areas over which there is no interesting variation with a few areas of real pattern. Slavish insistence on the significance of a global statistic will only mean that these local areas of interest are missed. Finally, it is often the case in GIS-based work that N itself is arbitrary. A good example occurs in the use of the raster data model where each sample object is a pixel. The ability to change the resolution means we can increase the N used (if not the real effective degrees of freedom) and make almost any result look significant. Quite what number should be taken as the sample size, for example, if we compute and model a contingency table of soil type cross-tabulated with land use using some form of categorical methods, I have not the faintest idea. Certainly, using such data to compute a global Moran's I as an index of spatial autocorrelation makes no sense at all. This is not, of course, to claim that statistical significance is irrelevant for studies involving correctly drawn samples of relatively small size from a well-understood population. Nor do I follow Openshaw in his declaration that almost all such formal approaches are irrelevant to GIS.

Rule 2.   Useful GISable analysis and modelling tools are study-region independent. This seems to me to be both self-evident and little understood. It is a clear consequence of the modifiable areal unit (MAU) problem and the word 'GISable' in the Rule is an irrelevance. Study boundary independence, in the sense of excluding any influence of arbitrary region edges, is a property that we should look for in virtually all the numbers we use in any geographical analysis. There are, perhaps, exceptions in studies that deal with computed values for spatial units based on administrative areas (county, state, nation, . .), but the general principle should be to look for numbers that are reasonably independent of any arbitrary boundaries. Many traditional spatial statistics depend on having a value for the area of the studied region. Many years ago, in an attempt to show that within drumlin fields the distribution of these landforms is spatially random, I misguidedly computed the Clark and Evans nearest neighbour statistic (my collaborator is blameless, see Smalley and Unwin, 1968). It should be abundantly clear that, simply by redefining the study area, I could have produced almost any R-index. As it was by total (mis)chance I seemed to hit on the scale of analysis that produced values close to the magical random expectation. This is an example of a very direct and obvious dependence on the study region selected which is taken through into the final global value by virtue of a calculation of the expected intensity of the point process using the study region area and the number of events. Similar, much more subtle dependencies occur in virtually all the work that we do. In the example given, the use of a measure of pattern based on an inter-event distance is reasonable (although there will be edge effect problems) and the technical fault would today easily be solved by randomisation.

Rather more difficult to solve are problems associated with area-enumerated data such as those obtained from a census of population as might be visualised using the familiar, if often misunderstood, choropleth map. Any density value or population ratio is highly dependent on the boundaries used and gives the classical MAU problem. Spatial statistical analysis can help here. If the mapped variable relates to a rate of incidence it can be estimated using a probability relative to an assumed process (Bailey and Gatrell, 1995) or by empirical Bayes (Langford, 1994). Alternatively, the problem can be transformed using a version of adaptive kernel estimation (Bracken and Martin, 1989) to estimate the underlying, continuous population-density function. Finally, using additional knowledge and an appropriate linear model (Flowerdew and Green, 1991) enables much better areal interpolation onto different sets of spatial units than can be obtained by the usual area weighting scheme used in most GIS.

Rule 3.   GIS relevant methods need to be sensitive to the special nature of spatial information. This seems self-evident but it remains unclear exactly what it is that is 'special' about spatial data. Anselin (1989) lists several characteristics that together differentiate it from other types of data. From the point of view of statistical analysis, the most salient fact is that they result not from a carefully controlled experimental design undertaken by the investigator specifically for the purpose of analysis but from messy surveys that, at least in the statistical sense, are 'unplanned'. As a consequence, in addition to the problems of spatial autocorrelation, modifiable areas, aggregation, and spatial scale, such data are often measured on inadequate and mixed measurement scales and contain spatially varying errors. In short, they are 'messy' and the methods used should recognise this.

Rule 4. The results should be mappable. This goes without saying, but see Unwin (1994), noting that the word 'mappable' should be taken to include almost any useful graphic display, including animation (Hearnshaw and Unwin, 1994).

Rule 5. GISable spatial analysis is generic. Existing GIS tool kits attempt to be truly generic, providing a set of functions in which very little is application specific. The commercial sense in this should be obvious but, equally, almost all of the methods listed in the previous section are equally generic, being applicable, for example, to any distribution of point, line, area or surface objects. As outlined above, this kind of generic utility may not extend into the area I have termed spatial modelling with the result that the GIS, or a slimmed-down version of it, becomes incorporated into the modelling software rather than vice versa.

Rule 6. GISable spatial analysis methods should be useful and valuable. Openshaw and Clarke (1996) see the motivation for this as 'luring' users into using spatial analysis and they draw the implication that the benefits of using it must outweigh the associated costs in the marketplace. Of course the methods used should be useful, but in academic use our benefits and costs might be seen rather differently from those accruing in commerce. Any cost/benefit equation, particularly if it involves costs which are IT related, is likely to have very limited validity over time. I would thus be less insistent on obvious immediate benefits than they are.

Rule 7. Interfacing issues are initially irrelevant and subsequently a problem for others to solve. Here Openshaw and Clarke (1996) argue that the important issue is gaining acceptance for the methods, not how they are implemented in today's computing environment. They point out that the computing world itself has changed enormously and that interfacing difficulties and compute intensiveness are no longer serious problems. If it can be done at all at present and there is a good case for doing it, then it will one day become common practice on affordable hardware located somewhere on a global network. As a not entirely irrelevant aside, they also point out that most of the available generic procedures incorporated into today's GIS typically run three or more *orders of magnitude* slower than they can be programmed in standalone software.

Rule 8. Ease of use and understandability are very important. My reservations concerning the wholesale incorporation of spatial statistical functions into GIS have already been listed. Typically, Openshaw and Clarke (1996) see this as a problem in technology, of making the functions easier to use and hiding 'the complexity of any relevant spatial analysis technology' behind 'suitable user interfaces'. As a dinosaur of a user who switches off Windows in favour of a DOS prompt and then programs most of what he does in FORTRAN, I am perhaps not the right person to comment on this, but as an educationalist (Unwin, 1991) my view is that, even if we accept its results, we cannot and should not rely on technology as a substitute for education and training.

Rule 9. GISable analysis should be safe technology. By 'safe' is meant that the results obtained should be reliable, robust, resilient and resistant to error and noise in both the data and the procedures used. This seems to me to be true for any technology, but I worry rather more about, for example, the safety of *fly-by-wire* technology than I do about GIS. The serious point is that as academics we have well-established quality-control mechanisms, such that the faults and problems inherent in any analytical methods are fairly rapidly discovered and corrected. With the exception, perhaps, of one or two real-time applications, most GIS results are not safety critical and I do not see why we should exclude methods simply because they can be misused.

Rule 10.   GIS methods should be useful in an applied sense. Finally, Openshaw and Clarke (1996) point out that GIS is primarily and predominantly an applied technology, with academic research falling into a different category with necessarily different criteria for usefulness. It is of interest to note that this utility of GIS in the marketplace, and of the concepts from quantitative/positivist geography that underpin it, is one of the major reasons for the current criticisms of GIS from the social theorists (Pickles, 1995). I suspect that in drafting this rule Openshaw and Clarke had in mind simply academic research into the underpinnings of GIS and using GIS within geography. If so, I would agree. These uses can take care of themselves and need not in any way influence the provision of facilities for the marketplace. What concerns me more is the use of GIS by our colleagues in other disciplines who have discovered 'geography' through the medium of GIS and whose only contact with the world of geography is through the same route. Clearly GIS is useful in these new applications, but it must give a strange view of what geography is all about and a very restricted notion of the potential of spatial information to help answer real scientific problems.

26.5 Conclusion: will it happen?

Thus far we have concluded that something called spatial analysis can be defined but that it can refer to several quite distinct approaches to knowledge creation. We have also demonstrated that, at least for the case of spatial statistical analysis, it can be useful in GIS. Our final question related to the assumption that it *will* be applied.

Clearly, spatial statistical analysis is not easy, and the eight impediments listed by Fotheringham and Rogerson (1993) remain as serious theoretical problems that, sooner or later, and no matter how user-friendly or powerful the technology adopted, all users of spatial data must face (these are the MAU and boundary problems, spatial interpolation, sampling and autocorrelation, goodness of fit, context-dependent results and non-stationarity, and aggregate or disaggregated models).

In deciding whether or not spatial statistical analysis will continue to be used over an ever-increasing range of applications in GIS, and I am by no means certain that it will, the basic problem has less to do with the availability of software, still less of software within a GIS framework, and more to do with disciplinary myopia and academic fashion. My experiences of the past year, during which I have visited, as a sort of inspector, a number of other geography departments in UK, convince me that not only do we teach less spatial statistical analysis to our undergraduates than we did 20 years ago before GIS came along, but also that by and large we teach it less thoroughly using techniques and materials appropriate to pre-computer days. Lots of people try to teach GIS as a technology without paying any attention to its theoretical underpinnings and the position is depressingly similar to the 1960s when statistical analysis became the fashionable item that displaced cartography and survey from geography curricula.

It is thus hardly surprising that research students do not cast their problems into frameworks for which modern spatial statistical analysis might be appropriate and that in GIS we remain ossified with a ragbag of functions that could have been (and were) implemented in software over 25 years ago. If we cannot get our act together on this, it seems unlikely that we will persuade colleagues in other disciplines who use GIS, still less the GIS vendors, to do anything about providing the necessary tools.

References

AGTERBERG, F. P. (1989). Computer programs for mineral exploration, *Science*, **245**, 76–81.

AITKIN, M., ANDERSON, D., FRANCIS, B. and HINDE, J. (1989). *Statistical Modelling in GLIM*, Oxford: Clarendon Press.

ANSELIN, L. (1989). 'What is special about spatial data? Alternative perspectives on spatial data analysis', National Center for Geographic Information and Analysis, Technical Paper, 89–4, Santa Barbara: NCGIA.

(1990). *SpaceStat: A Program For the Statistical Analysis of Spatial Data*, Santa Barbara: Department of Geography, University of California.

(1993). *The Moran Scatterplot as an ESDA Tool to assess Local Instability in Spatial Association*, Research Report 9330, Regional Research Institute, Morgantown, WV: West Virginia University.

ANSELIN, L. and GETIS, A. (1992). Spatial statistical analysis and geographic information systems, in Fischer, M. M. and Nijkamp, P. (Eds), *Geographic Information Systems, Spatial Modelling and Policy Evaluation*, pp. 35–49, Berlin: Springer-Verlag.

ANSELIN, L., DODSON, R. F. and HUDAK, S. (1993). Linking GIS and spatial data analysis, *Geographical Systems*, **1**(1), 3–23.

ATKINSON, P. M. (1995). A method for describing quantitatively the information, redundancy and error in digital spatial data, in Fisher, P. F. (Ed.), *Innovations in GIS 2*, pp. 85–96, London: Taylor & Francis.

BAILEY, T. C. (1990). GIS and simple systems for visual, interactive spatial analysis, *The Cartographic Journal*, **27**(2), 79–84.

BAILEY, T. C. and GATRELL, A. (1995). *Interactive Spatial Data Analysis*, Harlow: Longman.

BONHAM-CARTER, G. F. (1991). Integration of geoscientific data using GIS, in Maguire, D. M., Goodchild, M. F. and Rhind, D. W. (Eds), *Geographical Information Systems: Principles and Applications, vol. 2*, pp. 171–84, London: Longman.

BOOTS, B. N. (1994). Visualizing spatial autocorrelation in point data, *Geographical Systems*, **1**(4), 255–66.

BRACKEN, I. and MARTIN, D. (1989). The generation of spatial population distributions from census centroid data, *Environment and Planning, A*, **21**(4), 537–43.

BRUNSDON, C. and CHARLTON, M. (1995). Developing an exploratory spatial analysis system in XLisp-Stat, in Parker, D. (Ed.) *Innovations in GIS 3*, pp. 135–145, London: Taylor & Francis.

CLARK, M. (1990). Geographical information systems and model-based analysis: towards effective decision support systems, in Scholten, H. J. and Stilwell, J. C. H. (Eds), *Geographical Information Systems for Urban and Regional Planning*, pp. 165–75, Dordrecht: Kluwer.

CLARK, P. J. and EVANS, F. C. (1954). Distance to nearest neighbour as a measure of spatial relationships in populations, *Ecology*, **35**(2), 445–53.

CLIFF, A. D and ORD, J. K. (1973). *Spatial Autocorrelation*, London: Pion.

COMMISSION OF EUROPEAN COMMUNITIES (1995). DG XIII GI2000: Towards a European

Geographic Information Infrastructure, Draft, May 1995, Luxemburg: CEC DG XIII.

CRESSIE, N. A. C. (1991). *Statistics for Spatial Data*, New York: John Wiley & Sons.

DENSHAM, P. J. (1994). Integrating GIS and spatial modelling: visual interactive modelling and location selection, *Geographical Systems*, **1**(3), 203–19.

VAN DEURSEN, W. P. A. (1995). *Geographical Information Systems and Dynamic Models*, 190, Utrecht: Netherlands Geographical Studies.

DIGGLE, P. J. (1983). *Statistical Analysis of Spatial Point Patterns*, London: Academic Press.

DIGGLE, P. J. and ROWLINGSON, B. S. (1993). SPLANCS: spatial point pattern analysis code in S-Plus, *Computers and GeoSciences*, **19**(5), 627–55.

DYKES, J. A. (1995). Pushing maps past their established limits: a unified approach to cartographic visualization, in Parker, D. (Ed.) *Innovations in GIS 3*, pp. 177–187, London: Taylor & Francis.

EASTMAN, J. R. (1990). IDRISI: A grid-based geographic information system, Worcester, MA: Department of Geography, Clark University.

EASTMAN, J. R., KYEM, P. A. K., TOLEDANO, J. and WEIGEN, J. (1993). *GIS and Decision Making. Explorations in Geographic Information Systems Technology*, vol. 4, Geneva: United Nations Institute for Training and Research (UNITAR).

FISCHER, M., SCHOLTEN, H. J. and UNWIN, D. J. (Eds) (1996). *Spatial Analytical Perspectives in GIS*, London: Taylor & Francis.

FLOWERDEW, R. and GREEN, M. (1991). Data integration: methods for transferring data between zonal systems, in Masser, I. and Blakemore, M. (Eds), *Handling Geographic Information: Methodology and Potential Applications*, pp. 38–54, Harlow: Longman.

FOTHERINGHAM, A. S. (1992). Exploratory spatial data analysis and GIS: commentary, *Environment and Planning, A*, **24**(12), 1675–8.

FOTHERINGHAM, A. S. and ROGERSON, P. A. (1993). GIS and spatial analytical problems, *International Journal of Geographical Information Systems*, **7**(1), 3–19.

FOTHERINGHAM, A. S. and ROGERSON, P. A. (Eds) (1994). *Spatial Analysis and GIS*, London: Taylor & Francis.

GATRELL, A. C. (1994). Density estimation and the visualization of point patterns, in Hearnshaw, H. M. and Unwin D. J. (Eds) *Visualization in Geographical Information Systems*, pp. 65–76, Chichester: John Wiley.

GETIS, A. (1994). Spatial dependence and heterogeneity and proximal databases, in Fotheringham, S. and Rogerson, P. (Eds), *Spatial Analysis and GIS*, pp. 105–20, London: Taylor & Francis.

GOODCHILD, M. F. (1987). A spatial analytical perspective on geographical information systems, *International Journal of Geographical Information Systems*, **1**(4), 335–54.

(1989). Modelling error in objects and fields, in Goodchild, M. F. and Gopal, S. (Eds), *Accuracy of Spatial Databases*, pp. 107–13, London: Taylor & Francis.

GOODCHILD, M. F., GUOQING, S. and SHIREN, Y. (1992a). Development and test of an error model for categorical data, *International Journal of Geographical Information Systems*, **6**(2), 87–104.

GOODCHILD, M. F., HAINING, R. P. and WISE, S. M. (1992b). Integrating GIS and spatial data analysis: problems and possibilities, *International Journal of Geographical Information Systems*, **6**(5), 407–23.

GOODCHILD, M. F., PARKS, B. O. and STEYAERT, L. T. (Eds) (1993). *Environmental Modelling with GIS*, Oxford: Oxford University Press.

HAINING, R. P. (1990). *Spatial Data Analysis in the Social and Environmental Sciences*, Cambridge: Cambridge University Press.

HAINING, R. P. and WISE, S. M. (1991). *GIS and Spatial Data Analysis. Report on the Sheffield Workshop*, Discussion Paper 11, Regional Research Laboratory Initiative, Sheffield: Department of Town and Regional Planning, University of Sheffield.

HASLETT, J., WILLS, G. and UNWIN, A. (1990). SPIDER – an interactive statistical tool for the analysis of spatially distributed data, *International Journal of Geographical Information Systems*, **4**(3), 285–96.

HEARNSHAW, H. M. and UNWIN, D. J. (Eds) (1994). *Visualization in Geographical Information Systems*, Chichester: John Wiley.

HEUVELINK, G. B. M. (1993). *Error Propagation in Quantitative Spatial Modelling: Applications in Geographical Information Systems*, Technical Report No. 163, Utrecht: Netherlands Geographical Studies.

HEUVELINK, G. B. M. and BURROUGH, P. A. (1993). Error propagation in logical cartographic modelling using Boolean logic and continuous classification, *International Journal of Geographical Information Systems*, **7**(3), 231–46.

HEUVELINK, G. B. M., BURROUGH, P. A. and STEIN, A. (1989). Propagation of errors in spatial modelling with GIS, *International Journal of Geographical Information Systems*, **3**(4), 303–22.

LANGFORD, I. (1994). Using empirical Bayes estimates in the geographical analysis of disease risk, *Area*, **26**(2), 142–50.

OLIVER, M. A. and WEBSTER, R. (1990). Kriging: a method of interpolation for geographical information systems, *International Journal of Geographical Information Systems*, **4**(3), 313–32.

OPENSHAW, S. (1991). A spatial analysis research agenda, in Masser, I. and Blakemore, M. (Eds), *Handling Geographical Information*, pp. 18–37, London: Longman.

   (1994). Two exploratory space-time attribute pattern analysers relevant to GIS, in Fotheringham, S. and Rogerson, P. (Eds), *Spatial Analysis and GIS*, pp. 83–104, London: Taylor & Francis.

   (1995). Developing automated and smart spatial pattern exploration tools for geographical information systems applications, *The Statistician*, **44**(1), 3–16.

OPENSHAW, S. and CLARKE, G. (1996). Developing spatial analysis functions relevant to GIS environments, in Fischer, M., Scholten, H. J. and Unwin, D. J. (Eds), *Spatial Analytical Perspectives on GIS*, London: Taylor & Francis (in press).

PICKLES, J. (Ed.) (1995). *Ground Truth: The Social Implications of Geographical Information Systems*, New York: Guilford.

RIPLEY, B. D. (1981). *Spatial Statistics*, New York: John Wiley.

   (1988). *Statistical Inference for Spatial Processes*, Cambridge: Cambridge University Press.

ROWLINGSON, B. S. and DIGGLE, P. J. (1991). *SPLANCS: Spatial Point Pattern Analysis in a Geographical Information Systems Framework*, North West Regional Research Laboratory, Research Reports 23, Lancaster: Lancaster University.

SCHOLTEN, H. (1994). Specialist meeting report, GIS and spatial analysis, in *GISDATA Newsletter*, no. 3, 6–8.

SILVERMAN, B. W. (1986). *Density Estimation in Statistics and Data Analysis*, London: Chapman & Hall.

SMALLEY, I. J. and UNWIN, D. J. (1968). The formation and shape of drumlins and their distribution and orientation in drumlins fields, *Journal of Glaciology*, **7**, 377–90.

UNWIN, D. J. (1981). *Introductory Spatial Analysis*, London: Methuen.

   (1991). The academic setting of GIS, in Maguire, D. M., Goodchild, M. F. and Rhind, D. W. (Eds), *Geographical Information Systems: Principles and Applications, vol. 1*, pp. 81–90, London: Longman.

   (1994). ViSc, GIS and cartography, *Progress in Human Geography*, **18**(4), 516–22.

   (1995). Geographical information systems and the problem of error and uncertainty, *Progress in Human Geography*, **19**(4), 549–58.

   (1996). Integration through overlay analysis, in Fischer, M., Scholten, H. J. and Unwin, D. J. (Eds), *Spatial Analytical Perspectives on GIS*, London: Taylor & Francis (in press).

UPTON, G. J. and FINGLETON, B. (1985). *Spatial Statistics by Example, vol. 1: Point Patterns and Quantitative Data*, Chichester: John Wiley.

—— (1989). *Spatial Statistics by Example, vol. 2: Categorical and Directional Data*, Chichester, John Wiley.

WALDEN, A. T. and GUTTORP, P. (Eds) (1992). *Statistics in the Environmental and Earth Sciences*, London: Edward Arnold.

WANG, S.-Q. and UNWIN, D. J. (1992). Modelling landslide distribution on loess soils in China: an investigation using GIS techniques, *International Journal of Geographical Information Systems*, **6**(5), 391–405.

CHAPTER TWENTY-SEVEN

Linking GIS and Multilevel Modelling in the Spatial Analysis of House Prices

NINA BULLEN

27.1 Introduction: macro- and micro-scale analysis of house prices

Previous house price research has tended to focus either at the micro-scale of individual properties or the macro-scale of areal average prices (Jones and Bullen, 1993). Analysis based purely at the aggregate level or at the individual level can incur either the 'ecological' or the 'atomistic' fallacies (Dogan and Rokkan, 1969). These issues are discussed in the context of aggregating data on social and political behaviour by Olson (Chapter 30 in this volume). In the case of house price research, analysis at the individual property level often assumes that the same predictor/price relationships held everywhere. This approach misses the context in which housing markets operate and property transactions take place. Conversely, models specified exclusively at the aggregate level may incur the 'ecological' or 'aggregative' fallacy (Susser, 1973; Jones, 1995), the assumption that areal-level relationships apply uniformly at the individual level. In this case, contextual effects (the difference a place makes) are potentially confounded with compositional effects (due to variations in the mix of housing attributes within each place). Attempting to control for compositional effects by including an aggregate level variable, such as average house size, in a regression-based model cannot take account of the potentially varying relationships within an area (Jones and Bullen, 1993).

Another approach is to standardise the composition or 'mix' of housing characteristics sold in different areas and in different periods, so that comparisons are on a like-with-like basis. In the UK, for example, the Halifax Building Society publishes a house-price index based on a hedonic estimate of the cost associated with each property attribute (Fleming and Nellis, 1984). A major shortcoming of the hedonic approach is its treatment of contextual variation. Regional average prices are obtained by using a set of dummy variables to indicate the region in which each transaction occurs. In effect, separate regression equations are fitted to each region. However, several authors have argued that the assumption of uniform price coefficients is unrealistic even within a metropolitan area (Goodman, 1978, 1979, 1981; Palm, 1977), due to the existence of distinct submarkets (Straszheim, 1975). Again, the usual approach is to subdivide the data into smaller areal units, frequently called 'submarkets', and to perform a series of separate regressions. This is problematic on several counts. First, it reduces the amount of data from which to estimate the coefficients for each area. This may incur the 'small number problem' (Jones and Kirby, 1980). A single-level Ordinary Least Squares (OLS) analysis is likely to produce imprecise parameter estimates when the number of observations in each place is small and/or the range of the predictor variable is limited. For example, if all the property transactions in an area were three-bedroom properties, the slope term for that place could not be estimated at all (Jones and Bullen, 1994). Further, it is well known that observations within clusters (areas) tend to be more similar than a purely random sample. This is particularly likely with housing submarkets, which are generally defined to maximise homogeneity of housing characteristics. Regression-based analysis using OLS estimation can lead to substantially biased estimates of regression slope terms and their standard errors because it assumes that there is no residual autocorrelation within submarkets.

This brief synopsis of the methodological problems suggests that statistical modelling of the spatial variation in property prices needs to distinguish between compositional and contextual effects, to estimate geographical variation while simultaneously allowing for variations in individual property attributes, and be robust enough to deal with sampling fluctuations, unbalanced sampling design and the small-number problem. Multilevel modelling deals with these issues explicitly. Section 27.2 provides some background to the development of these models and their significance to the analysis of house price variation.

27.2 Contextualised quantitative modelling

Quantitative modelling within the social sciences has traditionally attempted to relate a response variable to a predictor by a universal constant. It is now increasingly recognised that model relationships may vary according to context, whether structural, institutional or geographical. Randolph (1991) stresses the social, historical and spatial contingency of housing market processes, arguing that 'although the major discontinuities in housing market structure can be understood at the aggregate level, the detailed segmentation of local housing markets is only definable through specific empirical analysis. The main implication of this is that the outcome of apparently similar sets of processes will vary between localities and over time because of the historical and social contexts within which they are embedded' (p. 35).

An important contribution to the development of contextualised quantitative modelling comes from the work of Casetti (1972, 1991, 1992) and Jones and Casetti (1992). The 'expansion method' is described both in terms of a research paradigm (Casetti, 1993; Jones, 1992) and a modelling technique that allows mathematical models to be modified to allow for contextual variation in parameter relationships.

Much of the literature on the 'expansion method' incorporates 'contextual drift' in model parameters by expanding the 'fixed' part of the model (see, for example, Can, 1990). Being equivalent to standard ANOVA and ANCOVA procedures, it is the most straightforward approach in computational terms because OLS can be used to estimate the parameters, using any standard statistical package. Models that involved an expansion of the 'random' part (Lindley and Smith, 1972) were more difficult to implement due to the difficulty of developing a valid estimation strategy. Developments during the 1980s, however, resulted in alternative modelling procedures that are variously known as 'multilevel modelling' (Goldstein, 1985, 1987); 'hierarchical linear modelling' (Bryk and Raudenbush, 1992; Raudenbush and Bryk, 1986) and 'variance components analysis' (Longford, 1986). Estimation of these models requires specialist software, such as the MLn package (Rasbash and Woodhouse, 1995); the HLM package (Bryk *et al.*, 1986) and the VARCL package (Aitken and Longford, 1986; Longford, 1986). For further details of these software packages and the range of model specifications which can be estimated, see Goldstein (1995), Jones (1991a, 1992a) and Kreft *et al.* (1990).

The next section gives a brief introduction to multilevel modelling, focusing on the two-level model which is estimated in the empirical case study that follows. There is a growing literature on multilevel modelling;[1] for more details on model specification and interpretation, see, for example, Goldstein (1987; 1995) and Jones (1991a and b, 1992b).

27.2.1 Introduction to multilevel modelling

Multilevel modelling is basically an extension of the general linear model to allow the development of models with several 'levels' or scales of analysis. There are many instances when the data structure is hierarchical, either intrinsically (for example, houses are situated within neighbourhoods), or due to the data collection procedure used (stratified or multistage sampling). Alternatively, the cross-classified structure (Goldstein, 1987) is ideal for data that are not purely nested. For example, properties at level 1 may belong to spatially overlapping units, such as census wards at level 2 and postal districts, also at level 2. An advantage of the cross-classified structure is that data collected for different spatial units can be readily incorporated into a multilevel model.

Midtbø (Chapter 15 in this volume) argues that the efficient handling of large spatial data sets requires appropriate data structures and better methods for spatial modelling. He discusses the spatial sampling of data points to extract representative information at the desired resolution. A multilevel data structure raises the issue of representative sampling to reflect the conceptual structure of the model, whether spatial or structural (for example, based on social groupings or institutions). The key point is that between-group differences cannot be estimated if the sample has hundreds of individuals nested within only a handful of groups; reliable estimates of between-place differences requires data from many places (for guidelines on sample sizes at each level, see Paterson and Goldstein, 1992; Mok, 1995).

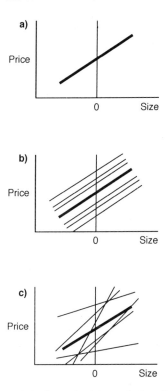

Figure 27.1 Alternative models for price/size relationships.

27.2.2 A two-level multilevel model of house prices

The usual single-level bivariate regression model is composed of two parts: fixed and random. The fixed part, shown in Fig. 27.1a, represents the underlying systematic relationship between a continuous response variable, y (the price of a property) and a continuous predictor variable, x (such as a measure of property size). The estimated values of the two 'fixed' parameters, the intercept and the slope terms, are assumed to apply to the whole sample. Fluctuations around this general relationship, after taking account of size, are captured by the random term. A major shortcoming of this single-level model is that any systematic place or contextual differences are subsumed within the general fixed relationship.

The simplest multilevel model structure is a two-level model with individual properties, i, nested within places, j. In addition to fitting the general relationship, the fixed slope and intercept terms vary between places/contexts. The intercept term is allowed to vary between places by specifying a between-place macro-model in which the price of an average-sized property in place j is seen as a function of the overall average price, plus a differential for each place. Rather than estimate a series of separate micro-models for each area, the multilevel model expands the fixed intercept at the place-level. The micro- and macro-models are then combined to form the 'random intercepts' multilevel model shown in Fig. 27.1b. For clarity, only six place-relationships are shown. The heavy line represents the overall average price/size relationship. Some areas attract a premium, whereas others are relatively inexpensive. Conceptually, the multilevel model does not view each place as a separate entity, but as part of a distribution of places. The two sets of random terms are assumed to be independent of each other. Under the usual assumptions, their distributions can be summarised by the variance terms at level 1 and at level 2 (Goldstein, 1987).

The parallel lines of Fig. 27.1b imply that the price/size relationship holds everywhere. This restrictive assumption is relaxed by specifying an additional macro-model in which the fixed slope term is allowed to vary between places. Combining the initial micro-model with the two macro-models produces the 'fully random' multilevel model in which both the slopes and the intercepts are allowed to vary between places (Fig. 27.1c). This represents a complex interaction between property size and place, with contextual variation in the effect of property size on price. There is no simple map of average property prices, but a complex geography of differential pricing, even after taking account of the housing mix or composition within each area.

27.2.3 Spatial units to reflect submarkets

Attention is now turned to the question of what constitutes the appropriate geographical context for a study of house price variation. Preliminary work has developed multilevel models in which individual properties nest within existing spatial units, such as UK counties (Jones and Bullen, 1993), London boroughs (Jones and Bullen, 1994) or Travel to Work Areas (Bullen and Jones, 1994). Although these higher-level units are clearly defined, they may represent 'chaotic conceptions' in terms of the operation of housing markets; there may be as much (or more) variation within these areas as between them. For example, TTWAs represent fairly self-contained labour market areas and have been used in previous house price research on the grounds that 'most house purchasers are constrained by commuting distance from their workplace' (Brunsdon et al., 1991, p. 232). However, the London and Heathrow TTWAs cover a large area because the commuting patterns are extensive and complex in the London area. In this case, errors of aggregation can arise: 'Pooling data over a wide geographic area implies that a single underlying price structure exists; that is, that the price premium for quality attributes is invariant across locations ... The size of the bias associated with such a procedure will depend on the true underlying structure of prices across markets' (Straszheim, 1975, p. 70). Straszheim (1975) suggests that if the true structural coefficients vary across submarkets, the data must either be stratified by submarket, or estimated such that particular coefficients are allowed to vary between j spatial and t temporal submarket divisions. This ties in closely with the multilevel modelling approach outlined above and suggests that individual property data should be aggregated to areas that reflect housing submarkets at the appropriate level 2 units.

However, although most urban economists agree that housing submarkets provide a useful working hypothesis, there is less consensus on how submarkets should be empirically defined and whether they can be described in spatial terms. Ball and Kirwan (1977) argued that the frequently observed homogeneity of housing characteristics within localities is not sufficient to establish the existence of separate submarkets. Their study of owner-occupation in Bristol identified spatial clustering of property types, but concluded that this did not constitute submarkets in the economic sense because they failed to find contextual variation in the implicit attribute prices. Spatial variation in property prices was due to the effects of different attribute mixes (composition effects). They suggest that a homogeneous housing market must include a uniform price structure for all the attributes of housing so that submarkets can be identified by differentiated attribute prices.

Goodman (1981) examined the residuals generated from hedonic price regressions for a sample of houses. He compared alternative submarket groupings using a set of criteria proposed by Cliff *et al.* (1975): *simplicity* – fewer submarkets are preferable to many submarkets; *similarity* – there should be a high degree of homogeneity of housing bundles within submarkets; *compactness* – contiguous areas should be grouped together within a submarket. This introduces an explicitly spatial issue: how to find the optimal zoning system or spatial partitioning that completely covers a study area so that each basic spatial unit (in this case, the property), is allocated to only one zone and the members of any zone are spatially contiguous. Openshaw (1977a) proposed an heuristic procedure for defining the optimal spatial partitioning based on Ward (1963), in which each basic spatial unit is initially allocated to a unique zone as sole member. The pair of zones whose aggregation produces the greatest improvement in the value of an objective function are then merged. This iterative procedure continues until either there is only one zone left (complete aggregation) or until topological discontinuities in the contiguity matrix prevent further aggregation of zones.

27.2.4 Using GIS to define housing market areas

Based on earlier work (Openshaw, 1977a and b), Openshaw and Rao (1994, 1995) have recently developed a zone design system, ZDES version 1.0, for use with ARC/INFO, although they note that the algorithms for the automatic zoning procedure could be linked to any geographical information system able to provide the appropriate contiguity information and to handle the zone aggregation process. ZDES consists of a set of AML and FORTRAN programs that provide a range of facilities for zone manipulation. In particular, the user can store zone systems as polygon coverages; output new zone systems as polygon coverages; aggregate selected spatial data to the new zone system, saving the output in the polygon attribute tables of the new coverage; and produce basic statistics on the input and output zone systems. Crucially, however, ZDES allows the user to design new zoning systems by specifying the function objectives and the optimisation methods. The current research will specify a function objective that produces zoning to optimise subregion homogeneity, based on the residuals from a two-level multilevel model of properties within TTWAs, in which all available property attributes are included as fixed terms. This will allow an empirical exploration of housing submarkets and define an intermediate zoning to reflect housing submarket areas. The new set of spatial units will provide an additional 'level' between properties and TTWAs, for further development of the multilevel models.

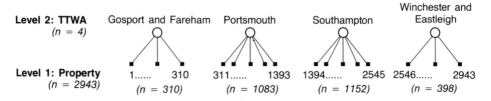

Figure 27.2   Two-level multilevel model for the Hampshire property data.

27.3 Case study: a two-level multilevel model for the Hampshire area

This section presents a preliminary empirical study designed mainly to pilot the use of the zone design algorithm, ZDES (Openshaw and Rao, 1994, 1995). Since the current UNIX workstation version of ZDES can only handle a few thousand zones, the structure is a two-level multilevel model of property transactions at level 1, grouped within four Travel to Work Areas around Hampshire, southern England (Fig. 27.2). The data on property transactions were obtained from the Nationwide Building Society. Table 27.1 provides further details, including a profile of the property attributes, which included both continuous and categorical variables (details on model specification for categorical variables are given in Bullen *et al.*, 1996).

Crucially, the database contained a spatial reference (postcode) which enabled the results from the statistical analysis to be linked with the GIS, adopting a 'loose coupling' approach via ASCII files (Goodchild *et al.*, 1992). The postcodes were linked to their 100m Ordnance Survey grid reference using the Central Postcode Directory (Raper *et al.*, 1992), maintained by the Post Office and available through the ESRC Data Archive. Randomising each point within its 100-m grid square gave each property a unique location (Fig. 27.3).

Since ZDES requires zone-based data, the point coverage of properties was converted to a zonal (polygon) coverage. In the absence of 'real' boundaries, the Voronoi technique identifies artificial boundaries based on the breakpoints between adjacent pairs of points (Clark, 1993). Arc/Info was used to generate a coverage of Thiessen polygons, based simply on the midpoint between adjacent points.

The multilevel models were estimated using MLn. The 'null' (variance components) model with no predictor variables provides a baseline for comparison (Table 27.2). The response variable is the log of the price, so that the fixed term (constant), when transformed, gives an estimate of the overall average price for all properties in the study area of £62 318. The random terms estimate the variance at each level. In this case, most (92 per cent) of the total variation is at level 1, between individual properties. The between-place (level 2) variation is small, but significant. This is partly because there are only four TTWAs (multilevel models usually require many higher level units to estimate between-place variation).

The model specification was then expanded to take account of compositional effects by including the property attributes as level 1 predictor variables. These are specified as 'fixed' terms and contribute to the overall, general relationship. The constant term (intercept) represents the price of the stereotypical property in the sample, which is a secondhand terraced house, built around 1945, freehold,

Table 27.1 Data summary: Hampshire and the surrounding area

Variable category	n	%	Variable category	n	%
Property type			*Month of approval*		
Terrace	1133	38.5	*1990*		
Semi-detached house	621	21.1	January	15	0.5
Detached house	465	15.7	February	81	2.8
Country cottage	4	0.1	March	245	8.3
Detached bungalow	125	4.2	April	32	1.1
Semi-detached bungalow	76	2.6	May	72	2.4
			June	150	5.1
Flat	470	16.0	July	150	5.1
Maisonette	49	1.7	August	117	4.0
			September	56	1.9
First-time purchase			October	125	4.2
No	1654	56.2	November	214	7.3
Yes	1289	43.8	December	120	4.1
New Property					
No	2384	96.3	*1991*		
Yes	109	3.7	January	127	4.3
			February	83	2.8
Tenure			March	98	3.3
Freehold	2379	80.8	April	127	4.3
Leasehold	564	19.2	May	167	5.7
Garage			June	197	6.7
No garage/parking space	1585	53.9	July	151	5.1
Single garage	1220	41.5	August	111	3.8
Double garage	138	4.7	September	139	4.7
			October	155	5.3
Central heating			November	177	6.0
Full central heating	2096	71.2	December	34	1.1
Partial central heating	269	9.1			
No central heating	578	19.6	**Total:**	2943	
Number of bathrooms					
1 bathroom	2708	92.0			
2 bathrooms	214	7.3			
3 bathrooms	9	0.3			
0 bathrooms	12	0.4			
Number of bedrooms					
1 bedroom	346	11.8			
2 bedrooms	875	29.7			
3 bedrooms	1339	45.5			
4 bedrooms	332	11.3			
5 bedrooms	47	1.6			
0 bedrooms	4	0.1			

Notes:
1 All mortgages were to ordinary borrowers, granted for house purchase by owner occupiers.
2 The response variable was the log of the price. Properties which sold for less than £25 000 (the minimum price after excluding properties sold for a nominal value of £1) were excluded.
3 Floor area: the data were centred around the average floor area of 747 units.
4 The age of the property was calculated from the date built: property built in 1991 was up to one year old at the beginning of 1992. Property age was modelled as a continuous variable, centred around the average age of 47 years.
5 The month of approval was modelled as a continuous variable, centred around January 1991.

with one bathroom, three bedrooms, full central heating and no garage or a parking space. It was sold in January 1991 to an existing owner-occupier (rather than a first-time buyer) for an estimated £52 052. Table 27.3 shows the fixed-term estimates, transformed to give the average premium/reduction associated with each property attribute, compared with the stereotypical property. Note that the additional fixed terms have had an effect on the random terms at each level. The

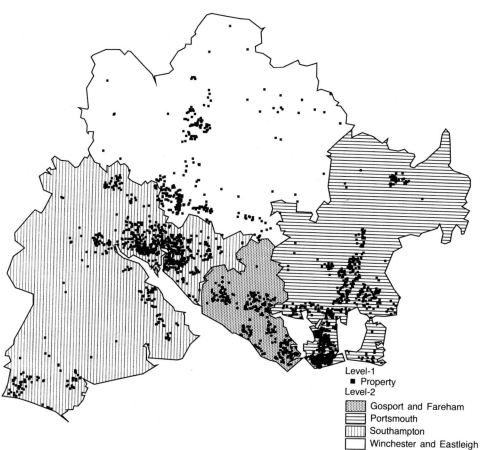

Figure 27.3   Spatial distribution of the Hampshire property data.

reduction in the house-level variance, compared to the previous 'null' model with
no predictor variables, reflects the fact that some level 1 heterogeneity was due
to measurable house characteristics. The level 2 variance term has also decreased
in absolute terms, suggesting that between-place price differentials are less
marked, once the characteristics of the properties sold within each TTWA are
allowed for. Tables 27.4 and 27.5 compare the predicted prices by TTWA, for each
model, based on all properties and on the stereotypical property respectively. Even
after taking account of the composition, there is some contextual effect on property
prices, which is highest within Winchester and Eastleigh TTWA.

The geographical variation in individual predicted prices was examined by
appending the level 1 residuals (obtained from MLn) to the correct zones in the
Arc/Info database. Figure 27.4 plots the level 1 residuals (positive/negative). The
patterning implies a contextual effect such that a uniform price structure does not
apply across the Hampshire area and suggests possible boundaries for housing
market areas.

The next stage will be to apply the ZDES regionalisation algorithm (Openshaw
and Rao, 1994, 1995) to aggregate individual properties to 'housing market areas'.
The input coverage of Thiessen polygons will be aggregated on the basis of the
residual values from the fully specified multilevel model using a function that

Table 27.2 The 'null' multilevel model

Parameter	Estimate	SE (U)	Transformed price (£)
Fixed part estimates			
Constant	11.04	0.05802	62 318
Random part estimates			
Level 2 variance	0.01319	0.009502	
Level 1 variance	0.1459	0.003806	

Table 27.3 Multilevel model with fully-specified fixed part

Parameter	Estimate	SE (U)	Transformed price (£)
Fixed part estimates			
Constant	10.86	0.03039	52 052
Semi-detached house	0.1048	0.01044	5 751
Detached house	0.3748	0.01415	23 668
Cottage	0.4754	0.09929	31 682
Bungalow (detached)	0.3692	0.01948	23 245
Bungalow (semi)	0.2065	0.02410	11 939
Flat	−0.0976	0.01414	−4 839
Maisonette	−0.0923	0.02940	−4 588
New property	0.1009	0.02036	5 526
1 garage	0.1251	0.00911	6 937
2 garage	0.3244	0.02055	19 946
Bathroom (2+)	0.1103	0.0155	6 070
1 bedroom	−0.2018	0.01692	−9 512
2 bedroom	−0.0679	0.00973	−3 416
4 bedroom	0.2237	0.01457	13 049
5 bedroom	0.3442	0.03158	21 386
Partial heating	−0.0257	0.0131	−1 319
No heating	−0.0937	0.00987	−4 657
Floor area	2.873e-06	7.76e-06	1.6
Floor area squared	−2.507e-05	7.949e-11	0.1
Property age	4.193e-04	7.877e-05	22
Month	−0.03504	0.00743	−1 792
Month squared	0.00206	6.574e-04	107
Month cubed	−4.306e-05	1.701e-05	−2
Random part estimates			
Level 2 variance	0.003305	0.002387	
Level 1 variance	0.03903	0.001018	

Table 27.4 Place-specific predicted prices for the 'null' multilevel model

TTWA	Level 2 residual	Rank	Predicted price (£)
Gosport and Fareham	0.0052919	2	62 648
Portsmouth	−0.13028	4	54 706
Southampton	−0.053451	3	59 074
Winchester and Eastleigh	0.17841	1	74 489

Table 27.5 Place-specific predicted prices for the model with fully-specified fixed part

TTWA	Level 2 residual	Rank	Predicted price (£)
Gosport and Fareham	−0.036814	4	50 171
Portsmouth	−0.036037	3	50 210
Southampton	−0.025285	2	50 752
Winchester and Eastleigh	0.098183	1	57 422

optimises region homogeneity. The outcome will be an Arc/Info coverage of new zones, to which the property data will be linked. At the time of writing, this stage is still in progress. By simply recoding the residual values to discrete values for positive/negative and 'dissolving' the Thiessen polygon coverage on this basis, it is possible to get an approximate idea of the likely housing market areas (Fig. 27.5).

The multilevel models will be developed further based on these empirically defined housing market zones. Possible extensions include the specification of complex heterogeneity (Bullen *et al.*, 1995), cross-level interactions and the inclusion of higher-level variables (Jones and Bullen, 1993). A cross-classified structure would allow for multiple contexts with spatially overlapping submarkets, which ties in with housing market processes since each segment of the housing market has its own spatial characteristics (Randolph, 1991).

27.4 Conclusions

Multilevel modelling can usefully be applied to the analysis of house prices in terms of taking account of micro- and macro-level sources of variation and allowing model parameters to vary according to context. This chapter has focused on a central issue: what is the appropriate context, in terms of spatial units, for an analysis of house price variation?

Although there is some debate as to whether housing submarkets can be spatially defined, this chapter has outlined a pilot study that explores the existence of submarkets based on a sample of properties within four TTWAs around Hampshire. The approach adopted involves a loose coupling (Goodchild *et al.*, 1992) between the specialist software package used to fit the multilevel models,

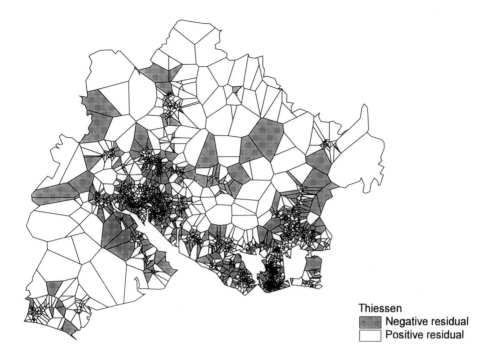

Thiessen
▨ Negative residual
☐ Positive residual

Figure 27.4 Plot of the multilevel model residuals.

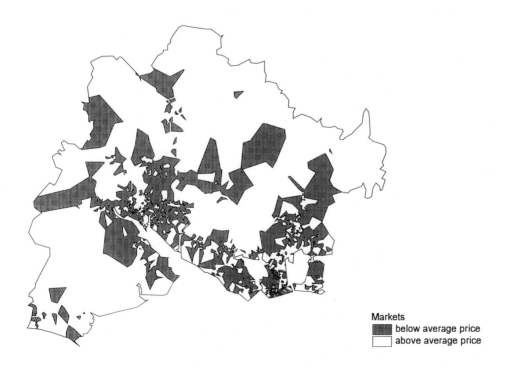

Markets
▨ below average price
☐ above average price

Figure 27.5 Preliminary view of housing market areas.

MLn (Rasbash and Woodhouse, 1995), and Arc/Info (ESRI, 1991). Although MLn includes some graphical capabilities for illustrating model outcomes, the GIS software enabled the predicted price differentials from the multilevel models to be mapped. The next stage will use these results to design a set of spatial units that optimise region homogeneity, using the zoning algorithm, ZDES (Openshaw and Rao, 1994, 1995). This approach follows the view that the selection of spatial units and the estimation of model parameters 'should occur simultaneously. Thus the end result of using an automatic zoning procedure is a set of spatial zones which represent both a mapping of a model on to the data and a mapping of the data on to a model' (Openshaw, 1977a, b).

Acknowledgements

A fellowship award from the European Science Foundation enabled me to participate in the ESF-GISDATA and NSF-NCGIA Summer Institute in Geographic Information. Thanks are due to the anonymous reviewers for their helpful comments and to Daniel Dorling for advice on the use of the property data, which was supplied by the Nationwide Building Society via the University of Newcastle upon Tyne, Department of Geography. The copyright for the Central Postcode Directory (postcode address file) is held by the Post Office, who bear no responsibility for the subsequent analysis based on the extracted data. Access to the CPD was provided by the Data Archive, University of Essex, via MIDAS, University of Manchester.

My PhD was funded by the Economic and Social Research Council's 'Analysis of Large and Complex Datasets' (ALCD) programme (Research grant No. R00429334048). The research on which this chapter is based was carried out at the Department of Geography, University of Portsmouth. I would especially like to thank my supervisor, Kelvyn Jones, Alistair Pearson and Paul Carter for help with using ARC/INFO. Rosemary Shearer kindly prepared Figs 27.1 and 27.2.

Note

1   The Multilevel Models Project also produces a regular newsletter which includes short technical papers and book reviews. Based at the Institute of Education, University of London, it can be contacted by email (temsmya@ioe.ac.uk) and via the Web site: http://www.ioe.ac.uk/hgoldstn/home.html.

References

AITKEN, M. and LONGFORD, N. (1986). Statistical modelling in school effectiveness studies (with discussion), *Journal of the Royal Statistical Society, Series A*, **149**, 1–43.

BALL, M. J. and KIRWAN, R. M. (1977). Accessibility and supply constraints in the urban housing market, *Urban Studies*, **14**, 11–32.

BRUNSDON, C., COOMBES, M., MUNRO, M. and SYMON, P. (1991). Housing and labour market interactions: an analysis of house price inflation in British LLMAs, 1983/87, in Satsangi, M. (Ed.), *Changing Housing Finance Systems*, Glasgow: Centre for Housing Research.

BRYK, A. S. and RAUDENBUSH, S. W. (1992). *Hierarchical Linear Models*, London: Sage.

BRYK, A. S., RAUDENBUSH, S. W., SELTZER, M. and CONGDON, R. T. (1986). *An Introduction to HLM: Computer Program and User's Guide*, Chicago: Department of Education, University of Chicago.

BULLEN, N. and JONES, K. (1994). Implications of a multilevel perspective for researching house prices, in Brown, S. (Ed.), *Proceedings of 'The Cutting Edge' Property Research Conference of The Royal Institution of Chartered Surveyors*, pp. 195–228, London: City University Business School.

BULLEN, N., JONES, K. and DUNCAN, C. (1996). Modelling complexity: analysing price variation between properties and between places, *Environment and Planning A*, (forthcoming).

CAN, A. (1990). The measurement of neighbourhood dynamics in urban house prices, *Economic Geography*, **66**, 254–72.

CASETTI, E. (1972). Generating models by the expansion method: applications to geographical research, *Geographical Analysis*, **4**, 81–91.

(1991). The investigation of parametric drift by expanded regression: generalities, and a family-planning example, *Environment and Planning A*, **23**, 1045–61.

(1992). The dual expansion method: an application for evaluating the effects of population growth on development, in Jones, J. P. III and Casetti, E. (Eds), *Applications of the Expansion Method*, pp. 10–41, New York: Routledge.

(1993). Spatial analysis: perspectives and prospects, *Urban Geography*, **14**, 526–37.

CLARK, M. J. (1993). Who needs boundaries? Paper presented at the AGI Annual Conference, Birmingham, 16–18 November (not in proceedings).

CLIFF, A. D., HAGGETT, P., ORD, J. K., BASSETT, K. A. and DAVIES, R. B. (1975). *Elements of Spatial Structure*, Cambridge: Cambridge University Press.

DOGAN, M. and ROKKAN, S. (1969). *Quantitative Ecological Analysis in the Social Sciences*, Massachusetts and London: MIT Press.

ESRI (1991). *Understanding GIS: The ARC/INFO Method*, Redlands, CA: Environmental Systems Research Institute.

FLEMING, M. C. and NELLIS, J. G. (1984). *The Halifax House Price Index: Technical Details*, Halifax: Halifax Building Society.

GOLDSTEIN, H. (1987). *Multilevel Models in Educational and Social Research*, London: Charles Griffin.

(1995). *Multilevel Statistical Models*, London: Edward Arnold, New York: Halstead.

GOODCHILD, M., HAINING, R. and WISE, S. (1992). Integrating GIS and spatial data analysis problems and possibilities, *International Journal of Geographical Information Systems*, **6**, 407–23.

GOODMAN, A. C. (1978). Hedonic prices, price indices and housing markets, *Journal of Urban Economics*, **5**, 471–84.

(1979). Externalities and non-monotonic price-distance functions, *Urban Studies*, **16**, 321–8.

(1981). Housing submarkets within urban areas: definitions and evidence, *Journal of Regional Science*, **21**, 175–85.

JONES, J. P. III (1992). Paradigmatic dimensions of the expansion method, in Jones, J. P. III and Casetti, E. (Eds), *Applications of the Expansion Method*, pp. 42–62, New York: Routledge.

JONES, J. P. III and CASETTI, E. (1992). *Applications of the Expansion Method*, New York: Routledge.

JONES, K. (1991a). Multilevel models for geographical research, *Concepts and Techniques in Modern Geography*, Norwich: Environmental Publications.

(1991b). Specifying and estimating multilevel models for geographical research, *Transactions, Institute of British Geographers*, NS 16, 148–60.

(1992a). Using multilevel models for survey analysis, in Westlake, A., Banks, R., Payne, C. and Orchard, T., *Survey and Statistical Computing*, pp. 231–42, Amsterdam: Elsevier.

(1992b). Amendment to 'Specifying and estimating multilevel models for geographical research', *Transactions, Institute of British Geographers*, NS **17**, 246.

(1995). 'Re-modelling a classic: a multilevel perspective on ecological correlations and the behaviour of individuals', mimeo, Portsmouth: University of Portsmouth.

JONES, K. and BULLEN, N. (1993). A multilevel analysis of the variations in domestic property prices: southern England, 1980–7, *Urban Studies*, **30**, 1409–26.

(1994). Contextual models of urban house prices: a comparison of fixed- and random-coefficient models developed by expansion, *Economic Geography*, **70**, 252–72.

JONES, K. and KIRBY, A. M. (1980). The use of a chi-square map in the analysis of census data, *Geoforum*, **11**, 409–17.

KREFT, I. G. G., LEEUW, J. DE and KIM, K.-S. (1990). *Comparing Four Different Statistical Packages for Hierarchical Linear Regression: GENMOD, HLM, ML2, VARCL*, Los Angeles, CA: UCLA Centre for Research on Evaluation.

LINDLEY, D. V. and SMITH, A. F. M. (1972). Bayes estimates for the linear model, *Journal of the Royal Statistical Society, Series B*, **34**, 1–41.

LONGFORD, N. (1986). VARCL: Interactive software for variance components analysis, *Professional Statistician*, **5**, 28–32.

MOK, M. (1995). Sample size requirements for 2-level designs in educational research, *Multilevel Modelling Newsletter*, **7**(2), 11–16.

OPENSHAW, S. (1977a). A geographical solution to scale and aggregation problems in region-building, partitioning, and spatial modelling, *Transactions of the Institute of British Geographers*, **2**, 459–72.

(1977b). Optimal zoning systems for spatial interaction models, *Environment and Planning A*, **9**, 169–84.

OPENSHAW, S. and RAO, L. (1994). A zone design system for ARC/INFO: methods for re-engineering census geography, Paper presented at GISRUK 94, Leicester, April.

(1995). Algorithms for re-engineering 1991 Census geography, *Environment and Planning A*, **27**, 425–46.

PALM, R. I. (1977). Homeownership cost trends, *Environment and Planning A*, **9**, 795–840.

PATERSON, L. and GOLDSTEIN, H. (1992). New statistical methods for analysing social structures: an introduction to multilevel models, *British Educational Research Journal*, **17**, 387–93.

RANDOLPH, B. (1991). Housing markets, labour markets and discontinuity theory, in Allen, J. and Hamnett, C. (Eds), *Housing and Labour Markets: Building the Connections*, pp. 16–51, London: Unwin Hyman.

RAPER, J., RHIND, D. and SHEPHERD, J. (1992). *Postcodes: the New Geography*, Essex: Longman.

RASBASH, J. and WOODHOUSE, G. (1995). *MLn Command Reference, Version 1.0*, University of London: Institute of Education.

RAUDENBUSH, S. W. and BRYK, A. S. (1986). A hierarchical model for studying school effects, *Sociology of Education*, **59**, 1–17.

STRASZHEIM, M. R. (1975). *An Econometric Analysis of the Urban Housing Market*, New York: National Bureau of Economic Research.

SUSSER, M. (1973). *Casual Thinking in the Health Sciences*, London: Oxford University Press.

WARD, J. H. (1963). Hierarchical grouping to optimize an objective function, *Journal of the American Statistical Association*, **58**, 236–44.

CHAPTER TWENTY-EIGHT

Modelling Urban Air Pollution using GIS

SUE COLLINS

28.1 Introduction

Concern regarding the effect of traffic-related air pollution on health continues to rise as the number of vehicles on the roads continues to increase. Particular concern relates to the prevalence of respiratory diseases such as asthma and its possible association with vehicle emissions. This has been reflected in the media with aptly named articles such as 'Gasping for Breath' – a report on the anxiety of the effect of exhaust fumes on children's health (*Independent on Sunday*, 10 October 1993).

Several studies have reported vehicle-related increases in the prevalence of asthma. Wichmann *et al.* (1989), for example, found increased levels of childhood asthma in streets subject to high levels of exhaust emissions, Wjst *et al.* (1993) reported an association between the prevalence of asthma and traffic volumes in Munich, Germany, and other articles (Ishizaki *et al.*, 1987; Nitta *et al.*, 1993; and Weiland *et al.*, 1994) have reported similar findings. Clinical reports into the role of air pollution in asthma, such as Wardlaw (1993), have concluded that 'air pollution could theoretically both increase the risk of developing asthma and exacerbate existing asthma' (p. 83).

In the SAVIAH study (Elliott *et al.*, 1995) 4500 children, between the ages of 7 and 11 in Huddersfield (UK) were interviewed by questionnaire. The prevalence of wheezing or whistling in the last 12 months was found to be as high as 17.8 per cent, with 30.9 per cent prevalence of occasional wheeze ever experienced (Kriz *et al.*, 1995). Similarly high values were recorded in the other European cities in the study, namely Amsterdam (The Netherlands), Prague (Czech Republic) and Poznan (Poland).

Epidemiologists therefore require accurate estimates of human exposure to air pollution in order to identify at-risk populations, as an aid to developing and monitoring control strategies. It has proved difficult, however, to obtain accurate estimates of exposure, especially in urban areas, where patterns of air pollution are very complex.

Consequently, in epidemiological studies, estimates of human exposure have been calculated at: (a) a regional level – exposure scores are estimated for whole areas such as counties or districts; (b) a local level – exposure scores are estimated for areas such as Census Enumeration Districts (EDs) or postal sectors; and (c) a personal level – where exposure is estimated for an individual either by personal monitoring or calculating the pollution concentration at their location of residence.

Personal monitoring is too costly, in terms of time and resources, to be used for large sample populations. Hence, there is a need to use a modelling approach. The difficulty, though, is to produce models which can give precise geographical estimates for a substantial number of points, over large and complex areas.

Regional pollution surfaces are traditionally generated by interpolating from monitoring sites, using various interpolation techniques such as trend surface analysis or kriging. Local patterns have been generated for particular roads or junctions, using Gaussian line-source dispersion models, such as CALINE (Benson, 1992), CAR International (Eerens et al., 1993) or Canyon Plume Model (Yamartino and Wiegand, 1986). The two approaches also require different data. Regional methods require concentrations at georeferenced points (x, y and z); local studies require more detailed data for example on surrounding building surface, traffic volumes and emission rates, road type and meteorology (wind direction, wind speed and atmospheric stability).

Near-source pollution is traditionally calculated using dispersion modelling techniques. These, however, do not have the capability to predict background levels of pollution and they can normally only predict accurately pollution concentrations within a relatively small zone adjacent to the emission sources. In the case of line-source models this zone ranges from 30 m (CAR) to 200 m (CALINE). Furthermore, near-source pollution exposure is well documented for simple configurations in open countryside, but exposure obtained in complex settings such as street canyons is poorly understood (Samson, 1988).

The problem with interpolating from monitoring sites is that changes in the pattern of air pollution occur over very small distances. In urban areas high degrees of local variation have been reported (Hewitt, 1991; Sexton and Ryan, 1988); this is supported by results from the SAVIAH study (Elliott et al., 1995) where, in some cases, twofold variations in levels of pollution were found over distances of less than 100 m. Interpolation thus tends to smooth the pollution surfaces and mask the high peaks of near-source pollution, especially where the sampling density is low.

Area-wide patterns of pollution can also be estimated by superimposing a grid over the study area and producing an estimate for each grid cell. This approach is commonly used for the development of emission inventories (Commission of the European Communities, 1995) or long-distance transport modelling (van Jaarsreld, 1994). It can also be used for pollution modelling in urban environments. The usual grid dimensions for urban-scale modelling are in the order of 1 to 10 km (Russell, 1988); an example of such an area-based study is that by Alexopoulos et al. (1993) who have developed a model to estimate emissions from traffic in Athens at a 1 km resolution. However, as Russell (1988) points out, there are problems associated with grid-based air quality modelling at such large resolutions, related primarily to 'a mismatch between the high concentrations that in fact do exist near the sources versus the lower concentrations computed by a model that immediately mixes those emissions throughout a grid cell of several kilometres' (p. 93). Exposure estimates are thus rendered homogenous across the grid cell. Recent advances in computer technology now afford the opportunity to describe the process of pollution mapping at a far more detailed resolution.

To this end, a methodology has been developed that employs a hybrid approach to pollution mapping. The method is based upon the principle that patterns of air pollution in urban environments can be described by two different, but related, components of variation (Collins *et al.* 1995):

- near-source variation (related to the dispersion processes associated with distinct point or line sources);
- background variation (reflecting differences in diffuse sources, broader – for example, local topographic – controls on dispersion and long-distance transport of pollutants across the study area)

The two components are predicted separately within a GIS environment using different techniques. The near-source air pollution is modelled by adapting the line dispersion model CALINE3 to operate within the GIS and the background variation is mapped utilising the spatial interpolation routines available in the GIS. The kriging interpolation technique was found to be the most applicable to this application (Collins *et al.*, 1995). The two separate components are additively combined to produce the final air-pollution map.

This chapter reports on the development of the near-source component, the application of the approach in Huddersfield (UK) and the validation of the final pollution map. The approach has been developed, applied and tested using the GIS Arc/Info version 6.2 running on a UNIX workstation, primarily in GRID, at a 10 m resolution.

28.2 Modelling near-source air pollution

Near-source pollution changes in response to local variations in vehicle emissions, road type, the porosity of the land adjacent to the roads and meteorology (wind direction, wind speed and atmospheric stability). The line source dispersion model CALINE3 was chosen in the first instance because of its capacity to include all these variables in the model. CALINE3 is based upon the Gaussian dispersion methodology, applying vertical and horizontal dispersion curves modified for the effects of surface roughness, averaging time and vehicle-induced turbulence (Benson, 1992).

CALINE3 predicts pollution concentrations for road links (road segments of uniform conditions) and can model up to 20 links and 20 receptors at any one time. Each link is a straight segment of constant width, traffic volume and emission rate. Real-world situations can be approximated by analysing multiple links, but the background pollution value must be specified by the user. Surface roughness, atmospheric stability, wind speed and direction are assumed to be constant over the study area. The links and the receptors are oriented within a coordinate reference system.

In area-wide detailed studies, road networks are extremely complex, with constantly changing conditions – reflecting the spatial variations in surface roughness, emission sources, background concentrations and the temporal variations in meteorology. In these situations a vast number of links would be created.

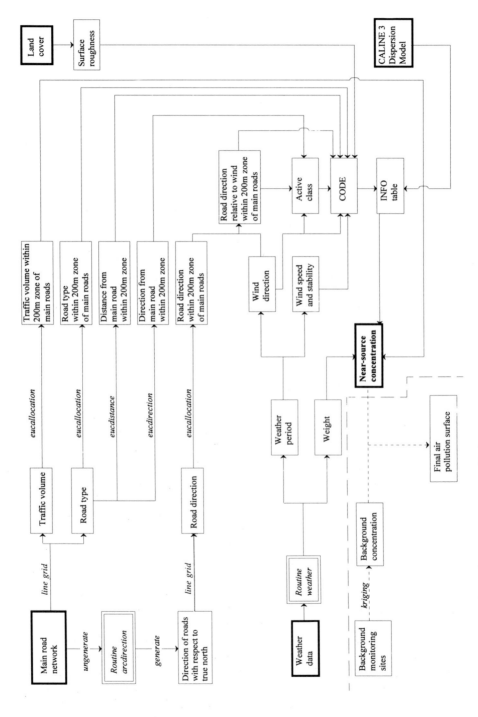

Figure 28.1   Modelling the near-source pollution.

Table 28.1 Code number and reference values

Code	Surface roughness (cm)	Mixing zone (m)	Wind direction (degrees)	Wind speed (m/s)	Stability (Pasquill scheme)
1	5	28	0	1	A
2	100	21	30	3	B
3	250	13.5	60	5	C
4	300	11.5	90	8	D
5	350		120	15	E
6			150		

Running each individual link through CALINE3 and then creating a pollution map would take a tremendous amount of time and effort.

To predict pollution concentrations efficiently in near-source areas (a 200 m band adjacent to the road links) an automatic approach to modelling pollution concentrations has been developed by adapting the CALINE3 model to operate within the GIS. This has been achieved by developing an Arc/Info routine using a combination of AMLs (Advanced Macro Language) and compiled FORTRAN programs. Outside the routine CALINE3 was run for a number of reference scenarios, each scenario reflecting a change in one of the input variables and the results were transferred to an Arc/Info data file. The structure of the near-source model is shown in Fig. 28.1.

28.2.1 Reference values for the scenarios

The input variables for the CALINE3 model include: surface roughness, mixing zone (width of traffic lanes plus 3 m on either side of the road), traffic volume, emission rate, atmospheric stability, wind direction and wind speed. The reference values for each of the input variables were chosen to reflect the full range of possible values available (Benson, 1979). A code number has been assigned to each reference value.

The code number and respective reference values for surface roughness, mixing zone, wind direction, wind speed and atmospheric stability are presented in Table 28.1. The scenarios were run with constant values for traffic volume (4000 vehicles/h) and emission rate (16 μg/vehicle km).

Pollution concentrations were calculated for receptors at distances 5, 20, 40, 65, 100 and 150 m perpendicular to the road links. Output from CALINE3 was presented as a series of tables in ASCII format. The FORTRAN program *concsext* was applied to extract the concentrations and write them to a new text file, in a format compatible with INFO. The information was then transferred to a defined table in INFO (*concs.exp*), a sample of which can be seen in Table 28.2.

CONCCODE is a unique code number. The five digits are formed from a combination of the codes for mixing zone, surface roughness, wind direction, wind speed and atmospheric stability class. RP# is the receptor location; receptors 1 to 6 are located on the opposite side of the road to receptors 7 to 12. Receptors 1 and 7 are located nearest to the roads.

Table 28.2 Sample data from the Concs.Exp.data file

CONCCODE	RP1	RP2	RP3	RP4	RP5	RP6	RP7	RP8	RP9	RP10	RP11	RP12
11111	0	0	0	0	0	0	152	94	62	41	26	17
11112	0	0	0	0	0	0	72	43	28	18	11	7
11113	0	0	0	0	0	0	47	28	18	12	7	5
~	~	~	~	~	~	~	~	~	~	~	~	~
23413	112	36	11	2	0	0	112	36	11	2	0	0
23423	49	15	5	1	0	0	49	15	5	1	0	0
~	~	~	~	~	~	~	~	~	~	~	~	~
45656	19	12	9	7	5	4	0	0	0	0	0	0

28.2.2 Description of the TRAFFPOL routine

Near-source pollution concentrations are generated with the command *traffpol* in Arc/Info at the ARC prompt. The *traffpol* routine requires three sources of input: a line coverage of roads in Arc/Info format, with road type code and traffic volume attributes; a polygon coverage for surface roughness in Arc/Info format; and an ASCII text file containing the weather data (wind direction, wind speed and atmospheric stability class). The form of the command line is:

&run traffpol <cover> <outgrid> <roadtype_item> <traffvol_item> <surface_rough> <weather_data>

arguments:
<table>
<tr><td> <cover></td><td>line coverage of roads to be analysed</td></tr>
<tr><td> <outgrid></td><td>output grid of near-source air pollution</td></tr>
<tr><td> <roadtype_item></td><td>item containing the code for road type</td></tr>
<tr><td> <traffvol_item></td><td>item containing the traffic volumes in vehicles/hour</td></tr>
<tr><td> <surface_rough></td><td>polygon coverage of classified surface roughness</td></tr>
<tr><td> <weather_data></td><td>text file containing the weather data (with extension)</td></tr>
</table>

The line coverage is converted to coordinates with the *ungenerate* command in ARC. The FORTRAN program *arcdirection* scans the arcs and calculates the bearing between successive vertices (pairs of coordinates). The maximum arc ID is identified and the individual sectors between vertices are classified according to direction. The arcs are scanned again and split for changes of class (that is, change of direction), assigning new IDs to new arcs. The segmented arcs are then written to a new file in arc ungenerated format and the arc IDs and direction classes written to a further text file for import into INFO.

The arcs are then transformed back to Arc/Info format with the *generate* command. The direction class is attached to the new arc coverage with the *joinitem* command. Three grids are then created with the *linegrid* command. The grids are created on items stored in the arc attribute table, namely, arc direction, road type and traffic volume. In GRID, for all cells within the near-source areas, the commands *eucdistance* and *eucallocation* are applied to calculate:

- the distance to the nearest road

- the value of the nearest road type
- the orientation of the nearest road
- the perpendicular direction to the nearest road
- the traffic volume of the nearest road.

Distances from the nearest road were then reclassified into six zones, with the central value of each zone corresponding to the location of the receptors. The orientation of the nearest road and the six wind-direction classes were used to create grids of relative road orientation. The wind direction, relative road orientation and perpendicular direction to the nearest road were then used to identify active cells, that is, those cells that will be in receipt of pollution for a particular wind direction.

The FORTRAN program *weather* was then applied to the weather data. In the first instance, the program classifies the wind directions and wind speeds into 12 and 5 classes respectively. The atmospheric stability class is based upon the Pasquill stability scheme and the classes A to F are reclassed 1 to 6. The program then calculates the frequency of each unique combination of wind-direction class, wind-speed class and atmospheric-stability class, referred to as a 'weather period' (Collins *et al.*, 1995).

Within the program, a new text file is generated containing the weather period and a weight for that weather period (based upon its frequency in relation to the total number of readings). The file is then converted into an INFO data file, where each record contains a unique weather period and its respective weight. The *cursor* functionality in Arc/Info is employed to select one record of the data file at a time. For each record the grid of active cells and the grid of relative road orientation, for that weather period, is selected.

The selected grids are used in combination with road-type allocation, surface roughness, classified distance from roads, wind-speed code and atmospheric stability class to give a seven-digit code for each cell. The seven-digit code is then used to extract the concentrations from the *concs.exp* table. The first two digits correspond to the receptor number (RP#) and the last five digits to the unique code CONCCODE (Table 28.2). The concentrations are then weighted for that weather period and the active cells updated to include the weighted concentrations.

The next record is then selected and the whole procedure is repeated. The new weighted concentrations are added to the previously calculated concentrations. The process is repeated until all the records have been selected. The final concentrations are then adjusted for traffic volume.

28.3 Case study

The method has been developed and tested in Huddersfield (UK) using data collected for the SAVIAH study (Elliott *et al.*, 1995). The study area (Figure 28.2) has a population of approximately 160 000 and covers an area 304 km². Altitude in the area ranges from 33 m in the east to 582 m above sea level in the west.

Nitrogen dioxide has been monitored as a proxy for traffic-related air pollution. The prime sources of outdoor, man-made NO_2 are vehicles (45 per cent) and

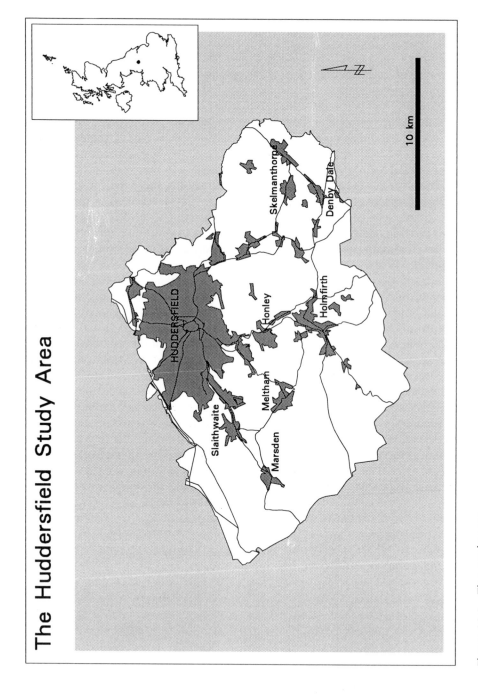

Figure 28.2   The study area.

combustion of fossil fuels in power generation (35 per cent); the remainder occurring as natural low background concentrations (Wardlaw, 1993). NO_2 data were collected at 80 permanent sites and 40 variable sites, during four two-week surveys:

1 – June 1993

2 – October 1993

3 – February 1994

4 – May 1994

Permanent sites have the same location for each survey (four readings per site) and the variables sites have a different location for each survey (one reading per site). The results from the June 1993 survey were subsequently dropped from the case study due to inconsistencies in the data. Data were also collected, on a two-weekly basis, for eight consecutive monitoring sites (November 1993–October 1994).

28.3.1 Input data

The road network was digitised from Ordnance Survey 1:10 000 topographic map sheets. Major emission sources were identified (that is, roads with day-time traffic flows > 250 vehicles/hour) and classified according to road type, as shown in Table 28.3.

Where possible, traffic flow-data were used from automatic traffic counts obtained from West Yorkshire Highways and Technical Services Joint Committee, Leeds. Where automatic traffic-count data were not available, data from manual traffic counts from Kirklees Highways, Huddersfield, were used. Traffic-flow data for the motorway were supplied by the Department of Transport, Leeds, based upon automatic traffic counts.

Data on land cover were obtained from classified, and then digitised, aerial photography at a scale of 1:10 000. The surface roughness coverage was generated using the classification scheme outlined in Table 28.4. Data on meteorology were obtained from a weather station in the study area.

The methodology was applied to the individual surveys 2, 3 and 4 and their estimated annual mean – adjusted for site and survey effect (Lebret *et al.*, 1995). The *traffpol* routine was run for all four cases to calculate the near-source variation. The background pollution surfaces were generated by applying the kriging interpolation technique, with the spherical model, in Arc/Info to the monitoring sites in the background areas (that is, beyond the 200 m near-source bands) for the three individual surveys and the mean. Results were then added to the near-source component of the model. The final pollution map for survey 4 can be seen in Fig. 28.3.

28.4 Validation of the results

The final pollution maps were validated to assess their ability to predict area-wide levels of NO_2. Results from the model were tested by comparing the estimated NO_2

Table 28.3 Mixing zone classification scheme

Mixing zone code	Road type
1	Motorway
2	Trunk roads
3	Main and secondary roads
4	Minor roads

Table 28.4 Surface roughness classification scheme

Surface roughness code	Land cover
1	Pasture, rough grass, reservoir, arble land, moor grass, peat sedge
2	Recreation, urban green space, very low and low density housing, quarries, disused and sequestered land
3	All tree types, public institutions, high density housing
4	Very high density housing
5	High density commercial, industry

concentrations with the monitored data from (a) sites located within the near-source band and (b) the consecutive monitoring sites, and were therefore not used in the kriging routine. Scatter plots of estimated concentrations against monitored data for the three individual surveys and the mean can be seen in Fig. 28.4. Correlation coefficients were calculated between the monitored and the estimated concentrations for the individual surveys and the mean. The correlations were found to be highly significant (< 0.001) for survey 3, survey 4 and the mean with coefficients as high as 0.7709.

28.5 Discussion

Analysis of the results indicated that the model failed to predict pollution values in areas where there were multiple sources of pollution, for example, at road junctions or where roads ran parallel and very close to each other. At these locations the pollution level is a function of all the emission sources in the near vicinity. However, the pollution levels estimated by the near-source model are only based upon the characteristics of the nearest road. There appears to be no apparent association between the monitored and the estimated concentrations in these locations, with almost zero correlation in all cases. Consequently, validation of the model was conducted in those areas where the data are considered to be reliable, that is one source of emission within a 200 m radius of the monitored site.

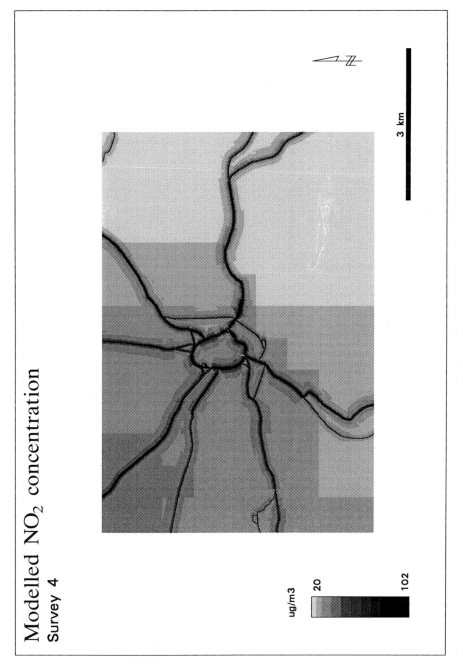

Figure 28.3 Pollution map – Survey 4 – Central Huddersfield.

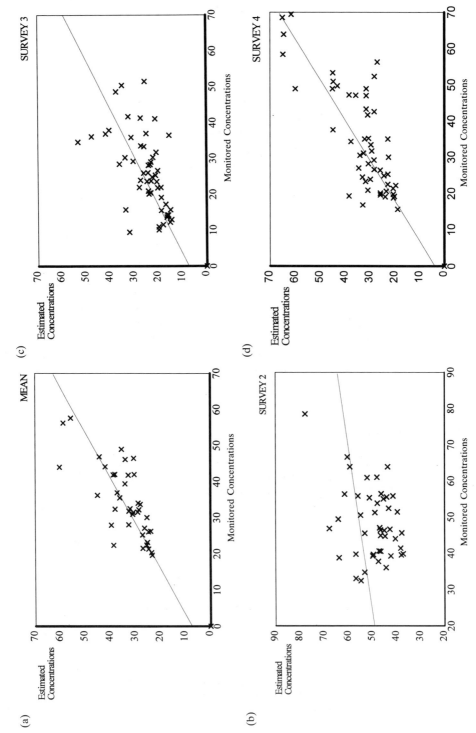

Figure 28.4 Scatter plots of estimated against monitored concentrations for (a) annual mean, (b) survey 2, (c) survey 3, and (d) survey 4.

During the development of the model a comparison of different kriging programs was undertaken to try to reduce uncertainty in the background pollution levels. The models produced by the different kriging programs were found to be very similar with no significant improvements to the overall accuracy of the approach.

28.6 Conclusion

The modelling technique provides reasonably good predictions of pollution for survey 3, survey 4 and the mean. There is no clear explanation why the results for survey 2 should be so poor – further monitoring and/or testing in other study areas may provide an explanation.

The next stage in the development of the model will be to attempt to resolve the problem of multiple sources of pollution by incorporating a multiple source option into the routine. Uncertainty in the background pollution is likely to be a result of the number and distribution of sites available for interpolation. Future work will therefore concentrate on improving the prediction of the background variation. In addition, as an aid to examining the applicability of the model, an analysis of variance will be undertaken to reveal any underlying patterns or trends that may prevail.

Constraints on the model could be imposed by the scarcity of the meteorological data. In the case of Huddersfield, there is only one meteorology station and it is assumed that the weather conditions are homogeneous across the study area. Naturally, this does not take into account the effect of terrain and building structure on patterns of meteorology. Future work will investigate the implication of meteorology on the model – including an in-depth study into the stability of the model.

Acknowledgements

Thanks are due to David Briggs for advice and to members of staff at the Institute of Environmental and Policy Analysis, The University of Huddersfield, UK, for the collection of data, especially Kirsty Smallbone.

References

ALEXOPOULOS, A., ASSIMAKOPOULOS, D. and MITSOULIS, E. (1993). Model for traffic emissions estimation, *Atmospheric Environment*, **27B**(4), 435–46.

BENSON, P. E. (1979). CALINE3 – *A Versatile Dispersion Model for Predicting Air Pollution Levels near Highways and Arterial Streets*, Report number FHWA-CA-TL-79-23, Washington, DC: Federal Administration.

(1992). A review of the development and application of the CALINE3 and CALINE4 Models, *Atmospheric Environment*, **26B**(3), 379–90.

COLLINS, S., SMALLBONE, K. and BRIGGS, D. (1995). A GIS approach to modelling small area variations in air pollution within a complex urban environment, in Fisher, P. (Ed.), *Innovations in GIS 2*, pp. 245–53, London: Taylor & Francis.

COMMISSION OF THE EUROPEAN COMMUNITIES (1995). '*CORINAIR*', Luxembourg: Office for Official Publications of the European Communities.

EERENS, H., SLIGGERS, C. and VAN DER HOUT, K. (1993). The CAR model: the Dutch method to determine city street air quality, *Atmospheric Environment*, **27B**(4), 389–99.

ELLIOTT, P., BRIGGS, D., LEBRET, E., GORYNSKI, P. and KRIZ, B. (1995). Small area variations in air quality and health, the SAVIAH study: Design and methods, Abstract in *Epidemiology*, **6**(4), S32.

HEWITT, C. N. (1991). Spatial variation in nitrogen dioxide concentrations in an urban area, *Atmospheric Environment*, **25B**(3), 429–34.

ISHIZAKI, T., KOIZUMI, K., IKEMORI, R., ISHIYAMA, Y. and KUSHIBIKI, E. (1987). Studies of prevalence of Japanese cedar pollinosis among the residents in a densely cultivated area, *Annals of Allergy*, **58**, 265–70.

KRIZ, B., BOBAK, M., MARTUZZI, M., BRIGGS, D., LIVESLEY, E., LEBRET, E., FISHER, P., WOJTYNIAK, B., GORYNSKI, P. and ELLIOTT, P. (1995). Respiratory health in the SAVIAH study, Abstract in *Epidemiology*, **6**(4), S31.

LEBRET, E., BRIGGS, D., SMALLBONE, K., REEUWIJK, H. VAN and FISHER, P. H. (1995). Small area variation in exposure to NO_2, Abstract in *Epidemiology*, **6**(4), S31.

NITTA, H., SATO, T., NAKI, S., MAEDA, K., AOKI, S. and ONO, M. (1993). Respiratory health associated with exposure to automobile exhaust: I. Results of cross-sectional study in 1979, 1982 and 1983, *Archives of Environmental Health*, **48**, 53–8.

RUSSELL, A. G. (1988). Mathematical modelling of the effect of emission sources on atmospheric pollutant concentrations, in Kennedy, D. (Ed.) *Air Pollution, the Automobile and Public Health*, pp. 161–205, Washington, DC: Health Effects Institute, National Academy Press.

SAMSON, P. J. (1988). Atmospheric transport of air pollutants associated with vehicular emissions, in Kennedy, D. (Ed.) *Air Pollution, the Automobile and Public Health*, pp. 77–97, Washington, DC: National Academy Press, Health Effects Institute.

SEXTON, K. and RYAN, P. B. (1988). Assessment of human exposure to air pollution: methods, measurements and models, in Kennedy, D. (Ed.) *Air Pollution, the Automobile and Public Health*, pp. 207–38, Washington, DC: National Academy Press, Health Effects Institute.

VAN JAARSRELD, J. A. (1994). *Atmospheric Deposition of Cadmium, Copper, Lead, Benzo(a)pyrene and Lindano over Europe and Surrounding Marine Areas*, Report Number 222401002, Bilthoven: RIVM.

WARDLAW, A. J. (1993). The role of air pollution in asthma, *Clinical and Experimental Allergy*, **23**(2), 81–96.

WEILAND, S. K., MUNDT, K. A., RUCKMANN, A. and KEIL, U. (1994). Reported wheezing and allergic rhinitis in children and traffic density on streets of residents, *AIR*, **4**(1), 79–84.

WICHMANN, H. E., HUBNER, H. R. and MALIN, E. (1989). The relevance of health risks by ambient air pollution, demonstrated by a cross-sectional study of Croup syndrome in Baden-Württemberg, *Off Gesundheitwes*, **51**, 414–20.

WJST, M., REITMEIR, P., DOLD, S., WULFF, A., NICOLAI, T., VON LEOFFELHOLZ-COLBURG, E. and VON MUTIUS, E. (1993). Road traffic and adverse effects on respiratory health in children, *British Medical Journal*, **307**, 596–600.

YAMARTINO, R. J. and WIEGAND, G. (1986). Development and evaluation of simple models for the flow, turbulence and pollutant concentration fields within an urban street canyon. *Atmospheric Environment*, **20**(11), 2137–56.

CHAPTER TWENTY-NINE

GIS in Quantitative Geomorphology – Assessment of Denudation Rates in the Central Kenya Rift

SIGRID ROESSNER

29.1 Introduction

Erosion processes are a cumulative expression of many environmental conditions. The fluctuation in character and intensity of these processes reflects changes in landscape factors, such as tectonism, climate, vegetation, and land use. The quantification of these erosion processes is important for complex modelling of landscape evolution at various spatial and temporal scales. Such models need calibration using realistic process rates and since advanced modelling approaches are often focused on the inclusion of factors controlling spatial and temporal process variations, it is important to provide high-resolution data for this calibration (Moore *et al.*, 1991; Summerfield, 1994). In this connection the rapidly advancing technologies of geographic information systems, digital terrain modelling and remote sensing allow spatially continuous quantitative evaluation of extensive areas and offer novel approaches in quantitative landscape analysis and modelling.

This chapter presents research on long-term average rates of mass removal for the last three million years for an extensive area within the East African Rift using GIS techniques to analyse digital-elevation data. The results of this study add to the existing worldwide database of regional rates of surface lowering (Einsele, 1992; Meybeck, 1976, 1987; Saunders and Young, 1983; Summerfield, 1994), which still has major gaps, especially within areas of limited hydrological survey including most of Africa.

This work contributes to a better understanding of the regional variation in magnitude of exogenic geomorphologic processes within a large but naturally uniform area. By using a quantitative approach based on geological evidence, natural average process rates are obtained which can be compared with present measurements of erosion. In this way estimates of the importance of human activity as a geomorphological factor are possible.

Different methods to quantify average rates of long-term surface lowering caused by erosional processes can be categorised as direct and indirect. The direct approaches use recent, short-term measurements of eroded material on slopes and in rivers (dissolved and suspended river loads) for extrapolation to longer time periods. The indirect approaches are based on the analysis of sedimentary basin fills or on the existence of dated relict erosional remnants within present surfaces which can be used to reconstruct previous topographical scenarios (Summerfield, 1991). The long-term averaging approaches avoid problems of direct short-term observations, like episodicity of measurements, sampling of atypical periods, and the necessity of detailed regional process understanding (Ahnert, 1970; Saunders and Young, 1983; Sutherland and Bryan, 1991). However, they also require special geological circumstances for conserving reliable geological evidence (Gale, 1992; Young and McDougall, 1993). Rates of surface lowering (denudation rates) are expressed in metres per million years (m/Ma) or millimetres per thousand years (mm/Ka). The term 'denudation' stands for mass removal as an integrative result of various erosional processes which are not subject to further process-oriented specifications.

The volcanic Bahati-Kinangop Plateau in the Central Kenya Rift (Figs 29.1 and 29.2) is an ideal site to assess long-term denudation rates using the elevation difference method based on relict surface reconstruction. Traditionally, this method is based on topographic maps in which the investigator defines a standard area unit and must manually determine the elevation difference within each unit. This chapter demonstrates how digital surface modelling techniques can be applied to obtain denudation rates with much higher spatial resolution and how, within this process, qualitative and descriptive knowledge are integrated and used for spatial analysis in a GIS.

29.2 The study area and its geological setting

The 710 km^2 study area (Fig. 29.1) includes a large part of the Bahati-Kinangop Plateau. The plateau is an integral part of the Central Kenya Rift between 1° 20 ' S and 0° 20' N. Like many other continental rifts, structures and volcanically created relief are dominant factors controlling the basin and drainage evolution of the Central Kenya Rift. Well-established stratigraphy and faulting chronology (Baker et al., 1988; Strecker et al., 1990), allow palaeotopographic reconstruction of an areally extensive terrain within the intrarift Bahati-Kinangop Plateau at the time when the plateau came into existence between 3.4 and 2.6 million years ago. The reconstruction is based on relict surface elements which form the uppermost parts of the present plateau surface.

The plateau belongs to the more than 3000 km^2 large Lake Naivasha catchment area (Fig. 29.1), which is one of the most important drainage and depositional environments in the Central Kenya Rift. This part of the rift is defined by boundary faults along the eastern Aberdare Range (4000 m) and the western Mau Escarpment (3000 m). The 60 km-wide rift floor is divided into the 1700–2000 m inner graben region in the West and the 30 km-wide intrarift Bahati-Kinangop

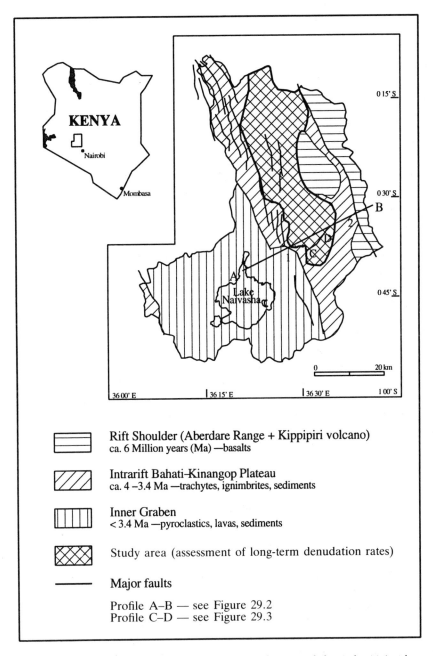

Figure 29.1 Major physiogeographic and geological units of the Lake Naivasha drainage basin in the Central Kenya Rift.

Plateau (2300–2500 m) in the East (Fig. 29.2). The inner graben hosts several endorrheic sedimentary basins characterised by alkaline (Lake Nakuru) and freshwater (Lake Naivasha) depositional systems.

The present structural and topographic configuration of the Lake Naivasha basin and adjacent areas was generated between 3.4 and 2.6 Ma, when major

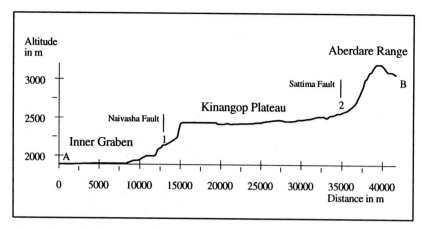

Figure 29.2 Topography of the Eastern Central Kenya Rift – derived from digital topography (cross-section profile, see Fig. 29.1).

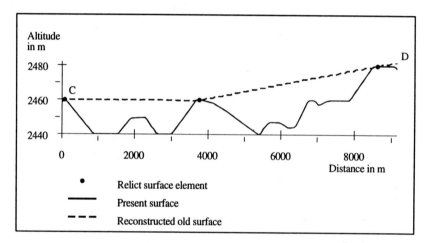

Figure 29.3 Principle of surface reconstruction based on erosional remnants (cross-section profile, see Fig. 29.1).

downward faulting occurred along the Sattima and Naivasha faults (Figs 29.1 and 29.2) as a result of extensional stress. These movements offset the areally extensive ashflow tuffs of the Kinangop Trachytes whose youngest (3.4 Ma) volcanic units form the present top surface of the plateau. Geological investigations (Strecker *et al.*, 1990) show that the deposition of the approximately 400-m-thick trachytes took place very rapidly between 3.7 and 3.4 Ma during the same time when the Sattima fault was active. Erosional remnants on interfluves within these youngest 3.4 Ma volcanic units represent the relict surface elements within the constructional volcanic relief that became segmented after downward faulting.

The dominant dendritic drainage patterns within the Aberdare Range – Bahati-Kinangop Plateau erosional system continue across the major Sattima fault boundary (Fig. 29.4), proving that the development of the drainage system was

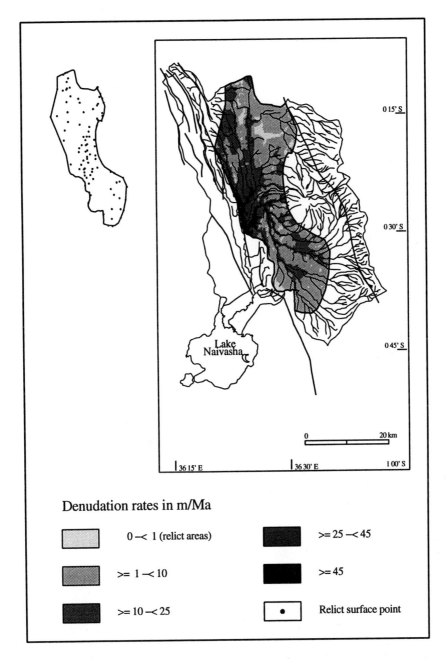

Figure 29.4 Long-term average denudation rates for the last 3 million years (in m/Ma) – Bahati-Kinangop Plateau (Central Kenya Rift).

superimposed on the tectonically created landscape. As the plateau and the Sattima Fault were not subject to pronounced subsequent tectonic movements the present drainage and state of erosion are thus a direct consequence of the climatic conditions and base-level processes in the Naivasha Basin. Since the extensive ashflows created an almost level landscape in the infilled rift valley, the remnants of this old landscape which dip by less than 3° can be utilised for reconstruction of the initial plateau landscape. The distribution of the erosional remnants (Fig. 29.4) makes it possible to decipher the character of this initial landscape in a spatially continuous way assuming an idealised non-eroded constructional surface prior to plateau formation.

Due to its equatorial position and pronounced relief, the present natural conditions of the Eastern Central Kenya Rift are characterised by vertical ecological zoning, including the submontane upland savannas of the rift valley (<2300 m), the montane grasslands and forest (>2300–<3200 m) and the (sub)alpine tree (bamboo) and shrublands (>3200 m) of the rift shoulder (Pratt *et al.*, 1966; Schmitt, 1991). Within the last 100 years, increased land-use intensity within the submontane and montane regions has led to extreme modifications of the original vegetation cover. Wide areas have been cleared and replaced by secondary grassland–bushland formations as well as by intensely used farmlands causing increased surface run-off and soil erosion.

29.3 Denudation assessment

Since the geological framework of the study area allows reconstruction of the idealised uneroded plateau surface at the time when the plateau came into existence (3.4–2.6 Ma ago), the location-specific elevation difference method (Fig. 29.3) can be applied.

In this study a GIS is used to reconstruct the topography of the initial plateau surface based on the results of digital modelling of the present eroded plateau topography. The GIS-based approach, including acquisition, organisation, analysis and output of the data, was performed with Arc/Info and consists of the following steps:

- Digital database creation
- Digital surface modelling
- Calculation and graphical presentation of denudation rates.

29.3.1 Digital database creation

The Kenya Rift has been the subject of detailed geoscientific investigations and land surveys for more than 50 years, which led to a detailed topographic database and a well-established documentation of landscape evolution including paleoecological aspects as well as geological stratigraphy and faulting chronology. The results of these investigations are stored in reports and conventional maps of different scales, reaching from 1:50 000 up to 1:250 000. Under these circumstances the creation of a digital database of sufficient resolution and accuracy was the first step within the GIS-based denudation analysis.

For the elevation difference approach, which is based on the analysis of the present plateau surface, digital topography holds the key position within the database. It was derived from the 1:50 000 topographic maps of the Survey of Kenya. Since these topographic maps are the most accurate source of spatial data which was available for this area, their UTM coordinates were used as the topographic reference system for the whole database. Landsat Thematic Mapper (TM) image data are the second main element of the digital database providing spatially continuous information. The geocoding of the image data was performed by an image–to–map registration. Spatial data of specific themes (mainly geological, vegetative) obtained from various 1:250 000 maps by manual digitisation represent the third type of information in the database. In contrast to topographic and image data, the process of the original creation of the spatially continuous thematic map information is difficult to evaluate because of the lack of supplementary descriptive documentation.

In order to manage these uncertainties and create a spatially consistent database for further analysis, the spatial matching of data layers was checked using the overlay capabilities of the GIS. The spatial accordance of topographic information and georeferenced image data was analysed with respect to the drainage network and showed very good correspondence. The overlay of thematic map information and image data was partly characterised by non-systematic artefacts which were eliminated by adjusting the geometry of the thematic map information to the image data. First, the course of faults was corrected, because they are very well detectable in the image data due to their pronounced morphological expression. In a second step these improved faults allowed a more reliable determination of the spatial extent of lithological units, because the major faults act as lithological boundaries. The correction of the lithological information was additionally supported by differences in the spectral reflectance characteristics between the lithological units. The geological features, selected in Fig. 29.1, are results of this process.

The approach described shows that GIS techniques make it possible to improve the original analogue database by a combined digital analysis of various spatial information sources which are generated independently by different technical methods.

29.3.2 Digital surface modelling

Digital modelling of the present eroded surface and the idealised uneroded initial plateau surface is based on 1:50 000 topographic maps which were generated by field surveys and photogrammetric measurements. These maps show a detailed representation of erosional features within the course of a single contour. The shape of the contours reflects favourable natural conditions for aerial survey within the Rift Valley and the Kinangop Plateau as well as a limited cartographic generalisation of the contours measured photogrammetrically.

Topography was assessed by manually digitising the whole Lake Naivasha drainage basin and its surroundings, reaching from 36° to 36° 45′ E and from 1° S to the equator (ca. 9200 km^2). Because of the large area and the labour-intensive digitising approach, zones requiring varying accuracy were determined. The discussion that follows refers to the zone with the highest accuracy covering the Lake Naivasha drainage basin east of the Naivasha fault (Fig. 29.1).

For the continuous surface representation (regular raster) a 30 × 30 m cell size was defined to maintain the positional accuracy of the digitised input data during the interpolation process and to match the spatial resolution of the Landsat-TM data. For that purpose high-information redundancy within areas of low input data density was accepted. Due to the use of the quadtree encoding scheme (Laurini and Thompson, 1992), the size of one raster dataset (45 × 75 km) was reduced from 15 to 4.5 Mbyte. This way the definition of the surface resolution was not limited by the resulting data volume.

29.3.2.1 *Present eroded surface*

Topographic data assessment was made with the goals of maintaining all topographic information pertaining to important erosional features while optimising the digitising effort. The equidistance of digitised contours varies with relief, which ranges from 20 m in flat areas, where each contour was digitised, to 200 m in areas with steep escarpments. Additionally, all spot elevations representing potential relict surface remnants were digitised. The density of the digitised points along one contour depends on the plan curvature of the contour in the map. Besides contours, the drainage network and outlines of the main ridges were also included in the topographic database.

To determine the most suitable approach for the continuous surface interpolation, experiments with mathematically defined surfaces and several regular and irregular input data distributions were carried out (Brehm, 1995; Utler, 1993). They showed that the direct raster interpolation approach requires more or less regular and densely distributed input data for a satisfactory interpolation. In contrast, the triangulated irregular network approach followed by a secondary raster interpolation is more robust against varying input data density and produces a surface with less noise, but has the disadvantage of an increased surface levelling tendency. Since in this study the use of contours for surface interpolation causes an extremely irregular distribution of input elevation information, the triangulated irregular network method followed by a secondary raster interpolation was chosen to handle the big differences in input data density.

Arc/Info allows the creation of a triangulated irregular network based on the Delaunay method from contours and/or elevation points, including additional surface-feature-type information such as breaklines of interpolation and erase polygons. Breaklines are often introduced in order to reduce the cut-and-fill effects that occur in connection with the interpolation of regular elevation rasters from triangulated irregular networks. Experiments with input data from this study have shown that the combination of contour elevation information and drainage network as breaklines of interpolation leads to a noisy surface after performing the secondary raster interpolation. This is due to the sensitive reaction of the bivariate fifth-degree polynomial-based algorithm to an unfavourable spatial distribution of the input data as well as the introduction of the drainage network as breaklines without elevation values. The bivariate interpolation method has to be used to maintain the effect of breaklines through the secondary raster interpolation. To avoid these artefacts which occur in the form of local pits and peaks, the triangulated irregular network was created without breaklines and the regular raster of mesh points was interpolated using the linear method.

The quality of the surface thus generated was assessed in two steps. First, the morphological correctness was visually checked by comparing the input contours with the contours interpolated from the digital surfaces. The best results were

Table 29.1 Present surface: elevation differences between control points and modelled surface values

Type of control points	Number of points	Mean (m)	Standard deviation (m)	Posit. max. (m)	Negat. min. (m)
Input data	2159	0.0	2.0	16.0	−15.0
Map points	51	2.5	8.5	35.0	−15.0
Different. GPS	47	0.2	9.9	24	−21

obtained in areas showing a similar spatial frequency of the relief in horizontal and vertical direction (for example, a dense and deeply eroded channel system along a steep fault scarp). In contrast, the coarse input data density in areas of isolated and pronounced relief elements caused cut-and-fill effects (for example, single, deeply eroded channels within the flat plateau area).

In a second step, a quantitative assessment of height accuracy was performed with reference data. For that purpose elevation values for control points were interpolated from the generated surfaces and compared with the original values (Table 29.1). The mean and standard deviation were calculated to describe the differences between control point and the modelled surface values.

This analysis was carried out for different control point types to evaluate the different stages of the surface generation process separately. A subset of input data points was analysed to assess the internal accuracy of the digital surface generation. The mean value of zero shows that this process was free of any systematic error. In contrast to the input data points, the two other control point types represent independent control information. Regularly spaced (5 km) map control points were used to characterise errors which occur in the conversion of analogue map contours into a digital continuous surface representation. A mean of 2.5 m shows a slight surface lowering tendency within the modelled surface. The standard deviation of less than half of the minimal equidistance of 20 m indicates a good quality representation of the available map information within the digital elevation model. Additionally, differential GPS measurements were carried out to evaluate the quality of the digital surface in comparison with the real surface to get an idea about the absolute accuracy of the topographic base used. Since most of the GPS measurements took place at pronounced relief positions to determine denudation rates directly in the field, the values obtained for mean and standard deviation show a good representation of the important erosional features in the digital surface model. Despite minor problems in areas of low and irregular input data density, the results of the accuracy assessment are in accordance with common quality expectations for digital surfaces which are derived from map contours.

29.3.2.2 *Idealised uneroded initial surface*

The geological framework allows reconstruction of the idealised uneroded plateau based on the recent plateau surface in selecting mesa-like relict surface elements.

Relicts of the old plateau surface occur in the form of erosional remnants on interfluves, isolated mesas and stratigraphic sections, including the 3.4 Ma. ash-flow tuff. Figure 29.3 illustrates the approach described of surface reconstruction on the basis of an example profile.

Before localising the relict surface elements, the final study area had to be determined. For that purpose the analytical capabilities of the GIS were used to perform spatial queries and select the suitable area for the denudation analysis which satisfies the required geological and topographical criteria. Additional boundary conditions are:

- Limitation of the study area to an elevation range between 2300 and 2500 m since field evidences of dated ash flows are only available for this part of the plateau
- Exclusion of the faulted area close to the reactivated western boundary fault (Fig. 29.1).

The final 710 km^2 study area shown in Fig. 29.1 represents the stable plateau area for which the initial plateau surface can be reconstructed on the basis of erosional remnants.

The localisation of points was performed by analysing the present digital topography (contours, perspective surface visualisations) in combination with overlays of thematic maps (lithology, faults), and georeferenced Landsat-TM data. Aerial photographs in the scale of 1:50 000 were also analysed. Only erosional remnants were selected which dip less than 3° locally. Within the study area 96 relict surface points could be determined (Fig. 29.4), which results in an average data density of one elevation point per 7.4 km^2.

The interpolation of a continuous, initial surface was performed in the same way as for the recent surface, so that the mesh points defining the initial surface are identical in number and position to the points of the present surface. Because of the low input data density the initial surface is highly oversampled.

Quality assessment was performed by interpolating secondary contours from the digital surface and by perspective visualisation. Fewer interpolation problems occurred due to the more regular and coarse distribution of the input data. Within the interpolated areas, a quality check based on independent control points was not possible because of the lack of independent control information. Quantitative accuracy assessment was carried out in the form of plausibility control to analyse if the modelled palaeosurface corresponds to the previously made assumptions. First, slope angles were calculated to ensure that the reconstructed surface does not dip more than 3° at any point. In a second step denudation rates were calculated for the relict points and for the regularly spaced map points. Additionally, the control-point approach was applied to assess the accuracy of the input data representation within the modelled surface (internal interpolation accuracy).

The results summarised in Table 29.2 show that for all relict surface elements the criterion of a nearly unchanged relief position is fulfilled, and the palaeosurface is higher or at least equal in elevation to the present surface. The few locations of a slightly higher present surface originated in the interpolation processes. In these cases the original palaeosurface values are replaced by the higher present ones.

Table 29.2 Palaeosurface: plausibility control and accuracy assessment

Type of control points	Number of points	Mean (m)	Standard deviation (m)	Posit. max. (m)	Negat. min. (m)
Relict points*	96	0.6	–	3.6	−0.5
Map points*	51	31.3	–	207.6	−1.4
Input data**	96	−0.1	0.4	0.9	−2.6

*Denudation (total elevation change) – normal distribution cannot be assumed
**Internal accuracy of palaeosurface modelling

29.3.3 Calculation of denudation rates

GRID, the cell-based modelling tool of the Arc/Info-software, was used to calculate and classify denudation rates. GRID requires a categorical interpretation of the raster data. This implies that the cell values apply to the entire cell (data type 'grid'), whereas in the case of a surface interpretation of the data the centre points of each cell are used as mesh points of the raster and values between them are interpolated from surrounding mesh points (data type 'lattice').

In the first step, two spatially identical surfaces were created by clipping both elevation data sets. This was achieved by defining the overlap between the lattices and a polygon coverage containing the study area boundaries. Secondly, the absolute elevation differences between the initial and the present surface were calculated for all mesh points. For further numerical analysis the elevation difference data (floating point lattice) were converted into integer grid data, which were accompanied by the creation of a value-attribute table (VAT) containing the two items 'value' and 'count' by default. The VATs are part of the Arc/Info georelational data model and allow the addition of new attribute information in relation to attribute information stored in other tables.

Using the VAT concept, overall, long-term average denudation rates (m(mm) per million or thousand years) were calculated in the following way.

1 Total values
   size of study area [a]: 710 km^2
   size of eroded area [a_E]: 693 km^2
   eroded volume [V_E]: 25.96 km^3
   plateau age [y_P]:
      Minimum: 2.6 Ma
      Maximum: 3.4 Ma

2 Overall denudation rate of the study area elevation change:

elevation change:

$$\Delta Z = \frac{V_E[m^3]}{a[m^2] y_P[Ma]}$$

for minimum plateau age (2.6 Ma):

$$\Delta Z = 14[m/Ma]$$

for maximum plateau age (3.4 Ma):

$$\Delta Z = 11[m/Ma]$$

Furthermore the high-resolution denudation rates for the minimum (2.6 Ma) and the maximum (3.4 Ma) plateau age were classified on the basis of histogram analysis (Table 29.3) to get a better picture of the frequency distribution of the denudation data.

The cartographic visualisation (Fig. 29.4) is based on denudation rates which represent the amount of surface lowering for the average time of assumed plateau existence, about 3 Ma.

29.4 Discussion and conclusions

The results represent values of average surface lowering since the time the plateau came into existence, between 3.4 and 2.6 Ma. During this period more than 80 per cent of the study area is characterised by denudation rates ranging between 0 (relict surface elements) and 15 to 20 m/Ma (Table 29.3). The real surface at the onset of faulting and plateau formation was probably higher than the present relict surface points. No information was available which allowed estimation of the amount of summit lowering down to the recent top level of the erosional remnants. Hence, only the minimum elevation change is assessed, leading to low denudation rates of less than 5 m/Ma for more than one-fourth of the study area. Within the eroded areas the observed elevation change reflects the surface lowering caused by fluvial erosion. Denudation rates mainly vary from 1–10 m/Ma for little-eroded (more than 50 per cent of the area) to 20–50 m/Ma for deeply eroded parts of the plateau which cover less than 15 per cent of the area. Highest rates of up to 100 m/Ma were obtained for the region close to the western boundary fault where the Malewa river exits the plateau and has caused pronounced retrogressive erosion and lateral channel cutting.

In the light of modern mechanical and chemical denudation rates from other regions (for example, Einsele, 1992) the results from the Central Kenya Rift are comparable to denudation rates in Mediterranean climates, where 10 m/Ma chemical denudation and 10 to 20 m/Ma and 40 to 50 m/Ma mechanical denudation have been determined for low relief river basins and mountain river basins, respectively. Although a wide range of denudation calculations for various environments for recent and palaeoecological settings exist, it is difficult to compare results due to different approaches of data generation.

The accuracy of the GIS-based approach presented in this study is determined by the reliability of geological evidence and the quality of the digital data analysis. Accuracy assessment based on independent control point information was only possible for digital modelling of the present topography. These investigations showed that the modelled, continuous surface represents the available input information (contours of 1:50 000 topographic maps) nearly free of systematic errors within the expected accuracy range (standard deviation of less than half the

Table 29.3 Classification of denudation rates

Classification (m/Ma)	% of total area	
	Minimum age (2.6 Ma)	Maximum age (3.4 Ma)
0 – <1	6.1	7.6
≥ 1 – <5	21.3	29.7
≥ 5 – <10	26.8	27.3
≥ 10 – <15	14.9	15.6
≥ 15 – <20	11.5	6.6
≥ 20 – <25	5.3	4.2
≥ 25 – <35	5.8	3.8
≥ 35 – <45	2.8	2.0
≥ 45 – <55	1.6	1.5
≥ 55 – <65	1.4	0.6
≥ 65 – <80	1.1	0.7
≥ 80	2.3	0.4

minimal equidistance). For palaeotopography, quality assessment was limited to internal accuracy and the check for major artefacts leading to contradictions with the basic assumptions for the palaeosurface reconstruction. This way a stable palaeosurface was generated which represents the relict surface elements in their correct position. Since priority was given to a fully reproducible surface-generation approach, no supporting elevation information was introduced. The resulting coarse and irregular distribution of relict elevation information (Fig. 29.4) led to a limited morphological correctness of the surface which was accepted because of its minor importance in the frame of the used elevation-difference method.

Nevertheless, previously mentioned uncertainties in geological evidence seem to be the main factor limiting the reliability of the denudation rates. In addition to the problem of the unknown amount of summit lowering, the accuracy of the denudation rates is mainly influenced by the 0.8 Ma time range of plateau formation which represents almost one-fourth of the maximal possible time span of its existence (3.4 Ma). Further work is still necessary in order to perform a more complex quantitative accuracy assessment (Goodchild, 1993; Heuvelink and Burrough, 1993; Maguire et al., 1991) for this multistage GIS analysis.

It is assumed that the interpolation character of the applied elevation-difference approach provides a higher reliability of the denudation data than most of the techniques based on long-term extrapolation of present short-term measurements. Due to the length of the analysed erosional period of about three million years, the rates obtained represent natural, average process characteristics which cannot be assessed any more in most parts of the world from present measurements due to the high intensity of human land-use activities. The results of this study are of special value in estimating the human impact on geomorphological processes.

This study shows that GIS techniques can be used for a reliable, quantitative assessment of denudation rates at high spatial resolution. The digital environment allows development of a sensible approach producing well-understood results which are the basis for further regional analysis of the denudation data including their relationship to other landscape parameters such as relief, climate, and surface cover.

Acknowledgements

This work was supported by the Deutsche Forschungsgemeinschaft (DFG), Sonderforschungsbereich Karlsruhe (SFB 108). I thank the government of Kenya for research permits; M. Strecker for commenting on earlier versions of this paper and the Summer Institute reviewers for critical comments and suggestions.

References

AHNERT, F. (1970). Functional relationships between denudation, relief, and uplift in large mid-latitude drainage basins, *American Journal of Science*, **286**(3), 243–63.

BAKER, B. H., MITCHELL, J. G. and WILLIAMS, L. A. J. (1988). Stratigraphy, geochronology and volcano-tectonic evolution of the Kedong-Naivasha-Kinangop region, Gregory Rift valley, Kenya, *Journal of the Geological Society of London*, **145**(1), 107–16.

BREHM, F. (1995). 'Untersuchungen zur Genauigkeit Digitaler Gelaendemodelle (Arc/Info und TASH)', unpublished diploma thesis, Karlsruhe: University of Karlsruhe.

EINSELE, G. (1992). *Sedimentary Basins – Evolution, Facies, and Sediment Budget*, Berlin: Springer-Verlag.

GALE, S. J. (1992). Long-term landscape evolution in Australia, *Earth Surface Processes and Landforms*, **17**(4), 323–43.

GOODCHILD, M. F. (1993). Data models and data quality – problems and prospects, in Goodchild, M. F. (Ed.), *Environmental Modeling with GIS*, pp. 94–103, New York, Oxford: Oxford University Press.

HEUVELINK, G. B. and BURROUGH, P. A. (1993). Error propagation in cartographic modelling using Boolean logic and continuous classification, *International Journal of Geographical Information Systems*, **7**(3), 231–46.

LAURINI, R. and THOMPSON, D. (1992). *Fundamentals of Spatial Information Systems*, London: Academic Press.

MAGUIRE, D. J., GOODCHILD, M. F. and RHIND, D. W. (1991). *Geographical Information Systems – Principles and Applications*, New York: Longman Scientific & Technical.

MEYBECK, M. (1976). Total mineral dissolved transport by world major rivers, *Hydrological Sciences – Bulletin des Sciences Hydrologiques*, **21**(2), 265–84.

(1987). Global chemical weathering of surficial rocks estimated from dissolved river loads, *American Journal of Science*, **287**(5), 401–28.

MOORE, I. D., GRAYSON, R. B. and LADSON, A. R. (1991). Digital terrain modelling – A review of hydrological, geomorphological, and biological applications, *Hydrological Processes*, **5**, 3–30.

PRATT, D. J., GREENWAY, P. J. and GWYNNE, M. D. (1966). A classification of the East African rangeland, with an appendix on terminology, *Journal of Applied Ecology*, **3**, 369–82.

SAUNDERS, I. and YOUNG, A. (1983). Rates of surface processes on slopes, slope retreat and denudation, *Earth Surfaces Processes and Landforms*, **8**(5), 473–501.

SCHMITT, K. (1991). The vegetation of the Aberdare National Park, Kenya, *Hochgebirgsforschung*, **8**, 1–259.

STRECKER, M. R., BLISNIUK, P. M. and EISBACHER, G. H. (1990). Rotation of extension direction in the Central Kenya Rift, *Geology*, **18**(4), 299–302.

SUMMERFIELD, M. A. (1991). *Global Geomorphology – An Introduction to the Study of Landforms*, Cambridge: Longman.

(1994). Natural controls of fluvial denudation rates in major world drainage basins, *Journal of Geophysical Research*, **99**(B7), 13.871–83.

SUTHERLAND, R. A. and BRYAN, R. B. (1991). Sediment budgeting – A case study in the Katiorin drainage basin, Kenya, *Earth Surface Processes and Landforms*, **16**(4), 383–98.

UTLER, E. (1993). 'Vergleich zweier Verfahren zur Interpolation digitaler Gelaendemodelle (Arc/Info und TASH)', unpublished diploma thesis, Karlsruhe: University of Karlsruhe.

YOUNG, R. and MCDOUGALL, I. (1993). Long-term landscape evolution – Early miocene and modern rivers in Southern New South Wales, Australia, *The Journal of Geology*, **101**, 35–49.

CHAPTER THIRTY

A Study of Voter Turnout in New York City Elections using Spatial Statistics

DAVID OLSON

30.1 Introduction

Many attempts have been made to explain why people vote, but theories which explain behaviour well in one context often fail in another. Political efficacy, for example, is strongly related to turnout at the individual level. But Americans express a high degree of political efficacy, even though they vote at low rates compared with citizens of European democracies.

Much is known about the attitudinal and demographic characteristics of voters and non-voters. For instance, voter turnout is strongly correlated with such socio-economic characteristics as income, education, and occupational status (Teixeira, 1992; Verba and Nie, 1972; Wolfinger and Rosenstone, 1980). Yet these characteristics of individuals cannot account for the wide variation which exists in turnout rates across time and space. The fact that variables which explain individual turnout behaviour cannot explain longitudinal turnout variation in recent years creates what Richard Brody (1978) has called the 'puzzle of political participation'.

'We really do not know', one study concludes after an extensive literature review, 'why people turn out to vote' (Aldrich and Simon, 1986, p. 277). Studying turnout at the local level should illuminate the factors which facilitate political participation. Because mobilisation is a product of an interaction between partisan élites and potential voters (and not merely the sum of individual decisions made in isolation), turnout results from activities which take place in the neighbourhoods in which citizens live. Rather than ask merely why individual A decides to vote while individual B decides to abstain from voting, we can now ask why a particular community or group participates at a different rate than another. This chapter will examine how geographically based aggregate social and political data can be gathered in common units for analysis and how such data can be used to test competing theories of voter turnout.

Many studies have stressed the fact that people do not vote because they lack resources to overcome the costs of voting. Hence, those who lack high levels of income and education are least likely to vote. This is known as the 'standard socio-economic model' of political participation (Zipp *et al.*, 1982). The implication is that the relationship between socio-economic status (SES) and turnout is a natural result of social stratification, although these relationships are generally not as strong in European democracies. A competing model of political participation which will be tested here stresses that voting is a social act and is influenced by neighbourhood characteristics as well as by individual socio-economic characteristics. This model can be termed the social structural model of turnout. Social structures of communities can be measured by mobility or home ownership rates. The hypothesis is that communities with stronger social structures are more likely to support high levels of political participation because residents of such communities will have more ties to their communities and to the residents who live in them.

Most studies of voter behaviour are based on survey data. There are numerous drawbacks to using such data to study voter turnout. Turnout is self-reported; surveys produce artificially inflated estimates of turnout. The sample size is generally small, so estimates of population characteristics often have large margins of error. Surveys usually do not capture local variation because they draw upon national samples. Yet one of the most interesting aspects of turnout is the fact that it varies enormously, both among different neighbourhoods and from election to election.

30.2 Building the database

The database used for this project contains all election districts (EDs) in New York City (the entire boroughs of Manhattan, Brooklyn, Queens, Staten Island, and the Bronx).[1] Election districts are the smallest reporting unit for political data in New York City. There were a total of 5671 election districts in New York City as of the 1994 election. They contain on average about 1400 persons, 1100 of whom are of voting age and 500 of whom are actually registered voters. All citizens over 17 are eligible to register and registration is a prerequisite for voting in the US.

The data used in this study are drawn both from the Board of Elections (registration data and election returns) and the US Census (individual and household demographic data). Once data from these two sources are transformed into common units it is possible to calculate turnout rates on a very small scale. Turnout is defined as the per cent of the eligible electorate (citizens over 17) casting ballots for a given office. The eligibility variable in the turnout denominator (from the US Census' Summary Tape File 3, Table P37) is reported in Census tract units and then is allocated to ED units.

Current 1993 election districts are used as the unit of analysis. Two types of transformations need to be made. Data in old, non-current election district units must be transformed into current election district units because ED boundaries are continuously being redrawn and many (about 49 per cent) were affected by the 1991 redistricting. Data in Census tract units also need to be transformed into current election district units. Both of these types of transformations can be made because of the existence of a 'link' or 'equivalency' file. This file links blocks and 'split-blocks' (Census blocks transected by an election district) with both current and former election districts. The data in this file list every block and split-block in New York City (35 952 cases). Every block or split-block is identified as being located in an old and a new ED. From this file we know how the process of redrawing election district boundaries has affected the underlying geography at the level of the block or split-block. In the file there are 5549 new election districts identified. The file tells us how many registered voters (as of 1990) live in each block or split-block. Blocks within EDs are often split by the Board of Elections when the population in them becomes too large.

To transform 1989 election data into 1993 election district units, the link file is sorted by old ED and an old ED per cent variable is created. This variable tells us what proportion of each old ED's registered voter population every block or split-block represents. We then multiply each of these percentages by the data (that is, votes reported by election district) we want to reallocate. The assumption must be made that voting patterns across an ED would be more or less evenly distributed. This block/split-block-level data is then summed up by new ED; the data have been transformed into new EDs. This makes mapping possible and makes direct comparisons possible between the 1989 and 1993 mayoral elections.

Of course, such data transformations require us to raise questions about the accuracy of the newly created data. It is important to notice that we must first disaggregate ED data into block/split-block data before reaggregating it. *If* different blocks or split-blocks in a common ED behave significantly differently politically and *if* that ED has been redistricted in some way, then the data will probably not be completely accurate. Since EDs are so tiny and there is remarkable consistency with how particular communities in New York City behave politically, errors in the estimates are expected to be small and randomly distributed. We may overestimate Rudolph Giuliani's vote in one ED, but it will probably similarly be underestimated somewhere else.

Basic demographic characteristics can be directly aggregated into ED units, as long as the data is available from the Census' Summary Table File 1 (STF1) and as long as the ED contains no split-blocks. Otherwise the data must be transformed. Total population estimates (from STF1) are assigned to each split-block on the basis of the proportion of the registered voters who reside there. Registered voters were geocoded into the split-blocks in which they live; total population figures are not available from the Census by split-block because it is political, not Census, geography. Voter registration is generally strongly correlated with total population.

Many interesting demographic variables are available in Census tract units (from Summary Tape Files 3 or 4). Tracts are larger than EDs, but they generally contain under three thousand people. Unlike EDs, tract boundaries rarely cut across blocks. As with the procedure used to transform data from 1989 to 1993 EDs, we need to weight the blocks and split-blocks before allocating data into them. Put simply, we need to know whether there are a lot of people in them so we know whether more people with a given characteristic are likely to reside there.

To allocate tract level data into ED units, we first determine what per cent of each tract's population is in each block or split-block using total population estimates. Simply multiplying the tract level data by this proportion will yield an estimate of how many people with a particular characteristic live in each block or split-block. These new variables are then summed up by ED.

Table 30.1 Variables defined with descriptive statistics

Variable	Mean	Std dev	Minimum	Maximum	N
Turnout93P	11.39	7.82	0	100	5497
Turnout in the 1993 Mayoral Primary Election					
Turnout93G	40.27	15.12	0	100	5527
Turnout in the 1993 Mayoral General Election					
Income	22.79	14.49	–	100	5462
Per cent household income $50–100 000 (from STF3/P60)					
Dropout	9.06	11.0	0	100	5409
Per cent 16–19 not graduates, not enrolled (from STF3/P60)					
Stability	64.49	9.50	0	100	5534
Per cent who lived in the same house in 1995 (from STF3/P43)					
Home ownership	30.49	24.36	0	100	5511
Per cent housing units owner occupied (from STF1B/H3)					
Reg90	66.72	16.66	0	100	5479
Per cent eligible registered in 1990 (from Districting Commission)					
Hispanic	22.71	23.05	0	100	5546
Per cent of Hispanic origin (from STF1B/P16)					
NHBlack	23.50	30.90	0	100	5546
Per cent of non-Hispanic Black (from STF1B/P16)					

Now election data and Census data is in common geographic units. Table 30.1 presents basis descriptive statistics of the variables which were used for this analysis; all were transformed into 1993 election district units. We can now calculate turnout with precision on a very small scale. Furthermore, we can identify variation in turnout and in voter behaviour over time and between communities in New York City. With a wide range of Census variables we can then explain what accounts for this variation in turnout and in the related issue of vote choice.

30.3 Ecological analyses

Since election districts are so tiny, many of the dangers of committing the ecological fallacy, or of attributing characteristics of groups to individuals, can arguably be averted.[2] For example, if we find that an ED has a high home ownership rate as well as a high turnout rate, we may assume that it is partly the participation of homeowners which accounts for that high turnout rate. Of course, this is not necessarily true (Goodman, 1953).

Many have assumed that statistical measures computed for aggregated units would have the same values as corresponding measures computed from individual level data. In 1950 William Robinson disproved this assumption. Robinson showed that the Pearson product-moment correlation coefficient (r) is not necessarily identical when individual versus grouped data are used. Using 1930 Census data, Robinson noted that the individual level correlation between being a native-born American and being able to read is 0.118. The same correlation between the same variables aggregated at the level of the state is −0.526. Thus state-level data would lead us to conclude, incorrectly, that foreign-born Americans are more literate than native-born Americans. Such a conclusion would comprise the commission of the 'ecological fallacy'.

Within a decade of Robinson's article, Leo Goodman published an article which focused on how regression, when carefully applied, could close the gap between individual and aggregate level statistics (Goodman, 1959). Laura Langbein and Allan Lichtman note, 'Aggregation bias will not bedevil a researcher who appropriately uses aggregate level information to explore the behaviour of groups themselves, rather than the individuals comprising the groups. An election analyst, for instance, might study the voting patterns of states or congressional districts' (Langbein and Lichtman, 1978, p. 10). For this study the interest is in the characteristics of election districts because this is how turnout is being defined.

The suggestion that the ecological fallacy identified by Robinson can be escaped simply by denying that we are making claims about individuals when ecological analyses are performed has been termed 'the holistic perspective' (Achen and Shively, 1995, p. 21). The problem with this seemingly simple solution, Achen and Shively write, is that 'the aggregate units ordinarily encountered in ecological inference lack the characteristics of unitary social actors. For example, they may be census enumeration tracts, voter precincts, or mosquito abatement districts – social units whose boundaries are unknown to the voters, whose correspondence to natural social groupings may be negligible, and whose impact on the social world is confined to a handful of non-resident clerks in a dusty and distant office' (Achen and Shively, 1995, p. 21).

It is difficult to know for sure what individual level processes underlie relationships which have been found with aggregate data. We always run the risk of falling into the trap of contextual fallacies (Kornhauser, 1978). But we can with ecological analyses demonstrate the effect of geography or place on voter behaviour.

30.4 Findings

Turnout does indeed vary tremendously, a fact which is often not adequately appreciated. In the 1993 general mayoral election in New York City, for example, turnout averaged about 40 per cent across all election districts (EDs). But the standard deviation was 15, suggesting that there are wide differences in turnout among EDs (Table 30.1).

Figure 30.1 shows the location of the five boroughs or counties which comprise the City of New York. Figure 30.2 should dispel any doubts that geography affects voter behaviour. New York City Mayor Rudolph Giuliani (the Republican-Liberal candidate) clearly received the most support from Staten Island, the Upper East Side of Manhattan, and the outer portions of the boroughs of Queens and Brooklyn. His major opponent, David Dinkins (the Democratic candidate), drew his support from northern Manhattan, the South Bronx and central Brooklyn. As expected, turnout also varies tremendously (Fig. 30.3). Turnout was noticeably high on Staten Island, on the Upper East and West Sides, Riverdale in the Bronx and in the outer portions of the boroughs of Queens and (to a lesser extent) Brooklyn. High turnout was not limited to high-income areas. Borough Park, for example, voted at high rates even though it is a relatively low-income area.

Figure 30.1   The five boroughs of New York City.

Commercial, non-residential areas such as midtown Manhattan, which are extremely high-income areas, nevertheless voted at very low rates. Turnout rates were also low in immigrant areas of Queens such as Jackson Heights and Elmhurst. Many minority areas in Brooklyn did not turn out to vote at very high rates; largely Hispanic areas of the Bronx voted at low rates.

The 1989 and 1993 New York City mayoral elections show how changing patterns of political participation can have a potentially significant impact on election outcome. In both of these contests the two major candidates were the same: David Dinkins and Rudolph Giuliani. Many of the campaign issues were the same and both elections were close. Yet the outcome was much different. Dinkins won in 1989, Giuliani in 1993. Turnout was a crucial element in these elections in at least three important respects. First, the referendum to secede from New York City which was placed on the ballot on Staten Island in 1993 increased turnout dramatically in that borough. As Fig. 30.4 illustrates visually, the number of votes cast increased by about 20 000 from 1989 to 1993. The votes were cast overwhelmingly (even more so than in 1989) for Rudolph Giuliani (by a margin of over five to one). Nearly half of Giuliani's election margin can be attributed to gains made with the Staten Island vote.

Second, demobilisation – defined simply as declining turnout – occurred among many of the election districts which contain significant minority (black

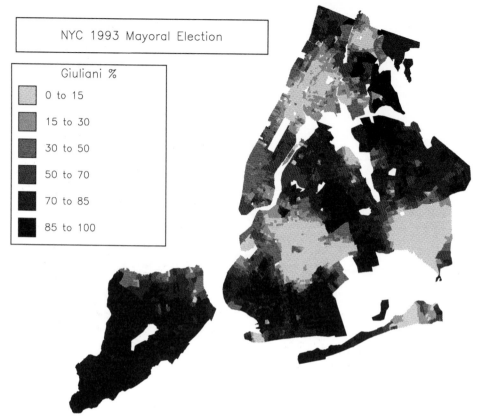

Figure 30.2 Giuliani's per cent of the 1993 mayoral vote.

and Hispanic) populations, including in Dinkins' home base in Harlem. This demobilisation contributed to electoral difficulties for Mayor Dinkins. Finally, turnout rose in some areas with large non-Hispanic white populations. This turnout was associated with support for Giuliani. Borough Park and Williamsburg in Brooklyn are two examples of such areas. Other factors explain overall variation in turnout as well.

The correlations (Pearson's r) presented in Table 30.2 show the relationships between turnout and some important variables which affect turnout rates. Registration is, of course, strongly correlated with turnout (Erikson, 1981), as only the registered are eligible to vote. Hispanic areas tend to vote at low rates. Socio-economic status, as measured by either income or education (high school dropout rates), correlates positively with turnout. Social structures of neighbourhoods, as measured by stability or home ownership, correlate positively with turnout. It is interesting to note that in the Democratic mayoral primary the relationship between income, education, or home ownership (which is correlated with income) and turnout is negative. Clearly a different sort of constituency, one from areas of much lower socio-economic status, was mobilised in the primary when Dinkins was chosen to run in the general election a few months later as the Democratic candidate. Such dramatic shifts in patterns of mobilisation are not generally captured in survey-based studies. It is not the case that poor, minority

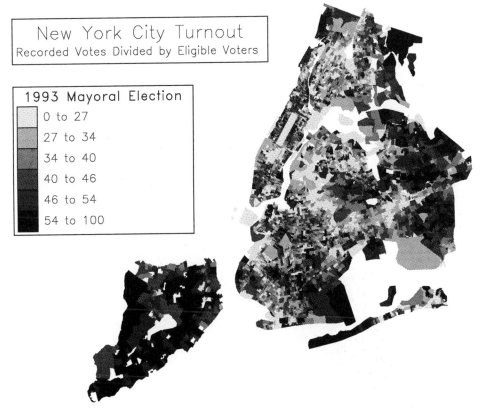

Figure 30.3 Turnout in the 1993 NYC Mayoral Election.

areas simply will not vote.

Within New York City, there are enormous differences in income levels among neighbourhoods. Figure 30.5 shows, for example, the percentage of the households in each ED which have incomes in the $50–100 000 range. Many such high-income households can be found in lower Manhattan around the financial district, in the more residential areas of Manhattan on the Upper East and Upper West sides, throughout most of Staten Island, in the eastern portion of Queens, and, to a lesser extent, in the far reaches of Brooklyn near the ocean. These areas tend to be middle- to upper-middle-class residential areas.

It is not simply the areas rich in material resources which vote (Fig. 30.5). Mobility also tends to depress turnout rates. The US Census asks whether residents lived in the same house in 1985 (about five years prior to when the Census was taken). The percentage who have moved is used as an estimate of mobility. Figure 30.6 illustrates that there are two types of areas which are highly mobile: upper-income areas in Manhattan and many low-income neighbourhoods. It is the middle-class, residential areas in Queens, Brooklyn, and Staten Island where mobility tends to be the lowest.

The more mobile a community, the lower will turnout rates tend to be. Although the exact relationships shift somewhat from election to election, this relationship is in general as strong and as consistent as the relationship between

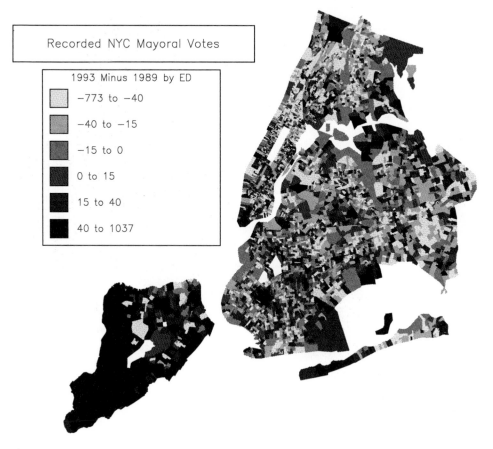

Figure 30.4   Shifts in NYC vote totals, 1989–93.

Table 30.2   Correlations of turnout with demographic variables

	Turnout 93 Primary	Turnout 93 General
Income	−0.14*	0.36
Dropout	0.04	−0.25
Stability	0.19*	0.32*
Home ownership	−0.13*	0.35*
Hispanic	−0.11*	−0.44*
NHBlack	0.45*	0.00
Reg90	0.60*	0.67*

Number of cases: 5153
* = 2-tailed significance < 0.001

income and turnout. Table 30.3 presents a multivariate regression equation using turnout in the 1993 general mayoral election as the dependent variable and income and stability as independent variables. Although a lot of variation is left unexplained (r square is just 0.15), each of these variables has approximately equal explanatory power, as the beta weights suggest. There is very little correlation

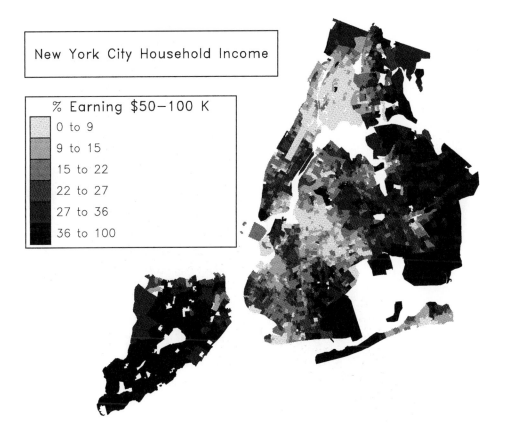

Figure 30.5 NYC household income.

between these two independent variables which measure distinct concepts (r is equal to 0.17), so multicollinearity should not be a major concern. It would be worth while exploring the extent to which there may be autocorrelation among the error terms. In any case, higher incomes and lower mobility are both associated with higher turnout, as both the socio-economic and social structural models of turnout predict. These relationships are statistically and substantively significant.

Mobility seems to have an effect on turnout which is independent of other variables such as income and education. To demonstrate this, Table 30.4 was constructed to divide EDs into a two-by-two matrix along the dimensions of mobility and education. The mean value of each variable was used to determine whether an ED is high or low along the dimensions of education and mobility. Highly mobile, poorly educated EDs clearly tend to have the lowest turnout rates. Areas of low mobility and high education have by far the highest turnout rates, as Table 30.4 makes clear. The other categories, as expected, fall in between. It is those areas with highly educated, non-mobile residents which turn out at the highest rates, with a 51 per cent turnout rate in general elections in the 1988 through 1993 elections. By way of contrast, low-education, high-mobility areas turn out at the lowest rates, with an average of only 35 per cent of the eligible electorate for these five elections, a difference of a considerable 16 per cent.

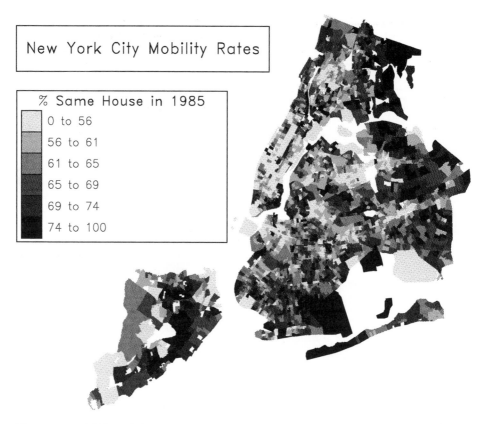

Figure 30.6 NYC mobility rates.

Moving has a destabilising effect which causes people to be less likely to vote because movers have less of a stake in the community. 'Urbanisation', Gunnar Myrdal wrote in his 1944 classic *An American Dilemma*, 'has always been associated with the weakening of traditional social structures' (Myrdal, 1944, p. xxviii). Social structures, however, are not necessarily uniformly weak, even in a city as large as New York. Mobility, which is a Census variable often used to measure social structure, is quite high in some areas, and this fact has a beneficial effect on turnout rates.

30.5 Conclusion

As surveys from the 1989 mayoral elections make clear, there is a strong relationship between socio-economic status and turnout. A CBS News/*New York Times* poll conducted 10–14 May, 1993, found, for example, that over 64 per cent of those who make under $15 000 annually admitted to not having voted in the 1989 mayoral election. This figure drops as income levels rise. For those who earn over $75 000, less than 32 per cent claimed not to have voted in the 1989 general mayoral election. A similar relationship is also found at the level of the election district, but it does not appear to be as strong, as correlation and regression

Table 30.3 Multivariate regression of turnout in the 1993 General Mayoral election (dependent variable) with income and stability (independent variables)

Variable	B	SE b	Beta	T	Sig T
Income	0.253953	0.012787	0.251517	19.860	0.0000
Stability	0.409636	0.020477	0.253347	20.005	0.0000
(Constant)	8.389207	1.314848		6.380	0.0000

Multiple R   0.38550
R Square   0.14861
Variables in the equation

Table 30.4 Matrix showing average ED turnouts sorted by mobility (per cent same house in 1985) and education (per cent college educated)

	High education (above mean)		Low education (below mean)	
High mobility	22.7	1988P	23.9	1988P
(above mean)	53.37	1988G	40.50	1988G
	23.85	1989P	21.66	1989P
	41.59	1989G	37.18	1989G
	27.25	1990G	21.00	1990G
	54.68	1992G	41.41	1992G
	10.56	1993P	10.22	1993P
	40.41	1993G	34.38	1993G
	$n = 1100$		$n = 1600$	
Low mobility	21.56	1988P	24.12	1988P
(below mean)	59.48	1988G	47.16	1988G
	27.99	1989P	26.29	1989P
	51.47	1989G	43.72	1989G
	35.13	1990G	26.52	1990G
	59.90	1992G	49.41	1992G
	12.90	1993G	13.57	1993P
	49.89	1993G	41.77	1993G
	$n = 800$		$n = 2100$	

P = Primary Election
G = General Election

analyses show (Tables 30.2 and 30.3). The frustration which often occurs when findings are altered as the unit of analysis is shifted is known as 'the modifiable areal unit problem'. It is troubling to discover that the meaning of findings can depend upon the unit of analysis which is used in a study. The smaller the unit of analysis, the less variation will be masked by the process of aggregating data. Turnout is rarely studied in units as small as election districts, and thus community social structure is often overlooked as a factor contributing to patterns of political participation.

In the Annual Address delivered at the Annual Meeting of the American

Political Science Association in 1967, Robert Dahl wondered what the ideal size of the city is for effective democratic governance. This inevitably led him to ponder the significance of the boundaries placed around cities. He concludes that it would be most useful if one could 'make the legal boundaries of a city coincide more closely with what might be called its sociological boundaries'. Dahl knew, however, that the boundaries humans create are not necessarily 'real'. Census tracts are sometimes used as substitute boundaries for neighbourhoods (Huckfeldt, 1986), but they are not drawn with any such sociological concept in mind. Similarly, a Congressional district, Fred Harris observes, 'has no natural coherence or intrinsic reason for being, except that it contains the requisite number of people' (Harris, 1995, p. 66). The unit of analysis matters. Dahl (1967) postulated that 'we may need different models of democracy for different kinds of units' (p. 968). He observed, 'Obviously different problems call for different boundaries' (p. 467).

The link between socio-economic status and turnout is well documented by survey researchers and is supported by the ED level data presented here. Even so, other factors, such as the mobility rates of a community, can have as powerful an effect on turnout. Stable communities in which residents have the strongest ties are most easily mobilised. But even these relationships are not necessarily fixed across time and space. Mapping data in small geographic units using GIS demonstrates this and provides the social scientist with a powerful tool for identifying and comparing social and political characteristics of areas, whether they are areas of New York City as presented here or areas of the UK as analysed by Daniel Dorling in Chapter 6 in this volume. Mobilisation and demobilisation are continuous processes, the results of which can be seen in Table 30.4. Shifting our analysis to the level of the election district forces us to question whether there is an inherent relationship between such variables as those which measure socio-economic status and political participation. Relationships which hold at one time in one place do not necessarily hold at another time or place, or at another unit of analysis.

Notes

1   This database was constructed with John Mollenkopf and Marta Fisch of CUNY Data Service. Much appreciation is extended to them. The original data were obtained from the US Census, the New York City Board of Elections, and the New York City Districting Commission. I am also appreciative of the contributions Todd Swanstrom from the University at Albany has made to help me conceptualise non-voting.

2   This is a point which Kristi Andersen has made. She anticipates the criticism she would later receive for using aggregate election data to make claims about the voting behaviour of individuals. Can you know whether *individuals* are switching party allegiances when data are in aggregated units? Changes in such voter behaviour could be caused either by changes in individual behaviour (conversion) or by changes in the composition of the populations (mobilisation). Andersen claims, 'As one moves down to smaller units of analysis . . . the danger of succumbing to the ecological fallacy is lessened. This occurs because we can couch our interpretations of the aggregate trends in a sounder knowledge of the partisan movements which are likely to occur in the same area (and which might well be obscured by analysing

units rather than individuals)' (Andersen, 1979, p. 84). Nevertheless, individual and aggregate level data are conceptually distinct types of data and it is necessary to keep the nature of the data in mind throughout the analysis.

References

ACHEN, C. H. and SHIVELY, W. H. (1995). *Cross-Level Inference*, Chicago: University of Chicago Press.

ALDRICH, J. H. and SIMON, D. M. (1986). Turnout in American national elections, in Long, S. L. (Ed.), *Research in Micropolitics*, vol. 1, pp. 271–301, Greenwich, CT: JAI Press.

ANDERSEN, K. (1979). *The Creation of a Democratic Majority 1928–1936*, Chicago: University of Chicago Press.

BRODY, R. A. (1978). The puzzle of political participation in America, in King, A. (Ed.), *The New American Political System*, pp. 287–324, Washington, DC: American Enterprise Institute.

DAHL, R. A. (1967). The city in the future of democracy, *American Political Science Review*, **61**, 953–70.

ERIKSON, R. S. (1981). Why do people vote? Because they are registered, *American Politics Quarterly*, **9**, 259–76.

GOODMAN, L. (1953). Ecological regression and the behavior of individuals, *American Sociological Review*, **18**, 663–4.

(1959). Some alternatives to ecological correlation, *American Journal of Sociology*, **64**, 610–25.

HARRIS, F. R. (1995). *In Defense of Congress*, New York: St Martin's Press.

HUCKFELDT, R. (1986). *Politics in Context: Assimilation and Conflict in Urban Neighborhoods*, New York: Agathon Press.

KORNHAUSER, R. R. (1978). *Social Sources of Delinquency*, Chicago: University of Chicago Press.

LANGBEIN, L. I. and LICHTMAN, A. J. (1978). *Ecological Inference*, London: Sage.

MYRDAL, G. (1944). *An American Dilemma*, New York: Harper & Row.

ROBINSON, W. S. (1950). Ecological correlations and the behavior of individuals, *American Sociological Review*, **15**, 351–7.

ROSENSTONE, S. J. and HANSEN, J. M. (1993). *Mobilization, Participation and Democracy in America*, New York: Macmillan.

TEIXEIRA, R. A. (1992). *The Disappearing American Voter*, Washington, DC: Brookings.

VERBA, S. and NIE, N. H. (1972). *Participation in America: Political Democracy and Social Equality*, Chicago: University of Chicago Press.

WOLFINGER, R. E. and ROSENSTONE, S. J. (1980). *Who Votes?*, New Haven: Yale University Press.

ZIPP, J. F., LANDERMAN, R. and LUEBKE, P. (1982). Political Parties and Political Participation: A Re-examination of the Standard Socioeconomic Model, *Social Forces*, **60**, 1140–53.

CHAPTER THIRTY-ONE

GIS Applications in the Arctic Environment

ALEXANDER BAKLANOV, TATJANA MAKAROVA
and LARISSA NAZARENKO

31.1 Introduction

The Arctic environment is in a state of constant change as a result of both natural and human influences. Current models suggest that environmental change, as a result of human activities in the earth-atmosphere system, will be more pronounced in the Arctic regions than elsewhere on Earth. Arctic ecosystems are fragile and susceptible to environmental change and human activities. As a result, the region is one of the most vulnerable, and, thus, may also be one of the most sensitive indicators of any changes. Furthermore, the Arctic environment is not isolated from neighbouring low-temperature regions, and there is a continuous interchange between them both in the transport of pollutants and in the movement of migratory species.

Economic development has taken place in the past in Russia with very little concern for its environmental consequences. Because of its low population density and abundance of mineral resources, the Arctic region has been a site of major industrial development based on extractive industries which have potentially a disastrous impact on the environment. In fact, there is already evidence that the limited ecological capacity of the northern regions of Russia and past unchecked developments have resulted in very large blocks of land now lying waste (Nickonov and Lukina, 1994). The current state of the Russian economy and the need to improve industrial efficiency and output mean that there is a grave risk of further environmental degradation unless steps are taken to prevent this. Against this background, this chapter describes the results of new environmental studies carried out in the Murmansk region by the Institute of Northern Ecology Problems to develop (1) a GIS database for radiation risk objects in the northern part of Russia, and (2) an ecological atlas of the Murmansk region. Both projects are part of a wider programme to create a regional ecological information system for the Kola North (Kalabin and Baklanov, 1992). It is felt that an environmental database will provide a mechanism for documenting the most important biological resources and the gravest threat to their survival. It will enable rapid access to information for Russia and collaborating companies from overseas to draw their attention on the major risks and plan their operations in a way to minimise their impact. Clearly, the research described in the following sections is made possible by the increasing availability of information technology such as GIS to collate and analyse environmental data and by the recent political changes which have given access to a large volume of information.

31.2 GIS database for radiation risk objects in northern Russia

The main objective of this project was to collate and assess all the available information (some of which is still protected for security reasons) on this sensitive topic and have a scientific base for evaluation rather than the cacophony of unreliable rumours on sources of radioactive pollution previously circulating. In addition to the threats created by human and industrial developments, the Arctic region is also strongly affected by nuclear power technologies and military activities, both on land and in the sea.

The seriousness of the threat to the Arctic and global environment from the accumulation, storage and deposition of industrially concentrated radioactive materials in northern Russia and adjacent Arctic water is evident. Many investigations and reports have provided a general description of radioactive materials from industrial and military activities in the Russian north and the adjacent ocean, and of their present distribution and deposition (Baklanov *et al.*, 1992, 1993; Foyn and Nikitin, 1993; Matishov, 1992; Matishov *et al.*, 1994; Nilsen and Bohmer, 1994).

Detailed information, however, is often lacking as available data have not been assembled and assessed. For example, in many critical areas there are no detailed data or information on the physical state, chemical composition and concentrations, and types of containers of the radioactive waste, nor on the production and management technologies used in the past. With this in mind the principal areas of concern at present lie in three main categories (Baklanov *et al.*, 1992; Matishov, 1992; Matishov *et al.*, 1994; Nilsen and Bohmer, 1994).

- The first category is radioactive waste (spent nuclear fuel, reprocessing plant residues, military materials, and contaminated equipment) from industrial and military activities, at present held in temporary storage or short-term dumps at various locations on or around the Kola Peninsula, Novaya Zemlya and the White Sea.

- Second, waste and contaminated equipment from the same sources as above, dumped in shallow to moderately deep waters in the Kara Sea, the Barents Sea, the Sea of Okhotsk and the North Pacific or Bering Sea.

- The third area of concern is radioactive waste and dispersed radioactive contaminants distributed in a large number of locations in inland northern and Arctic Russia as a result of military operations and associated weapons manufacturing, nuclear power station and its fuel cycle, mining, and the widespread use of nuclear explosions for civilian construction projects. In some of these locations the contamination of soil, biota and ground water is severe or critical, posing not only a present and future threat to resident human and local ecosystems but a prospect of serious future contamination of major north-flowing Russian rivers, with subsequent delivery of radioactive contaminants to the ocean.

In 1992 the Institute of Northern Ecology Problems (INEP) prepared the database and the map-scheme on radiation risk objects of the Arctic and Russian north (Baklanov et al., 1992, 1993). The next stage was the creation of a GIS database and a GIS map of radiation risk objects in Russian Arctic. Data were drawn from existing published information and fieldwork (Baklanov et al., 1992, 1993). The Soviet Union map at the scale 1:7 000 000 has been used. It was digitised by means of Arc/Info, PC version. The state borders, shore contours, lakes and river network, big cities and populated areas were geocoded. After that 16 types of radioactive risk objects were identified. The topology for these objects was generated (Fig. 31.1). The information about radiation risk objects is saved in the file in dbf-format organised with the help of dBASE-4 or PARADOX. More detailed information about each object is saved in the MEMO-file. This is a very convenient way of presentation of the textual information, commentary and interpretation of the data.

The history of radioactive waste in the ecosystems of seas, lands and archipelagos of the Arctic can be divided into four main periods of influence which often overlap and compound each other (Matishov et al., 1994). The first period relates to 1950–80 and is connected with the intensive outflow of radiation waste to the Kara Sea of West-Siberian rivers from the Ural and South-Siberian plutonium plants. The 30-year length of this period means that its cumulative effects are very dangerous to both local and trans-Arctic pollution.

The second period is from the middle 1950s to the beginning of the 1960s at the height of the Cold War when nuclear explosions in the atmosphere, underground and underwater were carried out in the nuclear test polygons of Novaya Zemlya.

From the middle 1970s to the beginning of the 1980s an additional source of nuclear pollution in the northern Arctic was from radiochemical plants located on the shores of France, England and Scotland into the Irish Sea and the English Channel. Then at the end of the 1980s the Chernobyl disaster released significant quantities of radioactive dust over a wide area. To add to the above sources of pollution, which have been accumulating into the Arctic region, the dumping of containers with solid radioactive materials into the Arctic seas took place between 1970 and 1990.

Of all of the above, by far the worst radioactive pollution of the Arctic sea and coast was that resulting from the atmospheric nuclear explosions on Novaya Zemlya (Cochran et al., 1989; Matishov et al., 1992). The Soviet Union performed 715 nuclear test explosions between 1949 and 1990. Nuclear bombs were used for the creation of water bodies, channels, mines and gas reservoirs. Nuclear bombs were also employed for extinguishing gas fires and for seismic research. There were 41 such nuclear explosions in the northern parts of Russia, most of them in Siberia.

On the Kola Peninsula, two nuclear bombs were exploded in the Kuelpor mine in the Khibiny Mountains, 15 km east of the town of Kirovsk, while at Novaya Zemlya there are four test sites (Baklanov et al., 1993), two at Chernaya Bay (southern field) on the southern island and two by the Matotchin Strait (northern field), dividing the northern and southern islands (Fig. 31.1). All atmospheric nuclear explosions were performed at the northern test site. The southern test site was in use for two years, from 1973 to 1975.

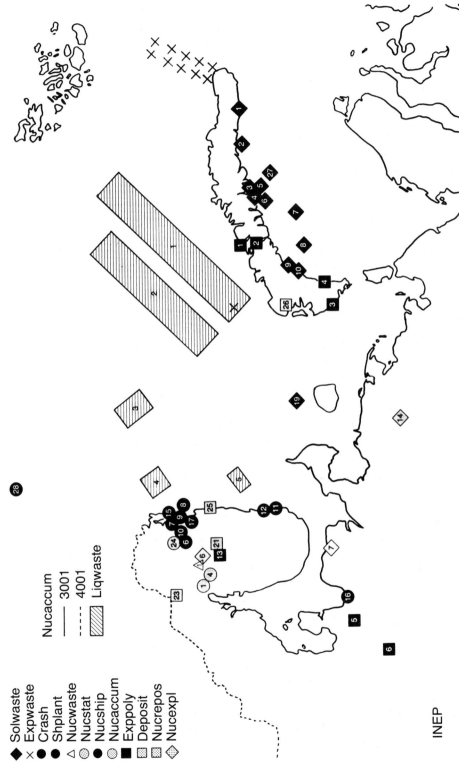

Figure 31.1 Radioactive risk objects in northern Russia.

It should be stressed that the test field at Novaya Zemlya has not been used since the demise of the Soviet Union. No nuclear explosion had taken place anywhere in Russia since 1990, as a part of the international freeze on nuclear testing.

One more problem of radioactive contamination is the dumping of radioactive waste. Radioactive waste has been dumped into the ocean ever since the Soviet Union started testing its first nuclear-powered submarines at the Severodvinsk shipyard in 1959. The Russian Northern Navy (formerly the Soviet) has, since 1960, dumped radioactive waste in the Barents Sea and Kara Sea on a regular basis. This comprises solid radioactive waste, liquid radioactive waste and nuclear reactors with and without fuel. The Northern Navy has sunk a total of 17 ships and lighters containing radioactive waste in these seas. The dumped containers are mostly filled with low- and medium-level radioactive waste, such as contaminated metal parts from the submarine's reactor sections and clothes and equipment used for work on the reactors. This kind of solid waste has been dumped in 10 different bays off the eastern coast of Novaya Zemlya and in the Kara Sea. The dump sites are shown in Fig. 31.1. The Navy has also dumped radioactive waste in the Japan Sea, Pacific Ocean and Baltic Sea.

Furthermore, radioactive waste has been dumped in the Barents Sea and Kara Sea from the civil state-run Murmansk Shipping Company's fleet of nuclear icebreakers. This includes liquid radioactive cooling waters from the ship's reactors and storage tanks for used fuel which have been dumped since 1959. The last dumping of liquid radioactive waste took place in November 1991, and this practice may be resumed if no alternative solutions are found.

All the available information on these events has now been stored in a GIS database and the various dumping fields in the Barents Sea are marked on Fig. 31.1. The liquid waste of the highest radioactive concentration has been dumped in three dumping fields, in the most remote part of the Barents Sea, while the less radioactive waste was dumped outside the shore of the Kola Peninsula (Baklanov et al., 1993).

The GIS database contains information about the amount and volumes of containers, radioactivity of wastes, years and precise site in coordinates of the radioactive waste dumping, providing a very valuable information base for monitoring the effects of these past activities on the local environment.

31.3 Digital mapping of the Murmansk region environment

The *Ecological Atlas of the Murmansk Region* is the scientific-reference edition, designed for scientists and managers connected with the problems of rational use of natural resources and nature protection. Leading specialists of the Murmansk Committee on Hydrometeorology and the Murmansk Marine Institute of the Kola Science Centre participated in this project. The leading institute was the Institute of North Industrial Ecology Problems. The maps in the *Atlas* describe the changes in the ecological situation during the last 70 years, for example, the period of most intensive industrial development of the territory. The findings of the extensive ecological investigations carried out in the Murmansk region for the last 30 years have provided the bases for compiling most of the specialised maps.

As a reference book, the *Ecological Atlas* will be of use for the residents of the

Murmansk region, who are interested in their ecological problems as well as for public health, local government institutions and organisations involved in ecological restoration and in education.

It has to be considered that the Murmansk region is now the most industrially developed and urbanised region in the circumpolar area. Some enterprises functioning here are the biggest in Europe: two nickel and copper smelters, the apatite and nepheline mining and processing companies and others. The whole territory of the Kola Peninsula is exposed to the effects connected with the air- and water-borne pollution of sulphur and heavy metals (Kryuchkov and Makarova, 1989).

Pollution creates ecological problems at various scales: regional – for example, the change of hydrochemical properties of the surface waters (the acidification, the content increase of heavy metals and sulphate); and local – for example, landscape devastation (the pollution of vegetation, soil, surface and groundwaters, the local climate changing and so on) – as well as affecting the human environment (increase of the morbidity and mortality rate of the population and decrease of the animal population) (Makarova, 1992).

The *Atlas* includes maps on the regional environmental pollution impact on the Murmansk region during 1990–3 elaborated by INEP and the Geography Department of Moscow State University which show pollutant emissions into the atmosphere, the heavy metal concentration in lichens, the deposition of anthropogenic sulphur, extreme ecosystem destruction resulting from industrial pollution, lake acidification, the nature-protected areas and the increase of Pinnipedia in colonies.

The Severonikel Company, a huge nickel producer, also yielding copper and cobalt, is located in the city of Monchegorsk. In the 1930s the enterprise was working with local ores and did not pollute much as the smelter was small and the ores were low-sulphur. But by the late 1970s the mines were depleted. To keep the smelter going – and to expand it – ores containing 30 per cent sulphur were shipped here from Norilsk in Siberia. As a result, today high-sulphur ore go to both Monchegorsk and the other Kola smelter at Nickel, from where sulphur dioxide plus moisture – acid rain – influences the environment.

Monchegorsk-generated poisons fall heavily on Laplandsky Zapovednik, the state nature biosphere preserve, created to protect reindeer. A quarter of the 1000 mile^2 preserve is damaged; some areas are dead. Metallurgical enterprises, heat power plants and transport also produce harmful emissions into the atmosphere. Figure 31.2 is an example of the rate of total local heavy metal accumulation in lichens, compared with the clean subarctic region. To create this figure, a map of the Kola Peninsula (scale 1:1 000 000) has been digitised together with polygons with different coefficients of the total local accumulation of heavy metal in lichens in comparison to the pure subarctic region. The highest coefficient, more than 50, of total local heavy metal accumulation in lichens is in the Nickel and Monchegorsk districts. The lowest coefficient (less than 2) relates to the eastern region of the Kola Peninsula. A similar situation occurs with the deposition of anthropogenic sulphur (Fig. 31.3). The deposition of the anthropogenic sulphur units is g/cm^2 per year. The highest deposition of anthropogenic sulphur is more than 2 g/cm^2 per year for the Nickel and Monchegorsk districts. The lowest deposition is less than 0.5 g/cm^2 per year for the eastern and western regions of the Kola Peninsula.

As shown, the serious ecological situation of the Murmansk region is the result

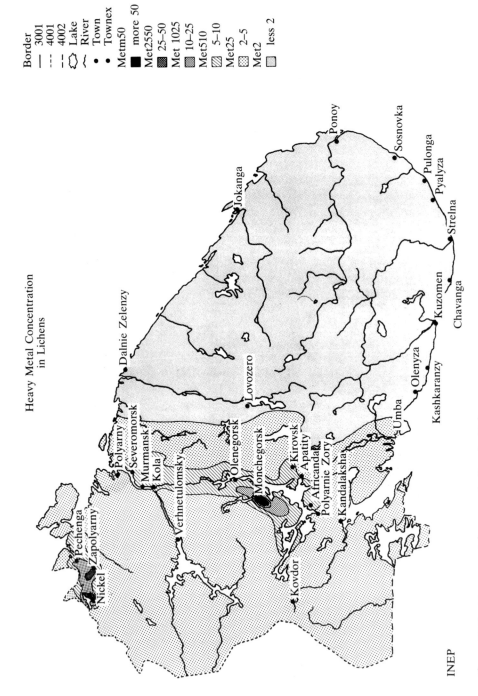

Figure 31.2 Heavy metal concentration in Lichens.

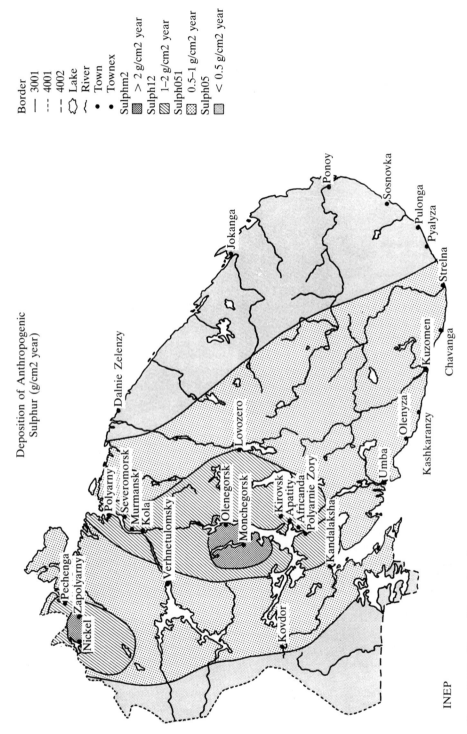

Figure 31.3 Deposition of anthropogenic sulphur.

of a variety of human activities. The *Atlas* developed with the aid of GIS, presents in an effective and comprehensive way the ecological data gathered and contains complex ecological descriptions, which could be applied as a base for decisions optimising environmental solutions, as well as estimates and prognosis maps. All the maps produced in the *Atlas* are based on the current investigations and complex ecological estimations of the territory integrated by geoinformation technologies.

There remains the following question: 'Is the self-regeneration of the northern ecosystem possible and under what conditions?'

It is impossible to answer this question simply. But if emissions in the air discharges in water bodies are stopped then the ecosystem degradation process could stabilise enabling gradual regeneration of the environment. It is clear, however, that it will be hard to achieve the level of the primordial ecosystem condition.

31.4 Conclusion

GIS is one of the best instruments for the presentation of large volumes of spatial environmental data. A GIS can help to perform many routines and even complex tasks. For example, a GIS can be used to analyse spatial relationships, to identify regions that meet multiple criteria, to model the impacts of policy options, and to measure and monitor dynamic processes.

There are many possibilities to perform qualitative and complex analysis with different types of data. With the help of GIS-technology it is much easier to combine geographical maps and high volumes of spatial data in various forms and to present them, leading hopefully to more informed decisions about the impact of current and proposed developments.

References

BAKLANOV, A., MOROZOV, S. V. and KLYUCHNIKOVA, E. M. (1992). The picking of risk zones and the elaboration of the extreme ecological radiation dangerous situation scenarios in North regions, in Kalabin, G. V. (Ed.), *Report of Scientific Researches*, p. 154, Apatity: Institute of North Ecology Problems, Kola Science Centre, Russian Academy of Sciences.

BAKLANOV, A., ZOLOTKOV, A. and KLYUCHNIKOVA, E. (1993). The map of risk objects of Arctic and Russian North, in *Econord-Inform*, 1, 9–16.

COCHRAN, T. B., ARKIN, W. M. and NORRIS, R. S. (1989). Nuclear weapons databook, *Soviet Nuclear Weapons*, 4, 332–82.

FOYN, L. and NIKITIN, A. (1993). Preliminary report of the joint Norwegian Russian expedition to the dumpsites for radioactive waste in the open Kara Sea, the Tsivolki Fjord and the Stepovogo Fjord September–October 1993, Norway: Joint Russian-Norwegian Expert Group for Investigation of Dumped Radioactive Waste in the Barents and Karma Seas.

KALABIN, G. and BAKLANOV, A. (1992). The creation principles of the regional information system of ecological monitoring, in Kalabin, G. V. (Ed.), *Ecological Geographical Problems of Kola North*, pp. 57–66, Apatity: Institute of North Ecology Problems, Kola Science Centre, Russian Academy of Sciences.

KRYUCHKOV, V. and MAKAROVA, T. (1989). *Airtechnogenic Impact to Ecosystems of Kola*

North, Apatity: Kola Science Centre, Russian Academy of Sciences.

MAKAROVA, T. (1992). Zonal and regional factors of ecosystem changing in the conditions of the anthropogenic pollution in Kola North, in *Ecological Geographical Problems of Kola North*, pp. 4–8, Apatity: Kola Science Centre, Russian Academy of Sciences.

MATISHOV, G. G. (1992). *Anthropogenic Destruction of Ecosystems of the Barents and Norwegian Seas*, Apatity: Kola Science Centre, Russian Academy of Sciences.

MATISHOV, G. G., MATISHOV, D. and PODOBEDOV, V. (1992). *Radio-nuclides on Kola Peninsula, Novaya Zemlya, Zemlya Frantsa-Iosifa and in Barents Sea*, Apatity: Kola Science Centre, Russian Academy of Sciences.

MATISHOV, G. G., MATISHOV, D., SZCZYRA, J. and RISSANEN, K. (1994). *Radio-nuclides in the Ecosystem of the Barents and Kara Seas Region*, Murmansk: Marine Biological Institute, Kola Science Centre, Russian Academy of Sciences.

NIKONOV, V. and LUKINA, N. (1994). *Forests Biogeochemical Function on the Northern Tree Line*, Apatity: Institute of North Ecology Problems, Kola Science Centre, Russian Academy of Sciences.

NILSEN, T. and BOHMER, N. (1994). Sources of radioactive contamination in Murmansk and Arkhangelsk counties, in *Bellona Report*, **1**, pp. 1–162, Oslo: The Bellona Foundation Books.

CHAPTER THIRTY-TWO

Ecologists, Farmers, Tourists – GIS Support Planning of Red Stone Park, China

KONGJIAN YU

32.1 Introduction: security patterns and SP approach

Landscape planning is a considered procedure of defence involving defenders of various processes. How can we defend the processes of our concern more effectively while maximising opportunities for change? This chapter tries to answer this question using the concept of security patterns (SPs) and demonstrates how GIS can be combined with the SP approach in landscape planning.

By definition, SPs are spatial patterns composed of strategic portions, positions, critical scales (sizes), numbers, shapes and inter-relationships that are associated with certain thresholds in the non-linear dynamics of processes in the landscapes. SPs have or potentially have a critical significance in safeguarding certain processes, for example, the process of species dispersal, spread of fire and other disturbances, visual perception and preference, and agricultural conversion.

In terms of their significance for the processes of our concern, security landscape components have three basic characteristics.

1 *Initiative*: the quality of a portion or position the occupation of which is likely to give the advantage of initiating certain processes.

2 *Efficiency*: the quality of a position or portion the occupation of which will give the advantage of less cost in energy and materials and be much more effective in promoting or controlling certain processes.

3 *Coordination*: the quality of a position or portion the occupation of which will give the advantage of effective spatial communication among neighbouring elements.

SPs are multilevelled. Each individual process in the landscape has its own security patterns (Fig. 32.1), and these individual SPs may compete and overlap spatially. Furthermore, each individual process has SPs at various security levels.

The concept of SPs is based on two assumptions concerning spatial patterns

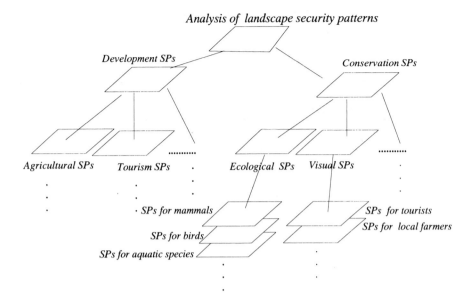

Figure 32.1 A presumed hierarchy of landscape security patterns.

and processes: (a) landscape patterns affect processes, and (b) there are strategic landscapes associated with some thresholds in the dynamics of certain processes.

Numerous observations suggest that the spatial patterns of a landscape influence various ecological processes such as species dispersal and population dynamics (Forman and Godron, 1986; Turner, 1989); human processes such as residential development and demographic dynamics (for example, Berry and Horton, 1970), and visual perceptual processes (Gibson, 1950; Lynch, 1960).

Not all portions and positions of the landscape are equally important in terms of their influence on individual processes. Some are more important than others, and some are strategically critical. Examples of such strategic portions and positions include the inlets and outlets of a basin and breaks in a corridor that have critical values for ecological processes (Forman and Godron, 1986; Merriam, 1984); the conspicuous landmarks, narrow defiles, gorges and bridges that have significant visual perceptual effects (Stein and Niederland, 1989; Tuan, 1974); as well as certain places that have a strategic significance for economic processes (Taaffe and Gauthier, 1973).

It is important to note, however, that in some cases various processes in the landscape may be controlled by spatial patterns that are not intuitively obvious nor visually apparent to a human observer. It is assumed that some kinds of thresholds exist in the trajectories of the dynamics of processes. At some points (in terms of number, size, shape and inter-distance of landscape elements), a slight change in landscape property produces sudden changes in the response of the process. Such thresholds have been recognised in urban development (Kozlowski, 1986). Similar to thresholds, other concepts have been proposed that may also be useful in understanding my ideas concerning the strategic landscape and security patterns such as safe minimum standards (SMS) (Bishop *et al.*, 1974; Ciriacy-Wantrup,

1968), carrying capacity, and ultimate environmental thresholds (UETs) (Kozlowski and Hill, 1993).

It is thus reasonable to assume that:

- landscape patterns associated with these critical thresholds or constraints are likely to be strategically critical in controlling or promoting certain processes;
- landscape design and management following these strategically critical patterns can more effectively safeguard or control the processes.

Therefore, it is worthwhile identifying and applying SPs in landscape planning. The following two aspects of exploration become the major focus of this chapter:

1 How can we define and identify SPs and what are they?
2 How can we apply SPs in landscape planning to achieve a less detrimental landscape, while at the same time maximising the changes acceptable to decision makers and/or developers?

These two aspects of inquiry compose an approach to landscape planning which I call the SP approach, or the approach of security patterns. It is an approach for defending various processes of our concern, aiming at a good balance of acceptable changes and a securer landscape by identifying and applying security patterns. The SP approach tries to establish 'stop signs' in the procedures of decision making for various landscape changes, and to safeguard the security of the processes at critical points. In a certain sense, defining SPs is a strategy of spatial defence, an operational weapon of negotiation aimed at a less harmful change by controlling critical points, or 'frontiers'. Defence by these SPs is expected to be more effective in safeguarding the landscape processes of our concern. GIS has great potential when combined with the SP approach in landscape planning and decision making (see Yu, 1995c for more detailed discussion on the SP concept).

A case study of the Red Stone National Park in south China, is used to illustrate the SP approach. This case is selected since it dramatically represents a defensible procedure of landscape change among defenders of three interacting, and often competing, processes in landscapes, including ecological, visual and agricultural conversion processes.

32.2 Defending the security of processes in Red Stone National Park: A case study

Red Stone National Park is 313 km^2 in size (Fig. 32.2). The dominant regional natural vegetation is composed of subtropical evergreen forests which have been seriously destroyed at the peripheral area with some isolated remnant patches scattered in the remote areas. The landscape is made up of hundreds of heavily eroded rocky hills, square with flat tops and steep slopes. This unique landform is the primary factor affecting the distribution of soil, vegetation, wildlife habitats, visual quality and agriculture. The remnant biological islands are extremely valuable in terms of biodiversity conservation and landscape restoration. The visual quality is extraordinary. It is one of the major tourist attractions in southern

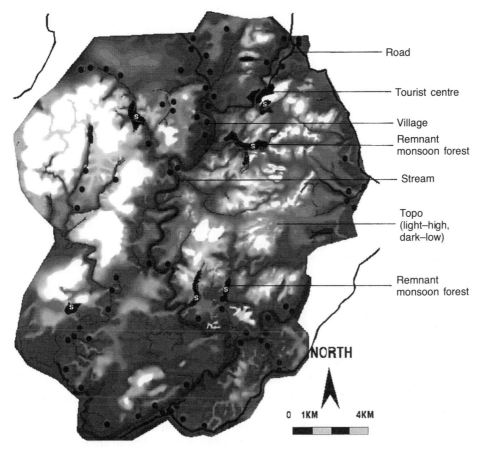

Road

Tourist centre

Village

Remnant
monsoon forest

Stream

Topo
(light–high,
dark–low)

Remnant
monsoon forest

NORTH

0 1KM   4KM

Figure 32.2 The landscape of the Red Stone National Park in south China.

China. The fertile soil and subtropical climate make this land one of the most productive agricultural areas. About 20 000 farmers live in 70 villages scattered in the small alluvial planes in this hilly landscape. The problems this national park now faces are typical of other national protected areas, namely, the conflicts between development, ecological and visual conservation. Landscape planning in this park is a defensible procedure taking place among defenders of three main processes: ecological, visual perceptual, and agricultural. The objectives of this case study are to explore an effective way of defending various landscape processes in this national park by identifying and applying SPs, and to demonstrate how GIS can be integrated into the defensible procedure of landscape change and decision-making.

32.2.1 Security patterns in the Red Stone National Park

32.2.1.1 Ecological SPs: Ecologists' defensive frontiers

Ecological processes concerned in this case are species dispersal and maintenance. Three groups of species are targeted: medium-sized mammals (Cervidae and

Viverridae families), pheasants (Phasianidae family) and amphibians (Crypto-
branchidae and Ranidae families). These species are native to this region and have
an endangered status. Ecological SPs are identified by analysing accessibility
surfaces that represent the potential coverage by the species of our concern.

Accessibility surfaces are developed using a minimum cumulative resistance
(MCR) model (Knaapen *et al.*, 1992; Yu, 1995b). This model conceives the
dynamics of species dispersal as a function of sources, distance and intermediate
landscapes. Native habitats of the target species are taken as sources of dispersal.
Intermediate landscapes are evaluated for their resistance to the dispersal of
species, and the dynamics of the dispersal process is simulated based on the
cumulative resistance to the dispersal of a certain species. Comparative resistance
values are assigned to various landscape attributes. Various factors such as cover,
slope, elevation and aspect may contribute to the resistance value of each cell of
the landscape. The probability of successful access to a cell by a species can be
expressed as:

$$\text{Accessibility} = f \operatorname*{Min} \sum_{i=1}^{i=n} (Di * Ri)$$

where f is some unknown but monotonically decreasing function. Di and Ri
represent the distances (number of cells) and resistance respectively when a species
travels across landscape type i. While f is some unknown function, the sum of the
weighted distance ($Di* Ri$), or the 'cumulative resistance' can be taken as an
indication of relative accessibility of the cell to the species through one possible
route. There are numerous routes from the sources to the cell, and the routes with
the lowest cumulative resistance, namely, minimal cumulative resistance (MCR),
can be used as the relative measurement of the accessibility of this cell from the
sources (habitats of target species).

Resistance classification (Ri) is based on individual cells of 25×25 m in size.
An interactive interface of the GIS model using Arc/Info is developed to allow the
processing of more precise data. Land use and land cover are the major factors
contributing to the resistance of the landscape. In our case, it is reasonable to
assume the more similar a cell to the natural habitats the less the resistance to the
target species of our concern. Eight major land-use and land-cover categories are
observed and they are closely associated with the degree of naturalness or the
intensity of human disturbances; these categories range from developed areas to
agricultural fields, grasslands, shrubs, coniferous forests, mixed forests, the
remnant subtropical forests and water. From developed area to the remnant
subtropical forests, the degree of human disturbance increases; therefore, in this
case it is assumed that the resistance to the dispersal of native target species
increases accordingly. The developed areas (including roads, housing and tourist
service centres) have the highest resistance to all target species (assigned a value
10) and the natural remnant forests the lowest resistance to the target species
(assigned value 0). Water bodies are assigned a high resistance value to the
medium-sized mammals but have a moderate resistance to pheasants and low
resistance to amphibians (here the quality of water is not considered).

The topographical factors including elevation and slope also contribute to the
resistance to some species. For the medium-sized mammals in this case, gentle

slope is considered to have less resistance than a steep slope, the extremely steep slope becoming a barrier to movement. For the pheasants, these topographical factors are not important. Amphibians are sensitive to the hydrological situation, which in this case is associated with the elevation because of the unique geological formation of this area.

Based on the resistance map, an accessibility surface can be developed using the function discussed above. The resultant accessibility surface resembles a topographic surface that is made up of equal-valued MCR contours. Following Warntz's model of surface interpretation (Warntz, 1966), it 'dips' at the sources, has 'peaks' that are least accessible to target species, has 'courses' with lower MCR value which run from 'pits' to 'pits', and has 'ridges' with higher MCR values which run from 'peak' to 'peak'. On each of the 'courses' or 'ridges' there is one 'pale' or 'pass'. From the MCR surface, we can also recognise the potential cliffs where values increase or decrease dramatically, and the potential flat planes where the species can spread quickly over the landscape.

The accessibility surface, therefore, reveals the potential patterns of coverage by the target species and the strategic values of landscape in terms of species dispersal and maintenance (Yu, 1995a and b). Based on the features of the accessibility surfaces, four structural components can be identified: buffer zones, inter-source linkages, radiating routes and strategic points. These four components, specified by certain quantitative and qualitative parameters, together with the identified sources (native habitats) compose a security pattern (Fig. 32.3). Changes in these components, quantitatively or qualitatively, will dramatically affect the security of the targeted processes.

Among others, three series of ecological SPs are identified respectively at high-, moderate- and low-security levels for different groups of species. For example, Figs 32.4 to 34.6 show the medium-sized mammals in this case. They could be combined into corresponding overall ecological SPs. These ecological SPs can be used by ecologists as defensive frontiers for the defence of the ecological processes at various security levels in the process of landscape planning and change.

32.2.1.2 *Visual SPs: Tourists' defensive frontiers*

The visual SPs are defined on the basis of critical landscape interpreted by visual sensitivity surfaces which are a combination of landscape visibility and preference evaluation. The calculation and mapping of landscape visibility were carried out using function that most GIS packages contain. GIS mapping of landscape preference is relatively more complicated. First, 572 individuals from China and USA were interviewed as to their preference evaluation for various landscapes in the case study area (for detailed discussion, see Yu, 1995c). Second, factor analysis and regression analysis were used to build the preference models. These preference models show the contributions of various spatial information to the visual quality of the landscape. This spatial information consists of landscape elements including water, rocks, vegetation, tourist service buildings, fields, weather conditions, and spatial dimensions associated with the position of viewers including foreground, midground and background. Both types of spatial information are classified, mapped and analysed using GIS. Finally, GIS was used to develop the landscape preference map based on the landscape preference model and the spatial information.

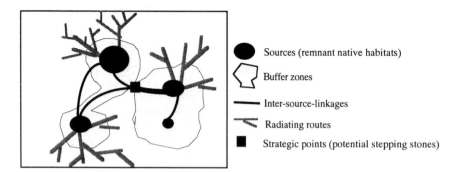

Figure 32.3 A schematic picture showing a typical ecological SP.

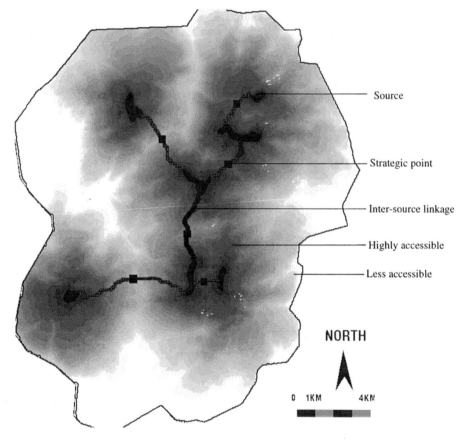

Figure 32.4 An accessibility surface for the medium-sized mammals and an ecological SP at a less secure level. The SP is composed of sources, strategic points and shortest inter-source linkages.

The visual security patterns (SPs) are defined in association with various security levels. Using the histograms of visibility and preference distribution patterns, some thresholds can be identified and used for the identification of visual security levels (Yu, 1995a). Three levels of SPs are identified: low, medium and high. These visual

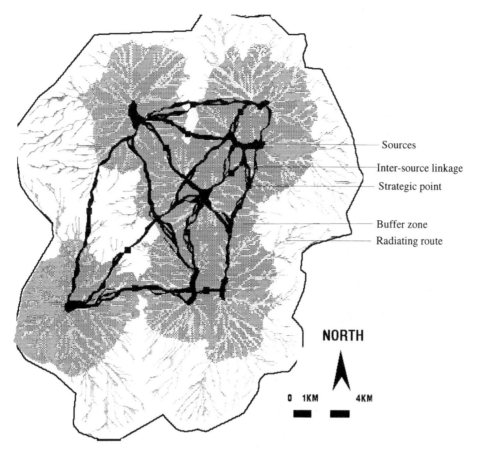

Figure 32.5 An ecological SP for the medium-sized mammals at a highly secure level. The SP is composed of sources, strategic points, all possible inter-source linkages, big buffer zones and some radiating routes.

SPs could be used as defensive frontiers by the defenders of visual perceptual processes and tourism during the procedure of spatial bartering and bargaining.

32.2.1.3 Agricultural SPs: Farmers' defensive frontiers

Local farmers have depended on their land for hundreds of years. Population growth requires more land to be converted into agricultural fields. In a certain sense, agricultural conversion in this case study is an issue of survival of the local people in this area. The potential security levels farmers want to achieve are normally determined by socio-economic analysis on local, regional and even national scales. This case study addresses the issue of agricultural SPs based on investigation of the landscape at the local scale and focuses on the issue as to where land conversion should pause or accelerate in terms of efficiency of productivity and impact on other processes in the landscapes.

The procedure of identifying agricultural SPs and ecological SPs is similar. Agricultural conversion is considered a process of disturbance with the 70 villages

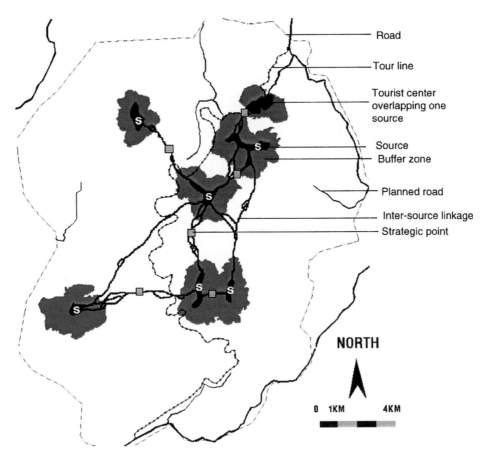

Figure 32.6 An ecological SP for the medium-sized mammals at a moderately secure level and the impact of tourist development. The SP is composed of sources, strategic points, some inter-source linkages and some buffer zones.

in the study area taken as the source of the spread of the disturbance. The intermediate landscapes are evaluated according to their resistance or cost of agricultural conversion. A convertibility surface (or the potential of conversion) is developed based on the cumulative resistance of intermediate landscapes. Based on this convertibility surface, agricultural SPs are identified at some thresholds or strategic values. Various agricultural SPs were identified corresponding to different security levels: low, medium and high. They could be used as defensive frontiers by defenders of the process of agricultural conversion in the spatial bargaining and bartering of landscape change and decision making.

32.2.2 Alternative changes based on SPs

Various alternative change models can be developed based on SPs. To illustrate this case four change alternatives are discussed as examples.

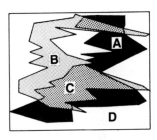

Category A: claimed as SPs exclusively for conservation, should be managed for maximum environmental qualities

Category B: claimed as SPs exclusively for agricultural conversion, should be managed for maximum agricultural production

Category C: overlapped SPs for both conservation and agricultural conversion, should be managed for an integration of both objectives

Category D: unclaimed areas for alternative changes vary according to decision-maker's preference, but the impact of change models should be evaluated

Figure 32.7 Management differentiation based on two general SPs of environmental conservation and agricultural conversion.

32.2.2.1 *Differentiation of management concentrations based on SPs*

At the highest level of the management hierarchy is the differentiation between landscape conservation, development (mainly agricultural conversion) and an integration of both (Fig. 32.7). The general management differentiation can be further subcategorised, when the environmental concern is specified into ecological and visual aspects. A weighing system has to be used when combining SPs of various ecological processes into an overall ecological SP. In this case, it is assumed all of the individual component SPs have the same weight ($= 1$).

32.2.2.2 *Strengthening landscape infrastructures based on SPs*

Taking ecological processes as examples, landscape ecological infrastructures can be strengthened by consolidating the SP components. These consolidations include, but are not limited to, the following aspects of improvement in quantity and quality:

1 increasing the buffer zones and making the intermediate landscapes more hospitable to native species;

2 having alternative linkages, widening the linkages and improving the connectivity of the linkages;

3 widening the radiating routes with native plants;

4 introducing native patches at the strategic points and expanding their dimensions.

By increasing the components of the ecological SPs, both in quality and quantity, the security of the landscape for the ecological processes can be improved proportionately. This gradual procedure, however, after reaching a certain threshold, will dramatically increase the security of the landscape and another security level for ecological processes will be achieved, for example from Fig. 32.4 to Fig. 32.5.

32.2.2.3 *Modifying introduced change models or trade-off SPs and exercising spatial bartering based on SPs*

Figure 32.6 shows a proposed tourist-development plan and its potential impact on the ecological SP for the dispersal of medium-sized mammals. As a result, the remaining native habitat at the upper-right corner and the immediate buffer zone will be destroyed by the expanding construction of tourist facilities and three ecological corridors will be negatively affected by the tour line.

One solution to reduce the negative impact of tourist development is to modify the plan of tourist development based on ecological SPs (Fig. 32.6). This solution suggests moving the tourist centre to the edge of the park and imposing a special management policy on the tour sections across the ecological corridors.

It is, however, possible that any modifications of the tourist-development plan may not be acceptable to the developers or local officials. In this case, the defender of ecological processes should consider the solution of spatial bartering to trade-off components of SPs for their overall consolidation. This solution of spatial bartering may include, but is not limited to, the tactics (Fig. 32.8).

Figure 32.9 shows an example of how the tourist-development plan can be adapted by trading-off some components of ecological SPs, but not security levels. It is suggested that the ecologist may abandon the native habitat at the upper-right corner of the map, but restore a native patch at the middle left and add a corridor to connect the two existing patches that will potentially be isolated because of the interruption of the corridors by the introduced tour line.

It should be noted that all the gains during the process of spatial bartering are

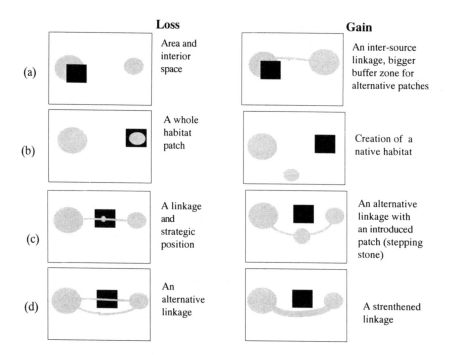

Figure 32.8 Possible spatial bartering tactics based on ecological SPs.

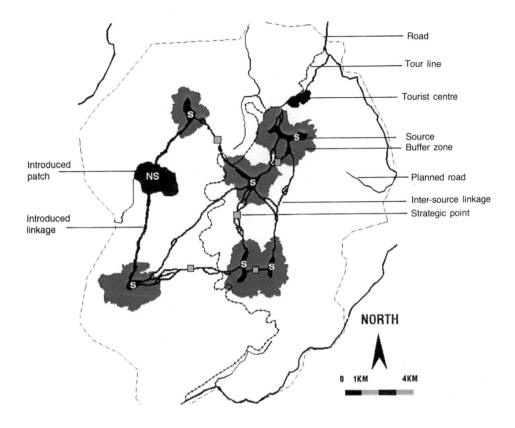

Figure 32.9 Tourist-development plan and spatial bartering on an ecological SP for the medium-sized mammals at a moderately secure level.

based on the identified ecological SPs and the surfaces that represent the accessibility of the landscape.

32.2.3 SP approach integrated with GIS in support of decision making for securer landscape changes

Those identified SPs at various security levels are the basis of strategies of spatial defence, or spatial bartering among defenders of ecological, visual and agricultural conversion processes, represented by ecologists, tourists and farmers.

Two situations have been simulated when using SPs in support of decision making.

Situation one. Negotiations for landscape changes within currently defined security levels. This is an optimistic result of the SP approach. Any of the four change models proposed in the last section may be acceptable.

Situation two. Negotiations for landscape changes beyond current security levels.

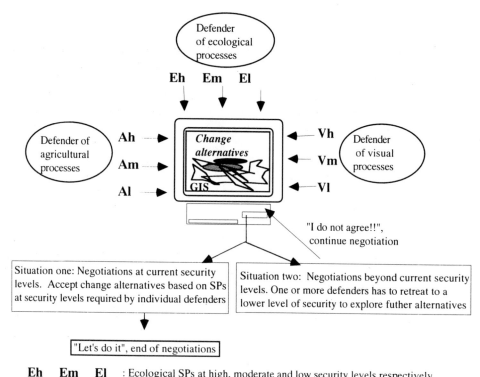

Figure 32.10 Strategies for negotiation and GIS-support decision making based on SPs.

Situation one can be taken as a special case of *situation two* where further solutions will be explored at other security levels. In *situation two*, one or more defenders has to give up his current requirement for the security of the concerned processes, and retreat to a lower level of security for further defence. Figure 32.10 shows the negotiation or gaming procedure and the strategies that reflect the decision-making process among the defenders of the three processes; each of the defenders has SPs of three security levels in mind. This procedure of redefining security levels may be repeated until solutions can be found.

As this case study shows, planners do not provide the optimum solutions or even any solutions at all to solve some projected problems. Instead, they are neutral consultants and moderators providing alternative strategies for each of the defenders in their defending various process in the landscape. These strategies are comparatively more efficient in achieving corresponding utility goals. The final solutions are the acceptable results achieved through negotiation among all defenders of the processes. GIS has great potential both in defining those spatial strategies based on the identification of SPs, and in the defensive procedure of negotiation among defenders of various processes.

32.3 Conclusion

The general conclusion of this research is that landscape security patterns could be very useful in landscape planning aimed at safeguarding various processes. Security patterns can be used as impact models guiding the modification or improvement of proposed change plans, as constraint criteria controlling the maximisation course of individual processes, as blueprints for the improvement of landscape structure, and as a basic reference frame for the procedure of spatial bartering.

The SP approach makes policy and management practice specify and concentrate on certain areas which can increase the efficiency of the decision-making procedure. Various landscape change alternatives are explored based on SPs within a certain security level for a certain process. Further change alternatives are developed at a lower security level only when none of the alternatives at a higher security level is acceptable to decision makers or defenders of various processes. SPs are 'stop signs' in the decision-making course that reduce the risk of the irreversibility of decision making and reduce the possibilities of catastrophes in landscape changes.

GIS plays an important role in simulating various processes, identifying security patterns, evaluating impact and developing landscape change models based on SPs. GIS shows its great potentials when combined with the SP approach in supporting landscape decision making.

Acknowledgements

Thanks are due to Carl Steinitz, Stephen Ervin and Richard T. T. Forman at Harvard University for their advice and support of this research, Hugh Keegan and other staff at the ESRI (Environment Systems Research Institute) for their support in the application of Arc/Info GIS, and Erin Crowley for her editing of the manuscript.

References

BERRY, B. J. L. and HORTON, F. E. (Eds) (1970). *Geographic Perspectives on Urban Systems*, 1st Edn, Englewood Cliffs, NJ: Prentice Hall.

BISHOP, A. B., FULLERTON, H. H., CRAWFORD, A. B., CHAMBERS, M. D. and McKEE, M. (1974). *Carrying Capacity in Regional Environmental Management*, Washington, DC: Office of Research and Development, US Environmental Protection Agency.

CIRIACY-WANTRUP, S. V. (1968). *Resource Conservation: Economics and Policies*, 3rd Edn, Berkeley: University of California Division of Agricultural Science.

FORMAN, R. T. T. and GODRON, M. (1986). *Landscape Ecology*, New York: John Wiley.

GIBSON, J. J. (1950). *The Perception of the Visual World*, Boston: Houghton Mifflin.

KNAAPEN, J. P., SCHEFFER, M. and HARMS, B. (1992). Estimating habitat isolation in landscape planning, *Landscape and Urban Planning*, **23**, 1–16.

KOZLOWSKI, J. (1986). *Threshold Approach in Urban, Regional and Environmental Planning: Theory and Practice*, St Lucia, Queensland, Australia: University of Queensland Press.

KOZLOWSKI, J. and HILL, G. (1993). *Towards Planning for Sustainable Development: A Guide for the Ultimate Environmental Threshold (UET) Method*, Vermont, USA: Avebury, Ashgate Publishing Company.

LYNCH, K. (1960). *Image of the City. Cognition and Environment: Functioning in an Uncertain World*, New York: Praeger.

MERRIAM, G. (1984). Connectivity: a fundamental characteristic of landscape pattern, in Brandt, J. and Agger, P. (Eds), *Proceedings of The First International Seminar on Methodology in Landscape Ecological Research and Planning*, **1**, pp. 5–15, Roskilde, Denmark: Roskilde Universitetsfolag GeoRuc.

STEIN, H. F. and NIEDERLAND, W. G. (Eds) (1989). *Maps from the Mind: Readings in Psychology*, Norman and London: University of Oklahoma Press.

TAAFFE, E. J. and GAUTHIER, H. L. (1973). *Geography of Transportation*, Englewood Cliffs: Prentice Hall.

TUAN, Y.-F. (1974). *Topophilia*, Englewood Cliffs: Prentice Hall.

TURNER, M. G. (1989). Landscape ecology: the effect of pattern on processes, *Annual Review of Ecology and Systematics*, **20**, 171–97.

WARNTZ, W. (1966). The topology of a social-economic terrain and spatial flows, in Thomas, M. D. (Ed.), *Papers of The Regional Science Association*, pp. 47–61, Philadelphia: University of Washington.

YU, K. (1995a). Cultural variations in landscape preference: comparisons among Chinese sub-groups and Western design experts, *Landscape and Urban Planning*, **32**, 107–26.

—— (1995b). Ecological security patterns of landscapes: concept, method and a case, in Fung, T. and Lin, H. (Eds), *The Proceedings of the International Symposium of Geoinformatics'95*, pp. 396–405, Hong Kong: The Chinese University of Hong Kong.

—— (1995c). 'Security Patterns in Landscape Planning: With a Case In South China', Doctoral thesis, Cambridge, MA: Harvard University.

CHAPTER THIRTY-THREE

A Generic Spatial Decision-support System for Planning Retail Facilities

THEO ARENTZE, ALOYS BORGERS and HARRY TIMMERMANS

33.1 Introduction

The planning of retail facilities that serve a spatially dispersed consumer population requires the analysis of demand-and-supply relationships in a geographic market area. Current GIS offers tools for storing, managing and visualising the spatial data involved. Furthermore, tools for network analysis and location analysis typically available in GIS are useful for retail planning. To support spatial decision making effectively the GIS tools should be complemented with models for analysis and decision support developed in the spatial and management sciences. The purpose of this study is the development of a spatial Decision Support System (DSS) that uses both GIS and complementary models. The system envisioned should support retail planning in both public (local or regional governments) and private sector (retailers) contexts.

Different approaches to spatial decision-support systems have been proposed dependent on whether emphasis is put on the generation of information (analytic tools), the visualisation of data and presentation of documents (media) or group processes (collective cognition) (Shiffer, 1992). Collaborative Decision Support Systems are a family of systems which are explicitly designed to facilitate group decision making. MacDonald describes an example of this approach in the field of solid waste planning in Chapter 34, in this volume. In other fields, the effectiveness of decision support may depend critically on the visualisation of spatial data and the presentation of documents or other forms of media. Recently, the use of multimedia in a GIS environment is receiving a lot of attention for this purpose. (See for example Chapter 36 in this volume where Bodum assesses this approach for local planning in the Danish context.) In The Netherlands, retail planning is one of the sectors in urban and regional planning with a relatively strong research input (Borgers and Timmermans, 1991; Oppewal, 1995; van der Heijden, 1986). Also in the private sector, planning analysis plays an important role (Beaumont, 1991; Breheney, 1988). Therefore, our approach is based on the assumption that DSS can improve the quality of decision making primarily by offering an interactive environment for analysis, provided that this environment is also suitable for users who do not have knowledge of the analytic techniques used. Compared with other approaches, the analytic capabilities of the system have higher priority than visualisation and group decision support.

The major stages of this project are the investigation of the information needs of planners and retailers, the integration of existing methods, techniques and models that match these needs and the design and development of a DSS. A group of representatives of retail companies, local governments and consultancies in retail planning participate in this project. The group regularly meets to give input from the side of users to the project. The system is developed by adding functions to the system and obtaining user feedback in an iterative process. At present, the system has reached a prototype stage. In the near future, the prototype will be applied in both a local government and a retailer context to test its usefulness and identify ways for further development. This chapter describes the results of each of the major stages of the project. The next section discusses the objectives in private and public sector retail planning and its implications for DSS design. The sections that follow introduce the major components of the proposed DSS and the way the system supports retail-planning decisions. Finally, we summarise the findings and discuss the implications of this study for improving the decision support capabilities of current GIS for location planning.

33.2 Retail planning and implications for DSS

The problems we focus on are planning the location (where?), type (what?) and the size (how much?) of retail facilities in a market area. Both retailers and local governments deal with these problems; retailers in the development of a marketing strategy for their chains and local and regional governments in the development of an active location policy regarding retail facilities. In The Netherlands, governments do not have the means to intervene directly in the retail system. They can, however, create conditions to influence retail developments by using regulation controls, financial instruments, investments in the public sector (for example, parking places) or forms of cooperating with developers/retailers. The aim of retail planning is to formulate plans for public spatial policies which attempt to guide processes in the retail system towards a set of objectives. The objectives involve a balancing between the interests of consumers, retailers and the community in general (Oppewal, 1995; van der Heijden, 1986, p. 165). The most important forms of analysis that relate to these objectives concern the accessibility and availability of facilities from residential locations, economic performance of facilities, and shopping-related mobility. Compared with planners, retailers normally pursue a smaller set of goals which are, moreover, not related to the retail network as a whole but to the subset of outlets that belong to their own chain. The goal is usually to utilise opportunities for improving the economic performance or the competitive strength (relative economic performance) of a chain, whereby economic performance may be defined in terms of turnover, profit or return on investment and competitive strength in terms of market share (Breheney, 1988).

Although the goals may differ, they can often be reduced to basic information categories, such as expected patronage of retail facilities as a function of plan alternatives. The difference in goals has consequences for the criteria and weights used to evaluate plans, but the underlying analysis is largely the same. Moreover, in The Netherlands retail plans are increasingly developed through forms of cooperation between planners and retailers. Both parties are increasingly aware that, for successful implementation, plans should meet the interests of both sides. As a consequence, (large) retailers also tend to analyse their location strategies in terms of public sector interests and, vice versa, planners also tend to consider the retailer's interests in plan alternatives. Given this convergence, it should be possible to develop a DSS that serves both parties, particularly if the DSS puts emphasis on analysis.

Although we find a common structure, retail planning problems may still differ in terms of data availability, preferences of decision makers and specific information needs. This diversity poses high demands on the flexibility of a generic DSS. To meet these demands, the system should allow users to specify the criterion variables to be monitored and the planning variables to be manipulated. Furthermore, the system should offer alternative models varying in sophistication and data needs so that users can trade-off accuracy of information and data requirements. To meet the different preferences of users, the system should also support both qualitative and quantitative forms of analysis in both user- and system-controlled problem-solving procedures.

Provided that the system can be adjusted to specific information needs, a DSS has the potential to improve the effectiveness of decision making. In general, an effective DSS improves information quality and facilitates the understanding and structuring of the problem. In the present field, DSS can potentially make an additional, more specific contribution. As van der Heijden (1986) argues, a DSS has the potential to improve the integration of the input of applied research into the planning process. At present, planners often consult researchers for questions arising in the planning process of the kind 'what will happen if . . ?' or 'what are the impacts of . . ?'. There is a time lag between the production of information and the needs for that information which creates a threshold for considering plan alternatives. Provided that the system is suitable for lay users, a DSS can reduce this time lag and thus support the inherently cyclic character of the planning process. Furthermore, DSS has the potential to improve the communication between planners and retailers by standardising analytic procedures. To realise these potentials, system properties such as flexibility, interactiveness and user-friendliness are essential.

Most current approaches in spatial DSS either use consumer choice or interaction modelling as the major technique for simulating decision scenarios (Birkin et al., 1994; Borgers and Timmermans, 1991; Roy and Anderson, 1988) or optimisation models for exploring the nature and possible solutions of location problems (Armstrong et al., 1990, 1991; Densham, 1991, 1994). The approaches proposed by Grothe and Scholten (1992) and Kohsaka (1993) use both types of techniques to support both the search for optimum locations and the simulation and analysis of decision scenarios. The contribution of this study is a system design that improves the interactive properties and flexibility of spatial DSS. The system environment should support the simulation of plan, population and goal scenarios and should be adaptable to a large variety of problems.

33.3 The DSS components

The spatial DSS described here is based on a GIS which offers functions for management, visualisation and elementary analysis of the spatial data involved.

The specific software used is called TransCAD (Caliper Corporation, 1992), which offers relatively advanced analytic tools particularly in the field of transportation. The second component of the DSS is a Microsoft Windows application currently under development, called Location Planner. Location Planner is based on a particular DSS-concept that we have developed in a former study (Arentze *et al.*, 1996a). This concept uses active and linked views of a spatial database to create a flexible and interactive environment for analysis and planning. Location Planner is a standalone program and is linked with the GIS through data files. Alternatively, it can be used in combination with general database management systems by means of dbf-files. This section gives an overview of the components of the system in terms of the database structure and the basic components of the knowledge base.

33.3.1 The structure of the database

The first step in developing a database is the definition of the study area. In most cases in the Dutch context, the study area is defined by the boundaries of a municipality. However, in cases where the types of facilities have a regional rather than local scale a larger study area is appropriate. Since the database differs from case to case, it is not part of the generic DSS developed here.

A database used in Location Planner consists of four main parts that describe the demand, supply, spatial structure and interactions between demand and supply in a market area. Figure 33.1 summarises the data involved in these parts.

The analytical tools available in TransCAD and most other GIS are useful for generating part of the data such as spatial clustering for delineating shopping areas; spatial aggregation for determining area attributes; localising geographic centroids of residential zones and shopping areas; generating distance data, possibly based on network analysis; locating physical barriers, such as, for example, waterways and railways; identifying candidate locations possibly using overlay, buffering analysis or spatial selections; and determining attributes of the environment of facilities, such as, for example, available parking space and distance to nearest bus

demand data	
residential zone (e.g. post code area)	population size
	demographic and socio-economic attributes

supply data	
individual facility or facility centre	floor space in each retail sector
	internal attributes (e.g. assortment, price level)
	external attributes (e.g. parking space, distance to bus stop)

interaction data
matrix: shopping trips or expenditure flows between demand and supply locations

distance data
matrix: distance or travel time between locations, possibly based on a transportation network

Figure 33.1 The components of a database in Location Planner (the data needs depend on the analysis purposes).

stop. It follows from this list that GIS and Location Planner are highly complementary in the sense that GIS tools can generate a substantial part of the database used in Location Planner.

It should be noted that the amount of data needed depends to a high degree on the optional methods of analysis one chooses to use. For example, using simple rule-based methods as approximates of more complex behavioural models requires only limited attribute and interaction data. Moreover, most of the demand and supply data needed are often available through internal secondary data sources or external data suppliers (Waters, 1995).

33.3.2 The basic components of the knowledge base

According to the model developed by Breuker and Wielinga (1989) there are four functional layers of knowledge involved in decision making in general. The domain knowledge, the lowest layer, describes the states and events of the system being planned. Inferential knowledge, the next layer, describes the relationships between these states and events, allowing one to draw inferences about the system. One layer higher we find task knowledge which prescribes ways to solve problems, when problems are well structured. Finally, the highest layer involves strategic knowledge which is the competence to structure complex problems. In a former study (Arentze et al., 1996c) we have used this multilayered model to describe and classify knowledge in facility location planning, whereby knowledge is defined broadly to include data, algebraic models, algorithms and heuristic knowledge. Again, Location Planner is a generic system and does not incorporate domain knowledge, which is case-specific. The knowledge base of Location Planner is composed of the three higher layers following the former study mentioned. The layers are hierarchically organised in the sense that knowledge in a higher layer builds upon knowledge in the lower layer.

The knowledge components in the inferential layer are the basic building-blocks of the knowledge base. To identify the inferential model components that are relevant for retail planning problems, we use the scheme shown in Fig. 33.2.

The boxes of the scheme represent states and events in the retail system and the arrows depict the dependency relationships. The future state has a central position in the scheme. It is the result of two types of changes of the present state in the plan period: developments that are autonomous in the sense that they are outside the control of the planner and actions that are planned. To be able to compare the future state with the goal state, an intermediate step is introduced in which the future state is described in terms of goal-variables. The discrepancy state which results from the comparison with the goal state indicates the effectiveness of the actions planned. The scheme shows the different functional components of the inferential model. For each component Location Planner offers various algebraic or heuristic methods. Figure 33.3 lists the methods that are available.

33.4 The levels of use of Location Planner

The user interface provides access to each of the layers so that users can choose between different levels of decision support dependent on characteristics of the

VIEWS AND LINKS

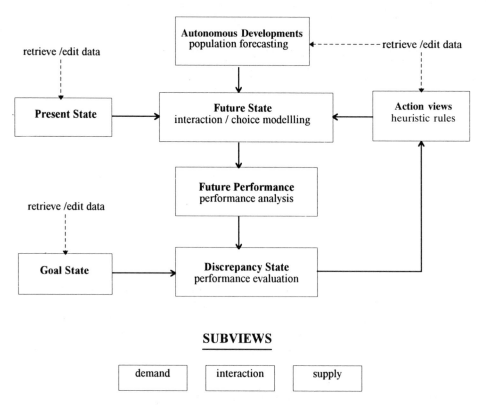

SUBVIEWS

Figure 33.2   The domain and inferential structure in Location Planner.

problem or preferences. This section describes the way Location Planner supports decision making at the different levels.

33.4.1 The domain and inferential level

At the lowest level Location Planner offers users the possibility of defining a model of the retail system and, next, to use the model for simulating decision, goal or population scenarios. The general structure of the model is fixed and corresponds to the scheme of Fig. 33.2. The boxes of the scheme represent different perspectives of the retail system, which are called views in Location Planner. As the first step, users specify within each view the data fields of interest for the problem under study. Internally, the user creates two sets of variables: directly observable attributes of spatial objects (population, facilities or interactions), which are called X-variables, and goal or criterion variables, which are called C-variables. The X-variables are used in the Present State, Future State, Autonomous Developments and Action views, and the C-variables are shared by the Goal State, Future Performance and Discrepancy State views. Within views there are three subviews

Autonomous developments	
forecasting demographic and socio-economic developments by residential zone	projection techniques

Future State	
predicting demand-supply interactions as a function of autonomous developments and actions	production constrained interaction model conventional MNL model model of multipurpose-trip shopping (Arentze *et al.*, 1993)

Future Performance	
analysing system performance at the level of residential zones or facility locations. Themes: service provision, market saturation, market penetration, accessibility, mobility, turnover, market share	interaction-based analysis (see Clarke and Wilson, 1994) catchment-area based analysis rule-based analysis distance-band-based analysis

Discrepancy State	
evaluating system performance by comparing observed and normative performance	classification methods

Actions	
defining actions that reduce observed discrepancies	heuristic rules (see Arentze *et al.*, 1996b)

Figure 33.3   Function and method of inferential knowledge components in Location Planner.

on residential zones (demand), facilities (supply) and interactions respectively. Not all subviews are meaningful in all cases. For example, Location Planner does not give an interaction view on the Goal State, since, normally, goals are not formulated at the level of interactions. As another example, if planners do not have the means to control population developments, the demand subview of the Action view is meaningless. A data field in a demand or supply view corresponds to a column of attribute data of residential zones or facilities. A data field in an interaction view, on the other hand, stores an origin-destination matrix. Figure 33.4 gives an example of a domain structure, whereby the problem is to monitor a market area for opportunities for expanding a retail chain.

The inferential structure, one level higher, is defined by specifying dynamic links between data fields. The scheme (Fig. 33.2) defines which views can be linked to each other and the types of links that are possible. The user specifies this general inferential structure by attaching to data fields an inferential method for updating the data dependent on data in linked views. This is done simply by selecting an inferential method available in the knowledge base of Location Planner (Fig. 33.3) and by specifying the data fields in linked views that are used as input. Figure 33.4 shows an example whereby a variety of different inferential models links the data fields in a model for monitoring a market area. Note that since each data field in interaction views corresponds to an O–D matrix, it is possible to use a set of these matrices for analysing future performance. A set of models is suitable when the consumer population is subdivided in segments that display different preference patterns (one model for each segment) or when supply is subdivided in sectors (one model for each sector). Depending on the information needs, one could also use the same model for different data fields (O–D matrices), for example, for allocating shopping trips (mobility analysis) and expenditure flows (economic performance analysis).

Having specified the dynamic links, the inferential structure is tailored to the

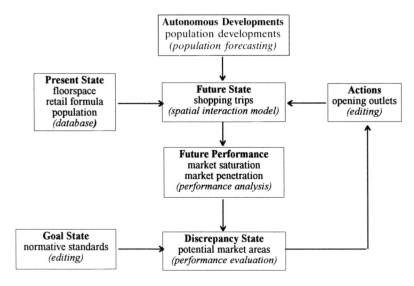

Figure 33.4   An example of an inferential model in Location Planner.

characteristics of the case such as the available data and preferences of the user. This structure is best viewed as a model (constructed by the user) of the retail system to be analysed. At this stage there are still no data to analyse. The data fields in the Present State and Goal State are external to the model – they cannot be defined based on the data in other views. Instead, they are defined by the methods selected for retrieving data from the database discussed earlier. The database is specified by the user as a list of TransCAD or DBase files. By specifying the methods for data retrieval, the user defines the links with the database. The data fields in the Actions and Autonomous Developments views are also often considered as exogenous. In general, views can be filled by editing or importing data.

At the domain level, the inferential model can be used to view (and edit) data. The user is able to view data within views in a table format or as a map of the area. A map shows the subdivision of the area in residential zones, the centroids of zones and the point locations of facilities (possibly centroids of shopping areas). Demand and supply data are shown on the map as labels attached to points. Interaction data, on the other hand, are represented by lines on a map connecting demand and supply locations. At the inferential level, the model can be used for simulating scenarios regarding actions, goals or autonomous developments, to obtain insights of the dependency relations in the retail system. Furthermore, at this level, the user can vary the specifications of the methods selected to assess the sensitivity of the results for arbitrary model settings.

An inference engine is built in the system for maintaining consistency in the model across the user's actions. Each time the user changes the specifications of the database or selects new methods the engine resets the dependent data fields. When the user gives an update command, the engine triggers the methods selected to update the fields that have been reset. Thus, it is possible to obtain immediate feedback information on the actions taken. To return to the example, a user could vary his or her assumptions on future population developments, actions of competitors (for example, opening or closing outlets), location strategies (for example, opening outlets of the own chain), goals (for example, normative market saturation levels), methods (for example, weights in the interaction model) and next view the impact of these changes, for example, in terms of future Discrepancies. Typically, at this stage users would use TransCAD for producing thematic maps of the results obtained in Location Planner.

In sum, at the inferential level Location Planner supports decision making by providing a dynamic environment for what-if analysis. It is important to note that the inferential model specified may be simple. Only a subset of the views available may be in use in a particular case. For example, a user may be interested in aspects of spatial opportunities of the population and ignore the interaction views. If there are no *a-priori* performance criteria, he or she will also ignore the Goal and Discrepancy State. Furthermore, we emphasise that, once the inferential model (data fields and links) has been specified, users can interact with the system without bothering about the underlying technical details. At this stage, therefore, using the system does not require any analytic knowledge (except in some cases where the interpretation of model outcomes is not straightforward). The system also assists lay users in the preceding stage of model specification, by providing a base of inferential models from which users can choose the one that suits the problem at hand best.

33.4.2 The task level

At the task level, Location Planner supports decision making by solving well-structured problems in retail planning. In contrast to the inferential methods, the methods at this level take the form of algorithms controlling the processes at the lower, inferential level, for achieving goals in the planning process. The task level in Location Planner is comparable with the Subproblem Solver module in the DSS approach developed by Armstrong *et al.* (1990). However, the specifications differ and are focused on retail planning problems. The Subproblem Solver in Location Planner generates plan alternatives, analyses the impact of scenarios, and compares/contrasts scenarios.

33.4.2.1 *Generating plan alternatives*

Several studies have shown the usefulness of location-allocation modelling for retail planning (Achabal *et al.*, 1982; Goodchild, 1984; Goodchild and Noronha, 1987). Location Planner uses these models to support the generation of optional strategies in locating new outlets/centres. The general procedure is as follows: the user defines the location problem by specifying (1) the maximum number of facilities to be located, (2) the optional types of new facilities and (3) the objective function and optionally a constraint function. The optional types are defined in terms of X-variables and the objective and constraint functions in terms of C-variables, which are dynamic data fields at the lower domain level. In combination with the set of candidate locations in the Present State, the maximum number and optional types of new facilities define the solutions that are possible. The constraint function identifies the ones that are feasible and the objective function measures their performance.

Location Planner uses the complete enumeration algorithm (described by Ghosh and McLafferty, 1987, p. 148) to find the best solutions. This algorithm systematically enumerates and evaluates all possible solutions using the inferential model at the lower level. An evaluation cycle involves implementing a solution in the Actions view and updating the relevant C-variables in the Future Performance view. Location Planner presents a selection of the solutions generated, dependent on the goals of the user. In many cases, the user will be interested in the list of, say, the N best solutions (in terms of the objective function). Alternatively, Location Planner can present the best solution for each so-called macro-strategy. A macro-strategy is defined by Ghosh and McLafferty (1987) as a combination of number and types of new facilities, which can be implemented in different ways, dependent on the chosen location strategy. The list of macro-strategies under optimum location strategy conditions are relevant for making investment decisions, since the costs of investment depend largely on how many and what type of outlets are developed. As another option, Location Planner can present the list of solutions that vary largely in terms of X-variables but perform all relatively high in terms of C-variables. Generally, the subset of solutions that are very different in geographical space, but similar in criterion space may help the decision maker to find creative solutions to the problem. Methods for selecting such sets are developed in the field of Modelling to Generate Alternatives (Brill *et al.*, 1990).

The general procedure based on complete enumeration can be used to solve a large variety of problems. The most simple and a very common problem is the selection of single candidate locations with high trading potential. This problem is defined by setting the maximum number of facilities to one, by specifying only one possible facility type and setting the objective function to measure economic performance. As another typical example we mention the problem of finding the number, location and type of new outlets that maximise the turnover or market share of a chain, whereby all outlets meet a minimum turnover. Furthermore, depending on the specification of the Present State, the procedure solves the problem of expanding an existing network or developing a network from scratch.

The complete enumeration procedure is able to find exact solutions for all location-allocation problems (that can be defined in terms of the inferential model), but it becomes intractable when the number of candidate locations is very large. For solving large problems, Location Planner offers the possibility to use the well-known interchange algorithm. This algorithm searches for the optimum solution step-wise by improving an initial solution through marginal changes (substitutions). The algorithm is designed to solve the p-median problem. However, in Location Planner it is used for solving the more general problem of finding the optimum location strategy given a macro-strategy. Therefore, in combination with a complete enumeration of macro-strategies, the algorithm can solve the general problem. Users should keep in mind, however, that the algorithm may fail to find the optimum solution especially in cases where the objective function has a very irregular shape (for example, in very restricted search spaces).

33.4.2.2 *Analysing the impact of scenarios*

Location Planner provides a module for analysing the separate effects (decreases, increases) of a scenario on the C-variables specified. Based on this information, decision makers can obtain insight in trade-off relationships between goal variables and can investigate the sensitivity of outcomes for reasonable variations in assumptions on the autonomous developments.

33.4.2.3 *Comparing and ranking scenarios*

Location Planner allows users to store Action or Autonomous Development scenarios for comparison. Having built a list of scenarios, users can load a scenario in the current inferential model to view and analyse its impact. Location Planner offers a set of methods developed in the field of discrete multicriteria decision-making, which supports the ranking of plan alternatives on the set of C-variables defined in the inferential model. The C-variables do not have to have a numeric scale. There are also techniques available for combining qualitative or mixed sets of criteria (Voogd, 1983).

33.4.3 **Strategic level**

At the *task level* Location Planner supports planning problems when they are well structured. At the higher strategic level Location Planner supports the structuring of complex problems in terms of task level goals. The strategic level is comparable with the Metaplanner module in the DSS approach developed by Armstrong *et al.* (1990). Again, in Location Planner the specification of strategic methods differs and is focused on retail planning problems.

The goal a planner or retailer wants to achieve may differ across retail planning problems as discussed earlier. Location Planner incorporates strategic knowledge developed in applied studies of retail planning to support the structuring of a set of standard problems. A strategic method is a kind of dynamic script that helps the user to develop a plan for achieving a certain goal. A script includes suggestions for specifying the inferential model (for example, loading a standard model) and a sequence of steps at the task level. A script is dynamic in that the suggestions are dependent on the characteristics of the problem, the data resources available, the preferences of the decision maker and results obtained in previous steps. A script not only helps to develop a strategic plan, but also monitors the planning process and suggests alternative specifications of methods or ways of analysis if impasses occur (Breuker and Wielinga, 1989). For example, a script may suggest that we re-specify the constraints if a spatial search algorithm does not generate an interesting set of possible location strategies.

If a given problem does not match any of the standard problems, then the Metaplanner of Location Planner applies general strategic knowledge for structuring the problem. In short, the Metaplanner acts as an expert regarding both the methodology in retail planning and the possibilities Location Planner offers. At present, the strategic layer has not yet been developed.

33.4.4 **Relations between layers**

Figure 33.5 summarises the methods available in the different layers and the way they are organised. Strategic methods control (in terms of guidelines) task methods to achieve the decision-maker's goal. Task methods in turn apply inferential

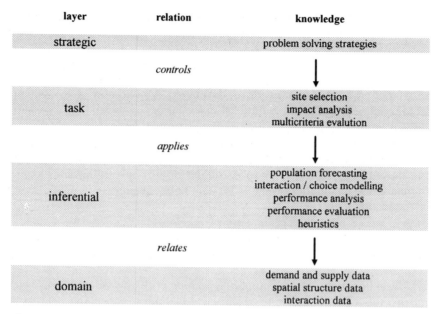

Figure 33.5 The knowledge layers in Location Planner.

methods to achieve subgoals. Finally, inferential methods relate data that describe the states and events in the retail system. The layers support different levels of usage including viewing data (domain level), simulating scenarios (inferential level), solving well-structured problems (task level) and structuring problems (strategic level).

The essential characteristic of this multilayered system is that all layers are accessible to the users, with the major advantage that they can easily switch between levels of decision support. For example, having generated an optimum network configuration at the task level, users can investigate the robustness of the solution by simulating possible population developments at the inferential level. Another advantage is that it improves the modelling capabilities of the system, provided that the specification of a method at a higher level is to some degree independent of the specification of its component methods. This is particularly significant for location-allocation algorithms at the task level. The complete enumeration algorithm imposes practically no restrictions on the specification of the inferential methods such as the objective and constraint function. The applicability of the interchange algorithm is more restrictive, but still allows various specifications of the location problem. Consequently, the layered structure makes many solution paths possible.

33.5 Discussion and conclusions

This chapter has introduced a spatial DSS for retail planning in both public and private sector contexts. As pointed out earlier, the information needs of local government and retailers show a large degree of overlap and tend to converge as cooperation in retail planning becomes more and more important. The described

DSS provides a generic problem-solving environment which can be tailored to specific information needs. Users can construct an environment for analysis (an inferential model) out of available component methods that suit the characteristics of the given problem. The environment can be made to fit problems that differ in complexity, varying from monitoring only a few possible qualitative variables to developing a plan considering a broad set of goals. The variety of methods available also supports data-poor forms of analysis. Because of this flexibility, the system is able to support a wide range of planning problems.

Once the system has been tailored, using the system does not require knowledge about the analytic techniques used. The lay user is able to vary easily and intuitively all kinds of conditions and receive immediate feedback on the consequences. The system supports the simulation of decision scenarios, goal scenarios and autonomous developments. Both the flexibility and interactivity of the analysis environment are realised by organising data and methods in active and linked views. The system supports decision making up to the strategic level, where the system incorporates expertise to help users in structuring complex problems and applying appropriate models.

The Location Planner software is currently under development. In the near future the usefulness of the system will be tested by applying the system to real-world cases in private and public sector planning. Location Planner is not a complete DSS in itself. It supports all the basic functions of a DSS only in combination with other GIS or general-purpose database software. The system is best viewed as a framework for analysing a spatial database that can be generated, managed and visualised in a GIS. The analytic tools (for example, for network analysis) typically available in a GIS are complementary to the analytic possibilities Location Planner offers. At a conceptual level, therefore, Location Planner shows how spatial data, GIS, and analytic modelling can be integrated. At the software level, however, the coupling with GIS is rather loose, namely through data files. Data generated in Location Planner can be used in a GIS and vice versa. A certain degree of independence has advantages. The disadvantage is, however, that users have to switch between environments within the same planning process (however, switching between programs is easy in a multitasking Windows environment).

Improving the decision-support features of current GIS would be a more fundamental answer to this problem of integration at the software level. This study has implications for developing GIS in that direction. Obviously, the analytic possibilities of current GIS should be enhanced to meet the information needs of decision makers. In addition the findings of this study suggest that there are structural improvements possible regarding the inferential and task level. Most typical analytic GIS functions are inferential in nature. The concept of active and linked views could be used to improve in particular the interactive properties of GIS at the inferential level and would be in line with the current development of object-oriented GIS. At the task level GIS supports spatial search. The tools involved should not be included as separate models, as in the traditional approach, but as algorithms controlling processes at the inferential level. As Location Planner shows, a hierarchical structure can improve the flexibility and modelling capabilities of the system. Finally, in current GIS, analytic capabilities are grouped and presented to users in a technique-oriented way. To improve the accessibility of system capabilities for end users, they should instead be linked to goals in modules that explicitly support the generation, analysis, and ranking of plan alternatives.

Acknowledgements

This research is supported by the Technology Foundation (STW).

References

ACHABAL, D. D., GORR, W. L. and MAHAJAN, V. (1982). MULTILOC: A multiple store location decision model, *Journal of Retailing*, **58**, 5–25.

ARENTZE, T. A., BORGERS, A. W. J. and TIMMERMANS, H. J. P. (1993). A model of multipurpose shopping trip behaviour, *Papers in Regional Science*, **72**(3), 239–56.

(1996a). Design of a view-based DSS for location planning, *International Journal of Geographical Information Systems*, **10**(2), 219–236.

(1996b). An efficient search strategy for site selection decisions in an expert system, *Geographical Analysis*, **28**(2), 124–146.

(1996c). The integration of expertise knowledge in decision support systems for facility location planning, *Computers, Environment and Urban Systems*, **19**, 227–247.

ARMSTRONG, M. P., DE, S., DENSHAM, P. J., LOLONIS, P., RUSHTON, G. and TEWARI, V. K. (1990). A knowledge-based approach for supporting locational decision making, *Environment and Planning B*, **17**(3), 341–64.

ARMSTRONG, M. P., RUSHTON, G., HOENEY, R., DALZIEL, B. T., LOLONIS, P., DE, S. and DENSHAM, P. J. (1991). Decision support for regionalisation: A spatial decision support system for regionalising service delivery systems, *Computer, Environment and Urban Systems*, **15**(1/2), 37–53.

BEAUMONT, J. R. (1991). Spatial decision support: some comments with regard to their use in market analysis, *Environment and Planning A*, **23**, 311–17.

BIRKIN, M., CLARKE, G. P., CLARKE, M. and WILSON, A. G. (1994). Applications of performance indicators in urban modelling: Subsystems framework, in Bertuglia, C. S., Clarke, G. P. and Wilson, A. G. (Eds), *Modelling the City: Performance, Policy and Planning*, London: Routledge, pp. 121–50.

BORGERS, A. W. J. and TIMMERMANS, H. J. P. (1991). A decision support and expert system for retail planning, *Computers, Environment and Urban Systems*, **15**(3), 179–88.

BREHENEY, M. J. (1988). Practical methods of retail location analysis: A review, in Wrigley, N. (Ed.), *Store Location, Store Choice and Market Analysis*, pp. 39–86, London: Routledge.

BREUKER, J. A. and WIELINGA, B. J. (1989). Models of expertise in knowledge acquisition, in Guida, G. and Tasso, C. (Eds), *Topics in Expert System Design: Methodologies and Tools*, pp. 265–95, New York: North-Holland.

BRILL, E. D., FLACH, J. M., HOPKINS, L. D. and RANJITHAN, S. (1990). MGA: A decision support system for complex, incompletely defined problems, *IEEE Transactions on Systems, Man and Cybernetics*, **20**(4), 745–57.

CALIPER CORPORATION (1992). *TransCAD Version 2.1*, Massachusetts: Newton.

CLARKE, G. P. and WILSON, A. G. (1994). A new geography of performance indicators for when planning, in Bertuglia, C. S., Clarke, G. P. and Wilson, A. G. (Eds) *Modelling the City, Performance, Policy and Planning*, pp. 55-81, London: Routledge.

DENSHAM, P. J. (1991). Spatial decision support systems, in Maguire, D. J., Goodchild, M. F. and Rhind, D. W. (Eds), *Geographical Information Systems: Principles and Applications, vol. 1*, pp. 403–12, Cambridge: Longman.

(1994). Integrating GIS and spatial modelling: Visual interactive modelling and location selection, *Geographical Systems*, **1**(1), 203–21.

GHOSH, A. and McLAFFERTY, S. L. (1987). *Location Strategies for Retail and Service Firms*, Massachusetts/Toronto: Lexington Books.

GOODCHILD, M. F. (1984). ILACS: A location-allocation model for retail site selection, *Journal of Retailing*, **60**, 84–100.

GOODCHILD, M. F. and NORONHA, V. T. (1987). Location-allocation and impulsive shopping: The case of gasoline retailing, in Ghosh, A. and Rushton, G. (Eds), *Spatial Analysis and Location-Allocation Models*, pp. 121–36, New York: Van Nostrand Reinhold.

GROTHE, M. and SCHOLTEN, H. J. (1992). Modelling catchment areas: Towards the development of spatial decision support systems for facility location problems, in Harp, J., Otten, H. and Scholten, H. (Eds), *Proceedings of the EGIS '92 Conference, Munich 23–26 March*, vol. 2, pp. 978–87, Utrecht: EGIS Foundation.

HEIJDEN, VAN DER, R. E. C. M. (1986). A decision-support system for the planning of retail facilities: Theory, methodology and application, PhD Dissertation, Eindhoven: Eindhoven University of Technology.

KOHSAKA, H. (1993). A monitoring and locational decision support system for retail activity, *Environment and Planning A*, **25**, 197–211.

OPPEWAL, H. (1995). Conjoint experiments and retail planning, PhD Dissertation, Eindhoven: Eindhoven University of Technology.

ROY, J. R. and ANDERSON, M. (1988). Assessing impacts of retail development and redevelopment, in Newton, P. W., Taylor, M. A. P. and Sharpe, R. (Eds), *Desktop Planning: Microcomputer Applications for Infrastructure and Services Planning and Management*, pp. 172–9, Melbourne: Hargreen Publishing Company.

SHIFFER, M. J. (1992). Towards a collaborative planning system, *Environment and Planning B*, **19**(6), 109–722.

VOOGD, H. (1983). *Multi-Criteria Evaluation for Urban and Regional Planning*, London: Pion.

WATERS, R. (1995). Data sources and their availability for business users across Europe, in Longley, P. and Clarke, G. (Eds), *GIS for Business and Service Planning*, pp. 33–47, Cambridge: GeoInformation International.

CHAPTER THIRTY-FOUR

A Spatial Decision-support System for Collaborative Solid Waste Planning

MARIANNE MacDONALD

34.1 Introduction

Urban and regional planning issues are integrally related to an area's spatial structure. Thus, geographic information systems (GIS) have been recognised as a valuable tool for planners. Solid waste planning exemplifies GIS technology's potential to assist in decision making. Many criteria must be measured and compared before a waste-management plan can be developed. These include resource requirements for land, energy, labour, capital and transportation infrastructure; environmental impacts from air and water pollution; concerns about equity; and political issues relating to locally undesirable land uses. Two traits that these criteria have in common are their large data and analysis requirements, and their spatial nature. GIS provides a suitable tool for managing and communicating information relating to the physical, political and environmental effects of solid waste facilities. However, additional tools are necessary for some analytical capabilities, especially for non-spatial data. For these reasons, solid waste planning provides a pertinent illustration of the usefulness of spatially oriented computerised decision support.

To take advantage of this potential, a prototype for a spatial decision-support system (SDSS) for solid waste planning was developed (MacDonald, 1994). This system, however, was designed for a single, technically knowledgeable user. Because the issues involved in making solid waste decisions are extensive and varied, decision making often requires input from a number of individuals and disciplines. Thus, it seems appropriate to extend the capabilities of this SDSS, and especially the GIS interface, to collaborative decision-making efforts. However, it is prudent to research the complexities involved in collaborative solid waste planning and the application of computers into a group decision-making environment before a technology design is established.

510

34.2 Collaborative solid waste planning

Solid waste planning is also collaborative effort. Many cities and regions have committees that meet regularly to resolve solid waste issues. These committees include representatives from local government, public interest and environmental groups, academics and business representatives to ensure that multiple viewpoints are considered. In general, members of these committees have a high level of knowledge of the field, but varying technical strengths. With so many ideologies and backgrounds involved in the planning process, however, it can be difficult to make decisions such as siting facilities or instituting recycling programmes.

Providing this type of committee with a spatial group decision-support system (SGDSS) may enhance their understanding of a plan's implications. The inclusion of GIS is a key factor. Having the SGDSS present multiple layers of information with the GIS enables a potentially wide range of value judgements and goals to be considered. It also provides a useful tool for communicating plans to other stakeholders.

There are four basic activities involved in group decision making: retrieving and generating information, sharing information among members, drafting policies and plans, and using information to reach a decision (Gray and Nunamaker, 1993). For solid waste management, an SGDSS with a geographic information system component can assist in all of these activities. However, group processes are complex and the outcomes of group efforts depend on a number of variables, of which technology is only one (Poole *et al.*, 1993). It may be naïve to think that a computer system can be developed to facilitate completely group decision-making. For this reason, it is important to understand the nature of the group and the variables involved in the task at hand before adapting the individual SDSS to collaborative situations. Then a system can be designed to address specific decision-making tasks without overwhelming the group with technology that may not improve decision making.

34.2.1 Group characteristics

The general goal of any solid waste plan is to provide efficient service to the public and/or private waste generators by removing and managing discarded materials in a manner that does not endanger public health. The planning group must study solid waste facilities and the flow of materials between them in order to address all relevant issues properly. This includes environmental, economic, technical, social, and political factors.

There are a few characteristics of solid waste groups that should be considered in developing an SGDSS.[1] These stem from the fact that members come from a number of organisations and their participation may be voluntary. It has been documented elsewhere that multi-organisational groups have some noteworthy attributes (Huxham, 1991). These include the absence of an authority to force participation or attendance of meetings. The only incentive is the desire to protect individual or constituent interests. Depending on the specific issue being discussed and the personalities of the individuals, this may lead to adversarial tensions between members. Although members may have worked together before, it may not have been in a setting of trust and open communication. For example, private business representatives usually deal with members from the Sanitation Department in the process of submitting competitive bids for services. They are often reluctant to divulge details of their business operations, and other members may be sceptical of private sector agendas.

Another characteristic relating to voluntary participation is the limited amount of time members devote to meeting preparation and tasks. There is no incentive for individuals to spend time or resources outside of meetings to investigate issues further, unless it is to protect a personal interest. Members may deal with solid waste issues regularly, so they remain salient. However, key features of group work, such as communicating, brainstorming and cooperating, are not necessarily practised outside of meetings.

As mentioned, members of the committee have a relatively high level of knowledge about solid waste systems. Because of the complex technical nature of solid waste systems, however, there is a great deal of variability and uncertainty about many factors such as pollution emissions and health risks. For this reason, as well as the lack of time, many issues requiring this information are not given thorough attention. Also, the ability of technical stakeholders to communicate verbally these variables to those without technical knowledge is limited. This affects the group's capabilities and may lead to insufficient analyses before decisions are made.

Even with these difficulties, solid waste groups have a great deal of potential. Although they may not participate in structured analyses or negotiations and consensus building as a group, they could. In fact, extending such a group to include these activities with other stakeholders could vastly improve the overall operation of solid waste systems. It has been noted elsewhere (Ozawa, 1993) that providing access to relevant data and analyses, in language that is understandable to all, gives disadvantaged stakeholders more influence over the process. This type of group could educate others by communicating its technical information. This would have two benefits. First, in order to communicate and legitimate its position to other stakeholders, the group would need to be more thorough and structured. This may also lead to greater idea generation and creative discussions of issues. The second benefit would be a better-informed public. This would empower them to participate in the decision-making process, thus fostering a proactive rather than reactive relationship.

34.3 Group decision-support systems

Much has been written about specific experiences with group decision-support systems (GDSS). During the 1980s a great deal of work was done on the development of such systems. In recent years, however, researchers have stepped back to question the true benefit of introducing computer support into group-decision situations. Researchers note that the purpose of GDSS is to direct and organise group processes better by keeping the group focused on the task, organising information, and documenting the process. This can provide members with greater procedural insight (Poole *et al.*, 1993). In addition, GDSS can be used to help deter undesirable group behaviours, such as dominance by specific members and 'group-think', by allowing the anonymous voting and posting of ideas. They can be used to promote desirable behaviour by providing communication media to assist in negotiation and consensus building (Finlay and Marples, 1992; Poole *et al.*, 1993). GDSSs have also been called upon to provide a prescriptive tool for group work, including the support of creative processes and idea integration (Broome and Chen, 1992).

With all of these positive goals, however, come drawbacks. Inclusion of computerised GDSSs has demonstrated increased start-up and mechanical friction. It has also been observed that some benefits, such as increased focus and structure, can be obtained by instituting manual procedures without the benefit of technology (Poole *et al.*, 1993). This same research also noted that there was less spoken communication with a GDSS and that there was no evidence of the ability of GDSS to balance participation in the group process. Finlay and Marples (1992) noted similar shortcomings, and suggested that the groups become too focused on the procedures of the system rather than on the issues.

These experiences highlight the difficulties of designing a system that meets the expectations that have been set up for GDSS. It is also important to realise that the technology is only one factor affecting the success of a group, and that measurement of success is not a simple matter (for example, see Pinto and Onsrud, Chapter 9 in this volume in respect to measuring the 'success' of GIS). For these reasons, system developers should be careful not to over-apply computer technology. In many cases, it may be likely that refinement in the choice of group members, the inclusion of an unbiased facilitator and the inclusion of manual decision-making procedures will vastly improve group decision-making situations without the difficulties and expense of computer support. These factors will guide the adaptation of the SDSS for solid waste planning for group processes.

34.4 The individual DSS

The original SDSS for solid waste planning was developed for a single user at a time. This user is an analyst or decision maker interested in assessing a plan and/or performing 'what-if' analyses. The system consists of a user interface, database system, analysis routines, and display and report generators. The GIS serves as part of the user interface, as well as a spatial data management and display tool. Spreadsheets, databases, mathematical programming systems and decision-making software supplementing the GIS provide more sophisticated analysis capabilities. Figure 34.1 presents the general flow of tasks performed with the SDSS and the technological tools used at each stage. A complete discussion of the technical details is provided by MacDonald (1994).

To accommodate the two general tasks of solid waste planners, plan development and evaluation, the SDSS can be applied in two ways. First, the system can evaluate actual or proposed policies for managing waste. The user supplies information about the facilities to be used, and the flow of waste between them, using the GIS and database system. Spreadsheets are then used to analyse the financial and environmental implications of the plan. Parameters that the user is unsure of can be supplied by expert systems where the user has final approval. The output consists of graphs, tables and GIS displays which demonstrate a plan's social, financial and environmental impacts. These can be used to compare options, inform decision makers, and legitimate decisions.

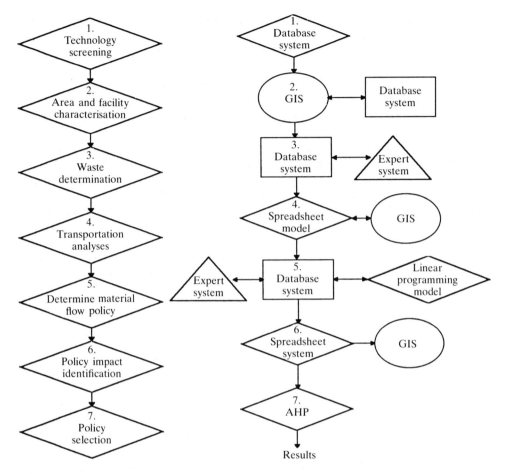

Figure 34.1   Stages of system analysis and corresponding software tools.

The second task the system can assist in is the development of solid waste management plans and policies. This is done by including linear and non-linear programming models which represent a variety of objectives and constraints. The user chooses from a set of objectives such as minimising cost or energy use, and specifies any political, financial, technical, social or environmental constraints on the program. This is all done simply by clicking on buttons or entering information into tables. The result of a model is a solid waste management scenario, including the facility technologies and locations, the types of materials to recycle, and the shadow costs of the constraints. The implications of this plan can then be determined using the evaluation path described above.

There are a number of spatial data layers that can be created by the user with the SDSS model results. These include the locations and types of facilities to be used in managing solid waste; buffer zones desired around solid waste facilities; flows of materials between facilities and generators along the region's transportation network; emissions of air and/or water pollutants from solid waste facilities; energy requirements for solid waste facilities; water demands of solid waste facilities; and financial costs to regions involved in the solid waste system. This information, along with the demographic and zoning data, can provide insights into potential political, land-use, social and environmental issues that may arise from the plan, especially if new facilities are being sited.

It is important to note that the data presented in the layers cannot be expressed using the same units of measure. While one planner may be concerned with financial costs, another may be interested in the social equity issues associated with locating solid waste facilities in low-income neighbourhoods. By presenting each attribute as a layer in the GIS, the implications of a plan can be expressed without imposing judgements about the importance of each. The users of the system can make those judgements themselves. In order to facilitate this, the multi-attribute decision-making procedure, the Analytical Hierarchy Process (Saaty, 1980), is included in the SDSS. Users indicate their preferences and judgements regarding the attributes or impacts that they feel are important. These judgements are then integrated with the measurements of the attributes determined by the analytical models of the SDSS. The result is a ranking of waste management programs. The user can then choose the one that best fits their specific situation.

34.5 The framework for the spatial GDSS

To summarise, the purpose of introducing technology into the group decision-making process is to provide tools to structure and focus group tasks; encourage idea generation and integration; help communication and consensus building; and to assist in data management and process documentation. This should all be done without inhibiting the creative benefits that result from open discussions of issues. This is a hefty task. The adaptation of the SDSS for solid waste planning to an SGDSS does not attempt to do all of these things. Rather, the design of the SGDSS addresses a subset of these requirements, and suggests the adoption of manual techniques to further improve the process. In fact, it may be appropriate to utilise only the most basic computational capabilities so as not to overwhelm or mislead the group.

In the specific case of collaborative decision making for solid waste planning, manual process changes can be used with technology to advance group processes. Broome and Chen (1992) suggest that there are three appropriate tasks for GDSS: recording ideas, structuring ideas, and graphical display. This specific system will include an additional task, providing necessary input data for analysis. With this in mind, the system is being designed around the communications and data management tool, the GIS. It has been recognised that GIS is not a complete enough tool to serve all of the analytical and decision-making needs of planning and policy-making (Harris and Batty, 1993; Scholten and Padding, 1990). However, it provides a valuable communication medium for group situations and is therefore potentially the most beneficial component of this GDSS. Communication difficulties present the largest obstacles for group work. Even the lack of resources for analysis can be improved by enhancing communications. Groups that focus on the issues, and communicate with each other about them should be better able to identify their information needs.

34.5.1 Physical framework

The individual SDSS sits atop an analyst's desk in a personal computer. In a group environment, however, the physical structure of the system becomes more complicated. DeSanctis and Gallupe (1985) have described four basic frameworks for GDSS. Considering the group's characteristics and the limited duration of their work, the decision-room structure is the most appropriate. With this framework, meetings occur in a single location equipped with a central personal computer and a large screen to present results at various stages in the decision-making process. The decision room for this system will be the room where meetings are traditionally held. This should help members feel more comfortable since their routine will not be disrupted. The technology will then be brought in for group meetings.

The preferred decision room has a keyboard and terminal for each member so that they can enter ideas and votes anonymously (Gray and Nunamaker, 1992). This encourages individuals who may be sceptical of expressing unpopular viewpoints to participate without fear of retribution. If the procedures enable members to express their beliefs anonymously, a more honest and complete solicitation of opinions and ideas will be achieved. However, to minimise the intrusion of the technology in the group process, manual procedures will be used to solicit anonymous ideas and votes. These will be entered by an impartial facilitator to encourage greater participation and discourage domination by select members. A potential drawback of manual procedures may be the increased time required for meeting procedures. Time savings with technology, however, are contingent upon members' proficiency in operating a keyboard.

The facilitator will operate the SGDSS to provide information, documentation and display capabilities for the group. The role of the facilitator is vital to the successful introduction of technology into this setting. He or she is responsible for encouraging communication and explaining the limitations of and gaps in the data and analyses. Because of the large amount of data used in solid waste planning, and the inability of the human mind to process many ideas at a time, there is a danger of information overload for members. To ensure that important ideas remain salient, Broome and Chen (1992) suggest providing hardcopy displays for information that is referred to regularly during the meeting. Thus, a printer is necessary to provide this hardcopy output.

34.5.2 Software framework

The individual SDSS flows through the tasks an analyst would follow in developing a solid waste plan. The decisions made by groups, however, are often less well defined. Therefore, the flow of the SGDSS should focus on managing and communicating information as it is required by the group. Thus, the system will centre on the component of the system that the members are likely to feel most comfortable with: the GIS. Figure 34.2 presents the stages of the group process with the SGDSS along with the tools used at each stage.

The most appropriate place to begin the discussion of an issue is by describing the status quo and identifying the problem or issue. For solid waste planning, this can be done with a GIS display of the solid waste facilities and the flow of materials

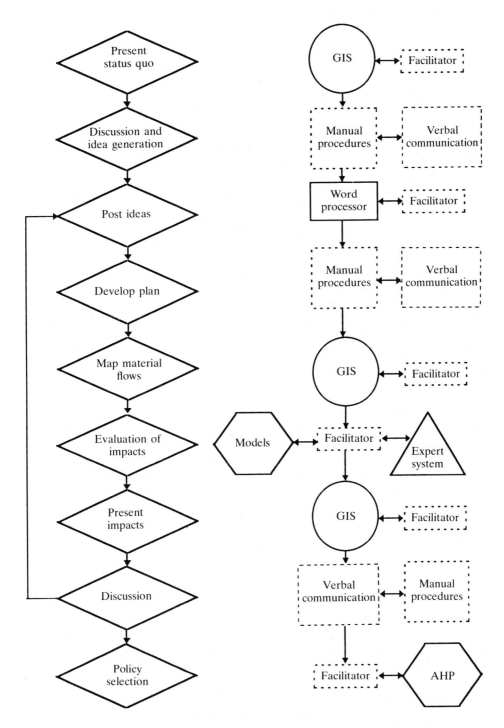

Figure 34.2   Stages of group process using the SGDSS and corresponding tools.

between them. Layers can be added by the facilitator to depict pollution sources and concentrations, demographic characteristics, or other relevant spatial variables depending on data availability. This can be supplemented by graphs, tables and charts describing the non-spatial variables such as economic costs or labour needs. It should be noted that this information will need to be compiled and entered into the system prior to the meeting. Therefore, members should have the ability to comment and/or vote on the accuracy of this representation before the discussion continues. If desired, corrections can be made on-site by the facilitator.

The facilitator will then present the issues at hand, and solicit comments and ideas for resolving them. At this point, the technology will be set aside and verbal and manual procedures will be employed to brainstorm and generate ideas. Once this has been completed, ideas will be entered into the system, posted for display, and output for later reference. In order to encourage creative thought and develop consensus, members should work to integrate and further refine these ideas. Again, this will be done using verbal and manual procedures to classify and extend ideas. The SGDSS will be used to display and document the results of this effort. The output of this task will be a set of waste flow scenarios to evaluate.

Before impacts can be evaluated, descriptions of scenarios must be converted into a mapping of material flows and facility usage in the GIS. This will be done by the facilitator with the guidance of the group. Next, the costs and environmental implications of the scenario will be estimated. Much data is necessary to perform these analyses. Although group members will have some knowledge of solid waste operations, rarely is complete technical, social and economic information accessible to all the individuals involved in the planning process. The SGDSS can enhance the gathering and sharing of these data.

As mentioned previously, the individual SDSS can assist in a number of tasks and is capable of performing complex analyses. However, due to the multiorganisational nature of solid waste groups, and the amount of uncertainty often involved in some relevant data, the group system should not attempt the same level of analytical sophistication. To do so could, at best, daunt some group members. At worst, it could mislead the group about the accuracy and meaning of results. It may be best for the group to work only with available data or information agreed upon by all members. This will ensure that members are comfortable with the decision-making process. The facilitator should work with the group to provide a greater understanding of the effect that data uncertainty will have on the decisions being made, perhaps by working through 'what-if' analyses for extreme cases.

The individual SDSS includes expert systems and databases so technical inputs can be retrieved quite simply. The extension of the SDSS to group decision making should also have this ability, but it should be enhanced to ensure group satisfaction with the data. This would be done by posting assumptions and data with high levels of uncertainty as they need to be entered into the system. Members can then vote on the accuracy of this information, and provide suggestions for inputs that they are familiar with. As mentioned earlier, members are often uncomfortable revealing specific knowledge about solid waste operations for proprietary reasons. The anonymous manual supply of this information, to be input by an impartial facilitator, may help encourage the sharing of valuable data.

If there are disputes about specific input values or assumptions, such as whether to consider maximum or average rates for pollution emissions, multiple scenarios can be evaluated to determine how the variation in input affects the overall implications of the system. Once data inputs are agreed upon, the models in the system will access group refined expert system and GIS data in order to summarise the implications of each scenario.

Implementation of the SGDSS involves data of varying levels of precision. Obviously, it is preferable to use data that is precise, accurate, and clear to users. This is possible for some classes of data such as points representing facility locations or emission points, and lines representing transportation networks. However, some of the spatially distributed information desired in decision making, such as pollution impacts and risk areas, results from sophisticated models with many assumptions. The use of these data should be carefully considered. The choice of models, if any, to include in a specific application of the SGDSS should be based on the decision being made, the quality of available data, and the members' understanding of models. The most important consideration is that models and assumptions be as simple as possible. If members do not understand the models, they will not trust the results. Complex models may not be appropriate for many group environments, especially in the formative stages.

With solid waste issues, much of the difficulty in planning involves differences in perceptions of factors such as risk or social and environmental impacts. Rather than attempting to model this, it may be more productive to use the system to facilitate the communication of these differences. This could be done by having members indicate their perceptions of impact areas using the GIS component, and then overlaying these areas to see how members' ideas compare. Some decision situations may not require modelling at all. It may be sufficient to inform group members of emission locations and concentrations, for example, rather than model health risks. In such a case, the SGDSS can use expert systems and GIS query tools to supply information to the group. If such models are desired in the system, the existence of an unbiased, scientifically knowledgeable facilitator is essential to explain the limitations of the models and to prevent misinterpretations of results.

The results of any models or information queries will be presented to members using tables and graphs for non-spatial variables, and GIS for spatial factors. The facilitator will need to organise the presentation and integration of data layers for the group. Displaying environmental and social data geographically allows for a greater understanding of potential implications of a plan or policy. Again, members will be provided with the opportunity to revise and comment on the results of the models. Group suggestions for changes can be made on-site by the facilitator.

Once a number of 'what-if' scenarios have been developed and analysed, their implications can be examined using GIS presentations and other graphical displays. Finally, the multiattribute decision-making procedure, the Analytical Hierarchy Process (Saaty, 1980), can be used to develop a ranking of alternatives based on the models' results, visual data and individuals' preferences. The values used for this process should be provided by each member anonymously, and results should be obtained for each member's opinion. By considering each member's values and having the group observe the variety of resulting scenario rankings, a greater understanding of the sensitivity to and effect of personal opinion will be gained.

34.6 Future research

One necessary area of future work is specific to GIS technology. For applications such as solid waste planning, where material flows are the primary concern, it would be highly beneficial for the GIS to include a graphical user interface (GUI) for specifying these flows. As currently designed, suggested waste flow scenarios are to be input into the GIS by the facilitator. This will require a facilitator that is technically proficient with GIS, and may substantially increase the time necessary for meetings. The optimal situation would be for the GIS to have a GUI that allows a user to indicate the source and sinks for materials simply by moving an icon and entering the amount of flow. This would simplify many GIS applications such as waste and water management, and product shipments.

Having designed the SGDSS for solid waste, the next stage is to apply the system to actual group-decision situations. The system should be tested in a variety of situations to determine how variables such as task, time, facilitator proficiency, group composition, and model complexity affect the process and its output. As noted earlier, it can be difficult to measure the success of GDSS in group processes. Does the system actually improve the decisions that are made? How can this be determined? If the members are more satisfied with their work is that an indication of success? More research needs to be done in this area if the real benefits of GDSS are to be determined. In dealing with technology that can be costly and time-consuming to develop, as well as disruptive to the status quo of group meetings, real benefits must be documented.

If this type of system is found to improve group-decision processes, and better inform stakeholders, it will be applied more frequently. As this technology becomes more widely accepted and understood, it may be extended to include more modelling capabilities, such as those used in the individual SDSS for developing planning scenarios. This may also enhance the ability of SGDSS to aid in negotiations over controversial plans (Dutton and Kraemer, 1985).

34.7 Conclusions

The potential of extending the prototype SDSS for solid waste-management planning to group decision making goes beyond helping local solid waste practitioners develop new programs. Once this system is established for waste planning, adapting it for other issues such as resource management, transportation, land-use and economic development would be possible by customising the compilation of data and models for these specific applications. If a complete regional base map is initially developed for the GIS, this same map can be used for other planning situations. Although obtaining and organising data is not a simple task, once the framework for the use and benefit of these data is specified, the task becomes more clearly defined and purposeful.

Integrating GIS with knowledge bases and models in a group decision-making environment provides a valuable visual and operational interface for soliciting ideas, analysing options and developing a consensus of opinions. The maps resulting from the SGDSS can provide a clear forum for presenting implications of potential decisions, such as environmental degradation or social equity. This enables group members to have equal access to information and to coordinate viewpoints so that the opinions of all members can be considered and analysed in a non-confrontational manner. This clearly facilitates planning processes, thus providing benefits for the group members and those affected by their decisions.

Note

1   This discussion is based on observations of the Solid Waste Advisory Committee in Philadelphia, Pennsylvania, and the Solid Waste Authority in Boulder, Colorado.

References

BROOME, B. and CHEN, M. (1992). Guidelines for computer-assisted group problem solving: Meeting the challenges of complex issues, *Small Group Research*, **23**(2), 216–36.

DESANCTIS, G. and GALLUPE, B. (1985). Group decision support systems: A new frontier, in Sprague, R. H. Jnr and Watson, H. J. (Eds) (1993) *Decision Support Systems: Putting Theory into Practice*, 3rd Edn, pp. 297–308, Englewood Cliffs: Prentice Hall.

DUTTON, W. and KRAEMER, K. (1985). *Modeling as Negotiating*, Norwood: Ablex Publishing Corporation.

FINLAY, P. N. and MARPLES, C. (1992). Strategic group decision support systems – a guide for the unwary, *Long Range Planning*, **25**(3), 98.

GRAY, P. and NUNAMAKER, J. F. (1992). Group decision support systems, in Sprague, R. H. Jnr and Watson, H. J. (Eds) (1993) *Decision Support Systems: Putting Theory into Practice*, 3rd Edn, pp. 309–26, Englewood Cliffs: Prentice Hall.

HARRIS, B. and BATTY, M. (1993). Locational models, geographic information and planning support systems, *Journal of Planning Education and Research*, **12**(3), 184–90.

HUXHAM, C. (1991). Facilitating collaboration: issues in multi-organizational group decision support in voluntary, informal collaborative settings, *Journal of Operations Research Society*, **42**(12), 1037–45.

MACDONALD, M. (1994). 'A Decision Support System for Integrated Solid Waste Management', unpublished PhD thesis, University of Pennsylvania.

OZAWA, C. P. (1993). Improving citizen participation in environmental decision making: the use of transformative mediator techniques, *Environment and Planning C: Government and Policy*, **11**(1), 103–17.

POOLE, M. S., HOLMES, M., WATSON, R. and DESANCTIS, G. (1993). Group decision-support systems and group communication: A comparison of decision making in computer-supported and non-supported groups, *Communication Research*, **20**(2), 176–213.

SAATY, T. L. (1980). *The Analytical Hierarchy Process*, New York: McGraw-Hill.

SCHOLTEN, H. J. and PADDING, P (1990). Working with geographic information systems in a policy environment, *Environment and Planning B: Planning and Design*, **17**(4), 405–28.

PART SIX

New Data Types: GIS and Multimedia

CHAPTER THIRTY-FIVE

Progress towards Spatial Multimedia

JONATHAN RAPER

35.1 Introduction

One of the application areas where developments in multimedia have made the greatest impact is in the manipulation and analysis of spatially referenced data. The availability of systems which can handle imagery, video and sound and organise them by webs of contextual links locally or across networks has proved especially attractive to geographers, planners, environmental scientists and others who commonly use spatial data in these forms. Recognising the importance of these developments, the European Science Foundation (ESF) 'GISDATA' Programme (Arnaud *et al.*, 1993) established a Task Force in 'Multimedia Geographical Information Systems' in 1993 to promote research in this area. A research meeting was held in Rostock, Germany, in May 1994 with representation from across Europe and the USA in order to draw together work spread over several disciplines and initiate an international dialogue on research in this area (work presented at this meeting is summarised in Craglia, 1994). This chapter aims to review the progress in spatial multimedia research and set out a preliminary agenda for research. Given the range of perspectives on multimedia technology and the rapid pace of developments, the chapter begins with a selection of published definitions and a review of literature available at July 1995. A short review of multimedia technology trends is included in order to identify particular hardware, software and communication tools which have been influential in initiating research in this area. Brief details of the emerging multimedia standards are given, as several of them deal with time and space representation which is a central concern in spatial multimedia research.

35.2 Definitions

Definitions are a key starting-point since the term 'multimedia' is used rather loosely in the geographical information systems (GIS) literature. 'Multimedia' has been defined by Laurini and Thompson (1992) as 'a variety of analogue and digital forms of data that come together via common channels of communication' (p. 595), thereby emphasising the information integration aspects. Bill (1994) considers that multimedia systems deal primarily with 'processing, storage, presentation, communication, creation and manipulation of independent information from multiple time-dependent and time-independent media' (p. 151) thereby emphasising its time-based nature. Raper (1995) defines multimedia GIS as 'the use of hypertext systems to create webs of multimedia resources organised by theme or location' (p. 94) thereby emphasising the information structuring issues. The term 'multimedia' GIS has also been introduced to the GIS literature with a wide range of meanings. The use of the term 'multimedia in GIS' has variously referred to visualisation toolset extensions, the addition of sound and video data types to GIS and the development of hypermedia spatial databases. No attempt is made here to restrict the use of the term.

35.3 Key multimedia technology for spatial multimedia

Multimedia technology can be embedded in any application program including a GIS or it can be supported at the operating system level (Hall, 1993). The key issues in GIS have been how to take advantage of new multimedia technology and whether to extend existing GIS or to add spatial functionality to multimedia.

35.3.1 Key hardware developments for spatial multimedia

The hardware developments reviewed here are divided into 'multimedia facilitating' and 'hardware–software integrated'.

Multimedia-facilitating developments. Multimedia technology is facilitated above all by a huge rise in the performance of low-cost processors such as Intel's Pentium or Apple/IBM's PowerPC and the availability of cheap memory, both of which have been incorporated into microcomputers, workstations, games machines and (now) television receivers. Dedicated processors have been developed which are capable of real-time compression, decompression, sampling, filtering, buffering of images and sound, and the availability of cheap fast memory has facilitated high-speed caching. These hardware technologies have made it possible to develop digital audio and video handling systems which are corruption-free and robust and which are incorporated in microcomputers as add-on boards. Recent experiments with digital TV signals and cameras preface a move towards fully digital systems from data capture through recording and broadcast to editing and compilation.

Based upon the review of applications discussed below, there are three main multimedia platforms which are in use in multimedia. GIS: the multimedia PC as defined by a manufacturers group led by Microsoft known as the Multimedia PC marketing council; the multimedia Apple Macintosh systems and the Silicon Graphics Indy Unix workstation. Spatial data sets are stored on large, fast, hard discs or compact discs (CD). Most current spatial data products on CD are based on the CD-ROM Yellow Book format (ISO 9660). Many users still have the original CD drives with a data transfer rate of 150 kbps with its relatively slow seek time compared with a hard disc. Few if any GIS data sets have been written in the White Book Video CD format allowing full motion video to be played from MPEG encoded video on an accelerated CD drive.

As yet there has been limited progress towards developing a library of spatial multimedia data gathered using digital audio and video capture techniques. However, experiments have been conducted with the capture of digital video from aerial videography (Green and Morton, 1994; Raper and McCarthy, 1994a) and by the use of touch screens or pen-based command systems to update maps stored in portable computers such as the MGIS system (Dixon *et al.*, 1993).

'Hardware–software integrated' developments. There is also a variety of new developments which integrate dedicated hardware and software (Cotton and Oliver, 1993). The main systems with spatial applications will be described briefly below.

- Compact Disc-Interactive (CD-I) from Philips (launched in 1991) uses the Green/White Book formats to allow the use of the CD as a source of alphanumeric data, images, animation, high-quality audio and video. Authoring is carried out, for example, using MediaMogul running under OS9. CD-I players are designed to plug into a television or VCR and be used with a remote controller. Typical spatial applications include museum guides and interactive golf.

- Data DiscMan from Sony is an extension of the audio DiscMan portable player. The Data DiscMan uses 8 cm discs to display graphics on an LCD screen and play sound for reference works and business information such as city maps.

- Kodak PhotoCD is a system for scanning colour prints and storing the images on a CD. To view the ImagePac format images requires PhotoCD hardware attached to a television which uses a further extension to the CD-ROM standard known as CD-ROM XA (Extended Architecture). Although it is intended to appeal to the non-scientific user who wants to develop colour print film and record it on CD, there are already spatial applications in the archiving of earth surface images and maps.

- Personal Digital Assistant (PDA) is a generic term for a hand-held computer with a small LCD screen and pen-based input using pointing or handwriting. A number of such PDAs have now appeared, such as Apple's Newton, which have been used for field data entry or update of stored maps.

- Ultra-small computers which can be worn by a user known as 'active badges' are being developed to track the location of individuals within buildings or sites and to route phone calls to the nearest phone. This work has involved the development of new forms of spatial databases.

- Games machines such as Sega Megadrive, Nintendo Gameboy and 3DO have developed simulated spatial scenarios which are based on dedicated hardware (sometimes incorporating a CD drive) with 'plug-in' games software cartridges. The VistaPro system allows the user to fly over any terrain input to the system in a predetermined path.

35.3.2 Key software architectures for spatial multimedia

Although multimedia developments are progressing through a series of self-reinforcing cycles of hardware and software innovation in which developments in one field fuel the other, the software developments appear to be pre-eminent in the multimedia GIS field at the moment. This is due to the need to serve a market only equipped with standard hardware systems. Hence, the development of software outside the dedicated hardware environments reviewed above has been based on extended operating systems offering support for the 'multimedia' data types, specifically video and sound. At the time of writing it is not clear whether hardware support for full motion video (FMV) will become a standard component of all hardware systems or whether the ever-accelerating pace of central processors will permit software-only support for FMV.

All the main operating systems for personal computers have been extended to offer support for specialised multimedia-supporting hardware such as video capture and for software-only handling of video and sound. Hence, Microsoft Windows has offered multimedia extensions since version 3.1 (Wodaski, 1994), IBM's OS/2 first offered the Multimedia Presentation Manager/2 extensions with version 2.1, Apple Macintosh System 7 has supported the Quicktime extensions since 1991 while the Silicon Graphics Irix operating system has offered support for video since the release of version 4.0. In addition to the operating system level support for multimedia, the software aspects of data storage have seen major developments, largely driven by 'information superhighway' requirements which have called for high-speed multimedia servers capable of handling 'media objects' as well as the traditional database data types. Further developments of this kind can be expected as the draft version of the relational database query language SQL3 is finalised. This will permit relational databases to store and query 'abstract' data types, especially multimedia GIS data (Raper and Bundock, 1993).

Built on top of these extended operating systems many new tools have been developed for visualising multimedia spatial data such as Exploratory (spatial) Data Analysis (EDA), animation of spatial processes and map-to-video 'hot-links' (which can be developed using tools such as Visual Basic). Designing user environments for the viewing and understanding of these new systems has involved considerable research into the human factors and ergonomics of computer use, especially for the handling of spatial data (Medyckyj-Scott and Hearnshaw, 1993). However, the key development in the visualisation of spatial data has been in the Virtual Reality (VR) field (Kalawsky, 1993). VR systems now enable a user to roam around a virtual terrain in real time either by use of head-mounted display (fully immersed VR) or by viewing the virtual environment on a high-quality colour monitor (through-the-window VR). Research is under way to define GIS-to-VR data translators and to adapt the VR control systems to constraints required in spatial applications involving terrains (Raper and McCarthy, 1994b).

Multimedia authoring in the field of spatial applications has been dominated by proprietary personal computer-based tools. These systems can be classified by the informational structures and metaphors they use for the development of their materials. The simplest informational structures are based on page-turning electronic books (such as StoryBook), scenario-building based on user interaction (such as Authorware) or movie-making (such as Director or Premiere) although the sophistication of some of the latter systems has been extended using scripting languages. More complex informational structures which allow the linking of multimedia data types grouped into abstractions in semantic nets are provided by systems using the hypertext model: these have proved popular in the development of spatial applications.

Hypertext or hypermedia systems have been used to develop hypermaps which are 'browsable hypermedia databases integrated with coordinate-based spatial referencing' (Raper and Livingstone, 1995, p. 681). The hypermedia model is particularly useful for heterogeneous collections of data where many abstractions have unique properties and cannot be typed or named in advance, or where arbitrary aggregations are meaningful such as in spatial applications.

Open hypermedia systems have begun to emerge in order to reduce the dependence on package-specific file structures and to provide operating system level support for associative linking of multimedia resources. Microcosm is an example of such a system developed for Windows 3.1, Macintosh and Unix (Hall, 1993). Microcosm was experimentally linked to the SPANSMAP GIS (Simmons et al., 1992) in order to link spatial and non-spatial data of a variety of forms referring to a development site. Work has also begun to design intelligent agents capable of locating and retrieving spatial information, generating spatial model templates, monitoring task execution in a GIS and managing collaboration between GIS and other systems (Rodrigues et al., 1995).

Digital data compression has also been of crucial importance to the effective storage and high-speed delivery of multimedia data types. Compression methods can be divided into lossless methods (images compressed are recovered exactly) and lossy methods (images compressed are recovered approximately). Lossless methods are appropriate for the compression of data which must be preserved exactly – such as earth surface images and maps – and can generally achieve an average of 3:1 reductions in data volume for single frames. However, lossy techniques can be used to compress images-to-be-viewed such as photographs since human perception of the high-frequency variations in an image is poor. Most systems now use the Joint Photographic Experts Group (JPEG) standard for still image compression which in lossy mode can achieve an average 20:1 compression while retaining excellent reconstruction characteristics. Many World-Wide Web (WWW) sites compress maps and images using the JPEG format.

For video, a further data reduction can be obtained over and above that of the individual image frames compressed with JPEG by using the Motion Picture Experts Group (MPEG) standard. The MPEG standard designates some frames of the video as reference frames and other frames are then encoded in terms of their differences from the nearest reference frames. Because of the limited implementation of MPEG in software and hardware at the time of writing, most work carried out with spatially referenced video has used proprietary video compression such as Apple's Quicktime format.

Although multimedia systems initially developed on standalone platforms, the development of the high bandwidth global Internet into a single heterogeneous Wide Area Network (WAN) has focused much development on the delivery of multimedia data across networks. Across the developed world governments, universities and corporations are developing digital networks while telecommunication utilities and cable television companies are developing access to homes and small businesses. This is achieved either by offering digital connections based on Integrated Services Digital Network (ISDN) transmitted over copper wires, or by replacing analogue services over copper wires with optical fibre networks capable of handling digital data. The huge increase in the transmission of multimedia data over the Internet has generated an urgent need to update existing infrastructures. Research is underway to revise the Open Systems Interconnect

(OSI) protocols and to develop multimedia conferencing using the International Telecommunication Union T.120 standards.

At the time of writing probably the most important implemented multimedia network 'service' is the World-Wide Web (WWW). This system allows any user with an Ethernet or modem, connection to the Internet and a browser such as Mosaic or Netscape to access data 'published' on a 'Home Page' anywhere in the world. Many spatial data resources have been 'published' on the WWW using JPEG/MPEG compression of images, graphics, video and sound. Shiffer (1995b) reviews some of the resources available and shows how this kind of spatially referenced data can be linked to the Collaborative Planning System developed for Washington's National Capital Planning Commission enabling spatially oriented browsing over the Internet. Raper and Livingstone (1995) describe a spatial data explorer called SMPViewer designed to facilitate extraction of spatial information from georeferenced image and map data resources obtained from the Internet, for example using the Xerox PARC Map Viewer application.

Telecommunication utilities and cable television providers are also experimenting with 'video-on-demand' services transmitted over existing analogue connections such as home shopping, videoconferencing, interactive television and long-distance collaborative game playing. High bandwidth networks also facilitate Computer-Supported Cooperative Work (CSCW) which permits physically separated colleagues to work together exchanging data and speaking to each other as necessary, perhaps using videoconferencing systems like CU-SeeMe.

To manage these huge and distributed resources of data many organisations are creating and maintaining 'digital libraries' which are designed to give network access to these collections of data. The first major attempt to establish a digital library for spatial data is the 'Alexandria Project' (Smith and Frew, 1995) which has been established by a consortium of US libraries, research groups and industrial corporations. The key aim of the Alexandria Project is 'To provide geographically dispersed users with access to a geographically dispersed set of library collections. Users will be able to access, browse and retrieve specific items from the data collections of the library by means of user-friendly interfaces that integrate visually based and text-based query languages' (p. 61). The Alexandria Project will initially include access to maps, orthophotos, AVHRR, SPOT and LANDSAT images as well as geodemographic data.

With the proliferation of software tools in the multimedia field, the need has been recognised for data exchange standards capable of translating one proprietary system into another. There are several standards at an early stage of development. HyTime is an international standard 10744 for the representation of information structures for multimedia resources through the specification of document type definitions and is based upon Standardised Generalised Mark-up Language (SGML) (ISO 8879). The Multimedia and Hypermedia information encoding Experts Group (MHEG) is a draft standard (DIS 13522) for representing hypermedia in a system independent form. It is optimised for run-time use where applications are distributed across different physical devices or on networks. The PResentation Environment for Multimedia Objects (PREMO) is a draft computer graphics standard due in 1997 (DIS 14478) which will define how different applications present and exchange multimedia data. It will define interfaces between applications so that different applications can each simultaneously input and output graphical data.

Each of these standards is developing a different concept of space and time (Hopgood, 1993). The PREMO and MHEG working groups are discussing the use of a four-dimensional space time which provides a comprehensive frame of reference for all events in a multimedia presentation. However, HyTime separates space and time permitting the definition of multiple time reference systems, for example permitting both 'video frame rate time' (30 frames per second) and 'user playback control time' describing how they are actually played by a user who is responsible for pauses and fast-forward events. The resolution of these debates and the concepts of space and time which are implemented have important implications for the spatio-temporal analysis of georeferenced multimedia.

35.4 Spatial multimedia: applications review

Today, the majority of GIS represent the world using geometry in vector or raster form linked to alphanumeric data arranged in tables. This kind of system is now used widely, notably in the maintenance of cadastral information and in facility management. GIS based on this design have been developed to handle very large continuous databases and can execute complex queries on the geometry and attributes. However, this kind of conventional system has limits to its representational ability since the spatial expression and behaviour of many dynamic phenomena cannot be handled; the qualitative data which the human eye or ear can easily recognise and interpret cannot be stored; and the support for visualisation of relationships between data in the spatial database is generally poor. The ability of multimedia technology to add to the representational scope of systems has proved attractive to users of spatial data: this has led to the development of multimedia GIS (Raper, 1995).

GIS developers seeking to extend the representational scope of spatial data-handling systems have adopted two main strategies: first, to bring multimedia handling within the GIS ('multimedia in GIS'); and, second, to develop GIS capabilities within a multimedia-authoring environment ('GIS in multimedia'). At the time of writing a number of GIS have addressed the first strategy by adding support for images, sound and video including GenaMap, Smallworld, GeoNavigator, Geo/SQL and SPANSMAP (Simmons, 1993). However, only a few GIS have also added support for associative hyper-linking of maps, images and multimedia resources: these include ESRI's Arcview with Geographic Hotlinks and Integraph's GeoMedia architecture. So far all of these 'multimedia in GIS' developments have isolated the multimedia data types from the 'standard' text and graphics data types and have not yet integrated them into the underlying data model. Some research work has been done to try to show how integration can be achieved by developing multimedia-extended spatial data models for GIS. For example, Yeorgaroudakis (1991) developed a generic data model for handling multimedia spatial data along with a set of query mechanisms; Kemp and Oxborrow (1992) developed a system to handle hospital- and ambulance journey-related multimedia data in the object-oriented Zenith environment; and Boursier (described in Craglia, 1994) has developed the HYPERGEO system which is an application generator for hypermedia geographic applications. Kemp (1995) evaluated the alternative implementations of the 'multimedia in GIS' approach and concluded that the georelational model falls short of existing aspirations since the multimedia data and their servers are not really integrated into the GIS. Instead Kemp (1995) suggests that 'multimedia in GIS' must be implemented using object-oriented languages constructed over database systems such as POSTGRES.

By contrast with this latter approach, systems based on existing multimedia development environments have integrated spatial data in a 'GIS in multimedia' design. This work is progressing rapidly on a variety of fronts and mostly uses established proprietary multimedia tools and their data models. This work is reviewed in the following sections.

35.4.1 Hypermedia spatial databases

A new development in spatial database design has been the creation of the hypermedia spatial database, as in, for example, the digital atlas. Creation of such databases has usually relied upon hypermedia-authoring tools such as Hypercard or Toolbook. Such databases usually contain a wide range of spatially referenced data which needs to be accessed by primarily spatial criteria. This means the defining and naming of spatial abstractions which may be aggregated into arbitrary collections of higher level abstractions. This structure is used, for example, for databases of tourist information where 'attractions' may be given the type 'museum' or 'art gallery' and then aggregated into 'day tour' objects at a higher level (for example, the National Gallery guide in London referred to by Buttenfield and Weber, 1993). These arbitrary abstractions may be transitory – such as the one-off route of a special road race like a marathon. Perhaps the earliest hypermedia spatial database developed was the Domesday System based on BBC microcomputers and the analogue Philips LaserVision discs (Rhind et al., 1988). The Domesday system was accessed by map gazetteers or tours and contained a huge amount of spatial data relating to population, employment and environment, complete with Ordnance Survey maps and photographs. Lewis and Rhind (1991), in one of the first reviews of multimedia in GIS, demonstrated how the Domesday LaserVision discs could be controlled by commands issued through the serial port of an Apple Macintosh once the BBC computer had become obsolete.

A number of hypermedia spatial databases have been developed for educational purposes: perhaps the earliest system was the BBC Ecodisc delivered on CD-ROM. Later examples include the Geonex/Ordnance Survey educational CD 'Discover York' developed using Windows Multimedia Extensions and the Open Mind/BBC Schools TV/Ordnance Survey 'Images of the Earth' CD-I project. Multimedia encyclopaedias such as *Encarta* from Microsoft and the *New Grolier Multimedia Encyclopaedia* (NGME) also incorporate digital maps and location videos: DiBiase (described in Craglia, 1994) showed how the multimedia resources of the NGME were planned with respect to the required interactive functions.

One of the most common motivations for the development of a hypermedia spatial database has been the need to visualise the urban environments for planning purposes. Examples of this have included Hypersnige (Camara et al., 1991b) for the exploration of regional planning information and SOFTANIS (Gomes et al., 1993) for exploring environmentally protected areas – both developed in Hypercard by the Centro Nacional Informação Geografica (CNIG), Portugal; the

London Covent Garden simulation developed by Parsons (1992) (developed in Hypercard) which uses an aerial photo base annotated with points of interest which are then illustrated by animation or video if activated; the Move-X system developed by Ertl *et al.* (1992) to merge video footage of vacant sites with architectural models and GIS data; the hypermedia system developed by Labbri for municipal building maintenance in Supercard (described in Craglia, 1994); and the Collaborative Planning System (CPS) based on Supercard (using Quicktime) designed by Shiffer (1993) to evaluate planning scenarios for Washington, DC which allows user navigation through a range of multimedia data.

Perhaps the best-known of such developments is the 'Great Cities of Europe' hypermedia spatial database developed as part of the European Union COMETT programme (Polydorides, 1993). GCE contains georeferenced planning information on a number of European cities organised by city, theme and image collection. The issue information is accessed by a thesaurus while the city information is accessed from a map base. The GCE is based on a multimedia GIS architecture developed by the University of Patras and Exodus Multimedia which uses Toolbook extended with C++ and integrated with the Windows multimedia extensions. It is to be delivered with data on 29 European cities on CD-ROM.

Hypermedia spatial databases have also begun to be used in facility management applications. Green and Kemp (1993) described the development of a pipeline management system developed in Toolbook on the PC which integrates a map base showing the pipeline at various different scales with detailed photographs (aerial and terrestrial) of the pipeline route. State (1993) developed a railway track information system based on video footage of the line and trackside equipment. By integrating inertial navigation systems with the video the distances were exported to Arc/Info and stored with details of the trackside equipment.

A popular application of the hypermedia spatial database is the digital atlas: an early example is the National Atlas Information System of The Netherlands which was created in Apple's Hypercard (Koop and Ormeling, 1990). Spatial enquiries may be made by clicking with the mouse pointer at any location on key maps and it is possible to link the selected location to various associated geodemographic data sets and analyse specified subsets of data. A similar system for the presentation of spatial information on the North American French-speaking communities was developed in Hypercard by Raveneau *et al.* (1991). Blat *et al.* (1995) describe the creation of the PARCBIT CD-ROM which contains a large quantity of planning-related data for the island of Mallorca. The user is presented with various methods of structured browsing which tests conducted on users showed had considerable time advantages over traditional approaches.

Other uses for hypermedia spatial databases have included the indexing of an image base such as the Portugal Interactivo project (Henriques *et al.*, 1992) where nationwide 1:15 000 stereo air photography is to be captured and compressed for access from optical discs. Mogorovitch *et al.* (1992) described a multimedia GIS architecture based on the integration of ARC/INFO with videodisc images and video for tourist-planning applications. The users can interactively access points of interest on the map and see images and video for that feature; they can search for all occurrences of a particular type of tourist sight or they can plan tourist routes. Craglia (1994) summarises work by Aurigi to develop a hypermedia spatial database for the public transport system of Florence which is implemented by spatially referenced integration between Hypercard and the GIS MapGrafix.

35.4.2 Hypermaps

When a hypermedia spatial database is integrated with coordinate-based spatial referencing such that each spatial 'object' has a stored location, the system can be defined as a hypermap. A hypermap is usually presented as a visualisation of a metric or topologic space which is accessed by browsing. Examples of hypermaps can be seen in the 'Great Cities of Europe' project described above.

Laurini and Thompson (1992) presented a data structure for a hypermap containing points, lines and areas (forming spatial features on a map) and suggested the use of R-trees to store spatial inter-relationships in a hypertext environment. They suggest that map (spatial) to document (non-spatial) relationships might also be structured using R-trees, although this requires the transformation of the documents into rectangles in 2-D coordinate space. The potential for hypermaps to become widely used depends on the ability of developers to conceive new spatial abstractions which can be implemented in the hypermap environment. Since the hypermap does not impose integrity constraints or planar enforcement of geometry in its design, there is a flexibility in implementation which is attractive when the overall size of the data set is not too great.

Fonseca *et al.* (1995) present a design and prototype of a multimedia GIS developed in Supercard which implements the hypermap concept for environmental impact assessment at the Expo '98 site in Lisbon. The user can browse and overlay images, maps and video of the site and can morph one site design into another to visualise the key differences between alternatives. Raper and Livingstone (1995) describe the design and implementation of a spatial data explorer called SMPviewer which allows the user to extract interpretations of images and maps digitised on a hypermap using the mouse. The geometry of the user-defined abstractions is stored as an object which itself can be linked to any other object in the SMP hypermedia database and exported using the Arc/Info 'ungenerate' format.

35.4.3 Cartographic visualisation and animation

A number of new publications have begun to address the difficulties of visualising complex spatial data sets and the opportunities provided by multimedia data and systems (Hearnshaw and Unwin, 1994; McEachren and Taylor, 1994). Some of this work focuses on the need to offer the tools of Exploratory (spatial) Data Analysis (ESDA) to the user. This will encourage them to focus on 'rebasing' the information by time, scale or unit of reporting and to attempt exploratory analysis of the relationship between spatially distributed variables. Exploratory Data Analysis has been realised for spatial data through systems such as SPIDER (Haslett *et al.*, 1990) and Exploremap (Egbert and Slocum, 1992). Unwin *et al.* (1994) show two applications of ESDA in the ARGUS project – a census data viewer and a three-dimensional data viewer. Map animation has also been proposed as a way of visualising complex variables, especially those organised by time (Dorling, 1992). Openshaw *et al.* (1994) suggested that animation of a temporal map series could be used to speed up time or to reorder the data within time periods. This approach has been used for applications ranging from cancer–cluster analysis to crime incidence. Weber and Buttenfield (1993) described a cartographic animation of average temperatures for the USA over the last 100 years which can be run in either direction in time. Schwarz von Raumer and Kickner (1994) and Ludäscher (described in Craglia, 1994) show how Toolbook and Arc/Info can be linked together to show the visualisation of pollution levels in real time. The Centennia interactive historical atlas of European history from Clock Software presents spatial animations of changing political boundaries, allowing movement forward and backwards through time and generalisation of the hierarchical level of the political units shown.

Animation has also been used to present views of points of interest in cities, for example in the 'Flying over Lisbon' presentation (Bacao *et al.*, 1993) produced in Director. Boytscheff (described in Craglia, 1994) showed how historical or future reconstructions of cities can be made using CAD tools and then visualised by 'fly-through' visualisation systems: several applications have been developed using this approach including the reconstruction of Mannheim in 1758 and the presentation of alternative plans for a new Stuttgart railway station.

Fisher *et al.* (1993) refer to the use of multimedia tools as means for improved map design. They specifically focus on the possibilities of using animation and sound to explore the error structure of a spatial data set. Cassettari and Parsons (1993) focus on the use of sound as a variable in GIS discussing how frequency, wavelength, intensity, spatial extent and duration could be mapped, giving airport sound pollution as an example. Fonseca *et al.* (1993) also look at the incorporation of sound (and video) into GIS and suggest which functional areas in a GIS might need to be enhanced to handle multimedia data.

However, Kraak (1996) has pointed out that while the technology of presentation has undoubtedly assisted visualisation, there have been few developments which codify the exploration and analysis of the new multimedia data types in the context of the theories of semiology and the 'grammar' of cartography.

35.5 Spatial user environment and usability

At the time of writing in 1995 most commercial GIS user interfaces have been ported to a window environment on PC (Macintosh or Windows) or workstation (X-Windows) making them much more visually attractive than the old command-driven text interfaces. Several GIS vendors have developed new interfaces such as GENIUS from GenaMap, SPANSMAP from SPANS and ARCVIEW from Arc/Info. However, as yet users have only seen enhancements of the screen designs rather than a deeper reorganisation of the way in which GIS are used. Although these products represent significant advances, they still suffer from a number of problems including poor database visualisation tools, complexity in the use of the analysis tools and poor customisability (Raper and Bundock, 1993).

Work has been undertaken on the human factors associated with GIS interaction (Medyckyj-Scott and Hearnshaw, 1993). The Usable Spatial Information Systems (USIS) project at Loughborough has attempted to assess the 'usability' of GIS using questionnaire surveys based on the Digital System Usability Scale (SUS) and observation of users. Preliminary conclusions (Medyckyj-Scott *et al.*, 1993) indicated that although users state that they have an overall positive impression of GIS, they frequently then go on to cite difficulties with jargon, file handling, lack of feedback, and inconsistencies between modules. These findings confirm the need to examine the GIS 'experience', especially since an international standard (IS9241) is under development for the 'Ergonomic requirements of office work with visual display systems' which will address the ergonomics of software systems.

Kraak (1989) carried out an early study on effectiveness of 3D perception in a spatial context. This has been followed by work such as that by Buttenfield and Weber (1993) who review some principles of visualisation for spatial hypermedia exploration. They specifically identify the need for navigation aids such as sonic and graphic cues.

It has been argued that tools such as hypermedia systems can be used to drive the 'front end' of GIS to improve the user experience. A first attempt made at Birkbeck to develop such a new GIS user interface was the HyperArc project to create a graphical 'front end' to the GIS Arc/Info (Linsey and Raper, 1993). Implemented using Hypercard on the Apple Macintosh and communicating over a network to a Vax or Unix host system running Arc/Info, HyperArc enabled a user to carry out some generic GIS tasks without specific knowledge of the Arc/Info command language. The creation of HyperArc was an extremely valuable exercise in identifying the important design factors for a GIS user interface of this kind. In particular it showed the difficulties inherent in using system terminology, and in coupling an interface directly to the underlying GIS.

A further attempt to develop a 'front end' to any GIS was the Universal Geographic Information eXecutive (UGIX) project (Raper and Bundock, 1993). UGIX is designed as a series of layers over the GIS: at the top, a graphical user interface (GUI) which has been designed around a building metaphor: functionality is expressed as furniture in rooms dedicated to specific real-world problems. Interaction with the room and furniture depicted in this layer generates an instruction in a fourth-generation language. This is then mapped into the host GIS using an application–interface model and transferred to the GIS via inter-process communication protocols. Embedded in the 4GL is the spatially extended SQL (SQL-SX) which gives the interface its expressive power in spatial terms. SQL-SX is spatially extended by adding support for spatial data types, operators, and functions using the facilities defined in the draft standard for SQL3 such as abstract data types. This overall design facilitates the assembly of tasks within a GIS environment, without the user needing to know how the underlying GIS actually works.

Simulated spatial environments multimedia systems have also been developed for spatial simulation. Amongst the most popular spatial multimedia systems are SimCity and SimEarth which are urban and global modelling programs presented as a game using colour simulations (Joffe and Wright, 1989). In the system the user must make decisions relating to city planning and welfare and global environments respectively: poor decisions are reflected in catastrophic outcomes!

More structured simulation and modelling environments are offered by programs such as Stella. This offers a visual programming interface to modelling systems in which complex mathematical expressions can be embedded. Output can be in the form of animated changes in the visualised system or conventional graphs and tables. Pictorial simulation models for spatial data sets such as those defined by Camara *et al.* (1991a and b) over a simple raster offer powerful new tools for spatial analysis in multimedia.

35.6 Virtual reality and three-dimensional GIS

The rapid development and decreasing price of VR systems have begun to make them attractive as visualisation environments for spatial data such as terrains. Jacobson (1991) and Edwards (1992) were the first to point out the possibilities open to users of spatial data for the exploration of complex spatial data sets. Raper *et al.* (1993) carried out a survey of awareness of VR in the spatial data community, finding that although knowledge of the technology was high in most sectors (especially the utilities) few organisations had any experience with systems due to the expense of VR systems and difficulties of data exchange with GIS. At the time of writing in early 1995 the cost of VR systems has fallen considerably and much progress has been made in their use with spatial data. Bagiana (1993) discussed the applications of VR at the European Space Agency (ESA) for simulations connected with the Columbus space station, the Mars Rover and the preparation of astronauts for space missions, and Smith (1994) described how the military are using VR to simulate battlefield conditions. Raper and McCarthy (1994b) discuss the development of GIS to VR translation routines in the VRGIS project showing how terrain data in triangulated irregular network form can be imported into the dVISE system from Division. Camara and Neves (1996) show how virtual ecosystems can be explored by using the VR system WorldToolkit to place the user in the physical position of any organism active within a simulated (coastal) ecosystem. They also give details of the development of the 'Virtual Tejo' project which allows the user to visualise pollution in the Tagus estuary using moving symbols within the water body.

Rapid developments have also taken place in the field of digital photogrammetry which has made it possible to stereocorrelate any two georeferenced raster images to generate digital-terrain models (DTMs) as sources of data for virtual reality simulations of terrain. Raper and McCarthy (1994a) show how a high-resolution video camera mounted on a light aircraft can be used to gather pseudo-vertically oriented imagery of the terrain at high quality and low cost. The DTM created by this method was then overlaid with the terrain image texture data to form a data set suitable for import into a VR system.

35.7 Spatial multimedia explorers and travel support

A number of systems have now been developed to allow users to explore large, complex spatial multimedia data sets usually stored on CD. Examples include the 'Global Explorer' from DeLorme which has tools to search 142 000 place names on its 1:1 million world atlas maps, with the facility to pan and zoom the maps; 'City Streets' from Road Scholar presents data on 180 US and 80 European cities and can be linked to GPS for in-car navigation; 'Explorer' from NextBase presents a digital world atlas and gazetteer complete with thematic maps and tables of statistics by country; and 'Euripides' containing a European road network database and small area demographic statistics. The explosion of available spatial data sets has led to a number of developments in the field of spatial metadata. In the US, the Federal Geographic Data Committee (FGDC) has proposed a draft Content Standard for Spatial Metadata which defines information which is regarded as essential in a metadata record for a spatial data set. In Europe the MEGRIN

group, representing 18 European National Mapping Groups, has drafted a
metadata standard for mapping between the scales of 1:25 000 and 1:1 million and
the European Union (EU) IMPACT II project Omega has developed a prototype
metadatabase on CD (Wood, 1994). Work is also under way to develop travel
advice and in-car navigation systems using multimedia GIS. Examples of projects
include the TITAN CD which contains travel and tourism information for Europe
in multimedia form (Power and Lay, 1994); PLEIADES, an automated system for
passing traffic conditions to cars by radio; and the EU DRIVE programme for
monitoring and managing traffic conditions called SCOPE (Stembridge, 1994).

35.8 Computer-based learning for spatial data

Hypermedia systems have proved ideal authoring environments for the creation
of interactive tutors in all subject areas, especially GIS. This is because there is
a range of alternative information structures available to organise the material,
visualisation tools such as animation are incorporated, and it is easy to import data
from other applications such as operational GIS. Hypermedia authoring tools
also have Computer-Aided Software Engineering (CASE) capabilities enabling
educators and trainers to develop their own applications quickly. These advantages, considered alongside the improved quality of the information presentation,
offer important new teaching methods.

One system which has been widely used is the GISTutor (Raper and Green,
1992). The GISTutor was developed as a means of introducing the generic
principles of GIS and some examples of their implementation. The intended users
were those with a basic knowledge of geography and cartography who wished to
understand how a GIS worked and what it was capable of doing. GISTutor was
developed in Hypercard on the Apple Macintosh, and version 2 has been ported
to Toolbook running under Windows. The system is organised as a hierarchical
set of concepts illustrated with graphics and animation which can be turned into
a web by hypertext linking. The system contains geometric algorithms, spatial
search and network shortest path simulations implementing real algorithms. A
tutor for GIS developed in France using Toolbook on the PC is GéoCube (Bernard
and Miellet, 1993) which contains sections on GIS concepts, organisations, systems
and digital spatial data available in France.

Another system for computer-based learning developed in a hypermedia
environment is the Scolt Multimedia Project (SMP) (Raper et al., 1992). Users
access the system using an 'organising metaphor' based on a time/space diagram,
carry out assignments using the spatial analysis tools and can export the geometric
interpretations made in a standard form. The system has been extended to use
Quicktime to incorporate video sequences in the databases linked to appropriate
places and times.

35.9 Research agenda

The volume of research which can be seen in the technology and application
reviews above indicates the sheer effort going into the (spatial) multimedia field
at the moment. However, the work undertaken in the spatial multimedia field

indicates an excessive concentration on the creation of hypermedia spatial databases and the visualisation of complex spatial data sets. Spatial multimedia researchers have barely examined the potential of a number of other fields mentioned in the technology review. This imbalance in the research activity in spatial multimedia has motivated the structure of the research agenda set out below. It is hoped that new work will address some of these topics. In no particular order, the following areas seem to require new work.

- *Data sourcing, access and pricing.* Where do the data come from at a physical level (discs or networks) and how do users find out about it (information navigation tools or intelligent 'spatial' agents)? How would spatial data be indexed, described, priced and made available, and what would the spatial, temporal and thematic descriptions and limits of these data sets be? How can spatial data be located on the World-Wide Web using resource discovery methods?

- *Data integration.* How can spatial multimedia data be integrated, for example, videos of the same place at different times? How can the data be tagged with metadata or knowledge to inform integration? How can space and time be represented in multimedia data sets in a consistent way?

- *Information structuring.* How can spatial multimedia be structured (time-space diagrams? hypermaps?)? Can spatial multimedia data be stored in conventional (relational) databases or will object-oriented databases be necessary?

- *User interaction.* What new usability issues are raised by spatial multimedia? Are sound and video as 'understandable' as symbols in conventional cartography? What user controls will be required to manipulate spatial multimedia (for example, camera angles over terrains)? Can immersive VR systems solve these problems? What is the potential for spatial analysis of multimedia data types and simulation? What manipulations can be defined over spatial multimedia data sets (for example, spatiotemporal range queries)? Can new forms of algorithms such as genetic algorithms and pictorial simulation help? What kind of simulation can be envisaged with multimedia data?

35.10 Conclusions

From the evidence in this review it is clear that this field has an important future. It is important now that researchers and vendors alike address the fundamental issues of (spatial) structuring at the data-model level, especially in the light of the potential for a distributed, networked information domain. Perhaps the data issues are even more important: how will confidentiality and copyright be preserved in this new environment; and, if preserved will (can?) spatial multimedia develop? There are more questions than answers at present: in many ways the ideal research environment!

Acknowledgements

Thanks to members of the ESF GISDATA Programme Task Force on 'GIS and Multimedia', in particular to Ralf Bill for his extensive comments on an earlier draft.

References

ARNAUD, A., CRAGLIA, M., MASSER, I., SALGÊ, F. and SCHOLTEN, H. (1993). The research agenda of the European Science Foundation's GISDATA Scientific Programme, *International Journal of GIS*, **7**(5), 463–70.

BACAO, F. J., MARQUES, J. M. and SILVA, R. G. (1993). Flying over Lisbon: a multimedia presentation, in *Proceedings of Workshop on Multimedia and GIS*.

BAGIANA, F. (1993). Tomorrow's space: journey to the virtual worlds in European Space Agency, *Le Bourget Air Show Guide*.

BERNARD, M. and MIELLET, P. (1993). Using hypertext for GIS training: the Gécube concept, in Harp, J., Ottens, H. and Scholten, H. (Eds), *Proceedings of EGIS Conference, Genoa* (**1**), pp. 322–7, Utrecht: EGIS Foundation.

BILL, R. (1994). Multimedia GIS definition, requirements and applications, in Shand, P. and Ireland, P. (Eds), *The 1994 European GIS Yearbook*, pp. 151–4, Hastings Hilton Publishers/NCC Blackwell.

BLAT, J. D., RUIZ, A. M. and SEGUI, J. M. (1995). Designing multimedia GIS for territorial planning: the ParcBIT case, *Environment and Planning B*, **22**, 665–78.

BUTTENFIELD, B and WEBER, C. R. (1993). Visualisation and hypermedia in GIS, in Medyckyj-Scott, D and Hearnshaw, H. (Eds), *Human Factors in GIS*, pp. 136–47, London: Bellhaven.

CÂMARA, A. and NEVES, N. (1996). Exploring virtual ecosystems, *IEEE Multimedia* (forthcoming).

CÂMARA, A., FERREIRA, F. C., FIALHO, J. E. and NOBRE, E. (1991a). Pictorial simulation applied to water quality modelling, *Water Science and Technology*, **24** (6), 275–81.

CÂMARA, A., GOMES, A. L., FONSECA, A. and LUCENA e VALE, M. J. (1991b). Hypersnige – a navigation system for geographic information, *Proceedings of EGIS '92 Conference*, pp. 175–9, Brussels, Utrecht: EGIS Foundation.

CASSETTARI, S. and PARSONS, E. (1993). Sound as a data type in a spatial information system, in Harp, J., Ottens, H. and Scholten, H. (Eds), *Proceedings of EGIS Conference, Genoa* (**A**), 194–202.

COTTON, R. and OLIVER, R. (1993). *Understanding Hypermedia*, London, Phaidon.

CRAGLIA, M. (1994). GIS and multimedia: report of the GISDATA Specialist Meeting, Rostock, 25–29 May 1994, *GISDATA Newsletter* (**4**), 11–16.

DIXON, P., SMALLWOOD, J. and DIXON, M. (1993). Development of a mobile GIS: field capture using a pen-based notepad computer system, *Proceedings of AGI '95 Conference*, Birmingham, **16**, pp. 1–6, London: Association of Geographic Information.

DORLING, D. (1992). Stretching space and splicing time: from cartographic animation to interactive visualisation, *Cartography and Geographic Information Systems*, **19**(4), 215–27.

DOZIER, J., GOODCHILD, M., IBARRA, O., MITRA, S. and SMITH, T. (1994). *The Alexandria Project: Towards a Distributed Digital Library with Comprehensive Services for Images and Spatially Referenced Information*, Santa Barbara: National Center of Geographic Information and Analysis.

EDWARDS, T. M. (1992). Virtual worlds technology as a means for human interaction with spatial problems, *Proceedings of GIS/LIS '92*, pp. 208–20.

EGBERT, S. L. and SLOCUM, T. A. (1992). EXPLOREMAP: an exploration system for choropleth maps, *Annals of the Association of American Geographers*, **82**(2), 275–88.

ERTL, G., GLEIXNER, G. and RANZIGER, M. (1992). Move-X: a system for combining video films, computer animations and GIS data, *Proceedings of 15th Urban Data Management Symposium, Lyon*, pp. 247–54.

FISHER, P., DYKES, J. and WOOD, J. (1993). Map design and visualisation, *Cartographic Journal*, **30**, 136–42.

FONSECA, A., GOUVEIA, C., CAMARA, A. S. and SILVA, J. P. (1995). Environmental impact assessment using multimedia spatial information systems, *Environment and Planning B*, **22**, 637–48.

FONSECA, A., GOUVEIA, C., RAPER, J. F., FERREIRA, F. and CAMARA, A. S. (1993). Adding video and sound to GIS, *Proceedings of EGIS '93 Conference* (**1**), pp. 187–93.

GOMES, A.-L., PEREZ, A. T. D., FONSECA, A. and GOUVEIA, C. (1993). SOFTANIS – a multimedia spatial information system for natural areas, in *Proceedings of Workshop on Multimedia and GIS*.

GREEN, D. and KEMP, A. (1993). A GIS may be the answer to every pipeline manager's dream, *Proceedings of AGI '93 Conference*, **1**(26), pp. 1–10.

GREEN, D. and MORTON, D. C. (1994). Acquiring environmental remotely sensed data from model aircraft for input to GIS, *Proceedings of AGI '94 Conference*, Birmingham, **15**(3), 1–27.

HALL, W. (1993). Hypermedia tools for multimedia information management, in Earnshaw, R. and Vince, J. A. (Eds), *Proceedings of British Computer Society Multimedia Systems and Applications Conference*, pp. 101–14, London: Academic Press.

HASLETT, J., WILLIS, G. and UNWIN, A. (1990). SPIDER: an interactive statistical tool for the analysis of spatially distributed data, *International Journal of GIS*, **4**(3), 285–96.

HEARNSHAW, H. and UNWIN, D. U. (1994). *Visualisation in GIS*, Chichester: John Wiley.

HENRIQUES, R., REIS, R. P., JOAO, E., FONSECA, A. and GOUVEIA, C. (1992). From forest fire simulation to the assessment of environmental impacts, *Mapping Awareness*, **6**(5), 13–17.

HOPGOOD, F. R. A. (1993). Use of time and space in multimedia systems, in Mumford, A. M. (Ed.), 'Multimedia in Higher Education: portability and networking'. Advisory Committee on Computer Graphics Technical Report 24, Univ. of Loughborough, pp. 1–5.

JACOBSON, R. (1991). Virtual worlds, inside and out, in Mark, D. M. and Frank, A. U. (Eds), *Cognitive and Linguistic Aspects of Geographic Space*, Dordrecht: Kluwer.

JOFFE, B. and WRIGHT, W. (1989). SIMCITY: thematic mapping + city management simulation – an entertaining, interactive gaming tool, in *Proceedings of GIS/LIS '89*, Falls Church, Virginia: American Society for Photogrammetry and Remote Sensing, pp. 591–600.

KALAWSKY, R. S. (1993). *The Science of Virtual Reality and Virtual Environments*, Reading, MA: Addison-Wesley.

KEMP, Z. (1995). Multimedia and spatial information systems, *IEEE Multimedia* (Winter issue, in press).

KEMP, Z. and OXBORROW, E. (1992). An object model for distributed geographic data, *Proceedings of EGIS '92 Conference*, pp. 1294–303.

KOOP, O. and ORMELING, F. J. (1990). New horizons in thematic cartography in The Netherlands. The National Atlas Information System, *Proceedings of EGIS '90*, pp. 614–23.

KRAAK, M. (1989). Computer-assisted cartographical 3D imaging techniques, in Raper, J. F. (Ed.), *Three-dimensional Applications in GIS*, pp. 99–114, London: Taylor & Francis.

(1996). GIS: issues in visualisation through multimedia, *IEEE Multimedia*, (forthcoming).

LAURINI, R. and THOMPSON, D. (1992). *Fundamentals of Spatial Information Systems*, London: Academic Press.

LEWIS, S. and RHIND, D. W. (1991). Multimedia GIS, *Proceedings of Mapping Awareness Conference*, pp. 311–22.

LINSEY, T. K. and RAPER, J. F. (1993). A task-oriented hypertext GIS interface, *International Journal of Geographical Information Systems*, **7**(5), 435–52.

MacEACHREN, M. A. and TAYLOR, D. R. F. (1994). *Progress in Cartographic Visualisation*, Oxford: Pergamon Press.

McEWAN, S. (1993). Future developments in information publishing, *Proceedings of British Computer Society Multimedia Systems and Applications Conference*.

MEDYCKYJ-SCOTT, D. and HEARNSHAW, H. (Eds) (1993). *Human Factors in GIS*, London: Bellhaven.

MEDYCKYJ-SCOTT, D., DAVIES, C. and BYRNE, V. (1993). Discovering the trials of GIS use: the USIS study of GIS usability, *Proceedings of GIS Research UK*, 170–9.

MOGOROVITCH, P., MAGNARAPA, M., MASSEROTTI, M. and MAZZOTTA, S. (1992). Merging GIS with multimedia technologies, *Proceedings of EGIS '92 Conference*, pp. 1085–94.

OPENSHAW, S., WAUGH, D. and CROSS, A. (1994). Some ideas about the use of map animation as a spatial analysis tool, in Hearnshaw, H. and Unwin, D. U. (Eds), *Visualisation in GIS*, pp. 131–8, Chichester: John Wiley.

PARSONS, E. (1992). The development of a multimedia hypermap, *Proceedings of AGI '92*, **2**(24), pp. 1–3.

POLYDORIDES, N. (1993). An experiment in multimedia GIS, *Proceedings of EGIS '93 Conference*, Genoa, pp. 203–12.

POWER, S. and LAY, M. (1994). TITAN CD: new horizons in GIS for electronic publication of travel information, *Proceedings of AGI '94 Conference*, **18**(2), pp. 1–8.

RAPER, J. F. (1995). Multimedia GIS from concepts to (virtual) reality, in Shand, P. and Ireland, P. (Eds), *The 1995 European GIS Yearbook*, pp. 94–7, Hastings: Hilton Publishers/NCC Blackwell.

RAPER, J. F. and BUNDOCK, M. S. (1993). Development of a generic spatial language interface for GIS, in Mather, P. M. (Ed.), *Geographical Information Handling – Research and Applications*, pp. 113–43, Chichester: John Wiley.

RAPER, J. F. and GREEN, N. P. A. (1992). Teaching the principles of GIS: lessons from the GISTutor project, *International Journal of GIS*, **6**(4), 279–90.

RAPER, J. F. and LIVINGSTONE, D. (1995). The development of a spatial data explorer within an environmental hyperdocument, *Environment and Planning B*, **22**, 679–87.

RAPER, J. F. and McCARTHY, T. (1994a). Using airborne videography to assess coastal evolution and hazards, *Proceedings of EGIS '94 Conference*, pp. 1224–8.

(1994b). Virtually GIS: the new media arrive, *Proceedings of AGI '94 Conference*, **18**(1), pp. 1–6.

RAPER, J. F., CONNOLLY, T. and LIVINGSTONE, D. (1992). Embedding spatial analysis in multimedia courseware, *Proceedings of EGIS '92 Conference*, (**1**), pp. 1232–7.

RAPER, J. F., McCARTHY, T. and LIVINGSTONE, D. (1993). Interfacing GIS with virtual reality technology, *Proceedings of AGI '93 Conference*, **3**(25), pp. 1–4.

RAVENEAU, J., MILLER, M., BROUSEAU, Y. and DUFOUR, C. (1991). Micro-atlases and the diffusion of geographic information: an experiment with Hypercard, in Taylor, D. R. F. (Ed.), *GIS: The Microcomputer and Modern Cartography*, pp. 201–23, Oxford: Pergamon Press.

RHIND, D. W., ARMSTRONG, P. and OPENSHAW, S. (1988). The Domesday machine: a nationwide GIS, *Geographical Journal*, **154**, 56–68.

RODRIGUES, M. A. S., RAPER, J. F. and CAPITCÃO, M. (1995). Implementing intelligent agents for spatial information, *Proceedings of Joint European Conference on GIS*, **1**, pp. 169–74, Bazel: AKM AG.

SCHWARZ VON RAUMER, H.-G. and KICKNER, S. (1994). *Konzeption und Entwicklung eines Geographischen Informations- und Plannungssystems für Regional- and Fälchennutzungsplanung*, Salzburg: Salzburger Geographische Materialen.

SHIFFER, M. (1993). Implementing multimedia collaborative planning techniques, *Proceedings of the Urban and Regional Information Systems Association Conference*, pp. 86–97.

SHIFFER, M. (1995a). Interactive multimedia planning support: moving from stand-alone systems to the World Wide Web, *Environment and Planning B*, **22**, 649–64.

(1995b). Environmental review with hypermedia systems, *Environment and Planning B*, **22**, 359–72.

SIMMONS, D. (1993). Multimedia SPANS MAP, *Second TYDAC European Users Conference*, Southampton, October 1993.

SIMMONS, D., HALL, W. and CLARK, M. (1992). Integrating GIS with multimedia for site management, *Proceedings of Mapping Awareness '92*, pp. 303–18.

SMITH, R. G. (1994). Current military simulations and the integration of virtual reality technologies, *Virtual Reality World*. March/April 1994.

SMITH, T. R. and FREW, J. (1995). Alexandria Digital Library, *Communications of the ACM*, **38**(4), 61–2.

STATE, R. (1993). A practical application of data capture techniques, *Proceedings of AGI '93 Conference*, **1**(14), pp. 1–3.

STEMBRIDGE, J. (1994). European roads: SCOPE for improvement? *GIS Europe*, pp. 28–30.

UNWIN, D., STYNES, K., DYKES, J., FISHER, P. and WOOD, J. (1994). WYSIWYG: visualisation in the spatial sciences, *Proceedings of the Association of GIS Conference*, Birmingham, November 1994.

WEBER, C. R. and BUTTENFIELD, B. P. (1993). A cartographic animation for average yearly annual temperatures for the 48 contiguous United States 1897–1986, *Cartography and GIS*, **20**, 141–50.

WODASKI, R. (1994). *Multimedia Madness*, Indianapolis: SAMS.

WOOD, T. (1994). Using GIS to develop the European market for geographic information: the Omega CD, *Proceedings of AGI '94 Conference*, **16**(1), pp. 1–9.

YEORGAROUDAKIS, Y. (1991). Visual representation of spatially referenced multimedia objects, *Proceedings of EGIS '92 Conference*, pp. 1250–60.

CHAPTER THIRTY-SIX

Towards Hypermedia-aided GIS in Local Planning

LARS BODUM

36.1 Introduction

Despite the potential of GIS for planning, there is substantial evidence that the diffusion of this technology in planning departments is a very slow process and that planners have not yet accepted GIS as one of the main tools in the planning process (Campbell, 1994; Kiib, 1996; Masser and Onsrud, 1993). Surveys conducted in the period 1992–4 have confirmed this (Budić, 1993; Craglia, 1992; Garcia and Perryman, 1992; Gordon Jr and Soubra, 1992; Masser *et al.*, 1996). Denmark is no exception to this trend. Although many municipalities in Denmark are aware of the possibilities and have used computers for many years, only a few of them are ready to use GIS for planning purposes. There can be many different explanations of this phenomenon, but the two most likely ones are that the possibilities offered by GIS technology are either too limited or too many. Maybe both of these explanations are true. The only way to find an answer is to get more knowledge about the way local planning is performed and how computers are used in planning offices today.

36.2 Local planning in Denmark

Denmark has one of the most decentralised systems of public administration in Europe. Local authorities administer more than 50 per cent of the public expenditure and each local and regional authority elects its own local and regional councillors to administer the revenues from local taxation levies. More and more fields of responsibility have been transferred from the central government to local authorities. The aim is that tasks should be devolved at the lowest possible level, and that there should be a coincidence between the level of responsibility for decision making and the level of responsibility for the financial consequences.

Each municipality has an obligation to prepare and maintain a structural plan and a set of local planning regulations, covering its whole area. Furthermore, the local authority has the right, and in some cases the obligation, to make more detailed local plans for parts of its area. In many cases, these local plans are made by the developers, followed by a public debate and the approval by the local authority. Local plans must be prepared and adopted before important or large developments can be carried out. This includes demolition of buildings, developments with more than 20 dwellings, schools, town halls, tourist hotels and so forth. The local authority may also prepare local and neighbourhood plans, when they wish to issue detailed planning regulations (see Fig. 36.1.)

The local plan may contain provisions about land use within the area, the size and delimitation of the contained properties, roads and paths, the alignment of tracks, pipes and transmission lines, the location of buildings and the size and external appearance of buildings. It can also contain provisions about the use of individual buildings, and require that existing buildings or vegetation must be retained. Finally, it can determine if areas should be transferred to urban, recreational residences or rural zones. The local plans are permanent and may only be changed by the adoption of another local plan that supersedes the earlier plan.

36.3 Geographical information technology in planning agencies

When a municipality decides to invest in new geographical information technologies, it is often a decision made without due consideration for the ability of the organisation to utilise the system (Campbell, 1994; Holmberg, 1994). Of course planners have been asked what kind of functionality they want from the system and they have probably had some concern about this new 'thing' and how it will affect their work. There is, however, no indication that planners have been integrated further into the implementation of this technology (see also Budić and Godschalk, 1994, for a recent American example).

Local planning is a very creative process, and one of the things we have learned about the computer is that it is a very rational tool (Dahlbom and Mathiassen, 1993; Dreyfus and Dreyfus, 1986; Winograd and Flores, 1986), so for the planners of today it is a question of how to combine the creativity of the process with the rationality of the tool. That is not an easy concept, when you think of implementing traditional GIS into local planning (Ottens, 1990).

Another problem with the use of standard GIS applications in local planning is the cultural and legal differences between the countries where the GIS is developed and the countries where it is used (Craglia, 1994). These conflicts have to do with traditions in working methods (for example, what type of spatial analysis is being done?), language problems (for example, most software is based on English as its language), geodetic differences (for example, map projection) and differences in the legal demands for visualisation of the planning documents. It is true that many of these problems are technically solved by some local developers, but I will still maintain that, for some countries, these problems can be very demanding to overcome.

It need not be so. In the recent decade, many system developers have shown that if we stop thinking of computers as machines with all the answers and look at them as comprehensive tools or media for number-crunching, data management, integrating, analysing, storing, and visualising graphic as well as non-graphic information, we will get a better understanding of what computers are and what they *can* do. This will eventually lead to better and more natural ways of

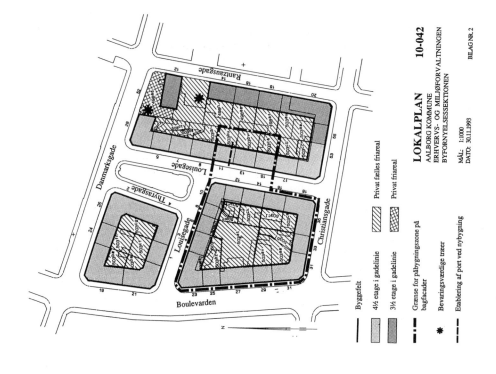

LOKALPLAN 10-042

AALBORG KOMMUNE
ERHVERVS- OG MILJØFORVALTNINGEN
BYFORNYELSESSEKTIONEN

MÅL: 1:1000
DATO: 30.11.1993   BILAG NR. 2

Byggefelt

4½ etage i gadelinie

3½ etage i gadelinie

Privat fælles friareal

Privat friareal

Grænse for påbygningszone på bagfacader

Bevaringsværdige træer

Etablering af port ved nybygning

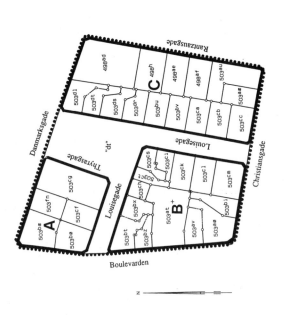

LOKALPLAN 10-042

AALBORG KOMMUNE
ERHVERVS- OG MILJØFORVALTNINGEN
BYFORNYELSESSEKTIONEN

MÅL: 1:1000
DATO: 30.11.1993   BILAG NR. 1

Lokalplangrense

Områdegrense

A B C   Delområder

Figure 36.1 Examples of Danish local plans.

communicating with computers (Dreyfus and Dreyfus, 1986; Winograd and Flores, 1986).

In the context of GIS and planning, planners need not only a tool for making analysis and presenting it on a static map. They also need the possibility of mixing other types of media and making other kinds of visualisations that involve spatial data.

36.4 New ideas

The working methods for making local plans in Denmark are guided by the Ministry of Planning in a very detailed way. The majority of planners have followed this guide for many years now without realising that a local plan can be different and more informative than what is presented. Although the last edition of the guide is from 1989 (Ministry of Planning, 1989), it ignores the fact that the term 'a document' can be something else than a pile of paper. As the integration of digital working methods continues, it is time to realise the change towards the digital document. There is no doubt that in a few years many municipalities will produce digital local plans, and these municipalities will have some excellent possibilities to use other digital working methods. The guide does not mention these possibilities. There is a need for an investigation of the possibilities in a situation where more and more information is only (or also) accessible in a digital format.

The main idea is that a standalone GIS application will not be able to fulfil these needs. What the planner really needs is a better handling of the digital communication process of the planning, and that is *not* one of the things GIS applications can handle very convincingly today. The aim of this study is to develop new interactive methods for the planner to evaluate and visualise projects (small or large) in the process of preparing the planning documents.

36.4.1 GIS and hypermedia

The future will involve the integration of technologies like GIS and interactive hypermedia. In contrast to the larger GIS and CAD applications, where the user interface and commands are pre-programmed, it should be the planner and the actual working process that decide how an application for design and decision support in a specific planning situation should look. It should be possible for the planner to customise the screen, to illustrate and visualise the changes in the environment and to set up the regulations for the actual plan in a dynamic way. Inside the framework given by the regulations and the means of communication, the planner should work in an interactive way with the geographical data. Interactivity is the essence of hypermedia provided by means of windows, icons, mouse and pointer interface. The hypermedia authoring tools give planners the possibility of making links between different aspects and views of the local plan. It allows the planner to experience the sight and sounds of their data, creating a 'virtual land base' of information (Kiib and Veirum, 1993; Nielsen, 1993; Shiffer, 1995).

With these tools it is possible to create scenes where different planning alternatives can be presented. That includes the import and editing of text, graphics, digital video, pictures and other data types. With the digital map as a central object of this planning process, it is natural to use the geographical reference as a way of organising this information. Every piece of information will be stored digitally and mapped (or linked to other objects) with a reference to the actual case. Eventually this type of digital 'map' will change into a hypermap. This hybrid between an ordinary digital map and a hypermedia document could still be connected with a GIS. The hypermedia-aided GIS enables the planner not only to collect, analyse and present attribute data as overlays in the digital map but also to act as an author of hypermedia documents. The author can link various quantitative and qualitative data to objects in the map, but he or she can also link the same information to various spots in the pictures or to words and graphics in the hypertext documents. This kind of design system requires that the term GIS is defined in a very special way. It is not enough to see GIS as a computer application or a set of tools alone. We must consider a definition that treats GIS as a medium (or integration of several media) (Shiffer, 1995).

36.5 Methods and results

The goal of the research described here is to outline and illustrate a possible working situation for the local planner in the future and facilitate the communication process of local planning in Denmark with the use of hypermedia and GIS.

The means used for reaching this goal are divided into two parts:

- a thorough analysis of the local planning process; and
- prototypes of hypermedia applications that will help the planners in a real-work situation and in their process of communicating the local planning.

36.5.1 Analysing the local planning process

The first part of the research included a survey among local planners in Denmark. This survey was mailed out to all 275 municipalities in Denmark in the autumn of 1993. With a response rate as high as 70 per cent it gave a very good impression of how and where planners used information technology in connection with the local planning process. Some of the areas covered in the survey were:

- Who are the local planners?
- How many plans do they make?
- What kind of computer tools do they use?
- Which information do they use in the process?
- How do they get the data?
- How do they visualise the local plan?
- How is the local plan published?
- How is the local plan presented?
- How is the local plan used in the administration afterwards?

After the collection and registration of all responses, a very intensive analysis of all the answers was performed. The main conclusion from the analysis of the survey has been that nearly all planners have access to computers but they are mainly used for word-processing. The amount of local plans is approximately 1000 plans per year for the 275 municipalities. That is an average of four local plans every year, but there is a big variation in this result. The majority of local planners have an architectural education (52 per cent) and only approximately one out of three (29 per cent) is a woman. The typical local planner in a municipality with less than 10 000 inhabitants is a male engineer, and the typical local planner in a municipality with more than 10 000 inhabitants is a male architect.

More than 60 per cent of the planning agencies have personal computers of some kind, and 23 per cent of them have workstations. A large part of the municipalities (98 per cent) still have text-based terminals for use with centralised mini-computers or mainframes. They play an important role in the planning process today, because many of the spatial referenced data used in the local planning are stored on these computers. If GIS were to be integrated into the planning process, a lot of these data could be made available in a readable and editable form. It would be very demanding to customise these data to a readable format. Although efforts are being made at governmental level for reconstructing these databases, this is a major obstacle for the implementation of GIS in local government.

Almost all agencies had a word processor of some kind and 50 per cent of them had spreadsheets. After these common software types, there is a major jump to Automated Mapping (AM), database and graphic tools. Only 10 per cent of the planning agencies say that they have some kind of GIS software today. To show just how few visual elements the local plans include today, one of the results of the survey was that the majority of planners do not use photography as a visualisation method. Other results from the survey showed the same pattern. Many planners used sketches and drawings in the planning process, but 90 per cent of them were made by hand. These results show that computers are not used much for illustration and visualisation purposes in the local planning.

Another result of the survey was the lack of analytical work accompanying the local planning. Very few planners made any kind of analysis in connection with the local plan and the ones that were performed were simple and did not require any type of GIS whatever. Most analyses were made in spreadsheets or with help from an electronic calculator. GIS would not help local planners in these situations.

From the results and from earlier research we found that:

- Planners spend much time producing reports and illustrations.
- Presentations and meetings are important in the daily work.
- The planner should be able to work with maps, texts, databases, graphics, photos and other kinds of media in an interactive way.
- The amount of geographical analysis in the local planning is overestimated.
- Administration of the regulations is an important task for the planner.
- The human–computer interface must be intuitive and easy to understand.

36.5.2 Prototyping

The second part of the research focused on the development of two hypermedia-aided prototypes that show how the relation between hypermedia and GIS could turn out in a real-life situation. With respect to the findings of both the survey and earlier research at Aalborg University, it was decided to sketch a model of the local planning situation as it is performed in Danish municipalities. This model is built around two different (although connected) interfaces, which represent the daily routines of the local planning process.

- TownPlanner
- Electronic local plan.

The first interface is the planner's desktop, and the second is a new electronic version of the local plan. The planner's desktop is a metaphor from the real-work situation of the planner, and the idea is to integrate the different kinds of information on the same desktop. The idea about the electronic local plan is to create a hypermedia-document that can be distributed broadly (Fig. 36.2).

36.5.3 TownPlanner: the planner's hypermedia-aided GIS desktop

The design of TownPlanner is developed with a work-oriented approach (Ehn, 1988). The desk is the working space of the planner where he or she prepares the local plan. To perform this task the planner needs various kinds of formal and informal information such as strategic plans, development plans, maps, statistics, cadastral data (maps and text), pictures, budget information and building information. Usually this is paper-based information stacked on the desk or side desks. The planner can easily get an overview of the different kinds of information looking at the label of each stack, but to get access to the contents he/she has to drag the stack to his desk, browse the contents and pick up the information needed for this local plan or permission.

From this basic perception of the working environment we created a small 'desktop control palette' with all the basic stacks of information represented as an icon. If you drag the icon of the map into the 'active area' of the desktop palette, the digital map of the local plan area will appear in a window at the screen. You may do the same with all the stacks and the screen would be totally covered with different windows, each providing a view of the 'real world'. If you activate the window of the cadastral information you will have the information displayed in a table. But you can also have the cadastral information displayed in the map window as an overlay, where each parcel is represented as an object (Fig. 36.3).

The planner looks at the 'real world' through a variety of views and windows. Each view gives the planner a certain description of the present situation, the future or the past. As in everyday planning situations, the system has incorporated the planner's links between these views. From one window, for instance, the cadastral, he or she may ask for more detailed information about the size of each parcel or who owns selected parcels, but you can also ask for supplementary information from different registers and have this displayed in tables as well as in the map. The cross-reference between the different kind of information is a key

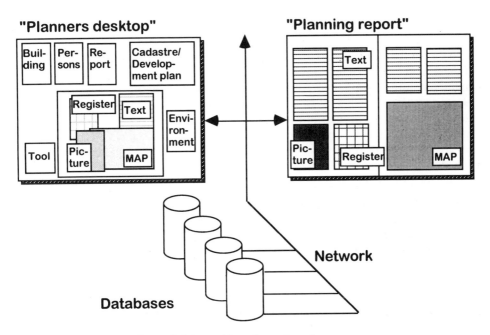

Figure 36.2 Concept for a division of interfaces for planners.

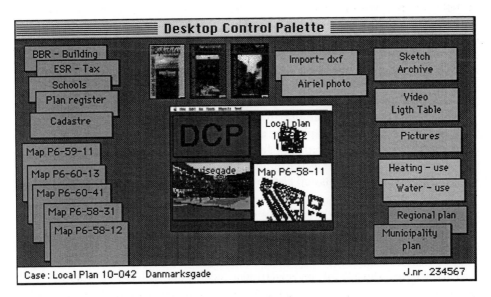

Figure 36.3 TownPlanner: the desktop control palette.

point in this implementation of these associated links. The planner can use the query facilities in the system to access the data as well as build his or her own links between different data (Fig. 36.4).

An important part of the TownPlanner is the ability to integrate pictures and video in connection with other traditional datatypes. Pictures are usually stored on shelves and put on a light-table when needed. This 'light-table' metaphor is also implemented in TownPlanner. The 'stack of pictures' icon on the desktop control palette represents a window with the video light-table. This is a tool that enables the user to browse through indexed lists of pictures and video sequences in the database having them displayed in a small window at the light table. The user is able to select the pictures he or she wants to link to his or her planning work and have them displayed in real size.

Each picture or video frame carries descriptions of its *content, context, light, colour, sound (if available), time code, address and coordinates*. This index enables the user to make queries about certain objects in a specific location and as a result a list is produced. All the pictures and video frames are geocoded. This feature gives the possibility of making a selection in the map window by making a selection polygon and after that specifying other criteria in the query line. Thus the result of the query not only produces a list of selected pictures but also gives dots on the map.

It is evident that one of the strong sides of TownPlanner is the possibility of storing and visualising data dynamically, such as navigational digital video and other time-dependent information. Scenarios that are difficult (or in most cases impossible) to create and show in normal paper-based local plans are instead possible in TownPlanner. It is important to continue research on the integration of digital video and GIS.

36.5.4 Electronic local plan: a hypermedia-aided document

A new and important alternative to the paper-based local plan is the concept of an electronic local plan. Eventually the idea is to integrate the flexibility of an electronic plan with the traditional local plan. This gives the municipality an opportunity to present the plan in a new, dynamic and customised way. The electronic local plan will become a hypertext/hypermedia document with all the possibilities this gives, such as the ability to dig into the archives and look at some of the background materials and also to view some of the many thematic maps that are produced.

It will be possible to comment on the plan on-line and in special cases (where data are available) to create one's own personal version of the local plan. The hope is that this new and more dynamic type of local plan could increase the participation in the local planning process.

This whole concept of an electronic local plan will be developed with considerations to the fact that a local plan is a legal document. That is why the electronic local plan can only be a supplement to the legally binding paper-based version. The prototype for this reporting tool will be developed as an HTML document and the media for displaying and interacting with this document will be one of the World-Wide Web browsers.

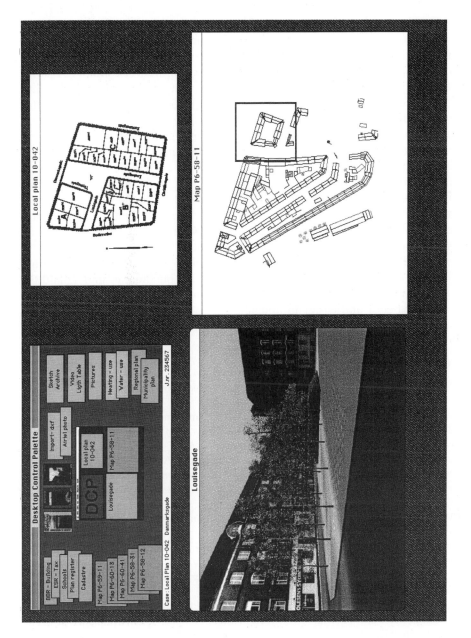

Figure 36.4 Views from TownPlanner.

36.6 Discussion and conclusion

The introduction of computing in planning agencies has mostly been conducted without considering the daily work situation of the planners. Although we have a tradition in Scandinavia for user-centred design projects (Ehn, 1988), there has (to my knowledge) never been any serious investigation of the planning process for the purpose of designing GIS-related applications. To succeed with the diffusion of GIS in the planning office, it is necessary to consider a new way of system development. This development should focus on the work situation and the planners' daily routines. It sounds like a banality, but local planners have some very basic needs when it comes to geographical information management that cannot be fulfilled by existing GIS technology despite its many functions.

The world of GIS is changing rapidly and so is the organisation of work and data connected to local planning in Denmark. New digital maps and attribute data will be available to the planner, and computers will become even more indispensable in the office. The demands within the organisation and from third parties will be higher. This means that for certain tasks new tools will be needed.

With these considerations in mind, there is a need for investigating opportunities and constraints of working environments where more and more information is (or also) only accessible in a digital format. This could be the chance to review the old working methods, and maybe give planners new inspiration. TownPlanner and the electronic local plan could help planners to understand the value of GIS and at the same time make the communication of documents like the local plan better and easier to update.

References

BUDIĆ, Z. D. (1993). GIS use among South-eastern Local Governments, *URISA Journal*, **5**, 4–17.

BUDIĆ, Z. D. and GODSCHALK, D. R. (1994). Implementation and management effectiveness in adoption of GIS technology in local governments, *Computers, Environment and Urban Systems*, **18**, 285–304.

CAMPBELL, H. (1994). How effective are GIS in Practice? A case study of British local government, *International Journal of Geographical Information Systems*, **8**, 309–25.

CRAGLIA, M. (1992). GIS in Italian urban planning, *Computers, Environment and Urban Systems*, **16**, 543–56.

—— (1994). Geographical Information Systems in Italian municipalities, *Computers Environment and Urban Systems*, **18**, 381–475.

DAHLBOM, B. and MATHIASSEN, L. (1993). *Computers in Context*, Oxford: Blackwell.

DREYFUS, H. L. and DREYFUS, S. E. (1986). *Mind over Machine*, Oxford: Blackwell.

EHN, P. (1988). *Work-Oriented Design of Computer Artefacts*, Stockholm: Arbetslivscentrum.

GARCIA, M. W. and PERRYMAN, M. (1992). Survey of the use of computers in planning by local governments of Arizona and New Mexico, *Computers, Environment and Urban Systems*, **16**, 517–29.

GORDON Jr, W. R. and SOUBRA, N. M. (1992). Geographical information systems and planning in the USA: Selected municipal adoption trends and educational concerns, *International Journal of Geographical Information Systems*, **6**, 267–78.

HOLMBERG, S. C. (1994). Geoinformatics for urban and regional planning, *Environment and Planning B*, **21**, 15–19.

KIIB, H. (1993). 'GIS technology in Danish local government', Skriftserien 119, Aalborg: Department of Development and Planning, Aalborg University.

—— (1996). Denmark: Local autonomy and register based information systems, in Masser, I., Campbell, H. and Craglia, M. (Eds), *Diffusion and Use of Geographical Information Technologies*, pp., Dordrecht: Kluwer.

KIIB, H. and VEIRUM, N. E. (1993). 'Hypermaps in urban planning', Skriftserien 109, Aalborg: Department of Development and Planning, Aalborg University.

LAUREL, B. (1993). *Computers as Theatre*, Reading MA: Addison-Wesley.

MASSER, I. and ONSRUD, H. J. (Eds) (1993). *Diffusion and Use of Geographical Information Technologies*, Dordrecht: Kluwer.

MASSER, I., CAMPBELL, H. and CRAGLIA, M. (1996). *GIS Diffusion: the Adoption and Use of GIS in Local Government in Europe*, London: Taylor & Francis.

MINISTRY OF PLANNING (1989). 'Lokalplanvejledning 2', Planstyrelsen, MiljØ-ministeriet.

NELSON, T. H. (1987). *Computer Lib/Dream Machines*, Redmond: Microsoft Press.

NIELSEN, J. (1993). *Hypertext and Hypermedia*, London: Academic Press.

OTTENS, H. F. L. (1990). 'The application of geographical information systems in urban and regional planning', in Scholten, H. J. and Stillwell, J. C. H. (Eds), *Geographical Information Systems for Urban and Regional Planning*, pp. 15–22, Dordrecht: Kluwer Academic Publishers.

SHIFFER, M. J. (1995). Interactive multimedia planning support: moving from stand-alone to the World-Wide Web, *Environment and Planning B*, **22**, 649–64.

SCHOLTEN, H. J. and STILLWELL, J. C. H. (Eds) (1990). *Geographical Information Systems for Urban and Regional Planning*, The GeoJournal Library, Dordrecht: Kluwer Academic Publishers.

WINOGRAD, T. and FLORES, F. (1986). *Understanding Computers and Cognition – A New Foundation for Design*, Reading MA: Addison-Wesley.

CHAPTER THIRTY-SEVEN

The Use of Multimedia Spatial Data Handling in Environmental Impact Assessment

ALEXANDRA FONSECA and ANTONIO CÂMARA

37.1 Introduction

Environmental Impact Assessment (EIA) 'is a systematic process that examines the environmental consequences of development actions in advance' (Glasson *et al.*, 1994, p. 3). It refers, according to Munn (1979), to the need to identify and predict the impact on the environment and on man's health and well-being, of legislative proposals, policies, programmes, projects and operational procedures, and to interpret and communicate information about the impacts. The assessment of environmental impacts is thus a multidimensional process, comprising several components, variables, parameters, and spatio-temporal scales. In essence, EIA is a process in which the assessment of the impacts of developments on the environment is made in a systematic, holistic and multidisciplinary way. On the other hand, this level of complexity has to be presented to the public in a simple and clear way. To achieve this objective, analytical and representational tools can be applied to the development of EIA studies. Tools such as environmental simulation models allowing the analysis and comprehension of the environmental processes, and the impacts evaluation of different scenarios, are widely used. These impacts often have a strong visual and audible component, and a spatio-temporal dependence. These features are not easily handled by the tools that are usually applied within EIA.

The potential of Spatial Information Systems (SIS) is being increasingly realised for use in Environmental Impact Assessment (Treweck and Veitch, 1992), although their full potential has not yet been fully explored. Often, when SIS are applied for environmental impact assessment, it is merely for the production and presentation of maps, despite the fact that the greatest value of SIS lies in their ability to analyse spatial data successfully (João, 1994).

Spatial analysis capabilities of these systems are not completely explored within the EIA process, because there is a lack of sufficiently developed modelling functionalities within SIS capable of dealing with environmental complexity. SIS

556

links to other software packages are also complicated and time consuming. Finally, large amounts of data are required for spatial environmental analysis and these data are usually not available at least in digital format.

Exploring multimedia capabilities with spatial data-handling functionalities might contribute to the development of systems where these problems can be partly overcome. These multimedia systems might improve the capabilities of visualisation and manipulation of the information. The spatial manipulation of videos, images and sounds would facilitate the analysis of environmental phenomena and give the possibility of using other environmental modelling approaches such as pictorial simulation and cellular automata modelling procedures.

The access to new data sources previously neglected (images, videos and sounds) that can be cheaper and easier to obtain and convert to digital format, may also contribute to bring a wider availability of information to use in EIA.

The incorporation and manipulation of these new data types with spatial data-handling functionalities, may indeed facilitate the consideration of the time dimension. At the same time, they adjust strongly to the visual and audible nature of most of the environmental impacts. The multimedia interfaces, when considering the design topics such as the ones used in video games, will allow a higher degree of interactivity (Crawford, 1990), which can be a highly significant achievement given the strong public participation component of the EIA process.

The current developments on network information-retrieval initiatives such as the World-Wide Web on the Internet have to be considered in this context. They will eliminate the need to integrate all these functions in a single system. The World-Wide Web provides a potentially much higher accessibility to the system and ease of updating the information within the EIA process. The developments associated with the World-Wide Web will also facilitate the access to different software, through the use of a common interface.

This chapter reviews the use of multimedia technologies in association with spatial information systems and discusses the concepts underlying the creation of a multimedia SIS. For this purpose, multimedia spatial information systems are first analysed in terms of data sources, information structuring and interface design. Then, the development of spatial data-handling multimedia functions for EIA is discussed, considering the possibilities arising from the manipulation of images, video and sound. The current developments on the World-Wide Web are analysed, considering that it can have relevant impacts on the creation of such a system. Finally, applications under development within this research project are described, giving the background to the development of a multimedia spatial information system for the EIA of the Expo'98 in Lisbon.

37.2 Multimedia and spatial information systems

Several multimedia applications have been developed during recent years, that include a spatial component (see, for example, Buford, 1994b; Raper, Chapter 35 in this volume). Some of these developments can be referred in this context such as the creation of hypermedia spatial databases, relying on the use of hypermedia-authoring tools such as Hypercard or Toolbook. In these systems a wide range of spatially referenced data can be accessed by spatial criteria. These are used, for example, in tourist information databases that explore the visual capabilities of multimedia technologies in association with the spatial information, to present the information to the user in an attractive way.

The concept of linking videos with maps was pioneered in the BBC Domesday project (Rhind *et al.*, 1988), a project that contributed to the spread of information in the UK, breaking the barrier between traditional geographic data and other types of data. Other applications of hypermedia spatial databases have planning purposes. Examples of this include Hypersnige (Câmara *et al.*, 1991) for the exploration of regional planning information, and 'Great Cities of Europe' (Polydorides, 1993), containing georeferenced planning information on a number of European Cities organised by issue, city and time.

Most of the work in the development of multimedia SIS has improved upon the concept of Domesday, using more modern technology (Câmara and Ferreira, 1993; Fonseca *et al.*, 1993; Scholten and Bijtelaar, 1993).

One of the most significant of this kind of development is the Collaborative Planning System (CPS), undertaken at MIT (Shiffer, 1993, 1995), designed to evaluate planning scenarios for Washington, DC, allowing navigation through a range of multimedia data, such as maps, aerial photographs, ground images, video clips and sounds. The system was designed to facilitate public participation and has considerable interactive capabilities. Other systems of this kind have been developed, such as those in the water-resources domain (Ashton and Simmons, 1993; Câmara and Ferreira, 1993), in geomorphological studies (Raper and Livingstone, 1995); in the development databases for retail analysis (Scholten and Bijtelaar, 1993), in the development of hypermedia systems for local planning (Bodum, Chapter 36 in this volume) and in the environmental impact of sound (Cassetari and Parsons, 1993).

The exploration of multimedia spatial data-handling capabilities within the Environmental Impact Assessment process is something that has not been done until now, to the best of the author's knowledge.

37.3 Environmental impact assessment and spatial multimedia tools

The central idea of the proposed multimedia integration with SIS is based on the assumption that the use of SIS for EIA can be significantly enhanced through the access to multiple types and sources of environmental data, and through the availability of new interactive visualisation and manipulation capabilities (Fonseca *et al.*, 1994). The research demonstrates that the use of images, video and sounds enhances the analytical capabilities of environmental spatial systems and the public participation in the EIA process.

This assumption is directly associated with the intrinsic nature of the EIA process. This process involves several steps ranging from the screening of projects and the scoping of alternatives and variables, to the analysis of impacts and its presentation to the public. EIA deals with large amounts of data, and comprises several variables and phenomena with complex inter-relationships varying in space and time. The impact results have also to be communicated to the public in a succinct and non-technical way. The transformation of a large number of variables describing the identified environmental effects into a reduced number of impacts, that have to be presented to the public, is a crucial problem. This requires the development of new mechanisms for presenting the information, allowing an effective integration of the different types of data and their dynamic and interactive exploration and manipulation.

The integration of multimedia capabilities with spatial data-handling functionalities may create an innovative framework for the development of new multimedia SIS functions, based on the availability of interactive data visualisation and manipulation capabilities and better interaction mechanisms. As a result it may help to deal with the crucial problem of transforming the effects in a small number of impacts. For example, impacts on water quality from a certain development might be much more easily transmitted to the public with images or videos showing the changes in water uses, than through the presentation of numerical tables or graphs of the water quality data and standards.

Multimedia Spatial Information Systems are analysed in the next sections in terms of data sources, information structuring and interface design. The development of spatial data-handling multimedia functions for EIA is analysed and the impact of the World-Wide Web is then discussed.

37.3.1 Multimedia spatial information systems

For the development of multimedia spatial information systems, three main topics are considered in this research: the data sources used, the access to the different data according to how the information is structured, and the user interface.

37.3.1.1 *Data sources*

Multimedia SIS integrate different types of data which include: maps, alphanumeric information, aerial photographs, text and graphics, images, video and sound. Some of these data types are traditionally used within SIS. Issues concerning the use of digital images, video and sound require further consideration.

Concerning the images, the resolution is a crucial issue as it defines the raster size (Fox, 1991). With video the most important issue is the enormous storage needed for digital requirements (Buford, 1994a). Associated with the necessity of compressing video data, low-level compression standards for data streams and hardware processing were developed (Fox, 1991).

Sound has a number of perceptual properties that correspond to its physical properties, being thus a medium for presenting data that has largely been neglected to date (Smith *et al.*, 1990). According to Cassetari and Parsons (1993) the use of sound in spatial information systems can be as part of the user interface or as a data item that may be manipulated as an integral part of the spatial database, either as an attribute to a spatial object or as a spatial object itself. Some of the principal elements which have relevance to the integration of sound into a spatial data system are: frequency and wavelength, intensity, spatial extent and duration.

37.3.1.2 *Information structure*

A non-linear structure is associated with a multimedia SIS, which provides a framework for accessing the non-standard data types. This non-linear structure comes from the typical representation of a multimedia document as a graph where each node contains information (text, images, video or sound). The nodes are connected by links, which help the user to move from one node to another. This is the classical structure associated with hypertext (Nielsen, 1990). Associated with the information structuring for multimedia applications, higher level standards for network and software operation were developed which include the MHEG and the Hytime standards (Buford, 1994a; Luther, 1994).

37.3.1.3 Interface design

The interface design of such a system should consider several topics: the functional requirements of the software according to the tasks to be performed; the model's adaptation to the user's cognitive representations; and the definition of the types of dialogue with the user.

Perceptual problems which are important in a multimedia environment are related to the resolution, colour and sound and to the response time. Some of these topics have been considered for video games (Crawford, 1990), a field where user-interface issues have been carefully taken into account.

Interactivity is another determinant characteristic of the user interface. Multimedia models have to be developed to support a high degree of interactivity. Interactivity may be measured by its frequency, range and significance (Laurel, 1990).

Based on the digital manipulation capabilities brought by multimedia technologies and considering the issues presented above, spatial data-handling multimedia functions for EIA are proposed and analysed in the next section.

37.3.2 Spatial data-handling multimedia functions for EIA

The field of interactive technologies has been rapidly growing, as previously presented, but little has been written about the conceptual aspects associated with the development of spatial data-handling multimedia functions. This has been one of the goals of this research.

To fulfil this objective, the benefits of adding video and sound to SIS were analysed in detail (Fonseca et al., 1993). According to this study, video and sound can be used within SIS across a wide range of different contexts ranging from the simple illustration and visualisation capabilities, to supporting the implementation of models within SIS and help in the evaluation of some spatial analysis results. According to these ideas, new functions for a multimedia SIS were proposed (Fonseca et al., 1993, 1994). A first approach to these functions applied to EIA is presented in Table 37.1, for digital video and images, and for sound.

Most of these functions can be classified according to their role in the EIA process: as improving data storage and access, the analysis capabilities or the communicability of results.

Data storage and access functionalities are associated to the availability of richer databases and to the possibility of performing multimedia dynamic queries based on real-world representations.

Analysis functionalities relate, for example, to the possibility of using interactive video and digital sound associated with maps, to help planners and decision makers visualise and better evaluate the impact of a new infrastructure development.

Table 37.1 Multimedia functions associated with the use of digital video, images and sound

Multimedia data types	Multimedia functions	Examples of application
Digital video and images	• Source of information	Ferreira (1993) is using digital video processing to extract the parameters to be used in atmospheric and water quality simulation models.
	• Show backgrounds, point scenes or transitions	It can give a more realistic and dynamic view of a study area.
	• Navigation images	It is possible to have 360° circular views for specific points and to access to video sequences in specific directions.
	• Fly over	To fly over an area, using the aerial photographs, the topographic information and the animation capabilities can give a more realistic and dynamic view of an area.
	• Synthesis of 3D images	3D graphics can improve realism and rendering techniques allow the representation of volumes or surface of objects.
	• Animated sequences	For example, animated sequences of temporal series of aerial photographs can give a dynamic picture of the evolution of an area.
	• Visual simulation models	This can allow a more natural approach to the analysis of a specific area and its future scenarios.
	• Multimedia dynamic queries	Allow a more natural and intuitive approach to the analysis of a specific area.
	• Superimposition of synthetic video on natural video images	Allow a more natural approach to the analysis of a specific area and its future scenarios.
	• Morphing	These operations can allow the visualisation of the transformations that have occurred in a specific situation or scenario.
	• 3D visualisation	Can be used to represent spatio-temporal variation of data, for example, data on pollutants.
Sound	• Stereo sound	To provide the notion of space.
	• Sound icons or music	To create movement or illustrate point scenes.
	• Audio simulation models	It can give, for example, a more realistic approach to the analysis of noise pollution.
	• Sound representing numeric data	The frequency and intensity can be associated with a certain range of values distributed in space.
	• Sound as annotations	Voice annotations in maps can be used to incorporate opinions during a meeting.

Other capabilities are related to the integration of spatial simulations associated with real images and to stereoscopic aerial photographs to get an improved visualisation of the phenomena and a more realistic evaluation.

Finally, results can be better communicated using multimedia functionalities. As previously discussed, the results of an EIA consist of compressed information which should synthesise in a small number of descriptors the complex and diversified part of the real world that was analysed. In a multimedia SIS the improvements in the communicability of the results are associated with the use of images which represent information in a compact way, and which is easier to understand and to manipulate.

A major characteristic of a multimedia SIS is the possibility of creating multiple representations of the same phenomena, thus offering the potential to generate alternative solutions to a problem: an important feature of the EIA process (Shiffer, 1993, 1995).

The current developments on the network information initiatives such as the World-Wide Web on the Internet will eliminate the need to integrate all the multimedia spatial functions in a single system. The integration of the Environmental Impact Statements (EIS) information in the World-Wide Web will make it possible to have individual browsing and group access to the information concerning the EIA studies. These aspects are considered in the following section.

37.3.3 Distributed spatial information systems for EIA

The current growing trend for the use of the World-Wide Web as a way to store and access (multimedia) information, will surely have an effect on the way EIS are presented to the community. This trend has to be taken into account when one considers the development of these multimedia spatial data-handling tools for the EIA process.

The access to the EIS through a hypermedia network based on the Internet, using the browsing tools available for the World-Wide Web, such as Mosaic and Netscape, will make environmental information available to a wider group of people (the globally networked community). This access is performed in an associative manner, where links are made through a distributed network, allowing access to a larger information structure and resulting in a facilitated updating of that information (Shiffer, 1995).

Besides these advantages, the availability of such a networked system makes it platform independent (Mosaic and other World-Wide Web browsing tools operate across the major computer platforms, such as UNIX, Ms Windows, Macintosh), which is another major advantage. Additionally, Mosaic and Netscape are based on a modular approach to the display of multimedia data, using several different help applications that allow access to the different media data files. One other major step forward is that World-Wide Web supports the spatial referencing of information. Thus, it is possible to click on a map and access a specific file associated to those coordinates on the map.

To test the applicability of these ideas to the EIA process, two applications are being developed within this project: a prototype including the multimedia spatial data-handling functions and a CD-ROM aiming to give users the possibility of exploring the environmental data of the Expo'98 area in Lisbon. The prototype will provide functions that can represent hypothetical multimedia spatial data-handling helpers on the World-Wide Web, through the use of its browsing tools (being thus the distributed version of the system). The CD-ROM represents an integrated multimedia spatial information system for the public (corresponding to the standalone version of the system).

37.4 Application

The Environmental Impact Assessment of the Expo'98 in Lisbon is being used as a case study to test these ideas. Expo'98 is a world exhibition on the theme of the Oceans, that will take place in the eastern part of Lisbon, during the Summer of 1998. This case study is an opportunity to test these ideas because the Expo'98 site is an area with several environmental problems. The environmental studies on the site have produced a considerable amount of information, including environmental, geographic and audio-visual data.

The eastern part of the city has for a long time been an industrial area adjoining Lisbon docks. Some of the facilities located in this area include oil refineries, gas deposits, military bases, the municipal land fill and the waste water treatment plant.

The exhibition will require major changes in the area, that will have to be evaluated in terms of their environmental implications. The exhibition site will be mainly located around one of the docks of that area of Lisbon. Almost all the existing buildings in this area are being demolished and new infrastructures and transportation facilities are being created.

The EIA of this project considers the impacts resulting from the reconversion of the area and the impacts of locating the Expo '98 infrastructures in the surrounding area.

The development of the multimedia applications for the EIA of the Expo'98 site required the analysis of this process to identify the tasks that will benefit from the addition of video and sound to SIS. This analysis was made for each of the stages of the EIA process: Screening/Scoping, EIA Study, Public Participation.

For the Screening/Scoping phase that deals more with the identification of determinant physical elements, environmental factors, and variables, most of the operations that can contribute to an improved 'picture' of the area were considered, such as the integrated access to video sequences, images, sounds in association to maps and graphics.

For the Environmental Impact Study, operations that can contribute more to the analysis and evaluation of the effects of the project were added, such as visual and audio simulation models, morphing and animation operations.

Finally, Public Participation might take advantage of the integrated illustration capabilities and interaction mechanisms (such as speech recognition and voice annotation), brought by these technologies.

37.4.1 EIA multimedia spatial information system prototype

The first application is a prototype developed using a hypermedia-authoring tool for the Macintosh, associated with the use of Quicktime® for the display of digital video and other tools for morphing, animation and 3D visualisation.

The major goal of this prototype is to provide a way to evaluate the use of multimedia SIS in the EIA process, either in supporting the development of the EIA studies or its presentation to the public.

The prototype is based on an iconic interface (associated to pull-down menus). The user can access different sources of information (maps, aerial photographs, alphanumeric data, images, videos, graphics and sounds) concerning the Expo'98

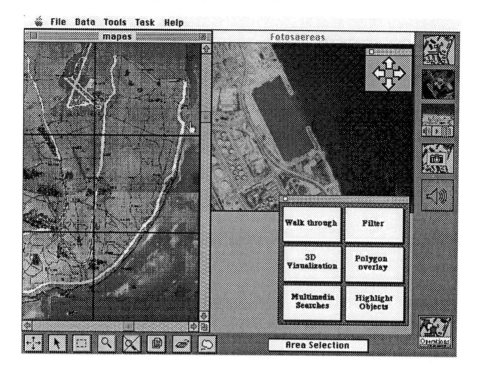

Figure 37.1   Multimedia operations for the area selection task.

area, and also some traditional SIS operations such as zooms and overlays. The multimedia spatial data-handling functions are accessed through a multimedia operations palette which contains different operations according to the selected EIA task: Area selection, Characterisation of the project and of the Environment, Analysis and evaluation, and Presentation of results. This association task/operations was made according to the different information and spatial manipulation requirements of each task.

A sequence of screens is presented to illustrate the use of the prototype within the different EIA tasks. For the *Area selection task* (Fig. 37.1) the multimedia operations available are mainly concerned with the visual exploration of a region and the manipulation of its components, to help delimit the study area. For the characterisation of the *Project and Environment task* (Fig. 37.2) some new operations are added that can help the characterisation of the study area and project. Data-visualisation techniques are included in this palette. The collection of soil pollutants data by a probe and the simultaneous visualisation of that data varying in depth, is one of the data-visualisation examples included.

In the *Analysis and evaluation task* (Fig. 37.3) the operations that can contribute to the assessment of visual and audible impacts are added, such as morphing operations, allowing us to evaluate visually the implementation of different objects in an existing environment.

In the *Presentation of results task* (Fig. 37.4) multimedia operations are mainly associated to the integration of information and the mechanisms of interaction with the public, such as speech recognition and voice annotation.

Figure 37.5 illustrates the performance of a multimedia query operation. It is

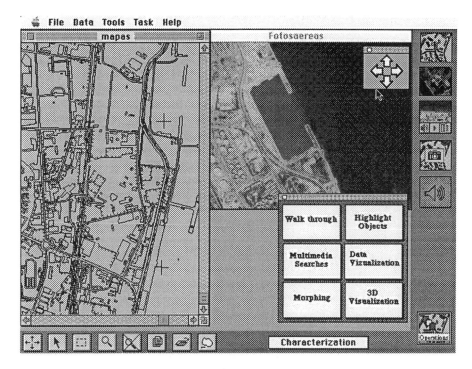

Figure 37.2   Multimedia operations for the characterisation of the project and of the environment task.

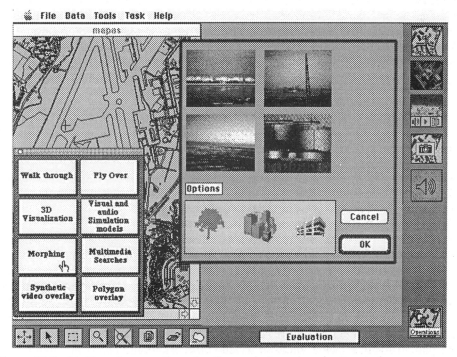

Figure 37.3   Multimedia operations for the analysis and evaluation task.

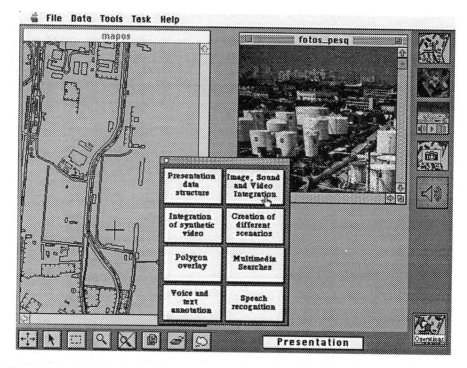

Figure 37.4 Multimedia operations for the presentation of results task.

possible to search for information either by text or by images, allowing a more intuitive way of retrieving information. In this case, a search by text of all the fuel deposits in that area is made. After the identification of their locations on the map, the user can access the images corresponding to the different fuel deposits, in the Expo'98 area.

Additionally, there are operations that are not associated with a specific task. Figure 37.6 presents a navigating operation that allows access to videos in specific directions or around specific points. In this example a 360° pan view around the dock can be activated, in association with the corresponding map and aerial photograph.

The next step for this work will be to integrate some of these capabilities in HTML documents and test its exploration for EIA within the World-Wide Web.

37.4.2 Expo '98 EIA multimedia spatial information system CD-ROM

The second application is a CD-ROM that is going to be distributed during the exhibition, intending to give to any visitor the opportunity to explore the environmental history of that area based on all the information of the EIA and exploring the spatial multimedia functionalities. In this approach, an object-oriented structure is being followed which considers three main components: the objects, models, and dialogue system.

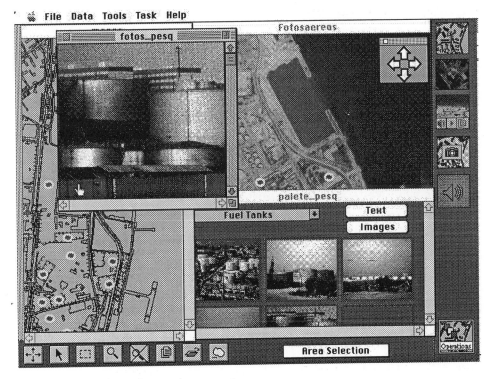

Figure 37.5 Multimedia queries operation.

The different objects include the background, physical objects and abstract objects. The background corresponds to the different geographical bases on which the navigation metaphor takes place. The physical objects are the existing natural or man-made elements within the Expo'98 area that have a physical existence. The abstract objects are the data that represent the state of the environment. These objects are stored in an object-oriented database from where they can be retrieved.

The available models include interpolation models (including morphing operations), simulation models (numerical and visual), static models (the database) and decision models.

The dialogue system is the interface that relies on the general metaphor over which all the application is based, and on the utilities that give users access to all the interactive navigation/exploration, visualisation/analysis operations.

According to these ideas, the system interface metaphor for the Expo '98 CD-ROM is based on the time/space selection. The navigation through the Expo '98 area relies upon a transparent vehicle, that flies, walks and dives all over the area. In this vehicle, there is a tablier where all the information concerning the baseline situation of the area, or the different scenarios, can be accessed. Through this tablier an environmental calculator – AMBICALC – and a visualiser can be used allowing for several operations on the different time/space scenes. Other navigation aids include agents, sounds and visual icons.

During the exploration of the environmental history of the Expo '98 area the

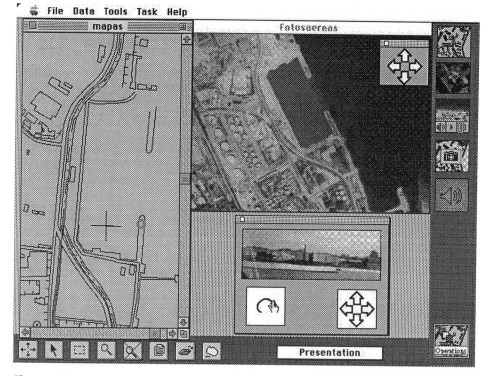

Figure 37.6 Access to videos in different directions and pan views.

visitor can view the area from different perspectives according to the different media available in the system. While in the vehicle, the user can choose to fly over an area in a certain time, land in some points accessing images or videos associated to specific directions. During the exploration one can change the spatial scales or the time unit.

For each time/scale combination, the information on the state of the environment can be accessed in different ways. Three-dimensional visualisations of the environmental descriptors considered will be available for each environmental theme: air, water, soil, and biota. These data are also available for the different scenarios considered in the EIA study. The CD-ROM will also include a sequential way to access that information, as an alternative to the interactive exploration described above.

Both the prototype and the CD-ROM will be tested not only with different people working in the EIA field but also with the general public, to evaluate the effectiveness and the value-added brought by the multimedia spatial data-handling functionalities, which is in essence the main goal of this research.

37.5 Conclusions

This chapter has renewed the concepts underlying the creation of a Multimedia Spatial Information System for Environmental Impact Assessment and presented a prototype to test these ideas and support the development of a CD-ROM with the environmental information of the Expo'98 area in Lisbon.

These developments aim to demonstrate that the use of multimedia technology capabilities within SIS may contribute to the development of more powerful environmental analysis due to the possibility of accessing additional types of data, the availability of new data visualisation and manipulation capabilities, and the use of multimedia interfaces with a higher degree of interactivity. These developments may contribute to the solution of crucial problem of transforming a large number of variables describing the environmental effects identified into a reduced number of impacts, that have to be presented to the public. The current developments on network information retrieval initiatives such as the World-Wide Web on the Internet will eliminate the need to integrate all these functions in a single system. This gives considerable advantages in terms of accessibility to the system and updating the information within the EIA process, as it makes it easier to add new capabilities and functionalities to the overall system.

Future research will concentrate on the development of analytical multimedia spatial data-handling capabilities for the Expo'98 EIA and the creation and testing of mechanisms of interaction with the public. The analysis of the integration of these mechanisms as World-Wide Web helpers will be another future research topic.

Acknowledgements

This work was partially supported by DGA and JNICT/DGA under research contracts number 86/91/J and number PEAM/C/TAI/267/93. We would like to acknowledge the special collaboration of Cristina Gouveia, and also the contributions of João Pedro Silva, João Pedro Fernandes, Elsa João and Andrew Thompson.

References

ASHTON, C. H. and SIMMONS, D. (1993). Multi-media: an integrated approach combining GIS within the framework of the Water Act 1989, *Proceedings of HIDROGIS, IAHS*, pp. 281–8, Wien.

BUFORD, J. F. K. (1994a). Multimedia file systems and information models, in Buford, J. F. K. (Ed.), *Multimedia Systems*, pp. 265–84, New York: Addison-Wesley.

(1994b). Uses of multimedia information, in Buford, J. F. K. (Ed.), *Multimedia Systems*, pp. 1–26, New York: Addison-Wesley.

CÂMARA, A. and FERREIRA, F. C. (1993). 'Simulation as an Exploratory Tool in Spatial Systems', presentation at the Workshop on Multimedia and GIS, CNIG, Lisbon, February.

CÂMARA, A., GOMES, A. L., FONSECA, A. and VALE, M. J. L. (1991). HyperSnige: a navigation system for geographic information, *Proceedings of EGIS'91 Conference*, pp. 175–9, Utrecht: EGIS Foundation.

CASSETARI, S. and PARSONS, E. (1993). Sound as data type in a geographical information system, in Harp, J., Ottens, H. and Schotten, H. (Eds), *Proceedings of EGIS'93 Conference*, **1**, pp. 203–12, Utrecht: EGIS Foundation.

CRAWFORD, C. (1990). Lessons from computer game design, in Laurel, B. (Ed.), *The Art of Human-Computer Interface Design*, pp. 103–11, Reading MA: Addison-Wesley.

FERREIRA, F. (1993). 'Aquisição de Informação Ambiental utilizando Vídeo', presentation at the Seminário de Verão, GASA/FCT-UNL, Monte da Caparica, Portugal, June.

Fonseca, A., Gouveia, C., Câmara, A. and Ferreira, F. C. (1994). Environmental impact assessment using multimedia GIS, *Proceedings of EGIS '94 Conference*, pp. 416–25, Utrecht: EGIS Foundation.

Fonseca, A., Gouveia, C., Ferreira, F. C., Raper, J. and Câmara, A. (1993). Adding video and sound into GIS, *Proceedings of EGIS'93 Conference Genoa*, vol. **1**, pp. 176–87, Utrecht: EGIS Foundation.

Fox, E. (1991). Advances in interactive digital multimedia systems, *IEEE Computer*, **24**(10), 99–121.

Glasson, J., Therivel, R. and Chadwick, A. (1994). *Introduction to Environmental Impact Assessment*, The Natural and Built Environment Series 1, London: UCL Press.

João, E. (1994). Using geographical information systems for environmental assessment, *Environmental Assessment*, **2**(3), 102–4.

Laurel, B. (1990). Interface agents, in Laurel, B. (Ed.), *The Art of Human-Computer Interface Design*, pp. 124–30, Reading MA: Addison-Wesley.

Luther, A. C. (1994). Digital video and image compression, in Buford, J. F. K. (Ed.), *Multimedia Systems*, pp. 143–74, New York: Addison-Wesley.

Munn, R. E. (1979). *Environmental Impact Assessment: Principles and Procedures*, 2nd Edn, New York: John Wiley.

Nielsen, J. (1990). The art of navigating through Hypertext, *Communications of ACM*, **33**(3), 297–30.

Polydorides, N. D. (1993). An experiment in multimedia GIS: Great cities of Europe, *Proceedings of EGIS'93 Conference*, vol. **1**, pp. 194–202.

Raper, J. F. and Livingstone, D. (1995). The development of a spatial data explorer within an environmental hyperdocument. *Environment and Planning B*, **22**, 679–87.

Rhind, D. W., Armstrong, P. and Openshaw, S. (1988). The Domesday machine: a nationwide GIS, *Geographical Journal*, **154**(1), 56–68.

Scholten, H. J. and Bijtelaar, B. (1993). 'GIS and multimedia: how to integrate. Technical aspects of a prototype', presentation at the Workshop on Multimedia and GIS, CNIG, Lisbon, February.

Shiffer, M. J. (1993). Augmenting geographic information with collaborative multimedia technologies, *Proceedings of Auto-Carto 11*, pp. 367–76.

(1995). Environmental review with Hypermedia systems, *Environment Planning B*, **22**, 649–64.

Smith, S., Bergeron, R. D. and Grinstein, G. (1990). Stereophonic and surface sound generation for exploratory data analysis, *Proceedings of CHI'90 Conference*, pp. 125–32, New York: ACM.

Treweck, J. and Veitch, N. (1992). GIS and its application to ecological assessment within EIA, *EIA Newsletter*, **7**, pp. 18–19.

CHAPTER THIRTY-EIGHT

An Open System for 3D Visualisation and Animation of Geographic Information

CHRISTOPHER GIERTSEN, ODDMAR SANDVIK
and RUNE TORKILDSEN

38.1 Introduction

Previous research in 3D visualisation and computer animation for GIS can be divided into two categories: pilot applications (McLaren and Kennie, 1989; Van Voris *et al.*, 1993) and more general considerations (Buttenfield and Ganter, 1990; MacEachren *et al.*, 1992). Most such research has so far been performed in fields related to GIS, while it has received little attention from the computer graphics and visualisation communities. Pilot application papers often describe an implementation in detail, but usually the scope is limited and the system design unsuited to being extended. Moreover, papers presenting general considerations seldom discuss implementation at a detailed level.

Current research activities in 3D GIS focus on spatial analysis as well as modelling and representing 3D objects and associated topological relationships (van Oosterom *et al.*, 1994). Generic tools or systems for 3D visualisation designed specifically for GIS are seldom considered in this context.

One way to view the shortcomings of previous research on visualisation functionality for GIS is that too few central requirements have been taken into consideration at the same time. For example, the scientist and the general public usually require different presentations of the same data. Furthermore, large repositories of 2D geographic information may be made more easily accessible by offering 3D visualisation functionality for 2D GIS data. The central problem is to identify and combine many such disparate requirements into a consistent and generic framework. This has been the main objective of our study.

The proposed solutions realise the conceptual framework proposed by Giertsen and Lucas (1994). In addition, the solutions are influenced by the authors' experiences from customer contact and the development of commercial 3D GIS-oriented applications aimed at visualisation and animation. Such applications include defence systems, television entertainment, marketing related to geographic areas, offshore surveying, news presentations and passenger information systems.

In our study, we have applied the following research methodology. First, we identified a set of requirements for 3D visualisation and animation in GIS. We then developed the proposed system in an iterative process, at each step evaluating the preliminary results with respect to these requirements.

38.2 General requirements for visualisation and animation in GIS

This section presents a mixture of user-oriented, functional and technical requirements. Throughout the chapter, we discuss how these requirements affect the design of a general visualisation system for GIS. We do not consider the quality and usefulness of the visualisation output, since the purpose of our study is to identify and implement generic visualisation and animation functionality for GIS. Requirements for visualisation output for GIS have been studied, for example, by Buttenfield and Ganter (1990) and MacEachren *et al.* (1992). In this section we do not discuss animation explicitly, but consider it as part of the field of visualisation.

1 *Visualisation systems for GIS must cover the needs of different users of geographic information.* Users of geographic information may require different presentations of the same data. Giertsen and Lucas (1994) identified three user categories and concluded that their needs span from abstract scientific visualisation to more intuitive and realistic visualisation.

2 *Visualisation systems for GIS must support different uses of GIS.* A GIS can be used for several purposes, each having specific requirements for visualisation. For example, visualisation in GIS is no longer simply for illustration, but has become part of the analysis process (Weibel and Buttenfield, 1992). Furthermore, decision support often implies comparing different scenarios (Fedra and Reitsma, 1990).

3 *Visualisation systems for GIS must be easy to use and support users with different skill levels.* Since a GIS may include more than 1000 functions (Openshaw, 1991), it is unlikely that advanced visualisation will find widespread use if all users have to learn all the details. Visualisation systems for GIS should therefore support multiple skill levels, be easy to use, and be interactive as far as possible.

4 *Visualisation systems for GIS must offer 3D presentations of both 2D and 3D GIS data.* Although much of the current research in GIS focuses on 3D data models and related analysis functions, the majority of the data in today's applications of GIS are 2D. As discussed in detail by Giertsen and Lucas (1994), it is possible to visualise many types of 2D GIS data using techniques for 3D visualisation.

5 *Visualisation systems for GIS must be easily extendible to support application-specific requirements.* Visualisation for GIS may involve different types of 3D modelling, particularly when creating 3D presentations of 2D data. It is, however, unreasonable to believe that it is possible to develop one general system that covers all modelling needs. In order to be sufficiently flexible, the visualisation system should therefore include a mechanism for adding new modelling programs according to needs.

6   *Visualisation systems for GIS must not be restricted to predefined data formats.*
Important information or features in GIS data can easily be lost if the data must
be converted to another format prior to visualisation. Furthermore, time-
varying data may exist in many different forms, which are difficult to
standardise. For these reasons, the visualisation system should be extendible
in order to accept any data format.

38.2.1 Discussion of requirements

This list of requirements affects two central issues: the choice of software
development tools and the overall design criteria for the system. When developing
a visualisation system, it is important to identify suitable software tools that
minimise implementation and maintenance efforts. Requirement (1) indicates that
the software tools should help produce both realistic and scientific visualisations
of the same data. Most current visualisation tools are designed to create scientific
data presentations, for example, contour plotting, isosurface plotting, volume
graphics, vector and tensor field visualisation. For an overview of such techniques
and research issues, see Rosenblum (1994). In order to meet requirement (1), the
development tool should include a graphical format with extensive support for
realism, that is, advanced object-hierarchy descriptions, material properties,
texture-mapping mechanisms, and so forth. To conclude, it is always possible to
represent a non-realistic object using a format for realistic objects, while the
opposite is usually difficult.

The need for user interaction (requirement 3) is another important issue
affecting the choice of a software development tool. In most scientific visualisation
systems, the user cannot edit objects produced by the visualisation algorithms since
they are designed to provide objective information based upon scientific data and
user-defined parameters. Viewing functions are usually the only kind of interaction
offered to the user. In the field of GIS, however, it may not be possible to produce
visual information which is entirely objective. Requirement (4) indicates that 3D
visualisations may be created from 2D data. Given that 2D data are incomplete
in this context, functions for interacting with objects in a scene can be useful. For
example, if 2D point data representing buildings are to be represented as 3D
symbols, it may be difficult to determine a natural orientation for each symbol.
Editing functions can then be used in combination with automatic procedures to
achieve acceptable results. Few development tools for 3D computer graphics
include functions for 3D object interaction. One exception is the Open Inventor[1]
toolkit, originally described by Strauss and Carey (1992). We chose this toolkit due
to its extensive support for realism *and* 3D object selection and editing.

Requirements (5) and (6) affect the overall design of the system in the sense
that the user must be able to include new data formats as well as new modelling
tools, hereafter called *model generators*. For this reason, our system design
includes an *open interface*, which is discussed in more detail in the following
sections. Due to requirement (3) we introduce the notion of a *template*, defined
as the coupling of a model generator and a set of parameters intended to emphasise
certain features in the data or to express information in a specific way. The set
of graphical objects generated by applying a template to a GIS data set is called
a *scene component*. Several templates may exist for the same model generator.

Users operate the system by defining templates, connecting data sets to templates, and creating scene components. The concept of templates allows the system to support multiple user skill levels, as described later.

38.2.2 Outline of system architecture

In order to obtain the desired flexibility in the visualisation system, we suggest that it should consist of an open framework, called the *integrator*, and an associated *kernel* containing general GIS-oriented visualisation functionality. To the user, the functionality in the integrator and the kernel will appear as one seamless system. Fig. 38.1 illustrates the main modules in the proposed system and the data flow between the integrator and the kernel. The GIS data referred to in the figure include any kind of results derived by spatial analysis tools (Unwin, Chapter 26, this volume), or data extracted by animation-oriented analysis tools, such as the query language for spatio-temporal information systems proposed by Rex (1995).

We suggest that the integrator should offer the following types of functionality:

- an *open interface* for adding new GIS data formats and model generators to the system;

- a *database manager* to keep track of all data and model generators known to the system and to combine scene components into a common data structure suitable for rendering; and

- a *scene component manager* with user interfaces for creating templates and for connecting GIS data sets to appropriate templates.

We suggest that the kernel should offer the following types of functionality:

- *view specification* (functions for specifying virtual camera parameters for single frames)

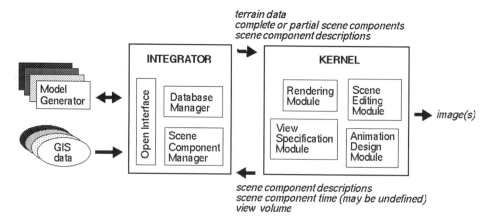

Figure 38.1 Outline of system architecture.

- *rendering* (functions for computing the visual display of scene components)
- *scene editing* (functions for changing features of objects in a scene component); and
- *animation design* (functions for composing, previewing, and computing animations).

The discussion of system architecture has so far focused on functionality. In order to ensure high performance in terms of response time and memory usage, it is necessary to focus on the flow of data between the integrator and the kernel. Model generators may create an enormous quantity of graphical objects, particularly for time-varying data. We therefore distinguish between *static scene components* for still pictures or backgrounds and *dynamic scene components* used for animation. Conceptually, a dynamic scene component can be viewed as a sequence of static scene components, where each item in the sequence is associated with a time value, hereafter called *scene component time*.

To ensure the best possible performance, we suggest that the kernel may receive scene components from the integrator in three different ways. The first option is a *scene component description*, containing a description of the data, model generator, and parameters used to create the actual scene component, but not the objects themselves. The second option, called a *partial scene component*, is a limited part of the actual scene component containing only the objects bounded by the virtual camera, and a time value if the scene component is dynamic. The third option is a *complete scene component* containing all objects created by applying a GIS data set to a template.

When the kernel requests partial or complete scene components, the integrator may receive the view volume defined by the virtual camera parameters, scene component time values, and scene component descriptions.

38.3 The integrator

This section describes in more detail the tasks typically performed by different types of model generators, as well as how new GIS data formats or model generators can be added to the integrator without recompiling any code. We then propose a user interface for the integrator supporting multiple user-skill levels.

38.3.1 Texture generator

A texture generator is a model generator producing 2D texture images to be mapped on to a terrain surface. Textures may be created from terrain data only, representing information such as steepness or height. Image data for the surface of a geographic area (for example, satellite data or digital airphotos) can be regarded either as textures themselves or as raw data. Texture generators operating on image data typically modify colours or emphasise objects in the image. Thematic data defined by points, polylines or polygons can optionally be superimposed on the textures.

38.3.2 Object generator

An object generator is a model generator producing 2D or 3D objects positioned relative to the terrain surface. The objects may either be geometrically defined or amorphous and created by means of procedural models (Ebert, 1994) or volume rendering methods (Kaufman, 1991). In order to meet the needs of different users of geographic information and different uses of GIS, the 3D objects may be either realistic or symbolic. The 3D representations of image data can be created by mapping values in the image to primitive 3D objects, for example, imported from a model library. In this case the object generator will typically be responsible for selecting primitive objects, for positioning the selected objects relative to the terrain surface, and for scaling them if this is needed. A terrain model is often input to an object generator in order to determine height values associated with 2D spatial data.

38.3.3 Implementing model generators

In order to easily add new model generators to the system, we have developed mechanisms for calling model generators, editing parameters and specifying input data. These mechanisms impose constraints on the implementation of model generators.

For each model generator, an associated *model generator description* must be created. This description includes a logical model generator name, file name for the executable program, an optional summary of functionality, and file format identifiers defining legal input to the model generator. The description can, for instance, be used to ensure that a model generator is linked to an appropriate data set.

In the prototype system a model generator is invoked with a system call. More efficient mechanisms, such as the use of interprocess communication, may be implemented in future versions. A command line parameter to the model generator defines its execution mode. In *production mode*, the model generator produces scene components. In *editing mode*, an interface panel for editing model generator parameters is displayed. In both modes, the model generator is called with at least three parameters: the name of a parameter file, the name of an input file, and the name of an output file. Optional parameters include a view volume or a scene component time if a partial scene component is to be created. We believe it is difficult to standardise the format of parameter files for model generators since they may be designed to serve highly different purposes. Instead we standardised user interaction with the parameters through the editing mode described above.

We define the *expert user* of the proposed system to be a person capable of implementing and using new model generators. The expert user is typically a person mostly concerned with extending visualisation functionality, such as an application programmer.

38.3.4 User interface

An import mechanism has been developed where a new model generator can be added to the system by entering the name of a model generator description file.

Similarly, a new GIS data set is included by giving a file name and file format identifier, which can be matched with the model generator description files. The user can assign logical names to model generators and data sets. For example, the logical name of a data set may represent several physical files.

We define the *advanced user* to be a person capable of importing new model generators and GIS data sets, defining templates and creating scene components. Examples of advanced users include application specialists and system programmers. The user interface panels shown in Figs 38.2 summarise core functionality in the scene component manager. All panels are designed according to the following principles:

- a *remove button* can be used to remove a selected item;
- an *edit button* can be used to edit existing parameters for a selected item
- an *add* button can be used to define new parameters for a new item
- a *description field* provides additional information about the selected item.

The upper interface panel in Fig. 38.2 shows the *template editor*, where the user defines a template by selecting a model generator, defining its parameters, and assigning a template name to this package of information. Parameters are defined by pressing the edit button to invoke the chosen model generator in editing mode. The lower panel in Fig. 38.2 shows the *scene component controller*, where the user can create partial or complete scene components. The example of the application shown in Fig. 38.2 is described later in the chapter (3D symbolic building generator).

All interface panels have been created using UIM/X (Hewlett-Packard Company, 1993).

38.4 The kernel

In our presentation of the kernel we distinguish between three categories of functionality:

1. standard visualisation functionality;
2. functionality available through Open Inventor, and
3. GIS-oriented visualisation functionality that must be developed and refined through further research.

We will briefly describe view specification, rendering, and scene editing, which fall into the first two categories. Animation functionality belongs to the third category and will be discussed in more detail.

We define the *high-level user* to be a person operating the proposed system through functions in the kernel. The high-level user is thus mostly concerned with composing existing scene components into a suitable presentation of the GIS data and creating animations.

38.4.1 View specification, rendering, and scene editing

View specification is handled through a *virtual camera model*, emulating a real camera. We have implemented a menu for editing lens parameters and use a

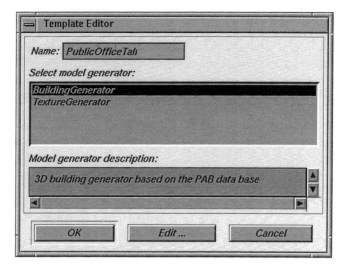

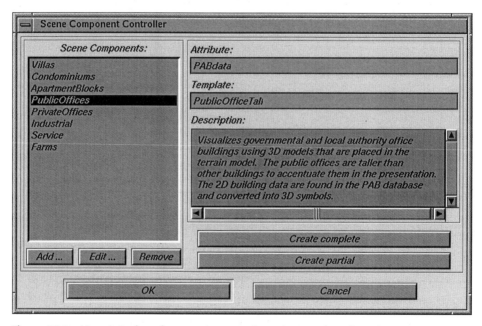

Figure 38.2   User interface for creating templates (upper panel) and scene components (lower panel).

position, pitch, yaw, and roll interface to specify the viewpoint and the line of sight. These camera parameters define the *view volume*, used for data selection and rendering. We have also developed an interactive 3D viewing function which responds to adjustments in position, pitch or yaw. This viewing function makes use of the rendering functionality described below. To provide detailed geographic references when specifying a view, we have implemented map presentations in a plan view including contours and/or thematic data such as roads, rivers, lakes, glaciers, coastline and urban areas.

Rendering has been implemented using Open Inventor's functionality for illumination, texture mapping, and hidden surface elimination (Wernecke, 1993). However, terrain data and scene components for large geographic areas require extensive memory resources and can dramatically degrade performance. To avoid this problem, we have implemented a demand paging system where the data are divided into patches that are read or discarded based on the current view volume. In addition, the user can select which scene components to display at any time. Each scenario in a GIS study can be represented by a set of scene components. By displaying different scene components in several windows, the visualisation system can be configured to support varying uses of GIS (see requirement 2).

Scene editing is available through ready-made user-interface mechanisms in Open Inventor. A picking mechanism based on a ray, or a cone, finds and highlights the frontmost 3D object behind the cursor. When an object has been selected, it can be manipulated (scaled, rotated, translated) or a material editor can be invoked to change its ambient, diffuse, specular, transparent, emissive and shininess properties. Open Inventor also offers general mechanisms for developing special-purpose editing functions, for example, to change texture-mapping parameters.

38.4.2 Animation design

This section describes GIS-oriented tools for designing an *animation sequence*, that is, tools for controlling virtual camera motion and synchronising dynamic scene components. The proposed animation module can handle any combination of static or dynamic virtual cameras and scenes, for example, a moving camera and a moving vehicle caught in an avalanche. Post-processing tools for editing and connecting animation sequences will not be discussed, since they are applicable to any field of study, including GIS.

Since the computing power for the majority of users is far below the performance required for complex real-time animations with high-quality rendering, we have chosen to design our system for a batch-oriented production of animations. When the necessary computing power becomes available, the kernel can be extended to include interactive control of real-time animations, for example, similarly to the virtual reality environment proposed by Koller *et al.* (1995). In the current design, we have included functions for real-time previewing based upon frame-selection mechanisms and quality reduction in rendering.

38.4.2.1 *Virtual camera motion*

When creating an animation sequence, the user must first define its duration. Points of time within this interval are referred to as *animation time*. If a static virtual camera is used, the duration must be specified explicitly. In the moving virtual camera case, the duration is implicit in a virtual camera path and a speed along this path. The remainder of this section focuses on techniques and problems related to controlling a moving virtual camera.

In the user interface for controlling a moving virtual camera, a position curve is defined in a plan–view map, thus becoming a natural extension of the interface for view specification. Height and speed curves as a function of the distance along the position curve must also be specified. All curves are B-splines with continuous second-order derivatives, ensuring smooth virtual camera movements. Muller *et al.* (1988) proposed a similar user interface, but with fewer degrees of freedom. Figure 38.3 shows motion curves for a flyby of southern Norway. A terrain profile is shown beneath the height curve in order to prevent the virtual camera from crashing into the terrain.

After these motion curves have been specified, an animation sequence can be computed by stepping along the position curve as a function of the speed curve and a user-defined number of frames per second. These computations are performed using numerical algorithms for integration and for solving non-linear equations available in the software development toolkit described by Press *et al.* (1992). When a position has been determined the corresponding height value is computed, defining the frame viewpoint. Next, the roll, pitch and yaw at the viewpoint are found by means of the tangent vector to the three-dimensional virtual camera motion curve. This tangent vector defines the line of sight. The animation frame is finally rendered, as described earlier.

For some applications, it is insufficient to control camera motion simply by means of user-defined curves. In order to provide richer functionality for camera control, we are currently developing functionality to automatically generate motion curves or adjust virtual camera parameters at computed viewpoints. Examples of this type of functionality and possible applications are listed below. From a research point of view, the challenge is to develop user interfaces and methodologies for controlling the combinations of different items in such a way that the resulting animation becomes predictable.

Functionality for generating virtual camera motion curves.

1 Position/height curve as a function of the position/shape of an object
    – for example, for moving the camera along a pipeline, or a moving car

2 Speed as a function of attribute data
    – for example, camera motion based on measured traffic density

3 Simulated motion curves
    – for example, creating a flyby along an ocean current

4 Forcing camera curves to match predefined views at key frames
    – for example, to easily blend animation and live video sequences.

Functionality for adjusting virtual camera parameters at computed viewpoints.

5 Simulation of roll as a function of the curvature of the virtual camera curve
    – for simulating airplane movement (roll > 0) or a helicopter movement (roll = 0)

6 Adding a value to the pitch
    – for example, for viewing details more directly beneath the camera

7 Changing the horizontal component of the line of sight (yaw)
    – for example, side window view from an airplane

8 Locking the line of sight to a point
    – for example, keeping the eyes fixed on an object while moving around it

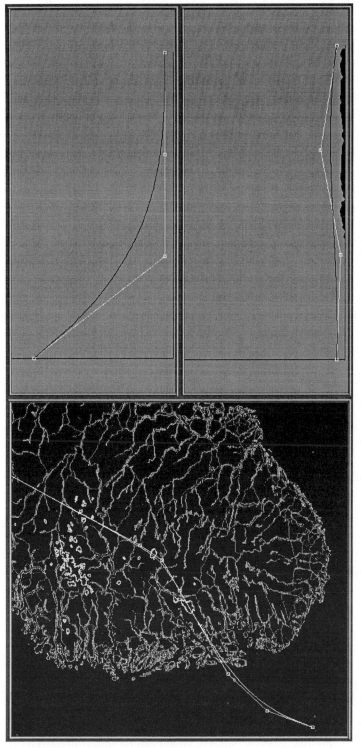

Figure 38.3   Spline curves defining position, height and speed for a virtual camera.

9   Simulating local camera movements in sharp turns
    – for example, a driver looking around a bend.

38.4.2.2 *Synchronising dynamic scene components*

There are several possible ways of synchronising camera and object motion in dynamic scene components. We are currently developing a framework, based upon the open system architecture and model generators, in order to fulfil the following goals: (1) to balance computational support in the synchronisation versus user-defined control, and (2) to utilise synchronisation functionality in modern graphics development tools such as Open Inventor.

Animations are computed based upon user-defined constraints, and thus the duration is known in advance. We therefore propose that the dynamic scene components of an animation are synchronised by mapping scene-component time to the animation time.

Goal 1. User-defined control can be achieved by composing and synchronising several less complex dynamic-scene components. On the other hand, sophisticated model generators with more internal synchronisation, specified in terms of constraints on the animation, provide a high level of computational support and simplify the tasks of the high-level user.

Goal 2. Toolkits such as Open Inventor include synchronisation functionality for animation in their graphical file format. This functionality is particularly useful in creating dynamic model generators where there are complex dependencies between the movements of different objects or subobjects. Open Inventor also supports functionality for device management and user interaction that can be linked to dynamic objects. This functionality allows the production of dynamic scene components where the user can select dynamic objects and change their dynamic features when viewing them.

Figure 38.4 shows the user interface panel for synchronisation of dynamic scene components, called the *animation manager*. In the left part of the panel, the user can select static scene components for the background and move dynamic scene component to and from the window for mapping scene component times to the animation time (middle right part of the interface panel). The upper right part of the interface panel contains buttons for creating an animation. In the lower part of the interface panel the user can load or save configurations of the animation manager.

38.5 Examples of applications

We have developed a terrain-based texture generator producing either realistic colours representing the landscape, or more abstract colour codes for steepness or height. A realistic summer landscape can be computed using green colours for low altitudes, brown colours for intermediate altitudes, and white for snow-covered mountains. Alternatively, a winter landscape can be created by assigning white to all altitudes. Steep mountain faces are assigned greyish colours and a noise function smoothes the transitions between different colours. The texture can optionally include shading. Map-attribute data defined by points (for example, houses, airports), polylines (for example, roads, rivers, lakes and power lines), and polygons (for example, glaciers and city areas) can be superimposed on the texture in user-defined colours and order.

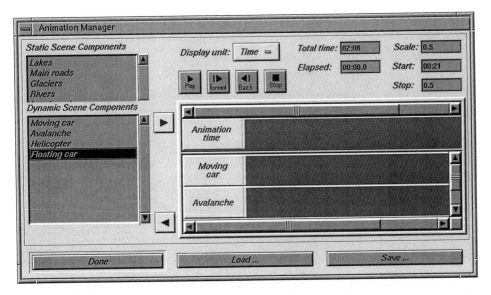

Figure 38.4   Interface panel for synchronising dynamic scene components.

Both examples described below[2] make use of the terrain-based texture generator.

38.5.1 Three-dimensional symbolic building generator

GIS data are often incomplete with respect to creating 3D visualisations. One such example is a public database containing information about all buildings in a country. Although detailed 3D CAD descriptions may exist for new buildings, the majority of buildings are typically represented simply by a 2D coordinate and a building code defining its type.

In many applications, it may be useful to create an intuitive visualisation of 2D building data, for example, in a news production to show residential areas affected by a pollutant. In order to explore such applications, we have developed an object generator performing the following three tasks for each building. First, the associated terrain model is sampled at the 2D coordinate to determine the altitude. Second, a 3D symbol is selected from a library based upon the building code. Third, the selected building symbol is positioned on the terrain surface using heuristics for determining a natural orientation.

The user interface panels of Fig. 38.2 demonstrates practical use of the proposed system for visualising building data. The upper panel shows how the user can select the building generator and define a template for creating public building symbols. The lower panel shows that the selected scene component named 'PublicOffices' is defined as the coupling between the 2D data set, called 'PAB' (properties, addresses, buildings) and the template called 'PublicOfficeTall'. To create the graphical objects in the 'PublicOffices' scene component, the user simply selects 'Create complete' or 'Create partial'. Figure 38.5 shows a picture created by means of the symbolic building generator.

Figure 38.5   Three-dimensional building symbols and terrain-based texture.

38.5.2 Three-dimensional pipeline generator

When a wide spectrum of professionals, politicians and managers are involved in a major industrial decision process, it is important that they have a common understanding of critical issues. During the planning phase of a pipeline from the Troll oil field in the North Sea to Mongstad refinery outside Bergen, Norway, we created a flyby animation along the planned pipeline. This animation was used in the decision-making process to develop a common understanding of critical sections of the sea floor in the fjords, where the pipeline could represent a threat to the environment.

Figure 38.6 shows a single frame in the Trollpipe animation, where we developed an object generator to create a 3D representation of the pipeline based upon a 2D polyline.

38.6 Conclusions

The proposed implementation contributes to clarify several non-trivial technical issues related to advanced visualisation and animation for GIS. First, we have drawn a line between core visualisation functionality that can be standardised and functionality which may be added depending upon needs and innovative ways to express information. Here we refer to the functionality in the kernel and the model generators respectively. Second, we have identified important requirements for the

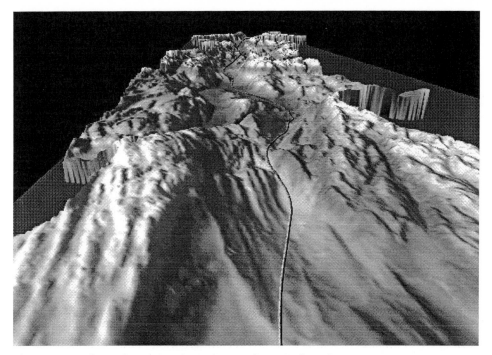

Figure 38.6   Three-dimensional pipeline and terrain-based texture.

'infrastructure' needed in a visualisation system for GIS. By using a software development tool with extensive support for realistic objects, the system can be configured by means of appropriate model generators to support both scientific and realistic visualisation. Furthermore, the system provides 3D visualisation from incomplete 2D data sets by using tools that support 3D object selection and editing. Third, the open system integrates GIS data from different sources by means of the model generators. This approach is flexible and facilitates the flow of data from a GIS into the visualisation system. In the prototype version of the system, all integration problems are solved using file transfers. In order to bring the proposed system beyond a research prototype, it is necessary to invest more efforts into data and process integration. Fourth, we have drawn a line between basic kernel functionality and open research areas, where we have stressed the need for constraint-based computational support for camera control and synchronisation of dynamic scene components.

From the user's point of view, the proposed system should be attractive since it supports multiple skill levels. It can also be configured to support the different uses of GIS and to cover the needs of different users of geographic information.

Acknowledgements

The Norwegian Mapping Authority and the Norwegian oil company STATOIL provided the data used to create Figs 38.5 and 38.6 respectively.

Notes

1 Open Inventor is a trademark for Silicon Graphics Inc.
2 Colour versions of the greyscale images shown in Figs 38.5 and 38.6 can be obtained in digital form from the authors, for example, by sending a request to: Christopher.Giertsen@cmr.no

References

BUTTENFIELD, B. P. and GANTER, J. H. (1990). Visualization and GIS: What should we see? What might we miss?, *Proceedings of the 4th International Symposium on Spatial Data Handling*, pp. 307–16, Zürich: University of Zürich, Department of Geography.

EBERT, D. (Ed.) (1994). *Texturing and Modeling – A Procedural Approach*, Massachusetts: Academic Press.

FEDRA, K. and REITSMA, R. F. (1990). Decision support and geographical information systems, in Scholten, H. J. and Stillwell, J. C. H. (Eds), *Geographical Information Systems for Urban and Regional Planning*, pp. 177–90, Dordrecht: Kluwer Academic Press.

GIERTSEN, C. and LUCAS, A. (1994). 3D Visualization for 2D GIS: an analysis of the users' needs and a review of techniques, *Computer Graphics Forum*, **13**(3), 1–12.

HEWLETT-PACKARD COMPANY (1993). *The UIM/X Developer's Guide*, Part No. B1183-90600, USA: Hewlett-Packard Company.

KAUFMAN, A. (Ed.), (1991). *Volume Visualization*, California: IEEE Computer Society Press.

KOLLER, D., LINDSTROM, P., RIBARSKY, W., HODGES, L. F., FAUST, N. and TURNER, G. (1995). Virtual GIS: A real-time 3D Geographic Information System, *Technical Report GVU-95-14*, USA: Georgia Institute of Technology.

MACEACHREN, A. M., BUTTENFIELD, B. P., CAMPBELL, J. B., DIBIASE, D. W. and MONMONIER, M. (1992). Visualization, in Abler, R. F., Marcus, M. G. and Olson, J. M. (Eds), *Geography's Inner Worlds – Pervasive Themes in Contemporary American Geography*, pp. 99–137, New Jersey: Rutgers University Press.

MCLAREN, R. A. and KENNIE, T. J. M. (1989). Visualization of digital terrain models: techniques and applications, in Raper, J. (Ed.), *Three-dimensional Applications in GIS*, pp. 79–97, London: Taylor & Francis.

MULLER, J.-P., DAY, T., KOLBUSZ, J., DALTON, M., RICHARDS, S. and PEARSON, J. C. (1988). Visualization of topographic data using video animation, *International Archives of Photogrammetry and Remote Sensing*, **27**(B4), 602–15.

VAN OOSTEROM, P., VERTEGAAL, W., VAN HEKKEN, M. and VILBRIEF, T. (1994). Integrated 3D modelling within a GIS, *Proceedings of AGDM'94 (Advanced Geographic Data Modelling)*, pp. 80–95, Delft, The Netherlands: Geodetic Commission.

OPENSHAW, S. (1991). Developing appropriate spatial analysis methods for GIS, in Maguire, D. J., Goodchild, M. F. and Rhind, D. W. (Eds) *Geographical Information Systems: Principles and Applications*, vol. 1, pp. 389–402, London: Longman.

PRESS, W. H., TEUKOLSKY, S. A., VETTERLING, W. T. and FLANNERY, B. P. (1992). *Numerical Recipes in C*, Chester, New York: Cambridge University Press.

REX, B. (1995). 'Animated Query Language for Spatio-Temporal Information Systems', paper presented at the ESF-GISDATA and NSF-NCGIA Summer Institute on Geographic Information, Wolfe's Neck, Maine, 26 July–2 August.

ROSENBLUM, L. J. (Ed.) (1994), Research issues in scientific visualization, *IEEE Computer Graphics and Applications*, **14**(2), 61–85.

STRAUSS, P. S. and CAREY, R. (1992). An object-oriented 3D graphics toolkit, *Computer Graphics*, **26**(2), 341–7.

VAN VORIS, P., MILLARD, W. D. and THOMAS, J. (1993). TERRA-Vision – the integration of scientific analysis into the decision-making process, *International Journal of Geographical Information Systems*, **7**(2), 143–64.

WEIBEL, R. and BUTTENFIELD, B. P. (1992). Improvement of GIS graphics for analysis and decision making, *International Journal of Geographical Information Systems*, **6**(3), 223–45.

WERNECKE, J. (1993). *The Inventor Mentor*, Reading, MA: Addison-Wesley.

CHAPTER THIRTY-NINE

Postscript:
New Directions for GIS Research

MICHAEL GOODCHILD

39.1 Introduction: research themes

The Organising Committee for the 1995 Summer Institute at Wolfe Neck identified six topics to serve as themes for the conference: Geographic data infrastructures; GIS diffusion and implementation; Generalisation; Concepts and Paradigms; Spatial Analysis; and GIS and Multimedia. The themes provided the organising framework for the conference, and all six are represented in this volume; together, they capture much of the active research in GIS and geodata on both sides of the Atlantic.

Almost 10 years ago, Ron Abler, then director of the Geography and Regional Science programme at the US National Science Foundation (NSF), was faced with a similar problem of identifying the research themes of GIS and geographic information analysis as the basis for NSF's solicitation for the new National Center for Geographic Information and Analysis. He picked five, after lengthy discussions with the research community: spatial analysis and spatial statistics; spatial relationships and database structures; artificial intelligence and expert systems; visualisation; and social, economic and institutional issues. The five make an interesting comparison with the Wolfe Neck six, and indicate how much priorities and thinking have changed in a decade. All but one of them are essentially technical, indicating the much higher and almost exclusive priority we gave to technical issues 10 years ago. Our understanding of social and institutional issues, a topic included almost as an afterthought in 1986 to cover a miscellany of issues that were then just beginning to be important, had become much more refined by 1995 and had expanded to include the public policy issues of national spatial data infrastructure.

For the second of these trans-Atlantic summer institutes to be held in Berlin in 1996, the organising committee has identified a further six topics: data quality; spatial models and GIS; remote sensing and urban change; spatio-temporal change in GIS; geographic information and the information society; and GIS and emergency management. 'Remote sensing and urban change' and 'GIS and emergency management' take us into domains of the application of GIS and geodata where the specific issues of the domain may be as important as the generic issues normally associated with GIS. This tension between generic and specific may also underlie other topics, such as 'spatial models and GIS', where the need to deal with the specifies of modelling hydrologic process, for example, may compete or even conflict with the need for generic modelling tools.

Of the six Berlin topics, the fifth, 'geographic information and the information society', may be the most significant indicator of deeper changes and trends in the field. In 1977 it was possible to assemble a substantial proportion of the entire research world of GIS in a conference of no more than 100 people, and for the discussion to focus solely on issues of software and hardware. The participants at the Endicott House meeting that year (Dutton, 1978) could hardly have imagined that by 1995 the field would have grown to the point where a legitimate topic of research interest would be the impact of geographic information technologies on society. Yet the attention given to the recent book *Ground Truth* (Pickles, 1995) is a clear indication that the time has come to take such issues seriously. One of the more interesting questions in this area concerns the nature of the geographies that will emerge in the information era, as telecommuting and the net bring us closer to a world in which 'there is no more "there"; everything will be "here"'.

The range of research issues associated with GIS and geodata has clearly expanded enormously in the past 20 years, as technical questions have been augmented and to some extent replaced by social ones. The pressures to work with experts in other fields have grown, as have the dangers of trying to build a new discipline in isolation. As GIS discovers new questions, there are often theoretical and conceptual frameworks already established in other disciplines to address them, and there is thus always a need to look carefully for appropriate analogues. Recently, researchers in GIS and geodata have begun to look to the library community, where structures and frameworks already exist for handling information search and resource discovery, as a way of cross-fertilising research in geodata cataloguing and metadata.

This apparently endless process of expansion of the GIS and geodata research agenda raises important and nagging questions: is the process potentially infinite, or limited; are there logical bounds to the field; does the field have a permanent core? What exactly is this field variously called GIS, geodata, geomatics, or geographic information science? If we can define it, then are there significant gaps in our current research programmes and emphases? Should we be doing more in some areas and less in others? How do the Wolfe Neck and Berlin themes fit within the broader framework? What choices should a young scholar in the field be making today to ensure a prosperous and productive future?

39.2 Defining the field: the limits

First and most clearly, the field is defined by geographic information, where 'information' is taken broadly to encompass facts, data, knowledge, and understanding, and 'geographic' implies a specific tie to locations on the Earth's surface. Because 'location' can refer to anything from the Atlantic Ocean to a position defined by coordinates to the nearest millimetre, the range of possibilities included in this definition is so enormous that it may be easier to define geographic information by its exceptions, which clearly include facts that are true everywhere on the Earth's surface, and facts that are true only at unspecified locations. Not surprisingly, it is often asserted that the majority of information is geographic in particular domains, such as local government.

'Geographic' is often replaced by 'spatial'. On the one hand this may mean a generalisation to any two- or three-dimensional frame, not only the Earth's surface; this would include medical imaging, where the frame is the human body; and it would also include engineering and architectural design problems where it is not necessary to tie the project frame to the Earth's geodetic frame. Clearly, many issues of geographic information extend to spatial information under this definition. On the other hand 'geographic' is often replaced by 'spatial' or even 'geospatial' for no other reason than the baggage the first term is believed to carry, although for many people 'spatial' probably carries stronger associations with outer space than with the Earth's surface.

So far so good. But why is geographic information worth singling out, rather than any of a myriad of other ways of partitioning the information pie? Why not a science of geriatric information, or genetic information? Are the issues of geographic information of sufficient importance, and sufficiently separable from issues of non-geographic information? Clearly we believe so, but we may have difficulty putting our collective finger on the reason, and may even feel a little nervous from time to time that there may not be anything particularly special about a database that happens to contain geographic coordinates. At the very least, many of the issues of geographic information also underlie spatial information, and there is little sense in separating research into the processing of raster information in GIS from research in image processing and pattern recognition. After all, geographic information in raster form may even fail the simplest test of being geographic: that of the presence of a geographic coordinate.

In summary, concern with geographic information is a necessary but not a sufficient condition for the field. There are issues of geographic information that may be better studied in other fields, because the geographic tie does not imply a strong enough separation. Many of the social and institutional issues of geographic information, for example, are not sufficiently different from similar issues of other information types to warrant their being studied separately, or to suggest that associated and distinct theoretical and conceptual frameworks need to be established.

39.3 Defining the field: the core

If the limits of the field are inevitably indistinct, then perhaps it is easier to define the core, by identifying the central, fundamental questions associated with GIS and geodata. Here again the presence of a geographic coordinate seems a weak basis on which to define a field. On the other hand, the concepts that are formalised and implemented in GIS go far beyond the description of point position, to include a much wider and more sophisticated set. They include the topological relationships of adjacency, connectedness, and intersection; geometric concepts of direction and distance; concepts of simple and complex geographic objects, from points, lines and areas to watersheds, viewsheds and anticlines; and the concepts embedded in models of the processes that affect the landscape.

Not all geographic concepts can be captured in a digital computer, in which everything must be expressed in terms of discrete binary representation. The Earth's surface is infinitely complex and continuous, so inevitably digital representations must be approximations or generalisations of the truth. This gives us one of the core defining problems of the field: that of finding useful approximations and representations, or data models for short, that capture significant aspects of real geographic phenomena in digital form. Moreover, the constraints on representations in digital computers are rather different from those of the human mind, leading to a second core problem: that of the tension between human cognition of geographic concepts and their digital representation. If these are not equal or at least equivalent, a digital database can be inherently difficult to comprehend and use. Finally, having built a digital database of geographic phenomena, we must then ask about the impact such a database has on the society that built it, and about the institutional problems inherent in building and managing it.

This line of argument leads to a vision of the core of the field: the study of geographic concepts; their role in human cognition; their formal representation in digital databases; their use in modelling, prediction and decision-making; the issues of the management and use of such databases; and their impact on society. Our earlier definition of geographic information, and the associated test of the presence of Earth coordinates, must be revised to allow for a more sophisticated range of geographic concepts that extends well beyond point location and simple geometric primitives to include all of the concepts used to represent, learn and reason about, model, and express our understanding of the patterns and processes found on the surface of the Earth. Some of these are generic, and some specific to particular domains and disciplines.

There are many contenders for an appropriate name for the field. 'GIS' is a popular contender, but it might suggest that one's research consisted of nothing more fundamental than the routine application of currently available software tools. The interpretation of the acronym seems to have drifted somewhat recently, from a type of information system, and therefore essentially software, to a shorthand for digital geographic information; hence the need to qualify 'GIS' further as in 'GIS system' and 'GIS data'. 'Geomatics' and 'geoinformatics' say little but are easily translated. My own preference is for 'geographic information science', which seems logical and informative, and plays nicely on the familiar acronym (Goodchild, 1992).

39.4 Where next?

If geographic concepts form the core of the field of geographic information science, then how well is the research community covering the core, exploring the periphery, and providing the basic research that will sustain new developments in GIS and geodata? How well do the topic sets of Wolfe Neck and Berlin address the fundamental needs of the field? These are large and challenging questions, so perhaps I can be excused for addressing them with a number of examples, rather than by attempting any kind of comprehensive analysis.

As sciences age they tend inevitably to institutionalise themselves, driven no doubt by the pressures of academic life, which tend to be centrifugal rather than centripetal. Small communities of like-minded scholars sustain themselves through the processes of peer review in publication, funding, and promotion, in ways that can seem almost completely detached from any larger agenda. Multidisciplinary fields like geographic information science are perhaps better equipped to avoid these patterns of behaviour, through a constant process of cross-fertilisation from other disciplines. At the same time it is important not to lose sight of the core, and to ask the kinds of questions outlined earlier is this topic best studied in this field; does its focus on geographic concepts provide a unique set of problems; and are there established theoretical or conceptual frameworks in other disciplines that might inform the problem?

One of the most challenging aspects of geographic information is its essential diversity. The problems of data quality, for example, are vastly different for raster images, vector representations of topographic features, and linear networks, and it is not at all obvious that there are benefits to be gained from studying them together. Arguments for the viability of geographic information science must look outward, as in the earlier discussion of limits, but also inward.

Research in geographic information science must also be informed by the essential nature of geographic data, which is so diverse and idiosyncratic as to make generalisation exceedingly difficult. It may be pointless, for example, to compare algorithms for performance when it is quite likely that for any given algorithm a set of geographic data exists for which that algorithm's performance is optimal. Geographic data sets do have generalisable characteristics: it is hard, for example, to find a data set that violates Tobler's 'first law of geography' that nearby places tend to be more similar than distant places; and it is also difficult to find a geographic data set that does not exhibit spatial heterogeneity (Anselin, 1989). The structure of geographic data is subtle and complex, as one can readily demonstrate by deleting all annotation from a map or image and asking questions such as 'What map is this?', 'Where is this?', or 'What scale is this?'

With these points in mind, I would like to suggest three areas where we are not yet doing enough geographic information science; as I noted earlier, this is very much a personal list and I make no claim to be inclusive or even fair to all of the GIS research community.

39.5 Metadata: describing data to others

The term 'metadata' has been called the least loved in science, in part I suspect out of frustration that solutions to such a seemingly simple problem are so hard to find. 'Data about data' allow us to locate useful information through processes of information resource discovery, much as one browses through a library; to describe the contents of data sets to others so that they can input and use them, or assess their fitness for use; and to capture technical information necessary to the successful packaging and handling of data in systems. In the case of geographic information, our current inability to do this is appropriately summed up in the acronym 'SAP', or spatially aware professional — that only someone with extensive experience in the field is capable of understanding someone else's description of a data set, or statement of needs. This is not to say the process is inefficient to a SAP — the information passed by the initials 'DCW' in a file header is useful and complex. To a non-SAP, however, the initials are meaningless and must be replaced by a detailed and exhaustive description. Our problem is that we have no accepted means of expressing that description. We lack a common, rigorous vocabulary that gives precise meaning to terms like 'layer'.

The momentum to find solutions to these issues is now very strong, through efforts like the US Federal Geographic Data Committee's Content Standards for Geospatial Metadata (http://fgdc.er.usgs.gov/metaover2.html), the Open Geodata Interoperability Specification (OGIS; http://www.ogis.org), and the efforts of Technical Committee 287 (CEN, 1995). What is missing at this point is a sound, general theory of geographic information on which a unified approach could be based, with appropriate extensions to the third spatial dimension, time, problems of multiple scales and accuracy. Such a general theory is a necessary precursor to a rigorous vocabulary and the implementation of a comprehensive approach to metadata. Further complicating the issue is the need to describe both analogue and digital data resources with the same vocabulary, since at least some information resources will continue to be in a traditional analogue form for some time to come.

39.6 Paradigms of spatial decision making

Not long ago, a computer was a machine that executed a series of instructions provided by a user, either in batches or interactively. The arrival of client–server architectures stimulated new ideas about how to interact with computers, leading most recently to the explosive impact of the World-Wide Web, with its metaphors of information publication and casual browsing. Peer-to-peer architectures will stimulate another round as they become more widely accessible with suitable software tools; one could argue that this is happening already in the 'chat-rooms' of the Net.

The same progression of architectures has stimulated new thoughts about spatial decision-making, and the role of computers in solving problems in areas traditionally reserved for professional planners. Batch computing allowed only the solution of problems that could be completely posed in advance, and reinforced the role of the professional as the custodian of an authoritative technology. Interactive computing, which began to appear in the late 1960s, led to the development of spatial decision-support systems, where ill- or incompletely posed problems could be refined through a human–computer dialogue, and allowed the technology to become a more intrinsic part of the decision-making process. It also promoted research on spatial reasoning, and other links between computing and spatial cognition. Client–server tools like the Web allow information about the decision-making process to be published for all to see, promoting citizen involvement and concepts of electronic democracy. Newer technologies like the Web are far easier to adopt and use than their precursors, which in part explains their explosive growth, but also puts great pressure on the research community to explore their implications.

39.7 Data fusion

The third example is more technical, but is similarly driven by changes that have occurred in computing and the increased interest in digital geographic information over the past two decades. As it becomes easier to share and exchange data, and greater and greater resources are invested in creating digital geographic data sets, it becomes increasingly likely that one will discover more than one source of the same geographic fact. The spurious polygon problem, a term given to the slivers that appear when two versions of the same line are overlaid, was an early example, occurring whenever a particular geographic line appeared in more than one layer or coverage. Software products have continued by and large to emphasise high geometric precision, and to deal with sliver polygons on a purely geometric basis rather than through a comprehensive analysis and tracking of positional accuracy.

Data fusion, or the successful merging of apparently conflicting geographic information, is increasingly necessary where GPS measurements disagree with positions digitised from topographic maps; when data must be updated with new measurements; when vector representations of features must be made consistent with raster backcloths; or where information must be reconciled across map boundaries. Besides purely technical issues of algorithms, it suggests the need for more comprehensive statistical models that are analogous to traditional methods for scalar measurements, and raises interesting questions of quality control, costs and benefits.

39.8 Discussion

The three areas described in the previous section as potentially fruitful for new research are all concerned with geographic information, and in each case it seems clear that the geographic dimension presents problems that are to some extent unique. In addition, they are all to some degree technology-driven; the need for research has been created to some extent by the technological advances of the past few years, and the impact these have had on society. The need for research on metadata, for example, is precipitated at least in part by the development of new search tools on the World-Wide Web, and the technologies of digital libraries, although one could argue that a much older motivation derives from the problems of finding maps and atlases in map libraries.

The idea that research should be driven by technology is itself somewhat new, and might have surprised the GIS research community of even 10 years ago, when computers and their applications were seen strictly as tools. Today, and particularly following the explosive interest in the Web of the past two years, it seems entirely reasonable that the research community should study, if not try to anticipate, some of the more significant effects of technology on society. In this sense our research priorities are changing, as computer technology's influence on late twentieth-century society continues to grow. Our new tools allow us to do things we never thought necessary, let alone possible, 10 years ago and, like it or not, many of the changes we now observe in society are indeed technology-driven.

This last seems an appropriate concluding point for this short postscript. The ESF/NSF Summer Institutes are one of the few series of meetings that have tried to achieve a comprehensive coverage of the field of GIS research in recent years, and they demonstrate both the field's vitality, and its essential coherence. The field has evolved from a relatively exotic speciality built around a small but powerful computer application, to something approaching a discipline in its own right, with a strong and well-defined intellectual core. It must continue to evolve rapidly if it is to keep up with the extraordinary changes that are occurring in society driven by computer technology, and we can expect to see evidence of that even in the short 12 months between the two Summer Institutes. By doing so, it will help ensure not only the future vitality of GIS, but also our understanding of the broader implications of the digital representation and manipulation of geographic concepts.

References

ANSELIN, L. (1989). 'What is special about spatial data?: alternative perspectives on spatial data analysis', NCGIA Technical Paper 89–4, Santa Barbara, CA: National Center for Geographic Information and Analysis.

COMITÉ EUROPÉEN DE NORMALISATION (1995). 'Geographic information – Data description – Metadata', Technical Committee 287, Brussels: CEN.

DUTTON, G. H. (Ed.) (1978). *Proceedings, First International Advanced Study Symposium on Topological Data Structures for Geographic Information Systems*, Cambridge, MA: Laboratory for Computer Graphics and Spatial Analysis, Graduate School of Design, Harvard University.

GOODCHILD, M. F. (1992). Geographical information science, *International Journal of Geographical Information Systems*, **6**(1), pp. 31–45.

PICKLES, J. (1995). *Ground Truth: The Social Implications of Geographic Information Systems*, New York: Guilford Press.

Index